World Scientific Proceedings Series on
Computer Engineering and Information Science 14

Intelligent Management of Data and Information in Decision Making

Proceedings of the 16th FLINS Conference on Computational Intelligence in Decision and Control & the 19th ISKE Conference on Intelligence Systems and Knowledge Engineering (FLINS-ISKE 2024)

World Scientific Proceedings Series on Computer Engineering and Information Science

Print ISSN: 1793-7868
Online ISSN: 2972-4465

Published:

Vol. 10 Uncertainty Modelling in Knowledge Engineering and Decision Making
edited by Xianyi Zeng, Jie Lu, Etienne E. Kerre, Luis Martinez and Ludovic Koehl

Vol. 11 Data Science and Knowledge Engineering for Sensing Decision Support
edited by Jun Liu, Jie Lu, Yang Xu, Luis Martinez and Etienne E. Kerre

Vol. 12 Developments of Artificial Intelligence Technologies in Computation and Robotics
edited by Zhong Li, Chunrong Yuan, Jie Lu and Etienne E. Kerre

Vol. 13 Machine Learning, Multi Agent and Cyber Physical Systems
Proceedings of the 15th International FLINS Conference (FLINS 2022)
edited by Qinglin Sun, Jie Lu, Xianyi Zeng, Etienne E. Kerre and Tianrui Li

Vol. 14 Intelligent Management of Data and Information in Decision Making
Proceedings of the 16th FLINS Conference on Computational Intelligence in Decision and Control & the 19th ISKE Conference on Intelligence Systems and Knowledge Engineering (FLINS-ISKE 2024)
edited by Etienne E. Kerre, Jie Lu, Luis Martínez, Tianrui Li, Javier Montero and Pablo Flores-Vidal

For the complete list of titles in this series, please visit
https://www.worldscientific.com/series/wspsceis

World Scientific Proceedings Series on
Computer Engineering and Information Science 14

Intelligent Management of Data and Information in Decision Making

Proceedings of the 16th FLINS Conference on Computational Intelligence in Decision and Control & the 19th ISKE Conference on Intelligence Systems and Knowledge Engineering (FLINS-ISKE 2024)

Madrid, Spain 16–21 July 2024

editors

Etienne E Kerre
University of Ghent, Belgium

Jie Lu
University of Technology Sydney, Australia

Luis Martínez
University of Jaén, Spain

Tianrui Li
Southwest Jiaotong University, China

Javier Montero
Complutense University of Madrid, Spain

Pablo Flores-Vidal
Complutense University of Madrid, Spain

World Scientific

NEW JERSEY • LONDON • SINGAPORE • BEIJING • SHANGHAI • HONG KONG • TAIPEI • CHENNAI

Published by

World Scientific Publishing Co. Pte. Ltd.
5 Toh Tuck Link, Singapore 596224
USA office: 27 Warren Street, Suite 401-402, Hackensack, NJ 07601
UK office: 57 Shelton Street, Covent Garden, London WC2H 9HE

Library of Congress Control Number: 2024025398

British Library Cataloguing-in-Publication Data
A catalogue record for this book is available from the British Library.

World Scientific Proceedings Series on Computer Engineering and Information Science — Vol. 14
INTELLIGENT MANAGEMENT OF DATA AND INFORMATION IN DECISION MAKING
Proceedings of the 16th FLINS Conference on Computational Intelligence in Decision and Control & the 19th ISKE Conference on Intelligence Systems and Knowledge Engineering (FLINS-ISKE 2024)

ISBN 978-981-12-9462-4 (hardcover)
ISBN 978-981-12-9463-1 (ebook for institutions)
ISBN 978-981-12-9464-8 (ebook for individuals)

For any available supplementary material, please visit
https://www.worldscientific.com/worldscibooks/10.1142/13882#t=suppl

Desk Editors: Nambirajan Karuppiah/Amanda Yun

Typeset by Stallion Press
Email: enquiries@stallionpress.com

Preface

FLINS and ISKE are two well-established international conferences, both launched by Da Ruan (1960–2011), that share a common objective: to link the theoretical foundations of computational intelligence with its applications to complex engineering and multidisciplinary decision-making problems, by putting together research institutions and industries, covering all the fields related to uncertainty management, operational research and artificial intelligence, including observation and experimentation, information processing, unavoidable ethical considerations and future trends in artificial neural networks, classification, clustering, computing with words, decision-making, evolutionary/genetic algorithms, expert systems, feature extraction, fuzzy systems, geographic information systems, intelligent systems, machine learning, optimization, soft computing and applications related to these topics as water, energy and environmental management, among others.

FLINS is the acronym introduced in 1994 standing for Fuzzy Logic and Intelligent Technologies in Nuclear Science, and ISKE is the acronym for Intelligent Systems and Knowledge Engineering, launched in 2006. Every two years, both conferences are organized jointly. This is the case of FLINS-ISKE 2024, held in Madrid (Spain), on July 16–21, 2024.

FLINS 2024 is the 16th in a series of conferences on applied computational intelligence. It follows the successful FLINS 1994 and 1996 in Mol, Belgium, FLINS 1998 in Antwerp, Belgium, FLINS 2000 in Bruges, Belgium, FLINS 2002 in Ghent, Belgium, FLINS 2004 in Blankenberge, Belgium, FLINS 2006 in Genova, Italy, FLINS 2008 in Madrid, Spain, FLINS 2010 in Chengdu, China, FLINS 2012 in Istanbul, Turkey, FLINS 2014 in Paraíba, Brazil, FLINS 2016 in Roubaix, France, FLINS 2018 in Belfast, North Ireland, FLINS 2020 in Cologne, Germany and FLINS 2022 in Tianjin, China.

ISKE 2024 is the 19th in a series of ISKE conferences, ISKE 2024 follows the successful ISKE 2006 in Shanghai, China, ISKE 2007 in Chengdu, China, ISKE 2008 in Xiamen, China, ISKE 2009 in Hasselt, Belgium, ISKE

2010 in Hangzhou, China, ISKE 2011 in Shanghai, China, ISKE 2012 in Beijing, China, ISKE 2013 in Shenzhen, China, ISKE 2014 in Joao Pessoa, Brazil, ISKE 2015 in Taipei, China, ISKA 2016 in Roubaix, France, ISKE 2017 in Nanjing, China, ISKE 2018 in Belfast, UK, ISKE 2019 in Dalian, China, ISKE 2020 in Cologne, Germany, ISKE 2021 in Chengdu, China, ISKE 2022 in Tianjin, China and ISKE 2023 in Fuzhou, China.

The joint FLINS-ISKE 2024 conference received 111 submissions, and the peer review led to the selection of the 65 papers included in this book.

The conference is organized in 20 sessions comprising the regular session plus the following 19 special sessions:

1. Industrial Applications of Artificial Intelligence, organized by Profs Kim Phuc Tran, Xianyi Zeng and Ludovic Koehl.
2. Information Fusion Techniques Based on Aggregation Functions, Pre-aggregation Functions, and Their Generalizations, organized by Profs. Benjamín Bedregal, Graçaliz Pereira Dimuro, Humberto Bustince, Javier Fernandez and Tiago Asmus.
3. Rough Sets: Theory, Applications, and Related Tools, organized by Henri Bollaert, Andrea Campagner, Richard Jensen, Jesús Medina and Adnan Theerens.
4. Dispersion, Consensus and Polarization Measures, organized by Profs. José Luis García-Lapresta and Daniel Gómez González.
5. Artificial Intelligence and Data-driven Techniques Applied to Operation, Control and Maintenance of Marine Energy Systems, organized by Profs. Matilde Santos and Payam Aboutalebi.
6. Synergy of Machine Learning and Soft Computing in Applications, organized by Profs. Irina Perfilieva and Vilem Novak.
7. Recommended Systems Supported by Computational Intelligence: Emerging Topics and Applications, organized by Profs. Luis Martínez, Raciel Yera and Rosa Mª Rodríguez.
8. Advances in Reasoning Based Rational Decision Making, organized by Profs. Shuwei Chen, Li Zou and Jun Liu.
9. Data-Driven Approaches in Large-Scale Group Decision-Making: Innovations, Challenges, and Applications, organized by Profs. Luis Martínez, Bapi Dutta and Diego García-Zamora.
10. Artificial Intelligence in Marketing, organized by Profs. Ramón Alberto Carrasco González and Jesús Serrano Guerrero.
11. Multicriteria Decision Making, organized by Prof. Begoña Vitoriano.

12. Autonomous Learning in Uncertain Decision Situations, organized by Profs. Kun Wang, Hang Yu, Zhen Fang, Guangquan Zhang and Jie Lu.
13. Evolving Deep and Transfer Learning Models for Computer Vision and Medical Imaging, organized by Profs. Li Zhang, Chee Peng Lim, Haoqian Huang and Hang Ruan.
14. Trustworthy Federated Continual Learning, organized by Profs. Xin Yang, Hao Wang, Jielei Chu and Wei Huang.
15. Efficient and Scalable Incremental learning, organized by Profs. Chuan Luo, Hongmei Chen, Yanyong Huang and Jihong Wan.
16. Data and Knowledge Enhance Learning, organized by Profs. Shengdong Du, Pengfei Zhang, Bing Wang and Jia Liu.
17. Three-way Decision and Granular Computing, organized by Profs. Dun Liu, Xiaoqing Ye and Decui Liang.
18. Logic and Intelligent Agent, organized by Profs. Ningning Zhou and Long Hong.
19. Explainable Artificial Intelligence under Small Data, organized by Profs. Long-Hao Yang, Fei-Fei Ye, Haitian Lu and Haibo Hu.

We are sure that readers will enjoy selected papers of FLINS-ISKE 2024 proceedings and plan to attend future editions of these conferences.

We offer our sincere gratitude to all authors, referees, special session organizers, session chairs and members of the international scientific committee, who through their merit have contributed to the success of FLINS-ISKE 2024.

Editors
Etienne Kerre
Jie Lu
Luis Martínez
Tianrui Li
Javier Montero
Pablo Flores

About the Editors

Etienne Kerre obtained a master's in Mathematics from Ghent University in 1967 and at the same university a PhD in 1970 on LEED (Low Energy Electron Diffraction) theory supported by a research grant. In the early 1970s, he discovered a Xerox copy of Zadeh's seminal paper "Fuzzy Sets". Immediately, he fell in love and started research on fuzzy topology. Already, in 1976, Etienne started some optional courses on fuzzy set theory and its applications. In the same year, he launched a research group called Fuzziness and Uncertainty Modelling that hosted many local and foreign researchers. Till now, 33 students obtained a PhD in mathematics or computer science under his guidance from Belgium, China, Bulgaria, Spain, Hungary, and Egypt. Most of them became professors and continued fuzzy research. Professor Kerre's research was very diversified ranging from pure mathematical to applied. He co-authored or co-edited more than 500 papers in international journals and proceedings of international conferences. He became an IFSA fellowship, and in 2021, he was awarded the prestigious IFSA award for significant impact on fuzzy logic-related research and applications.

Distinguished Professor Jie Lu is a world-renowned scientist in the field of computational intelligence, primarily known for her work in fuzzy transfer learning, concept drift, recommender systems, and decision support systems. She is an IEEE Fellow, IFSA Fellow, Australian Computer Society Fellow, and Australian Laureate Fellow. Professor Lu is the Director of the Australian Artificial Intelligence Institute (AAII) at the University of Technology Sydney (UTS), Australia. She has published

6 research books and over 500 papers in leading journals and conferences, won 10 Australian Research Council (ARC) Discovery Projects and over 20 industry projects as leading chief investigator, and has supervised 50 PhD students to completion. Prof Lu serves as Editor-In-Chief for *Knowledge-Based Systems* and the *International Journal of Computational Intelligence Systems.* She is a recognized keynote speaker, delivering over 40 keynote speeches at international conferences. She is the recipient of two IEEE Transactions on Fuzzy Systems Outstanding Paper Awards (2019 and 2022), NeurIPS Outstanding Paper Award (2022), Australasian Artificial Intelligence Distinguished Research Contribution Award (2022), Australian NSW Premier's Prize on Excellence in Engineering or Information & Communication Technology (2023), and the Officer of the Order of Australia (AO) in the Australia Day 2023.

Luis Martínez is a Full Professor at the University of Jaén, Jaén, Spain. He has been the main researcher in 19 R&D projects and has also published more than 290 JCR papers and more than 300 contributions in inter/national conferences related to his areas. His current research interests include intelligent decision support, fuzzy logic, and computing with words. He was a recipient of the IEEE Transactions on Fuzzy Systems Outstanding Paper Award 2008 and 2012. He is the Co-Editor-in-Chief of the *International Journal of Computational Intelligence Systems.* He is an IFSA Fellow 2021. Eventually, he was appointed as a Highly Cited Researcher 2017–2023 in Computer Sciences.

Dr Tianrui Li is a Professor and the Director of the Key Lab of Cloud Computing and Intelligent Technique of Sichuan Province, School of Computing and Artificial Intelligence, Southwest Jiaotong University, China. Since 2000, he has co-edited 10 books and 12 special issues of international journals, received 28 Chinese invention patents, and published over 500 research papers (e.g., AI, IEEE TPAMI, IEEE TKDE, IEEE TEC, IEEE TFS, IEEE TIFS, IEEE ASLP, IEEE TIE, IEEE TC, and IEEE TVT) in refereed journals and conferences (e.g., ACL, CVPR, ICCV, ICDE,

ICML, IJCAI, KDD, UbiComp, WWW, ICDM, CIKM, and EMNLP). Five papers were ESI Hot Papers and 23 papers were ESI Highly Cited Papers. He serves as the Editor-in-Chief of *Human-Centric Intelligent Systems* (Springer), area editor of the *International Journal of Computational Intelligence Systems*, editor of *Information Fusion* and *WIREs Data Mining and Knowledge Discovery*, Associate Editor of *ACM Transactions on Intelligent Systems and Technology*, etc. He is an IRSS Fellow and Steering Committee Chair (2023–2024), IEEE CIS Emergent Technologies Technical Committee (ETTC) member (2019–2020), IEEE CIS Senior Members Committee member (2018–2020), a senior member of ACM and IEEE, ACM SIGKDD member, Chair of IEEE CIS Chengdu Chapter (2013–2018), Treasurer (2016–2021), and Secretary (2022 till present) of ACM SIGKDD China Chapter.

Javier Montero is a Full Professor at the Department of Statistics and Operational Research, Faculty of Mathematics, Complutense University of Madrid (Spain). He holds a PhD in Mathematics from Complutense University since 1982 and has been leading research projects since 1987. He is the author of more than 150 research papers in refereed journals. He has published a similar number of refereed papers as book chapters. His research interests are in Aggregation Operators, Preference Representation, Multicriteria Decision Aid, Group Decision Making, System Reliability Theory, and Classification problems, mainly viewed as applications of Fuzzy Sets Theory. He has been a scientific visitor to more than 10 universities, has served the scientific community as the area editor of several international journals, and has been a member of more than 100 conference program committees. He has been the President of the European Association for Fuzzy Logic and Technology (EUSFLAT) and the President of the International Fuzzy Systems Association (IFSA). He has been the Vice-Rector (twice) at Complutense University, Dean of the Faculty of Mathematics at this university, and Head of its Department of Statistics and Operational Research, with more than 20 years in different academic management positions. Javier Montero has been also acknowledged as an IFSA Fellow. He is also a member of the European Academy of Sciences and Arts.

Pablo Flores-Vidal holds a PhD. in Data Analysis, specializing in Image Processing, from the Complutense University of Madrid (UCM, 2019, Extraordinary PhD Award). Currently, he serves as an Associate Professor in the Department of Statistics and Data Science at the Faculty of Statistics (UCM, since 2022). Dr. Flores-Vidal has accumulated a nine-year research career and has authored 12 articles as the first author and contributed as a member to three research projects alongside the group "Data Science and Soft Computing for Social Analytics and Decision Aid". His thesis memoir, *Image Processing Problems based on the incorporation of human characteristics*, showcases two novel image detection algorithms. Dr. Flores-Vidal further expanded his research horizons during a stay at the University of California, Berkeley, working on projects related to tree recognition using LiDAR images. Dr. Flores-Vidal has actively participated in numerous international conferences, including ISKE, FLINS, EUSFLAT, and IPMU. Beyond research and academia, Dr. Flores-Vidal has co-organized five editions of FuzzyMad. Additionally, Dr. Flores-Vidal has shared his expertise with students at the ESIC Business School and UNIR as a professor. Currently, Dr. Flores-Vidal has served as Mobility and Erasmus Coordinator at the FEE since 2022 and holds the role of co-main organizer for the FLINS-ISKE 2024 Conference.

Contents

An Empirical Study of the Impact of CEO Dominance on Corporate Performance Based on Facial Physical Feature Recognition

Lingyun Zhou*, Yao Mu†, Baojun Ma‡, Yi Chen§ and Jian Zhang¶

Key Laboratory of Brain-Machine Intelligence for Information Behavior (Ministry of Education and Shanghai), School of Business and Management, Shanghai International Studies University, Shanghai 201620, China

**zhoulingyun@shisu.edu*

†muy@shisu.edu

‡mabaojun@shisu.edu

§cheny@shisu.edu

¶jianzhang@shisu.edu

Unlike earlier studies that fail to measure CEO dominance accurately, our research develops a quantitative measure for CEO dominance based on a deep learning method, which can recognize the physical features from CEO's facial photographs. Taking more than 1800 corporates' data from 1992 to 2022 as a sample, this chapter aims to investigate the effects of CEO dominance on corporate performance. The results suggest that higher CEO dominance is associated with lower corporate performance. Furthermore, both CEO's high non-salary compensation and large firm size significantly mitigate the negative impact of CEO dominance on corporate performance.

Keywords: Facial physical feature recognition; CEO dominance; corporate performance; non-salary compensation; firm size.

1. Introduction

Chief Executive Officer (CEO) is the core of the management team in a corporate, possessing the legal power and responsibility to lead and control. In terms of operation, the CEO holds significant influence on business activities such as investments, financing, mergers and acquisitions, thereby impacting the corporate performance [1]. Upper echelons theory suggests that the personal characteristics of the CEO can be an important factor in firm performance [2]. Furthermore, compared with demographic features such as age, gender and education, the CEO' s personality is rather crucial to lead to various subjective or irrational behaviors, which subsequently affects the performance of corporate.

† Corresponding author.

Dominance is an important personality trait that refers to an individual's leadership, desire for control and assertiveness and is often manifested by a strong style of actions that exert control and influence on the surroundings [3]. Within the top management team of a corporate, there is naturally a game of power and cooperation between the CEO and other executives. Therefore, a highly dominant CEO may impose his own views and seek greater influence on corporate decisions, which would hurt the firm performance [4]. However, the empirical results regarding the impact of CEO dominance on corporate performance remain inconclusive in previous studies [5, 6], due to the significant challenges in measuring CEO dominance and especially obtaining large samples. On the one hand, CEOs often refuse to participate in invasive psychological questionnaires, while on the other hand, non-invasive methods such as proxy variables (e.g., CEO compensation) may only reflect the advantage of the CEO over other executive members rather than directly measuring the dominance.

In addition, research in cognitive psychology has shown that rich facial cues can be used to infer an individual's dominance, so some studies recruit raters to evaluate CEO dominance based on facial photographs [7]. However, such methods are susceptible to interference from the raters' prior knowledge, decreasing the reliability of the ratings. With the advantages of deep learning models in feature learning and large-scale computing, data-driven approaches that utilize deep learning may shed light on the accurate measurement of CEO dominance by identifying facial cues. Vernon *et al.* [8]77 designed a neural network model, which is trained on a huge number of facial photographs, to obtain a dominance calculation weight matrix based on 65 facial physical features, and the experiments have shown a high correlation between the machine-learned dominance scores and the manual ratings.

This study calculates CEO dominance based on facial physical feature recognition with a deep learning method and empirically investigates its impact on corporate performance, dealing with the existing limitations. To validate the relationship between CEO dominance and firm performance, we collected annual financial data of more than 1800 publicly listed corporates in the United States from 1992 to 2022, along with the corresponding demographic features and facial images of CEOs during their tenure. Furthermore, considering both the CEO' s non-salary compensation and firm size can to some extent influence the corporate's decision-making and strategic choices; this chapter also focuses on the moderating effects of these two factors so as to further explore the boundary conditions of CEO dominance effects.

2. Literature and Hypotheses

Dominance is an important personality trait that refers to an individual's leadership, desire for control, and proactiveness in social situations. The degree of leaders' dominance determines their willingness to acquire and maintain power, which can affect the extent of control they have over subordinates [3]. Studies have found that, a CEO with high dominance may dominate various strategic choices and business decisions of the corporate, restrict the work of other members in the executive team, and make other executives fear to express ideas that contradict those of the CEO [4]. Furthermore, researchers have discovered that as CEO dominance increases, the ability of shareholders as well as corporate regulatory bodies to check the CEO would diminish, thereby amplifying the CEO's influence on corporate decisions[9]. In this context, the increasingly monopolized power of the CEO may pose greater risks to the corporate. For instance, CEOs with high dominance tend to exhibit opportunistic behaviors and are more likely to be inclined to explore new technologies or ventures, which, however, are often time-consuming and entail uncertain returns, resulting in potential losses for the corporate [10].

Hypothesis 1: *The degree of CEO's dominance is negatively related to the corporate performance.*

According to principal-agent theory, there is a natural conflict of interests between investors and managers of a corporate. To protect their own interests, investors usually influence the motivation and stimulate the expected behaviors of managers with compensation [11]. A higher compensation intensifies CEOs' aversion to risk and thus, based on the principle of loss aversion, suppresses their tendency to make highly risky decisions. To this regard, CEOs, faced with threats such as bankruptcy, unemployment, and personal loss of interests associated with the corporate, would actively restrict the risks for weakening the negative impacts on firm performance [12]. In addition, CEO's compensation can be mainly divided into salary and non-salary components, and the later can account for around 90% of the total compensation. Existing studies suggest that compared to salary, non-salary compensation depends more on the firm performance, which may better motivate CEOs to make efforts [13].

Hypothesis 2: *Non-salary compensation of CEO negatively moderates the reationship between CEO dominance and corporate performance.*

Firm size is one of the most commonly used contextual variables of organizations. Research has found that firm size is an important factor influencing the organization's inertia [14]. In particular, larger-scale corporates possess more established structures and decision-making processes, which leads to more challenges in exploring changes and also limits managers' discretion [15]. Furthermore, larger firms are believed to have more redundant resources, larger market shares, and stronger competitiveness in the market. The buffering effect of these redundant resources significantly reduces corporates' sensitivity to poor performance, and consequently, can mitigate CEOs' risky decisions so as to improve firm performance [16].

Hypothesis 3: *Firm size negatively moderates the reationship between CEO dominance and corporate performance.*

3. Data and Measurement

3.1. *Dataset*

This study matched the Execucomp dataset[1] containing CEO data with the Compustat dataset[2] containing annual accounting data of corporates, and selected the successfully matched data from 1992 to 2022 as the initial sample. The following processing was also carried out according to the research context: 1) All CEOs were picked out using the TITLE and CEOANN labels in the Execucomp dataset; 2) Based on the derived CEO list, CEOs' facial photos were collected through Google; 3) The collected photos were positioned and cropped before calculating the CEO dominance; 4) Financial corporates (i.e., with SIC code 6000-6999) and those with missing control variables were excluded; 5) Winsorization was carried out on the continuous variables of corporates at the 1% and 99% percentiles. After these processing, a total of 22274 records were obtained, which contained 1834 corporates and 2872 CEOs.

3.2. *CEO dominace based on facial physical features recognition*

This study employed facial physical features recognition to measure CEO dominance, that is, constructing the key explanatory variable for corporate performace. The deep learning based methods generally train their parameters on a huge number of CEO facial photos, which may decrease the bias in manual scoring and thus enable a more reliable and accurate measurement of CEO dominance. In this regard, the CEO dominance calculation process was implemented as follows. First, this study preprocessed the facial photos of CEOs collected from Google, used the MTCNN (Multi-Task Convolutional Neural Network) model to automatically locate the face of each CEO, and cropped the image to a unified size with the face at the center. Next, the IBUG facial recognition software was used to locate 68 key facial landmarks on each photo, and 65 facial features were constructed based on the coordinates of these key landmarks. Then, the trained weight matrix for dominance with facial physical features [8] was used to calculate the degree of CEO dominance.

---

1 https://wrds-www.wharton.upenn.edu/pages/grid-items/compustat-execucomp-basics

2 https://wrds-www.wharton.upenn.edu/pages/get-data/compustat-capital-iq-standard-poors

3.3. *Model*

To assess the impact of CEO dominance on corporate performance as well as the moderating effects of CEO's non-salary compension and firm size, a regression model is constructed as in Eq. (1).

$$ROA_{i,t} = \alpha + \beta_1 DOM_{i,t} + \beta_2 MOD_{i,t} + \beta_3 DOM_{i,t} \times MOD_{i,t} + \sum \beta_i Control_{i,t} + \gamma_t + \delta_i + \varepsilon_{i,t} \quad (1)$$

Here, *i* represents the firm, *t* represents the year, *ROA* is the corporate's return on assets, *DOM* is the degree of CEO dominance, *MOD* is the moderator (i.e., CEO's non-salary compension or firm size), and *Control* denotes the control variables. Furthermore, α is the intercept, $\beta_1/\beta_2/\beta_3$ are the impact coefficients of focus, γ_t and δ_i represents the fixed effects of year and firm, respectively, and $\varepsilon_{i,t}$ represents the error. This study utilizes the cluster-robust standard errors with firm ID as the clustering variable to construct a two-way fixed effects model, and thus to control for potential issues from cross-sectional correlations and time-series autocorrelations.

4. Results

4.1. *Descriptive statistics and correlation analysis*

The definition and descriptive statistics of variables are shown in Table 1.

Table 1. Definition and descriptive statistics of variables.

Type	Variable	Definition and calculation	Avg.	Std.
Independent	ROA	Return on assets (=Net income/Total assets)	0.04	0.19
Dependent	DOM	Normalized degree of CEO dominance in range [0,1]	0.46	0.12
	Non-salary	CEO's non-salary compensation (in millions) (=Annual income-Salary) and adjusted with US GDP indices	7.58	16.39
	Size	1 when above the average annual assets and 0 otherwise	0.50	0.50
Control (CEO)	Gender	1 when famale and 0 when male	0.04	0.19
	Age	Actual age	55.25	7.35
	Tenure	Time duration (in year) in this position	8.33	7.45
	Salary	Salary (in millions) adjusted with US GDP indices	1.08	0.58
(corporate)	AIR	Asset intensity (=Property, plant and equipment/Sales)	0.52	0.90
	GROWTH	Growth choice (=Capital expenditure/Property, plant and equipment)	0.23	0.17
	BLEV	Book leverage (=Total liabilities/Assets)	0.26	0.98
	MLEV	Market leverage (=Total liabilities/Equity book value)	0.17	0.15

According to the Pearson correlation analysis of variables, the coefficients between CEO dominance, non-salary compensation, and firm size are all below 0.5, indicating that the empirical analysis is little affected by multicollinearity.

In addition, the average VIF value for all variables is 1.22, with a maximum value of 1.47, which is far below the common threshold of 10. Therefore, it can be inferred that the selected variables do not exhibit serious multicollinearity problems and are suitable for the multivariate regression analysis.

4.2. *Regression results and discussions*

The results of regression analysis are summarized in Table 2.

Table 2. Regression results of the empircal study.

	Control only	Main effect	Main+Moderator 1	Main+Moderator 2
DOM		-0.028***(-2.72)	-0.036***(-3.31)	-0.049***(-2.98)
DOM×Non-salary			0.001**(-2.55)	
DOM×Size				0.034**(-1.98)
Non-salary	0.000(-1.34)	0.000(-1.26)	-0.000**(-2.22)	0.000(-1.29)
Size	0.006**(-2.17)	0.007**(-2.21)	0.007**(-2.23)	-0.009(-1.09)
Gender	-0.004(-0.84)	0.000(-0.04)	-0.001(-0.14)	0.000(-0.04)
Age	-0.000*(-1.71)	-0.000*(-1.84)	-0.000*(-1.87)	-0.000*(-1.89)
Tenure	0.000***(-2.66)	0.001***(-2.83)	0.001***(-2.83)	0.001***(-2.86)
Salary	0.006***(-2.96)	0.006***(-2.99)	0.007***(-3.02)	0.007***(-3.06)
AIR	-0.022***(-6.47)	-0.022***(-6.48)	-0.022***(-6.49)	-0.022***(-6.54)
GROWTH	0.078***(-9.49)	0.078***(-9.51)	0.078***(-9.52)	0.078***(-9.52)
BLEV	0.053***(-3.86)	0.053***(-3.87)	0.053***(-3.84)	0.053***(-3.85)
MLEV	-0.318***(-18.66)	-0.318***(-18.70)	0.318***(-18.70)	0.318***(-18.68)
Intercept	0.082***(-7.96)	0.096***(-8.04)	0.100***(-8.28)	0.106***(-7.82)
Fixed effects	Controled	Controled	Controled	Controled
R^2	0.510	0.511	0.511	0.511

Note: *, **, and *** represents p-value smaller than 0.1, 0.05, and 0.01, respectively.

As illustrated, first, the nagative effect of CEO dominance on corporate performance is significant at the 0.01 level, supporting Hypothesis 1. This indicates that highly dominant CEOs tend to dominate strategic decisions, reduce the balance imposed by shareholders and regulatory bodies, and increase the likelihood of risky decisions, thereby negatively affecting firm performance. Second, the interaction term DOM×Non-salary is significantly positively related to ROA at the 0.05 level, implicating a negative moderating effect of CEO's non-salary compensation that validates Hypothesis 2. This suggests that highly dominant CEOs, due to their higher non-salary compensation, exhibit a greater aversion to risk, which suppresses their preference for risky decisions and mitigates the negative impact on firm performance. Third, the coefficient of DOM×Size is also significantly positive at the 0.05 level, supporting Hypothesis 3. This implies that highly dominant CEOs, when working in larger firms, face restrictions on their discretion and benefit from the buffering effect of redundant resources, thus reducing the risk-seeking decisions and mitigating the negative impact on corporate performance.

For robustness checks purpose, this study: 1) adopted profitability, market value, and return on investment, respectively, to measure corporate performance instead of ROA; 2) retained the original CEO dominance degree as the main independent variable rather than that after normalization; and 3) changed the cluster-robust standard errors with CEO and industry ID as the clustering variable. The results of robustness checks reveal the consistent findings, that is, CEO dominance has negative impact on corporate performance, and CEO's non-salary compensation and firm size both negatively moderate the effect.

5. Conclusion

The impact of CEO personality on firm performance has always been a hot topic in the field of corporate governance, and dominance, as an important personality trait, has received considerable attention. This study utilizes a deep learning method to analyze CEOs' facial photos and obtains CEO dominance based on the recognition of facial physical features. Then, we investigate the impact of CEO dominance on corporate performance using real-world financial data. The empirical analysis reveals the following: (1) Higher degree of CEO dominance is associated with lower corporate performance. (2) The higher the non-salary compensation of the CEO, the weaker the negative impact of CEO dominance on corporate performance. (3) The larger the firm size, the weaker the negative impact of CEO dominance on corporate performance.

From the theoretical viewpoint, this study measures CEO dominance using a facial physical features recognition-based approach, so as to deepen and enrich understanding of the impact of CEO personality on corporate performance through empirical tests with a large sample size. Furthermore, this chapter explores the moderating effects of contextual factors, that is, CEO' s non-salary compensation and firm size. From the practical viewpoint, rather than solely focusing on demographic characteristics, this study conducts a deeper analysis of the personality reasons behind CEOs' decision-making and strategic choices. The findings can provide valuable insights for investors in making decisions on corporate investment, risk warning and even CEO selection.

Future efforts may employ advanced deep learning methods to further improve the calculation of CEO dominance and, probably, enrich the methods with CEOs' voice or video data for enhancement. Another direction is to explore the impact of CEO dominance on corporate performance in other aspects (e.g., social responsibility) and the corresponding mechanisms.

Acknowledgments

Work supported by the National Natural Science Foundation of China (72172092/72302153).

References

1. Y. L. Jaw and W. T. Lin, Corporate elite characteristics and firm's internationalization: CEO-level and TMT-level roles, Int. J. Human Resource Mgt. **20**, 220 (2009).
2. D. C. Hambrick and P. A. Mason, Upper echelons: the organization as a reflection of its top managers, Acad. Mgt. Rev. **9**, 193 (1984).
3. R. A. Emmons and D. P. McAdams, Personal strivings and motive dispositions: exploring the links, Personality Social Psyc. Bulletin **17**, 648 (1991).
4. D. C. Hambrick and R. A. D'Aveni, Top team deterioration as part of the downward spiral of large corporate bankruptcies, Mgt. Sci. **38**, 1445 (1992).
5. D. E. Re and N. O. Rule, Predicting firm success from the facial appearance of chief executive officers of non-profit organizations, Perception **45**, 1137 (2016).
6. N. O. Rule and N. Ambady, The face of success: inferences from chief executive officers' appearance predict company profits, Psyc. Sci. **19**, 109 (2008).
7. T. G. Canace, A. M. Cianci, X. K. Liu and G. T. Tsakumis, Paid for looks when others are looking: CEO facial traits, compensation, and corporate visibility, J. Biz. Res. **115**, 85 (2020).
8. R. J. W. Vernon, C. A. M. Sutherland, A. W. Young and T. Hartley, Modeling first impressions from highly variable facial images, PNAS **111**, E3353 (2014).
9. K. B. Lewellyn and M. I. Muller-Kahle, CEO power and risk taking: evidence from the subprime lending industry, Corp. Gov. Int. Rev. **20**, 289 (2012).
10. J. Tang, M. Crossan and W. G. Rowe, Dominant CEO, deviant strategy, and extreme performance: the moderating role of a powerful board, J. Mgt. Stu. **48**, 1479 (2011).
11. E. P. Jansen, K. A. Merchant and W. A. Van der Stede, National differences in incentive compensation practices: the differing roles of financial performance measurement in the United States and the Netherlands, Acc. Organizations Soci. **34**, 58 (2009).
12. T. A. Gormley, D. A. Matsa and T. Milbourn, CEO compensation and corporate risk: evidence from a natural experiment, J. Acc. Econ. **56**, 79 (2013).
13. J. S. Demski and G. A. Feltham, Economic incentives in budgetary control systems, Acc. Rev. **53**, 336 (1978).
14. D. C. Hambrick and S. Finkelstein, Managerial discretion: a bridge between polar views of organizational outcomes, Res. Organizational Behavior **9**, 369 (1987).
15. P. G. Audia and H. R. Greve, Less likely to fail: low performance, firm size, and factory expansion in the shipbuilding industry, Mgt. Sci. **52**, 83 (2006).
16. D. A. Levinthal, Random walks and organizational mortality, Admin. Sci. Q. **36**, 397 (1991).

Toward Sustainability Optimization in Touristic Route Recommendation

Pablo Sánchez[*,‡] and Alejandro Bellogín[†,§]

[*]*Instituto de Investigación Tecnológica (IIT), Universidad Pontificia Comillas, C. Alberto Aguilera 25, E28015, Madrid, Spain*

[†]*Universidad Autónoma de Madrid, Madrid, Spain*

[‡]*psperez@icai.comillas.edu*

[§]*alejandro.bellogin@uam.es*

While route recommendation plays a vital role in digital tourism services, conventional methods tend to be inherently complex, as they need to consider several constraints at once: user preferences, geographical and scheduling information, etc. Besides, recent interests from society and institutions have motivated a greater emphasis on sustainability across multiple sectors, including tourism. This work proposes to use reranking techniques as a plain and efficient method to generate personalized routes for users while considering sustainable goals in the procedure. We believe that these results could lead to new research directions regarding sustainable route recommendation techniques.

Keywords: Recommender systems; route recommendation; sustainability.

1. Introduction and Motivation

For several years, tourism has witnessed an exponential growth, becoming a vital economic force for many countries.[1] As Location-Based Social Networks (LBSNs) such as Foursquare or Gowalla gain traction, users can easily share their trips and check-ins at the different Points-of-Interest (POIs) they have visited. This abundance of geotagged content on LBSNs has gained considerable attention among users who are interested in finding new venues to visit when they are in a specific city.[2]

When recommending POIs to users, most current studies tend to focus on the accuracy of recommendations, by analyzing how many relevant venues we are recommending to the target user.[3] However, in the classical recommendation domain, many researchers have alerted about the necessity of evaluating recommendations on other aspects. Novelty and diversity are examples of beyond accuracy dimensions, and although they have been explored in the traditional recommendation problem, they could be especially relevant in the POI recommendation domain (and other connected problems). Exploring these additional dimensions could greatly benefit users, as it allows them to uncover less popular POIs, and also for the venue owners, as they might benefit financially from these recommendations. By exploring these beyond accuracy aspects, both route and POI recommenders could offer users a wider array of candidate venues to visit, favoring a more complete exploration of the city.

Besides, in the Information Retrieval, Recommender Systems, and Machine Learning communities, there is a growing emphasis on concepts, such as novelty, diversity, and fairness. Nevertheless, this is normally limited to statistical biases that can be obtained from data, like popularity bias,[4] or linked to measurements on sensitive attributes.[5] In any case, these trends match some of the Sustainable Development Goals (SDGs)[a] stated by the United Nations in 2015, such as reaching gender equality (SDG 5), lessen inequalities (SDG 10), or responsible production and consumption (SDG 12). Thus, SDGs have attracted attention to particular societal problems that require collective action. In the context of tourism, besides those previously mentioned concerns, SDG 11 is especially relevant: "make cities and human settlements inclusive, safe, resilient and sustainable". Certain studies have already tackled these and related topics within the realm of the so-called *green tourism*[6–8] or *sustainable tourism.*[9] However, as far as we know, there is no link with the development of personalized routes.

In this context, this work considers the route recommendation problem from a sustainable viewpoint. More precisely, our approach involves the use of reranking techniques to generate route recommendations based on traditional POI recommendation methods. These strategies are capable of optimizing different criteria, including considerations such as the proximity of recommended venues and their categories, in order to encourage sustainable tourism by suggesting eco-friendly destinations. Our approach offers several benefits, such as straightforward implementation, effectiveness, and clarity, which could benefit both users and other stakeholders involved in the process. Moreover, the following evaluation includes both classical and novelty metrics, providing a thorough comprehension of the recommendation effectiveness when addressing the varied needs of the users.

This chapter is structured as follows: in Section 2, we present our proposal based on rerankers. Section 3 shows the performed experiments and, finally, in Section 4, we discuss the conclusions and future work of our study.

[a]`https://sdgs.un.org/goals`

2. Exploiting Reranking Methods for Sustainable Optimization

In this work, we propose to use item reranking techniques in order to create valuable sequences of venues. Our approach offers two main advantages with respect to the use of complex route generation algorithms. First, we can generate different rerankers depending on the dimension we aim to maximize (or minimize). Secondly, the rerankers enable us to derive routes even from recommendation models that may not inherently incorporate any sequential component. Now, Section 2.1 presents a reranking framework for route recommendation that will be used in our work. Then, Section 2.2 introduces our proposal for a sustainable reranker in this domain.

2.1. *Reranking framework*

We follow the same formulation as in our previous work,[10] based on the original formulation from.[11] Beginning with an initial pool of candidate items for a particular user, generated by any recommendation model, our objective is to reorder them to identify a route that best satisfies the user's interests. As we select each candidate item and include it within the final recommendation list R, we simultaneously delete it from the candidate pool. The formulation is shown in Equation 1.

$$f_{obj}(u, i, R) = \lambda \cdot f_{rec}(u, i) + (1 - \lambda) \cdot f_{seq}(u, i, R) \tag{1}$$

Equation 1 defines the objective function $f_{obj}(u, i, R)$ considered in our greedy reranking process. Based on the score obtained in Equation 1, we select the venue i that obtains the highest value from the current set of candidate venues. These are the venues present in the original recommendation list that have not been chosen yet for the final itinerary recommended to the user.

In this case, as it is typical in the field,[11] the objective function combines the score assigned to the user u and venue i pair by a recommendation algorithm, represented by the $f_{rec}(u, i)$ component, and the score provided by the sequence-aware reranker component, represented by $f_{seq}(u, i, R)$. Here, R represents the final recommendation list being generated. The sequence-aware reranker can take into account, if necessary, the previously selected candidate items that already belong to the final recommended route for user u.

Finally, a parameter λ is used to balance the trade-off between the original score of the recommender and the sequential reranker. A lower value of λ indicates a greater influence of the reranker, and therefore more possible changes to the original recommendation provided.

Our previous work[10] defined seven different formulations for the sequence-aware reranker component f_{seq}, classified in three major families: independent reranker (the score is solely determined by the user-item pair under consideration), for example, based on scores produced by Random or Recommender-based strategies (`rec`); based on the preceding item (the score assigned to item i is only influenced by the last item included in list R, denoted as i_{n-1}), such as the rerankers based on Distance (`dist`), Feature-based Markov Chain (`feat`), or Item-based Markov Chain (`item`); and dependent on the entire sequence (the score assigned to an item is determined by considering the entire sequence that has been generated so far, that is, the current list R and the potential next item i), using algorithms based on Longest Common Subsequence (`lcs`) or Suffix Tree (`stree`).

2.2. *A reranker for sustainable route recommendation*

We propose a new reranker that corresponds to the last family described before (dependent on the whole sequence), named **Outdoor Reranker**. For this reranker, we identify a target distribution T_{Dist} of the categories of the venues to be recommended to the user, so the next candidate to be included in the route should be the one minimizing the difference between the target category distribution T_{Dist} and the categories of the route being generated so far, that is, the categories of R.

To promote more sustainable recommendations, the target distribution T_{Dist} is defined based on categories tailored for sustainability, such as parks and botanical gardens or, in general, outdoor POIs.[8] By doing so, we aim to provide a more balanced and nuanced set of recommendations that align with the user's preferences (by considering the recommendations coming from the original algorithms) while also promoting the exploration of more sustainable categories (through the use of the Outdoor Reranker, `outd`).

3. Experiments

3.1. *Settings*

In this section, we will describe the evaluation methodology followed in our experiments. We use the Trip builder dataset from,[12] that combines two data sources, Flickr and Wikipedia, by building trajectories using the geotagged photos taken by the users in three Italian cities, although we focus on the city with a higher number of check-ins, Rome. This dataset, once processed, provides 7,954 users who created 27,252 routes (trajectories) on 394 venues, producing 61,330 interactions between the users and the venues.

We chose the last itinerary of users with a minimum of three distinct routes as long as they have at least 4 POIs in the last route to create the test set. The remaining data will be retained for training the recommenders. The evaluation methodology, in line with recommendations from the research community,[13] selects the candidate items from the training set, to ensure a more realistic setting. This means that for each user in the test set, we will consider as candidate POIs all venues that appear but not been visited by the user in the training set.

3.2. *Recommendation approaches*

To cover classical algorithms and methods more tailored for the tourism domain, we have included in our experiments a varied combination of methods, which shall be classified as classic and geographical approaches. In the first family, we include a popularity recommender (Pop), two nearest-neighbor methods[14] (one based on users: UB, and another on items: IB), and two matrix factorization models (one based on Alternate Least Squares: HKV,[15] and another on Bayesian Personalized Ranking from[16]: BPRMF). In the second family, the algorithms exploit the geographical coordinates of the items, such as AvgDis (that suggests the closest POIs to the user's average location), IRenMF (a weighted Matrix Factorization technique proposed in[17] that considers the geographical influence between the target POI with respect to its neighboring venues and the region it belongs to), and RankGeoFM (a matrix factorization model proposed in[18] that incorporates the geographical influence between neighboring POIs while also capturing user geographical preferences by integrating an extra latent matrix).

3.3. *Results*

We now present the performance results obtained when applying the reranking framework described before to the recommendation algorithms listed in the previous section. Specifically, for accuracy, we report nDCG_s, which stands for nDCG incorporating the sequential component with LCS penalization, as defined in.[10] For novelty, we use the Expected Popularity Complement (EPC), which assumes that a POIs is novel if its popularity is low.[11] Higher values in all metrics indicate a better performance of the

Table 1. Results of standard recommenders. **Bold** denotes the best results for each metric, in *italic* the (next) best result in each family (classical/geographical) is emphasized.

	Pop	UB	IB	HKV	BPRMF	AvgDis	IRenMF	RankGeoFM
nDCG	**0.508**	0.469	0.473	0.425	*0.508*	0.362	0.401	*0.486*
$nDCG_s$	**0.447**	0.420	0.419	0.386	*0.447*	0.334	0.367	*0.427*
EPC	0.848	0.865	0.864	*0.894*	0.848	**0.965**	*0.937*	0.865

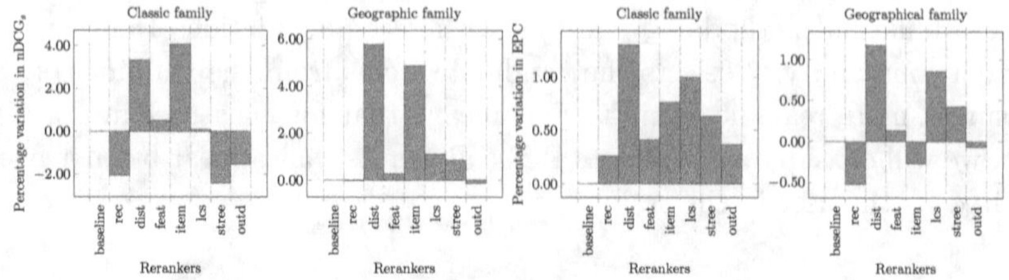

Fig. 1. Variation in $nDCG_s$ (left) and EPC (right) with respect to no reranker (see reranker abbreviations in Sections 2.1 and 2.2).

models, indicating a higher accuracy or novelty. Moreover, unless specified otherwise, these metrics will be reported on rankings of length 10.

Table 1 shows a strong popularity bias, since the popularity recommender outperforms other models in terms of accuracy (nDCG). This is a well-known fact in the community for other domains,[4] but in this case, factors such as high sparsity and a concentration of check-ins in a short number of venues, reinforce this behavior. The trend in performance stays the same when incorporating sequence in accuracy measurement ($nDCG_s$), where Pop and BPRMF obtain the best results among the classical approaches, whereas RankGeoFM is the best geographical algorithm in terms of accuracy. EPC, as a measure of novelty, prompts a balance with the accuracy metrics, as the best methods now (AvgDis and HKV) correspond to the worst methods according to $nDCG_s$ or nDCG.

Let us now analyze the impact of the reranking framework. Figure 1 shows the performance variation when using such framework on the best method of each family. These figures take as baseline that method, which is then used by the reranker to reorder its first 20 recommended items, and apply Equation 1 with $\lambda = 0.5$ exploiting each reranker capabilities. We observe that distance-based and Item-based Markov Chain rerankers are the only ones that consistently improve accuracy and novelty in both recommendation families. Regarding the outdoor reranker, it is remarkable that, by promoting more sustainable venues, it is able to gain novelty at the expense of accuracy in the classic family, whereas for the geographic family the observed changes are negligible.

Hence, we conclude that, by optimizing for sustainability through reranking, recommendation performance either stays the same, or, at least, one of the analyzed evaluation dimensions (accuracy or novelty) improves.

4. Conclusions and Future Work

In this chapter, we have extended a reranking framework proposed previously and analyzed how sensitive the recommendation methods were when promoting outdoor activities. By doing so, we have observed which algorithms and rerankers are more effective in promoting more sustainable suggestions and healthier recommendations for users. In particular, our proposed reranker has shown promising results in these directions. However, promoting only outdoor POIs suggests the necessity of combining our approach with other models to maximize its usefulness.

A promising line of future work would involve extending the reranking framework for multi-objective route recommendation. This would allow us to combine and optimize several rerankers at once, each of them tailored to a different goal. Such an approach holds particular interest in tour planning, where it could be seamlessly integrated.[19] Related to the previous idea, we believe the proposed reranker could be used to further sustainability objectives – such as reducing CO_2 emissions[8] – and, in general, other principles of fair recommendation with an emphasis on sustainability.[9]

Acknowledgments

Work supported by Grant PID2022-139131NB-I00 funded by MCIN/AEI/ 10.13039/501100011033 and by "ERDF A way of making Europe."

References

1. H. Rasool, S. Maqbool and M. Tarique, The relationship between tourism and economic growth among brics countries: a panel cointegration analysis, *Future Business Journal* **7**, p. 1 (Jan 2021).
2. L. W. Dietz, A. Sen, R. Roy and W. Wörndl, Mining trips from location-based social networks for clustering travelers and destinations, *J. Inf. Technol. Tour.* **22**, 131 (2020).
3. P. Sánchez and A. Bellogín, Point-of-interest recommender systems based on location-based social networks: A survey from an experimental perspective, *ACM Comput. Surv.* **54**, 223:1 (2022).
4. A. Bellogín, P. Castells and I. Cantador, Statistical biases in information retrieval metrics for recommender systems, *Inf. Retr. Journal* **20**, 606 (2017).
5. E. Amigó, Y. Deldjoo, S. Mizzaro and A. Bellogín, A unifying and general account of fairness measurement in recommender systems, *Inf. Process. Manag.* **60**, p. 103115 (2023).

6. S. kwon Hong, S. il Kim and J. hyun Kim, Implications of potential green tourism development, *Annals of Tourism Research* **30**, 323 (2003).
7. G. Divsalar, A. Divsalar, A. Jabbarzadeh and H. Sahebi, An optimization approach for green tourist trip design, *Soft Comput.* **26**, 4303 (2022).
8. A. Susanty, N. B. Puspitasari, S. Saptadi and S. Prasetyo, Implementation of green tourism concept through a dynamic programming algorithm to select the best route of tourist travel, *IOP Conference Series: Earth and Environmental Science* **195**, p. 012035 (dec 2018).
9. D. Streimikiene, B. Svagzdiene, E. Jasinskas and A. Simanavicius, Sustainable tourism development and competitiveness: The systematic literature review, *Sustainable Development* **29**, 259 (2021).
10. P. Sánchez and A. Bellogín, Applying reranking strategies to route recommendation using sequence-aware evaluation, *User Model. User Adapt. Interact.* **30**, 659 (2020).
11. M. Kaminskas and D. Bridge, Diversity, serendipity, novelty, and coverage: A survey and empirical analysis of beyond-accuracy objectives in recommender systems, *TiiS* **7**, 2:1 (2017).
12. D. Monti, E. Palumbo, G. Rizzo and M. Morisio, Sequeval: A framework to assess and benchmark sequence-based recommender systems, *CoRR* **abs/1810.04956** (2018).
13. A. Said and A. Bellogín, Comparative recommender system evaluation: benchmarking recommendation frameworks, in *Eighth ACM Conference on Recommender Systems, RecSys '14, CA, USA - October, 2014*, (ACM, 2014).
14. X. Ning, C. Desrosiers and G. Karypis, A comprehensive survey of neighborhood-based recommendation methods, in *Recommender Systems Handbook*, (Springer, 2015) pp. 37–76.
15. Y. Hu, Y. Koren and C. Volinsky, Collaborative filtering for implicit feedback datasets, in *Proceedings of the 8th IEEE International Conference on Data Mining, December 15-19, Pisa, Italy*, (IEEE Computer Society, 2008).
16. S. Rendle, C. Freudenthaler, Z. Gantner and L. Schmidt-Thieme, BPR: bayesian personalized ranking from implicit feedback, in *Proceedings of the 25th Conference on Uncertainty in Artificial Intelligence, Montreal, QC, Canada, June 18-21, 2009*, (AUAI Press, 2009).
17. Y. Liu, W. Wei, A. Sun and C. Miao, Exploiting geographical neighborhood characteristics for location recommendation, in *Proceedings of the 23rd International ACM CIKM Conference, Shanghai, China, November 3-7, 2014*, (ACM, 2014).
18. X. Li, G. Cong, X. Li, T. N. Pham and S. Krishnaswamy, Rank-geofm: A ranking based geographical factorization method for point of interest recommendation, in *Proceedings of the 38th International ACM SIGIR Conference, Santiago, Chile, August 9-13, 2015*, (ACM, 2015).
19. S. Choachaicharoenkul, D. W. Coit and N. Wattanapongsakorn, Multiobjective trip planning with solution ranking based on user preference and restaurant selection, *IEEE Access* **10**, 10688 (2022).

Optimizing Influence Maximization in Social Networks Using Seed Node Update Strategy

Xinjie Bai[*,†], Hongmei Chen[*,†,§], Chuan Luo[‡] and Tianrui Li[*,†]

[*]*School of Computing and Artificial Intelligence, Southwest Jiaotong University, Chengdu 611756, China*
[†]*National Engineering Laboratory of Integrated Transportation Big Data Application Technology, Southwest Jiaotong University, Chengdu 611756, China*
[‡]*College of Computer Science, Sichuan University, Chengdu 610065, China*
[§]*hmchen@swjtu.edu.cn*

Influence maximization is an important research topic in complex networks, aiming to find a seed node set with the most significant ultimate influence in social networks. Recently, the I_WVoteRank algorithm proposed a semi-local extended neighborhood concept based on the WVoteRank algorithm, considering the 1-hop and 2-hop neighbors in the voting process to calculate the voting scores of nodes. However, it initializes the voting ability of all nodes to 1. In contrast, we propose a voting method called WVoteRank++, which initializes the voting ability of nodes to their degree. This approach is significant as it considers the node's connectivity, thus providing a more accurate representation of its influence. To reduce the time complexity, we only update the voting scores of seed nodes and their influenced neighbors within three hops. Additionally, we introduce the concept of voting proportion, which measures the percentage of votes a node casts to its neighboring nodes. This concept is crucial in our algorithm as it helps determine the influence of a node on its neighbors. We use the Susceptible-Infected-Recovered and Linear Threshold models to simulate the spreading progress on six real and synthetic datasets. The experimental results show that, in most cases, the proposed algorithm outperforms other baseline methods regarding the final infected scale, the infection scale, and the LT model.

Keywords: Influence maximization; social networks; voting approach; SIR model.

1. Introduction

As network data and Internet technology continue to evolve rapidly, the emergence of complex networks and social networks is a prime example. Social network research focuses on community detection, influence maximization, and network propagation (network node importance ranking, network robustness analysis, etc.) [1]. In these networks, information spreads astonishingly, potentially reaching the entire network. This makes the influence maximization

---

[§]Corresponding author.

(IM) problem particularly relevant in practical fields, such as marketing, advertisement deployment, and public sentiment monitoring. The IM problem seeks to identify k seed nodes within a social network that can achieve the widest impact radius. The ranking of node importance and the identification of key nodes are of paramount importance. The IM problem is considered an NP-hard problem, and greedy algorithms are often considered approximate solutions. However, the high computational cost limits their applicability to small networks. Kumar *et al.* [2] proposed an improved voting method based on an extended neighborhoods concept, which considers 1-hop and 2-hop neighbors during the voting process to determine influential nodes. However, the algorithm assigns an identical initial value to all nodes, which does not effectively determine the initial node set. Selecting the initial node set is a formidable task as it dramatically impacts the network structure and maximizes its influence during propagation [3].

Liu *et al.* [4] proposed a voting method named VoteRank++, which considers the varying degrees of closeness between nodes and reduces the voting ability of 2-hop neighbors of the selected nodes. This method updates only the scores of nodes whose voting scores may change, avoiding the cost of calculating the voting scores of all nodes in each iteration. However, this approach has its limitations. Despite nodes of different degrees having different voting capabilities, the algorithm only considers the 1-hop neighbors for the voting process, neglecting the broader neighborhood. It is important to note that nodes beyond direct neighbors also play a significant role, a factor that our proposed method, WVoteRank++, addresses.

This chapter introduces a novel approach to the influence maximization problem — the WVoteRank++ algorithm. This innovative method initializes the voting scores of nodes based on their degree, taking into account both the node itself and its neighboring nodes. We also introduce the concept of voting proportion to evaluate the percentage of votes a node gives its neighbors, considering the voting scores of 1-hop and 2-hop neighbors [2]. We only update the voting scores of seed nodes and their neighbors within three hops to optimize time efficiency. We validate the effectiveness of our method through simulations using the Susceptible-Infected-Recovered [5] and Linear Threshold models [6]. The results clearly show that WVoteRank++ surpasses the baseline methods in spreading speed and infected scale.

The rest of this chapter is structured as follows. Section 2 describes the proposed algorithm, and Section 3 shows the analysis of experimental data and the experimental results. Finally, we discuss some conclusions and future research directions in Section 4.

2. Proposed Algorithm

In our research, we have proposed a method to enhance the I_WVoteRank [2] algorithm to identify a subset of influential nodes. We have improved the computation methods for both the voting score and voting ability while reducing the time complexity of node updates. For each node $v \in V$ in the network, there exists a tuple (VS_v, Va_v), where VS_v and Va_v is the voting score of obtained from neighbor nodes up to two hops of node v and voting ability of node v, respectively. The detailed procedure of WVoteRank++ is shown in Algorithm 1. We will now provide a comprehensive overview of the algorithm.

Algorithm 1: WVoteRank++

```
Input: A network G(V,E), the number of spreaders l.
Output: A set S including l influential nodes.
S ← ∅
foreach v ∈ V do
    Calculating the initial voting ability of each node by Eq. (1).
while |S| < l do
    H ← Γ_v
    foreach u ∈ Γ_v do
        Updating the voting ability of the 1-hop neighbors of seed nodes by Eq. (4).
        foreach w ∈ Γ_u do
            Updating the voting ability of the 2-hop neighbors of seed nodes by Eq. (4).
    foreach v ∈ H do
        foreach i ∈ Γ_v do
            foreach j ∈ Γ_i do
                Determining the voting proportion VP_uv of node v by Eq. (2).
        Calculating the voting score of each node by Eq. (3).
```

2.1. *Initializing voting ability and score*

The initial node set has the potential to significantly impact the special nodes in the network structure, enabling them to achieve maximum influence during the propagation process. We assign $Vs_v = 0$ to each node and define the initial value of Va_v as its degree. Mathematically, the Va_v of node v is calculated as

$$Va_v = log\left(1 + \frac{k_v}{k_{max}}\right), \tag{1}$$

where k_v is the degree of node v, k_{max} is the highest degree among all nodes.

2.2. *Calculating voting score*

Due to variations in the similarity and attractiveness between nodes [4], different neighboring nodes receive varying percentages of votes from the same node. For instance, given a node u, the voting proportion VP_{uv} of u towards its neighbor node v can be calculated using the following formula

$$VP_{uv} = \frac{k_v}{\sum_{w \in \tau(u)} k_w}, \tag{2}$$

where k_w is the degree of node w, w is the direct neighbor of u.

I_WVoteRank [2] introduced a novel approach called the semi-local extended neighborhood concept, which considers both the immediate neighbors (1-hop) and secondary neighbors (2-hop) to determine the voting score of a node. The voting score Vs_v is represented by the following equation

$$Vs_v = \sqrt{|\tau(v)| \left(\sum_{i \in \tau(v)} V_{a_i} \times VP_{iv} \right) + \sum_{j \in \tau(i)} V_{a_j} \times VP_{ji}}, \tag{3}$$

where $\tau(v)$ is the one-hop neighbor collection set of vertex v, and $|\tau(v)|$ denotes the number of one-hop neighbors of v. VP_{iv} stands for the voting proportion between v and i, where i is the direct neighbor of v. And j is the direct neighbor of i and the two-hop neighbor of v in VP_{ji}.

After obtaining the scores of all nodes, the node with the highest score is selected as the seed node. Its voting ability is updated to 0 to ensure that the selected node does not participate in the next iteration of the voting process.

2.3. *Suppressing neighbors and updating*

In order to achieve coverage of the entire network by the seed node set and reduce the probability of selecting neighboring nodes within two hops of the chosen nodes during the voting phase, we propose to decrease the voting ability of their neighbor nodes. Since the voting score of seed nodes is derived from their neighbors up to two hops away, the following equation indicates the updated voting ability of the 1-hop and 2-hop neighbors

$$V_{a_v} = \begin{cases} V_{a_v} - \delta & if\ V_{a_v} - \delta > 0 \\ 0 & otherwise \end{cases}, \tag{4}$$

$$\delta = \frac{1}{\langle k \rangle d}, \tag{5}$$

where $\langle k \rangle$ is the average degree of the nodes. If a node is a one-hop neighbor, its degree is d=1; for a two-hop neighbor, d=2.

Since the remaining nodes have not been altered, in order to reduce time complexity, only the seed nodes and their voting scores up to three-hop neighbors are updated.

3. Experimental Study

To validate the effectiveness of the algorithm, we will compare the proposed method with popular baseline methods, namely Degree [7], K-shell [8], IKS [9], H-index [10], VoteRank [11], EnRenew [12], MCDE [13], ECRM [14] and VoteRank++ [4] on various real and synthetic datasets. The fundamental characteristics of these experimental networks are presented in Table 1.

Table 1. Fundamental characteristics of 6 experimental networks

Network	Nodes	Links	k_max	<k>
CENew	453	2025	237	8.94
Router	5022	6258	106	2.49
Email	1133	5451	71	9.62
USAir2010	1574	17215	314	21.87
Hamster	2426	16631	273	13.71
Condmat	23133	93497	281	8.08

The susceptible infected recovered (SIR) model, a crucial tool in network science, is employed to simulate the spreading progress. It involves three types of nodes: susceptible nodes (S), which have not yet received information from neighboring nodes; infected nodes (I), which propagate the information; recovered nodes (R), which have already propagated the information and no longer spread it. The initial nodes infect susceptible nodes with a probability μ and recover to become recovered nodes with a recovery rate β, where recovered nodes cannot be infected again. The infection rate μ has a significant impact on the propagation process. When μ is large, the information can rapidly spread on the network [4]. We considered the infection rate $\mu = 1.7$ and the ratio of initial infected nodes $\rho = 0.03$ in the subsequent experiments.

The final infection scale $\mathrm{F(t_c)}$, a key metric representing the extent of information propagation in the network, is the number of nodes that are infected and then recover in the next iteration in the SIR model. Assuming the SIR simulation is completed at time $\mathrm{t_c}$, $\mathrm{F(t_c)}$ enumerates all nodes that were infected during the information dissemination process and then recovered at time t_c. Figure 1 illustrates the $\mathrm{F(t_c)}$ values resulting from different initial infection ratios ρ after 1000 iterations using different methods. Notably, when the initial infection rate is the same, the larger the proportion, the wider the final infection scale of WVoteRank++, especially on Condmat, Email, Hamster, and Router networks. As ρ gradually increases, the influence range of WVoteRank++ rapidly expands compared to other methods, indicating the potential for the information to spread to a greater number of nodes in subsequent iterations.

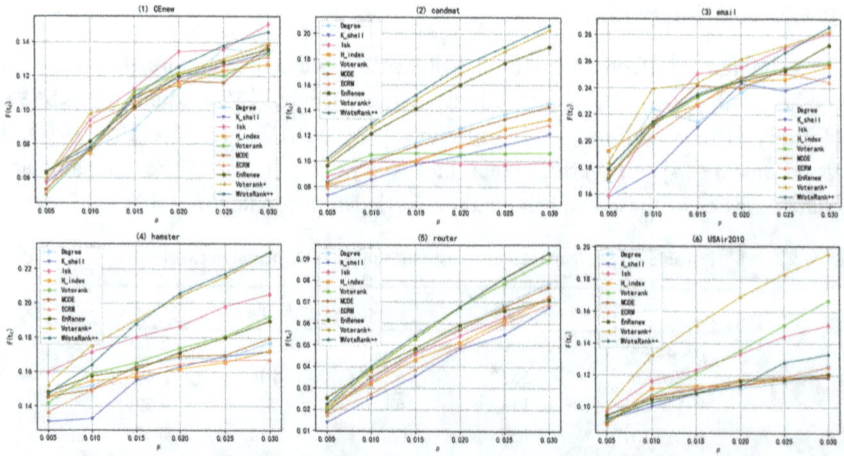

Fig. 1. $F(t_c)$ resulting from different initial infection ratios ρ after 1000 iterations

The infection scale provides a temporal scope of information from spreading nodes in the network, serving as an indicator of the propagation speed. Figure 2 illustrates the changes in infection scale over time, after 1000 iterations, under different methods. Among the compared algorithms, WVoteRank++ consistently outperforms others regarding infection scale during the stable phase across Condmat, Email, Hamster, and Router networks. Furthermore, regarding propagation speed, WVoteRank++ exhibits a shorter duration in influencing a similar number of nodes in the CEnew, Condmat, and Router networks. This implies that the influential nodes selected by our method possess more robust dissemination capabilities.

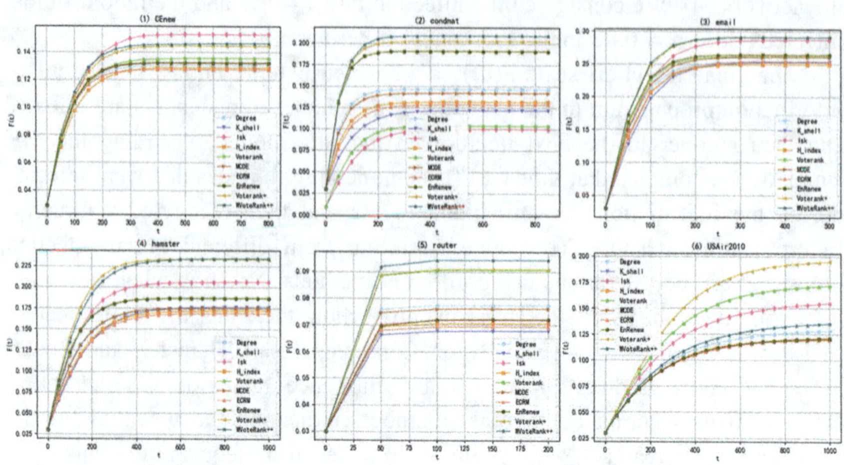

Fig. 2. The infection scale $F(t)$ at various time t after 1000 iterations

In addition to the SIR model, the LT model [6] is also employed to measure the number of stable activated nodes by influencing the spread of nodes. The final number of infected nodes activated by different initial infection ratios ρ is shown in Figure 3. WVoteRank++ performs well on most networks, particularly CEnew, Condmat, Email, Hamster, and USAir2010. Despite potentially less performance on the Router network, according to the SIR model, there is a noticeable enhancement in $F(t_c)$, thus demonstrating the versatility of the proposed method in most cases measured by the SIR and LT models.

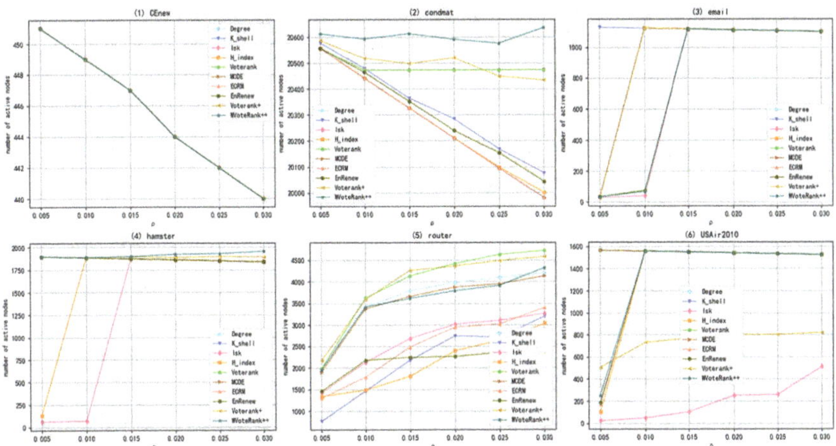

Fig. 3. The final number of infected nodes activated by different initial infection ratios ρ

4. Conclusions and Future Work

This chapter proposes an algorithm based on WVoteRank++ to identify the most influential nodes in complex networks. The algorithm utilizes node degree to initialize the voting ability, introduces the concept of voting proportion to measure the percentage of votes that a node gives to its neighbors, and calculates the voting score of each node by summing the voting ability of its 1-hop and 2-hop neighbors. Finally, the algorithm iteratively selects a set of high-impact seed nodes by updating the voting scores of the seed nodes and their neighbors within three hops. The SIR model simulates the information propagation process on six real complex networks. The experimental results of $F(t_c)$, $F(t)$, and LT model demonstrate that the WVoteRank++ outperforms other baseline methods. In future research, we will consider the mutual influence between nodes when initializing the voting capability. As an exceptional form of social network, researching how to utilize voting algorithms to identify the most influential papers or authors in academic networks is also a promising area for investigation.

Acknowledgments

This work was supported by the National Natural Science Foundation of China (Nos. 62376230, 61976182, 62076171, and 62306196), and the Natural Science Foundation of Sichuan Province (2022NSFSC0898).

References

1. G. Facchetti, G. Iacono and C. Altafini. Computing Global Structural Balance in Large-Scale Signed Social Networks[C]. Proceedings of the National Academy of Sciences of the United States of America, 108(52):20953-20958, 2011.
2. S. Kumar and A. Panda. Identifying influential nodes in weighted complex networks using an improved WVoteRank approach[J]. Applied Intelligence, 52:1838-1852, 2022.
3. E. Yu, Y. Fu, Q. Tang, J. Y. Zhao and D. B. Chen. A Re-Ranking Algorithm for Identifying Influential Nodes in Complex Networks[J]. IEEE Access, 8:211281-211290, 2020.
4. P. Liu, L. Li, S. Fang and Y. Yao. Identifying influential nodes in social networks: A voting approach[J]. Chaos Solitons Fractals, 152:111309, 2021.
5. R. PastorSatorras and A. Vespignani. Epidemic dynamics and endemic states in complex networks[J]. Physical Review E, 63(6):066117, 2001.
6. D. Kempe, J. M. Kleinberg and É. Tardos. Maximizing the Spread of Influence through a Social Network[J]. Theory of Computing, 11:105-147, 2015.
7. LC. Freeman. Centrality in social networks conceptual clarification[J]. Social networks, 1(3):215-239, 1979.
8. G. Sabidussi. The centrality index of a graph[J]. Psychometrika, 31(4):581-603, 1966.
9. Z. Li and X. Huang. Identifying influential spreaders in complex networks by an improved gravity model[J]. Scientific Reports, 11 (1):22194-22194, 2021.
10. L. Lü, T. Zhou, Q. Zhang and H. E. Stanley. The H-index of a network node and its relation to degree and coreness[J]. Nature communications, 7(1):10168, 2016.
11. J. Zhang, D. Chen, Q. Dong and Z. Zhao. Identifying a set of influential spreaders in complex networks[J]. Scientific reports, 6(1):27823, 2016.
12. C. Guo, L. Yang, X. Chen, D. Chen, H. Gao and J. Ma. Influential Nodes Identification in Complex Networks via Information Entropy[J]. Entropy, 22(2):242-242, 2020.
13. A. Sheikhahmadi and M. A. Nematbakhsh. Identification of multi-spreader users in social networks for viral marketing[J]. Journal of Information Science, 43(3):412-423, 2017.
14. A. Zareie, A. Sheikhahmadi, M. Jalili, M. S. K. Fasaei. Finding influential nodes in social networks based on neighborhood correlation coefficient[J]. Knowledge-based systems, 194:105580, 2020.

A Design Method of Surveying Robotics for Transmission Line Construction

Wenzheng Wang[*,‡], Xiangqing Wen[†,‡,§], Bin Zhang[†,‡], Liangmin Zhong[*,‡], Jianguo Hou[†,‡] and Xuan Jiang[†,‡]

[*]*Shenzhen NARI Technology Co., Ltd, Shenzhen, Guangdong, China*

[†]*NARI Technology Co., Ltd, Nanjing, JiangShu, China*

[‡]*NARI Group Corporation (State Grid Electric Power Research Institute) Nanjing, JiangShu, China*

[§]*wenxiangqing@sgepri.sgcc.com.cn*

Transmission line construction is a work with high safety risks and heavy tasks. The traditional manual method for measuring line sag is influenced by objective factors, such as terrain, weather, and span, and has certain limitations in extreme environments. In addition, spacer rods are the most hardware components in transmission line construction, and the manual measurement of the installation position of conductor spacer consumes a lot of manpower. This chapter proposes a robot that can move directly on the transmission line, which can achieve measure sag value during the process of transmission line stringing construction and also can mark the installation position of the conductor spacer during the installation process. This chapter introduces the design of the structure, control system, sag measurement method, and implementation of the conductor spacer marking function . Finally, the test results of the robot on site and subsequent work are introduced.

Keywords: Transmission line construction; sag measurement; spacer installation; robot.

1. Introduction

Transmission line stringing construction has the highest safety risk in transmission line construction tasks and is also the most difficult part to control safety and quality. The key is to measure and control the sag value of the line, not any positive or negative deviation are allowed[1]. Positive deviation could cause insufficient distance to the crossing object, leading to discharge tripping accidents. Negative deviation could cause a geometric increase in the tension of the wire, posing a safety risk of overloading the tower and the grounding wire itself. At present, transmission line sag measurement is mainly carried out through manual observation[2]. Real-time monitoring methods for transmission line sag based on different principles such as tension, inclination, wire temperature, stress, images[3,4], and unmanned aerial vehicles equipped with optical

[§]Corresponding author.

scanners [5,6] have also been proposed. However, these methods are either inconvenient to operate or have low measurement accuracy that does not meet the requirements.

In addition, the installation workload of spacer rods during transmission line construction is huge. At present, the installation position of spacer rods is mainly measured and marked manually, which has problems such as high-altitude operations, high labor intensity, and low efficiency[9,10].

This chapter proposes a robot that can autonomously move on the overhead transmission line. The robot can measure the altitude of the line's sag point, calculate and calibrate the data, and feed back the measurement results to the control personnel to achieve precise control of the sag value. During the installation process of the spacer rods, the robot uses the tower as the base point to measure the installation position of the spacer rods one by one, and complete the marking, which saves a lot of work.

2. Design Method

2.1. *The structural design of the robot*

The main structure of the robot includes four parts: driving mechanism, pressing mechanism, marking mechanism, control box. The driving mechanism consists of two sets of walking wheels that are directly driven by dual DC motors. Each set of walking wheels is equipped with an encoder and brake to ensure the walking speed, climbing ability, and motion reliability of the robot.

Located in the middle of the driving mechanism is the marking mechanism, which can mark the installation position of the spacer rod measured by the robot. The pressing mechanism is located below the marking mechanism, which can improve the contact between the walking wheel and the ground wire, prevent the walking wheel from deviating, and also improve the climbing ability of the walking wheel; The clamping mechanism is installed on the guide shaft, which can ensure that the trajectory of the clamping mechanism's up and down movement is fixed to achieve reliable operation. The bottom of the robot is a control box, responsible for energy supply, motion control, information communication, human-machine interaction, information perception and other functions. In addition, located at the top right end, there is a single line radar that can measure the height difference between the surrounding wire and the wire where the robot is located, thereby calculating the sag value of the surrounding wire[11-13]. The RTK antenna is located at the top left end where provides the best signal. As shown in Fig.1.

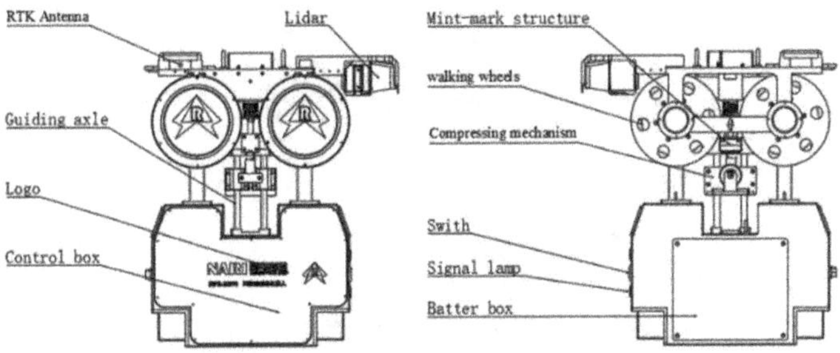

Fig. 1. The structural design of the robot

2.2. *Control system design scheme*

The control system of the transmission line surveying robot is divided into motion control platform and operation management platform. The motion control unit adopts the STM32H743ZIT6 chip based on the Cortex-M7 architecture as the main motion control core, serving as the "cerebellum" of the robot; The operation management unit adopts NVIDIA Pascal™, the architecture of the artificial intelligence supercomputer module Jetson TX2 builds the "intelligent brain" of robots. The hardware and network architecture are shown in the following Fig. 2.

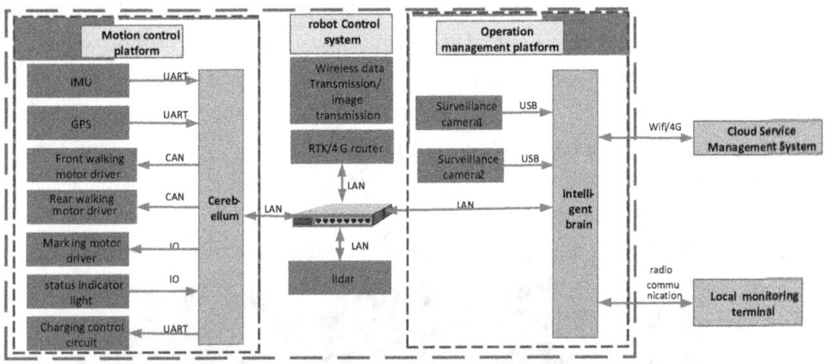

Fig. 2. The control system and network architecture

The software function is shown in Fig 3. According to the robot's function requirements, three modules are deployed in the robot's "intelligent brain": robot task planning module, line sag value calculation, and robot operation control, to achieve intelligent control of robot operations; As the real-time controller of the robot, the "cerebellum" is responsible for managing the robot's motion status, collecting sensor information, and controlling the robot's walking and marking actions.

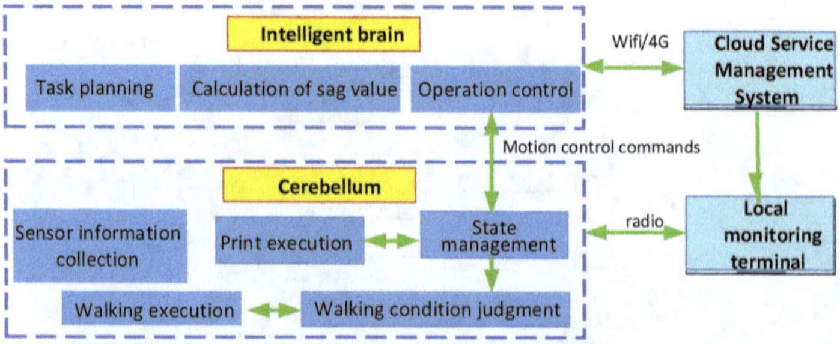

Fig. 3. Control System Software Functional Framework.

3. Sag Value Measurement and Correction Method

3.1. *The workflow of the robot*

The following figure shows the dynamic display interface of sag values, where the green line represents the design value and the red line represents the measured value. The workflow of the robot is as follows:

(i) The robot moves to the sag measurement point P and issues a tightening command to the tightening operator to start the tightening machine.

(ii) After the tightening machine is started, the position of the robot will also change with the movement of the wire. At this time, the robot will move autonomously to maintain its position near the sag measurement point P.

(iii) The robot sends real-time sag measurement values to the tightening machine operator. When the operator of the tightening machine observes that the sag value is close to the design value, a deceleration command is issued.

(iv) When the sag value reaches the design value (the red line coincides with the green line), the tightening machine operator stops the tightening operation.

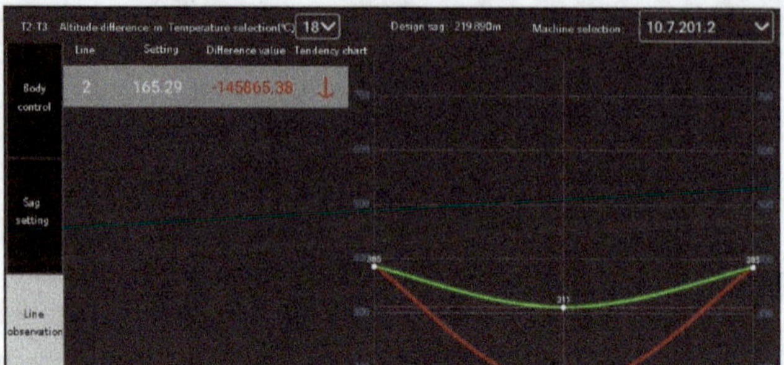

Fig. 4. Sag value display interface

3.2. *Calculation and correction of sag values*

Unlike indirect measurement methods, robots can directly measure the height information of sag points to obtain sag values, The maximum sag point of the line occurs at the center of the line span[14], so the calculation method for sag value f is as follows:

$$f = \frac{h_A + h_B}{2} - h_C \tag{1}$$

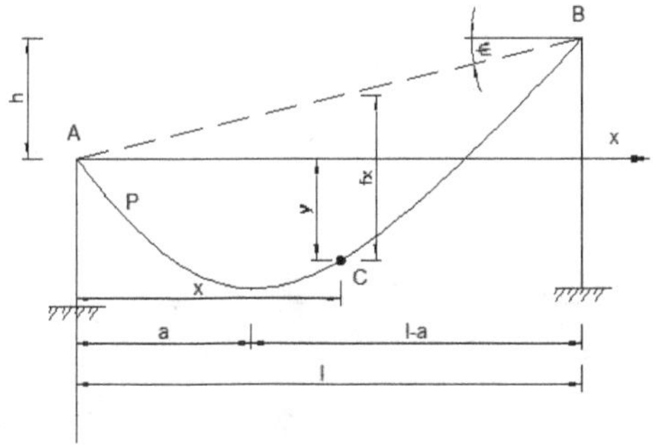

Fig. 5. Schematic diagram of sag value calculation

The measured sag values need to consider the influence of the robot's own weight on the sag measurement values. Due to the own weight of the robot, the measured sag value is greater than the actual sag value, resulting in a negative deviation of the sag value. By using the principle of calculating line sag under concentrated load, the correction value can be calculated.

Based on the parabolic model principle of overhead line conductor under a single concentrated load, the Maximum sag value of the overhead line under its own weight and a single concentrated load during robot operation is[14]:

$$f_{c\max} = \frac{l^2}{8H}\left(\frac{\omega}{\cos\psi} + \frac{2Q}{l}\right) \tag{2}$$

In the formula, Q represents the robot's own weight, l represents the line span, and H represents the horizontal tension of the grounding wire. ω represents self gravity per unit length of ground wire ψ represents the elevation angle of the height difference between two towers.

According to the parabolic model principle of overhead line conductors, the formula for the midpoint sag value of overhead lines under no load is[14]:

$$f_m = \frac{l^2\omega}{8H\cos\omega} \tag{3}$$

Based on equations (2) and (3), the correction value for the influence of the robot's self weight load on the sag value of the grounding wire at the midpoint of the line span is obtained as follows:

$$f_{\Delta} = f_{c\max} - f_m = \frac{Ql}{4H} \tag{4}$$

According to equations (1) and (4), obtain the corrected sag value f'.

$$f' = f - f_{\Delta} = f - \frac{Ql}{4H} \tag{5}$$

4. Spacer marking function

The robot is equipped with a odometer and an inertial navigation module, which can calculate the installation distance and position of the line spacer based on the length of the line span. The marking of the installation position of the spacer is also completed through a marking mechanism, as shown in Fig. 6.

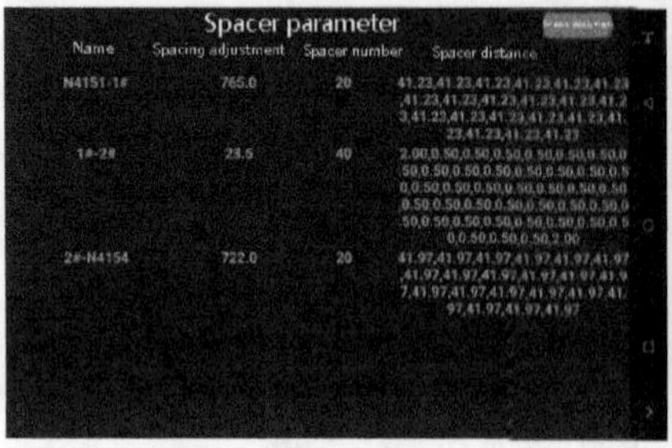

Fig. 6. Spacer mark position configuration interface

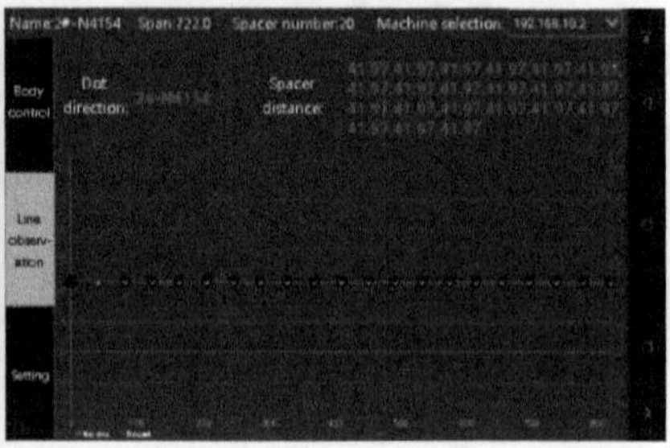

Fig. 7. Spacer mark display interface

During the robot marking operation, the system can display the real-time working status of the robot, including the direction of the robot marking operation, the spacing between the spacer rods, the progress of the operation completion, and other information. as shown in Fig. 7.

5. Test and Verification

Our testing environment is the Luoshan Yangtze River Crossing Project, a 1000kV ultra-high voltage AC project located in Yueyang, Hunan, China, and the 500kV Fengcheng-Meili transmission line Yangtze River Crossing Project located in Jiangsu, China. We conducted tests on the marking function of the spacer in the previous project and sag measurement in the latter project. The height of the two towers in 500kV Fengcheng-Meili transmission line Yangtze River Crossing Project is 333.46 meters. The sag point is located in the middle of the two towers. The designed sag value is 244.2 meters. The measured sag height and calculated sag values of the robot are shown in Table I, which show that the robot is getting closer to the design value of sag in the process of tightening, and the final construction value of sag point is very close to the design value, which can meet the construction requirements. The following figure show the test photos of the robot on site and effect of marking the installation position of the spacer in Luoshan Yangtze River Crossing Project.

Fig. 8. Robot field work picture

Fig. 9. Robot field marking effect

Table 1. Sag measuring record sheet (part)

Serial number	Height(m)	Sag value(m)
1	89.1	244.36
2	89	244.26
3	88.8	244.06
……	……	……

The SAG measuring robot scheme proposed in this paper is proved to be feasible by our field test, and it can help the overhead transmission line builders to improve work efficiency, reduce construction time and improve construction quality. In the future, our main work is to further enrich the 4G communication methods, and to ensure the stability of robot communication in any situation, the other is to further reduce the weight of the robot, making it easy to carry and place it on the ground wire.

References

1. Yadong Shang, Zhangpeng Zhou, Sag measurement of the 750 kV transmission lines, Advances of Power System & Hydroelectric Engineering. 24(5): 39-41(2008).
2. Chenguo Yao, Lei Zhang,Chengxiang Li, Yu Li, Zuo Zhou, Measurement method of conductor ice covered thickness based on mechanical analysis and sag measurement, High Voltage Engineering. 39(5): 1204-1207(2013).
3. Xingyu Ma, Fei Hou, Research and Application for transmission line sag Measurement Method Based on Aerial Image Sequence, 2016 3rd International Conference on Engineering Technology and Application, pp.390-394, (Kyoto,Japan,2016).
4. XinBo Huang, XiaoXia Zhang, LiCheng Li,Bing Luo, Measurement of Transmission Lines Conductor Sag Using Image Processing, High Voltage Engineering , 37(8): 1961-1965(2011).

5. P. Greeshma Shivani, Somani Harshit, Ch. Vamsi Varma, and R. Mahalakshmi, Detection of Icing and Calculation of Sag of Transmission Line Through Computer Vision， 2020 Third International Conference on Smart Systems and Inventive Technology. (Tirunelveli, India, 2020) , pp. 689–694.
6. Fang Ye, Junke Guo, Meng Tian, Zhangqi Wang, Xing Zhang, Xiangying Yu, Sag Measurement of Overhead Lines Based on 3D Direct Linear Transformation Theory, Journal of Noah China Electric Power Universit,44(4):71-75 (2017).
7. Shen Jinqing, Yao Hongwei , Chen Hua, et al. Research and Application of Spacer Installation Robot, E3S Web of Conferences. EDP Sciences(2021).
8. Liang Liang, Zhengqing Wu, Zhengyang Qian, et al. Reliability Analysis of Structure Strength based on Spacer Installation Robot, Journal of Physics: Conference Series, (Guilin, China,2021).
9. Filip Zori´c , Stjepan Flegari´c , Goran Vasiljevi´c , et al. Autonomous Installation of Electrical Spacers on Power Lines Using Magnetic Localization and Special End Effector, Machines, 11(5),510(2023).
10. Jonathan Cacace, Santos M. Orozco-Soto , Alejandro Suarez ,et al. Safe Local Aerial Manipulation for the Installation of Devices on Power Lines: AERIAL-CORE First Year Results and Designs, Applied Sciences. 11(13):6220(2021).
11. Kirti Shankar Sharma and P. V. Manivannan, Stereo Image Partitioning based Fuzzy Logic Controller for Real-time Obstacle Detection and Avoidance, International Journal of Mechanical Engineering and Robotics Research, Vol. 9, No. 8, pp. 1158-1169 (August 2020).
12. Eftichia Badeka, Theofanis Kalampokas, Eleni Vrochidou, Konstantinos Tziridis, George A. Papakostas, Theodore P. Pachidis, Vassilis G. Kaburlasos, Vision-based Vineyard Trunk Detection and its Integration into a Grapes Harvesting Robot, International Journal of Mechanical Engineering and Robotics Research, Vol.10, No. 7, pp.374-385(July 2021).
13. Deshinta Arrova Dewi, Elankovan Sundararajan, Anton Satria Prabuwono, and Lau Meng Cheng, Object Detection without Color Feature: Case Study Autonomous Robot, International Journal of Mechanical Engineering and Robotics Research, Vol. 8, No. 4, pp. 646-650(July 2019).
14. Bozhi Li, Construction calculation principle of high voltage overhead transmission line , 2nd edn. (China Electric Power Press, Beijing, 2008).

Integrating Social Information Learned through Auto-encoder into Matrix Factorization for Recommender Systems

Tianyou Li[*,†], Hongmei Chen[*,†,§], Chuan Luo[‡] and Tianrui Li[*,†]

[*]*School of Computing and Artificial Intelligence, Southwest Jiaotong University, Chengdu 611756, China*

[†]*National Engineering Laboratory of Integrated Transportation Big Data Application Technology, Southwest Jiaotong University, Chengdu 611756, China*

[‡]*College of Computer Science, Sichuan University, Chengdu 610065, China*

[§]*hmchen@swjtu.edu.cn*

Cold start and data sparsity issues often plague traditional recommendation systems, as they only rely on rating matrices. Social recommendation systems address these issues by considering users' social trust relationships. However, trust data is often sparse because users need more information on these supplementary resources. To overcome these challenges, we propose a new recommendation model that models the trust relationship matrix using deep neural networks. In this process, we use the sparse auto-encoder to learn latent features of the user's trust relationships and use the learned latent features to construct the user's neighbors. Finally, we combine the original and learned latent social relationships with the matrix factorization model. The experimental results on two datasets have validated the efficiency of this method.

Keywords: Recommender systems; deep neural networks; sparse auto-encoder; matrix factorization.

1. Introduction

Due to the continuous increase in Internet users, the online world constantly expands with information. For this phenomenon, we can use machine learning algorithms to help users quickly identify intelligent systems for the information they want. The most effective and widely used of these systems is the recommender system (RS).[1] RS serves to understand users' past behavior, model their preferences, and use the model to predict potential ratings for items. However, because many items are only rated by a few users, the rating matrix is often heavily sparse, which causes cold start and data sparsity issues in recommendation systems.[2]

[§]Corresponding author.

Additional information can be considered for recommendation to address the data sparsity problem in RS.[3] Social recommendation systems successfully improve recommendation quality by utilizing trust relationships or tag information between users as additional information.[4,5] The social recommendation system assumes that two users with a trust relationship usually have similar opinions.[6] Integrating social information can tackle the problem of data sparsity, but there are still some problems. Social relationships are often sparse, and many users only have relatively few clear social relationships. The effectiveness of social recommendation systems is very limited when users lack social relationships.

In order to solve the problem mentioned above, this chapter proposes a recommender system that combines deep neural networks and matrix decomposition. The proposed method first utilizes the sparse auto-encoder to extract latent features from the trust relationship matrix. The extracted low-dimensional latent features are then used to extend the MF method and generate recommendations by predicting unknown items.

The remaining part of this chapter is structured as follows. Section 2 presents the proposed method. Section 3 presents the setup of experiments and experimental results. Finally, some conclusions and future research directions are presented.

2. Proposed Method

This section mainly describes the proposed recommender system that integrates the latent feature vectors learned by the sparse auto-encoder into the matrix factorization model.

2.1. *Sparse auto-encoder*

Auto-encoder is a neural network model that can be used to learn low dimensional representations of input data in an unsupervised way[7]. It mainly consists of an encoder and a decoder. The encoder includes an input layer and a hidden layer, through which the input data can be represented as low dimensional latent features. The decoder includes a hidden layer and an output layer, which are used to convert feature vectors back to the original input space.

For a set of input samples $X = \{x_1, x_2, \dots, x_m\}$ containing m input vectors , the encoder obtains its new feature vector $h(x_i) \in R^k$, where k is the number of latent features. In the decoder section, the obtained feature vectors can be mapped back to the reconstructed version of the original data. The entire process can be described as:

$$h(x_i) = \delta(Wx_i + b_1), \tag{1}$$

$$y(h(x_i)) = \delta(W^T h(x_i) + b_2), \tag{2}$$

where W and W^T are the connect weight matrix, b_1 and b_2 represent the bias of the input layer and the output layer, respectively.

The objective of an auto-encoder is to minimize the reconstruction error between the input vector x_i and the output vector $y(h(x_i))$. The objective function of the auto-encoder is defined as

$$J(W,b,X) = \min_{W,b,c} \sum_{i=1}^{m} \left\| y\big(h(x_i)\big) - x_i \right\|^2. \tag{3}$$

Sparse auto-encoder optimizes traditional auto-encoders by introducing sparsity constraints in hidden layers. The sparse auto-encoder imposes constraints on the average activation value of hidden layer neurons and uses KL divergence to approach the default value to achieve a sparsity effect, which will be included as a penalty term in the objective function. The objective function of a sparse auto-encoder is defined as

$$J_{sparse}(W,b,X) = \min_{W,b,c} \sum_{i=1}^{m} ||y\big(h(x_i)\big) - x_i||^2 + \gamma \sum_{j=1}^{m} KL(\rho||\hat{\rho}), \tag{4}$$

where β is a random value in $[0,1]$ used to control the extent of the sparsity penalty factor. $\sum_{j=1}^{m} KL(\rho||\hat{\rho})$ is defined as

$$\sum_{j=1}^{m} KL(\rho||\hat{\rho}) = \sum_{j=1}^{m} \rho log \frac{\rho}{\widehat{\rho_i}} + (1-\rho) log \frac{1-\rho}{1-\widehat{\rho_i}}, \tag{5}$$

where $\rho = \sum_{j=1}^{m}(a_j(x_i))$ represents the average activation of the jth hidden layer neuron, m represents the number of hidden units in each layer, and a_j is the activation value of the neurons.

Due to many users lacking sufficient trust relationships, the trust relationship matrix is usually highly sparse. For this issue, the sparse auto-encode[8] is used to learn latent features of trust relationships. In order to learn latent features using the auto-encoder, the input value should be a one-dimensional vector. Therefore, for each user u in the user set U, the vector $S_u \in R^{|U|}$ of user u can be constructed based on the trust relationship matrix S, which corresponds to the rows in S. In order to learn latent features, these vectors will serve as inputs for the sparse auto-encoder. The sparse auto-encoder has a sparsity limitation, where the network selects k neurons with the highest activation value and ignores most other neurons. So, the corresponding latent feature $\widehat{S_u} \in R^k$ $(k \ll |U|)$ can be obtained through a sparse auto-encoder.

After obtaining latent features through sparse auto-encoders, each user's trust relationship will have a low dimensional representation. These acquired representations can serve as supplements to enrich social relationships.

We can use these representations to calculate the similarity between users to obtain the user's nearest neighbor. The cosine similarity between users is

$$CosSim(u_1, u_2) = \frac{\widehat{S_1} \cdot \widehat{S_2}}{|\widehat{S_1}||\widehat{S_2}|}, \tag{6}$$

where $\widehat{S_1} \in R^k$ and $\widehat{S_2} \in R^k$ are the latent features vector of the user u_1 and user u_2, respectively. Users with relatively high similarity values are considered to have latent trust relationships.

2.2. *Matrix factorization*

Matrix factorization (MF) is a classic method applied in recommendation systems, which factorizes the user-item rating matrix into a user latent factor $p_u \in R^d$ and an item latent factor $q_i \in R^d$ to predict the score. On the basis of SoReg[9], we incorporate trust relationships into the regularization term to constrain the optimization of the objective function. The objective function is defined as

$$L = \sum_{u,i \in R}\left(r_{u,i} - p_u^T q_i\right) + \lambda(||p_u||_F^2 + ||q_i||_F^2) + \alpha \sum_u \sum_{v \in F(u)} t_{u,v} ||p_u - p_v||_F^2$$
$$+ \beta \sum_u \sum_{v \in S(u)} s_{u,v} ||p_u - p_v||_F^2, \tag{7}$$

where $r_{u,i}$ represents the rating of user u on item i, while $||p_u||_F^2$ and $||q_i||_F^2$ are regularization terms used to prevent overfitting, λ is a regularization parameter. $||p_u - p_v||_F^2$ is the social regularization term used to reduce the latent factor distance between user u and user v. $F(u)$ denotes the original set of trusted users by user u, $t_{u,v}$ is the original trust value between users u and v. Besides, $S(u)$ denotes a set of users with latent trust relationships for user u, and $s_{u,v}$ is the trust similarity between users u and v obtained through Eq. (6). α and β are parameter to control the intensity of social constraints. This paper chooses the stochastic gradient descent (SGD) method to learn the parameters in Eq. (7).

3. Experimental Study

3.1. *Datasets and evaluation metrics*

This method was tested on the FilmTrust and CiaoDVD datasets. The FilmTrust dataset contains 35497 ratings, 1642 users, 2071 items, and 1853 trust relationships. The CiaoDVD dataset contains 278483 ratings, 7375 users, 99746 items, and 111781 trust relationships. Table 1 shows the specific information of these datasets.

Table 1. Datasets

	user	item	trust
FilmTrust	1642	2071	1853
CiaoDVD	7375	99746	111781

Two metrics, including the mean absolute error (MAE) and the root mean square error (RMSE), are employed to evaluate the effectiveness of the proposed method.

3.2. *Experimental results*

In our meticulously conducted experiment, we employ the fivefold cross-validation method. Parameters are carefully determined to ensure the optimal performance of the algorithm. We utilize the sparse auto-encoder with three hidden layers on the FilmTrust and CiaoDVD datasets to learn latent features. The number of neurons of the sparse auto-encoder is precisely set to 1200 in the first hidden layer, 600 in the second layer, and 300 in the third layer.

Moreover, for the FilmTrust and CiaoDVD datasets, the number of latent features k learned from the trust matrix is set to $k = 60$ and $k = 70$, respectively. γ is used in Eq. (4) to adjust the weight of sparsity penalty, and is set to $\gamma = 0.1$. The dimensions d of factor p_u and q_i are set to 10, α and β are the weights used to control social constraints and are both set to 0.01, and the regularization parameter λ is set to 0.001. The number of $S(u)$ is set to $|S(u)| = 5$ because having too many users in the user's neighbor can reduce the accuracy of the recommendation.

To test the performance of our algorithm, we choose the SVD++[10] algorithm, the TrustSVD[11] algorithm, the SocialReg algorithm, and the CUNE[12] algorithm as comparative experiments. SVD++ is a classic matrix factorization model. SocialReg and TrustSVD are two mainstream matrix factorization-based social recommendation models that optimize the matrix factorization model by integrating social information. CUNE extracts implicit social information from product user feedback and integrates it into matrix factorization models. Compared with the above methods, the effectiveness of our model is shown in Table 2 and Table 3.

The experimental results show that the proposed algorithm has certain improvements in MAE and RMSE metrics compared to the comparative algorithm. For instance, on the FilmTrust dataset, the MAE value of our model is

0.6493, and the RMSE value is 0.8342 with the %60 training data. The MAE value of our model is 0.6415, and the RMSE value is 0.8295 with the %80 training data. When the training data is sparser, the performance improvement of our model is more significant.

Table 2. Experimental results on Filmtrust

Train	%60		%80	
Metrics	MAE	RMSE	MAE	RMSE
SVD++	0.6624	0.8645	0.6494	0.8613
TrustSVD	0.6588	0.8553	0.6472	0.8581
SocialReg	0.6647	0.8595	0.6431	0.8542
CUNE	0.6612	0.8439	0.6510	0.8426
Our model	**0.6493**	**0.8342**	**0.6415**	**0.8295**

Table 3. Experimental results on CiaoDVD

Train	%60		%80	
Metrics	MAE	RMSE	MAE	RMSE
SVD++	0.8424	1.1315	0.8218	1.1105
TrustSVD	0.8376	1.0921	0.7896	1.0818
SocialReg	0.8051	1.0457	0.7688	1.0053
CUNE	0.7967	1.0392	0.7651	0.9943
Our model	**0.7783**	**1.0181**	**0.7502**	**0.9857**

4. Conclusions and Future Work

This chapter unveils a method that fuses deep neural network-based trust relationship mining with matrix factorization. Integrating rating matrices, original social trust relationships, and latent trust relationships into the recommendation system opens up new possibilities for overcoming the data sparsity problem and enhancing recommendation performance. The potential of this approach, which also incorporates trust relationships, is bound to pique curiosity and anticipation.

In future work, we plan to incorporate more auxiliary information into recommendation systems to enhance recommendation performance and develop auxiliary information using other deep learning models.

Acknowledgments

This work was supported by the National Natural Science Foundation of China (Nos. 62376230, 61976182, 62076171, and 62306196), and the Natural Science Foundation of Sichuan Province (2022NSFSC0898).

References

1. G. Adomavicius and A. Tuzhilin, Toward the next generation of recommender systems: A survey of the state-of-the-art and possible extensions, *IEEE transactions on knowledge and data engineering* **17**, 734 (2005).
2. T. Anwar, V. Uma, M.I. Hussain and M. Pantula, Collaborative filtering and kNN based recommendation to overcome cold start and sparsity issues: A comparative analysis, *Multimedia Tools and Applications* **81**, 35693 (2022).
3. M. Jamali and M. Ester, A matrix factorization technique with trust propagation for recommendation in social networks, in *Proceedings of the fourth ACM conference on Recommender systems*, 2010.
4. D. Rafailidis and F. Crestani, Recommendation with social relationships via deep learning, in *Proceedings of the ACM SIGIR International Conference on Theory of Information Retrieval*, 2017.
5. E. Zheng, G.Y. Kondo, S. Zilora and Q.i. Yu, Tag-aware dynamic music recommendation, *Expert Systems with Applications* **106**, 244 (2018).
6. S. Ahmadian, M. Meghdadi and M. Afsharchi, Incorporating reliable virtual ratings into social recommendation systems, *Applied Intelligence* **48**, 4448 (2018).
7. Y. Bengio, A. Courville and P. Vincent, *Representation learning: A review and new perspectives, IEEE Transactions on pattern analysis and machine intelligence* **35** ,1798 (2013).
8. N. Jiang, W. Rong, B. Peng, Y. Nie and Z. Xiong, An empirical analysis of different sparse penalties for autoencoder in unsupervised feature learning, *2015 international joint conference on neural networks (IJCNN). IEEE*, 1, (2015).
9. H. Ma, D. Zhou, C. Liu, Recommender systems with social regularization, in *Proceedings of the fourth ACM international conference on Web search and data mining*, 2011.
10. Y. Koren, Factorization meets the neighborhood: a multifaceted collaborative filtering model, in *Proceedings of the 14th ACM SIGKDD international conference on Knowledge discovery and data mining*, 2008.

11. G. Guo, J. Zhang and N. Yorke-Smith, Trustsvd: Collaborative filtering with both the explicit and implicit influence of user trust and of item ratings, *in Proceedings of the AAAI conference on artificial intelligence*, 2015.
12. C. Zhang, L. Yu, Y. Wang, C. Shah and X. Zhang, Collaborative user network embedding for social recommender systems, in *Proceedings of the 2017 SIAM international conference on data mining*, 2017.

Voice-based Mood Recognition: An Application to Mental Health

Ricardo Hidalgo*, Victoria López and Diego Urgelés

Quantitative Methods, CUNEF University,
Madrid 28040, Spain
**ricardo.hidalgo@cunef.edu*

Voice recordings constitute information that enriches previous studies on mood disorders in which sensor data that measure physical activity are used. The analysis of all these data is applied to mental health to prevent crises due to depression or mania mainly. This chapter analyzes voice records and the application of machine learning for the first classification of emotions (sadness, happiness, fear, etc.). Subsequently, the emotions detected will be associated with mood states (mania, depression, mixed, euthymia, etc.). The usefulness of voice recordings to improve diagnostic prediction in the classification of mood disorders is concluded.

Keywords: Mental health; voice recording analysis; data analytics; support vector machines; mood disorders; fuzzy logic.

1. Introduction

The continuous advancement of technology allows individual behavior records to be monitored in real time through activity sensors. These data are especially useful in medicine applications[1] and more specifically on emotional disorders because of the lack of clinical data.[2]

In recent years, related research has been aimed at preventive and personalized medicine where sensor data is analyzed to determine when a manic or depressive episode will occur.[3,4]

The analysis of clinical data using machine learning techniques to aid in diagnosis has been widely used[5,6] and psychological and physiological indicators are related to emotional states.[7]

This chapter is integrated into the Bip4cast project[8] where sensor data is integrated with other clinical data sources. In this chapter, patient voice records are analyzed as a new source of information, and a proposal for integration in Bip4Cast is made based on the classification results of the emotional states detected.

The structure of this chapter is as follows. Section 2 shows the methodology of the study. It includes data collection and processing, visualization, and integration within the main project Bip4Cast. Section 3 is devoted to the results related to training from Spanish audios and training from two language (German and Spanish) audios. This chapter ends with the conclusions and future work in Section 4.

2. Methodology

Identifying emotional crises is complex. This approach personalizes patient care, uses a knowledge base for classifying emotional crises using accurate data, and weighs key variables. It includes a procedure for classifying emotional states from voice samples.

2.1. *Data collection*

We use two public anonymized audio databases. One of them contains about 500 audios in German.[9] This database was recorded by 10 persons (5 men and 5 women) between 25 and 35 years of age. The classification is done with 7 emotions (happy, anxious, fearful, disgusted, angry, and neutral). To make the recordings, the participants had to record 10 sentences in each of the mentioned moods. The second database used contains about 1500 audios in Spanish.[10] It was recorded by a woman and a man reading a series of sentences with different mood states (happy, anxious, fearful, disgusted, angry, and different kinds of neutral, such as fast or slow neutral). The total duration of this audios is about 8 hours. The information of interest used in this work is as follows: Acoustic amplitude (mean, min, max, std, range): measures the amplitude of the signal (Pascals). Pitch (mean, min, max, std, range): Refers to the fundamental frequency (first harmonic) (Hz). Harmonicity (mean, min, max, std, range): Represents the degree of acoustic periodicity (dB). Intensity (mean, min, max, std, range, quantile): Represents the intensity at linearly separated time points, (dB). Speech rate (syllables/second): Measures the speed and density of speech per unit of time. Articulation rate (syllables/second): Obtains a measure of speech rate where all syllables are excluded from the calculation. Energy (mean, min, max, std, range): Energy transmitted by sound waves, which comes from the vibration of the sound source, (Pascals). MFCC 1-12 (mean, min, max, std, range): They represent Mel's frequency cepstral coefficients as a function of time. Formants 1-5 (mean, min, max, std, range, quantile): Represents the spectral structure as a function of time,(Hz).

In addition, both databases classify these audios according to the emotions expressed in them. As a result, after processing we obtain the metadata sets that will be used for training and testing.

2.2. *Data processing*

The data processing in this study has two distinct but extremely linked parts, on the one hand, the signal processing to be performed on the sound files, and the other hand the processing of the characteristic metadata of each sample that must train the classification system that will indicate which categorical variable corresponding to a mood corresponds to the analyzed sample.

2.2.1. *Signal processing*

To create the sound data processing pipeline we have used Praat[11] through Parselmouth[12] as an approach interface to Praat from Python. For the automatic processing of the databases, we need a specific directory structure to facilitate the correct labeling. The audio must be in WAV format, and in this work at 256kb/s, 16kHz, 1 channel, PCM (Little/Signed). Once the metadata has been extracted with the interest variables, we may save the information externally and continue with the classification system.

2.2.2. *Classification system*

The classification system used was the support vector machine (SVM), which has already been used with satisfactory results in the classification of emotional states[13] and has demonstrated higher performance than CNN.[14]

The most important equation for the Support Vector Machine (SVM) algorithm is the one that defines the optimization problem it tries to solve.

When data are linearly separable, the optimization problem can be formulated as follows.

- Objective Function: being $\bar{w}$ =control variables vector

$$Minimize\ \frac{||\bar{w}||^2}{2}$$

- Constrains; For each data point (x_i, y_i) where x_i is the feature vector and y_i is the class label (+1 or −1), the following must hold: $y_i(w.x_i + b) >= 1$. This ensures that each data point is on the correct side of the margin.

For this paper, a slack variable ξ_i is introduced for each data point to allow for misclassifications, and the objective function becomes to minimize:

$$\frac{||\bar{w}||^2}{2} + C\sum_i \xi_i \qquad \gamma = \sum_i \xi_i$$

Either *gamma* or C, which is a regularization parameter that controls the trade-off between maximizing margin and minimizing misclassifications. They are calculated using a comparative system that seeks the best results for both parameters.

We used the RBF (Radial Basis Function) kernel because of its flexibility, the RBF kernel can go from a linear to a very complex classifier, depending on the values of its hyperparameters. Simplicity of the decision surface. The parameter C in the RBF kernel compensates for the misclassification of the training examples with the simplicity of the decision surface. A low value of C makes the decision surface smooth, while a high value of C aims to correctly classify all training examples, and influence of the training examples. The parameter γ defines how much influence a single training example has. The higher the γ, the closer other examples must be to be affected. The proper choice of C and γ is critical for SVM performance.

The Extra Trees classifier technique has been employed which is useful in determining the most important features of the SVM model.

2.3. *Visualization*

The tools and techniques used have been the following.

- Matplotlib and Seaborn; are used in this work to visualize the data and decision boundaries of the SVM model.
- Confusion matrix; to describe the classification's performance.

2.4. *Bip4Cast data integration*

This work analyzes emotions detected through voice recordings: happiness, sadness, rage, anger, irritability, panic, anxiety, exaltation, and fear. The Bip4Cast project aims to predict mood states, that is, depression, mania, mixed, or euthymia (neutral state). Emotions and mood states are related through indicators constructed ad hoc and based on previous publications.[15] After a classification of voice recordings according to the emotion detected, an integration of these results is due in the Bip4Cast project. The aim is to assign to each record a soft data vector of dimension 3 indicating the Depressive, Manic, or Euthymic degree in a fuzzy logic sense. For example, (.1; .7,.2) represents a grade 0.1 in depression, a grade 0.7 in mania, and a grade 0.2 in euthymia, and can be interpreted as a state closer to mania than to depression. These soft variables will be used in future work to provide a computational treatment with fuzzy logic.

3. Results

We should consider this work useful to complement the daily and nocturnal motor activity information that can be extracted and analyzed by additional pathways. This section shows the results of the training with Spanish audios, the training combining Spanish and German audios, and a first approximation of the table for mapping emotions to mood states as the source of the aggregation process.

3.1. *Trainning from Spanish audios*

The normality and uniformity of the variables have been checked. Fig.1 shows the distribution of two of the selected variables from the Spanish data set. Fig.2a shows the training results of the model for this data set.

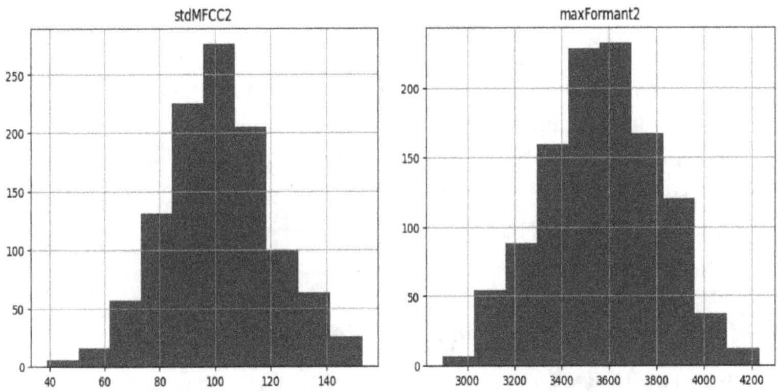

(a) stdMFCC2 (b) MaxFormant2

Fig. 1: Histograms for two selected variables

We can highlight how the training and testing for this data set obtain very significant values, highlighting how for Restless, Nervous, and Anxious.

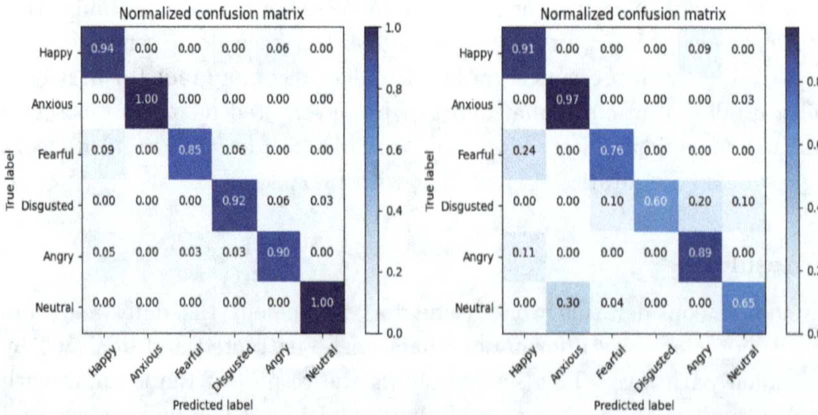

(a) Spanish audios (b) Mixed Audios

Fig. 2: Result details

A well as Neutral has been correctly classified in all test cases. Equally noteworthy is the high level of correctness for Happy, Disgusted, and Angry. As well as a very reasonable value for the Fearful mood.

3.2. *Trainning from mixed audios*

The normality and uniformity of the variables have been checked. Fig.2b shows the training results for this data set. The Anxious mood is classified independently of these languages with a very high level of accuracy. The same can be considered for Happy and Angry. However, important differences are observed for Fearfull, Disgusted, and Neutral. The significant drop in the latter mood is certainly striking.

3.3. *Mapping emotions into mood states*

In the analyzed database, 6 emotions have been recognized: happiness, sadness, irritability, panic, anxiety, and anger. Table 1 shows the degrees (unnormalized[15]) of membership in the tree clusters under study: manic, depressive, and mixed. The depression cluster presents the highest values in the emotions Sadness (2.29) and Anxiety (1.88) while the Mania cluster presents the highest values in the emotions Irritability (2.24) and Anxiety (2.14), these data being consistent with the opinion of verified experts.

Table 1: Clustering emotions into mood states.

Emotion	depressive	maniac	mixed
happiness	0,5	2,0	1,12
sadness	2,29	2,05	2,30
irritability	0,93	2,24	1,92
panic	1,04	1,19	1,62
anxiety	1,88	2,14	2,50
anger	0,67	1,81	1,34

4. Conclusions and Future Work

This work analyzes the usefulness of voice recordings in predicting crises due to emotional disorders. It is concluded that audio records are useful in determining the emotion expressed by individuals since these records can be successfully classified using machine learning techniques. In this chapter, the SVM technique has been used but other techniques can also give good results. As a consequence, this source of information (voice records) is useful to enrich the information collected about an individual with other techniques in previous projects, such as the Bip4cast project where motion sensor data and clinical data are used. Including alternative and independent sources of information improves knowledge of the individual's emotional state and therefore results in crisis prevention. This source of information is easily integrated into the previous project and is therefore proposed for future work. Fuzzy modeling of the voice variables is also proposed as future work.

Acknowledgments

This work is supported by the grant ACERTA Computational analysis of risk state in mood disorders from the Spanish MICINN.

References

1. V. López, G. Valverde, J. Anchiraico and D. Urgelés, Specification of a cad prediction system for bipolar disorder, in *Uncertainty Modelling in Knowledge Engineering and Decision Making: Proceedings of the 12th International FLINS Conference*, 2016.

2. M. Čukić and V. López, Progress in objective detection of depression and online monitoring of patients based on physiological complexity., *Frontiers in Psychiatry* **13**, p. doi: 10.3389/fpsyt.2022.828773 (2022).
3. P. Llamocca, V. López, M. Santos and M. Cukić, Personalized characterization of emotional states in patients with bipolar disorder, *Mathematics* **9**, p. 1174 (2021).
4. M. Čukić, V. López and J. Pavón, Classification of depression through resting-state electroencephalogram as a novel practice in psychiatry, *Journal of Medical Internet Research* **22**, p. e19548 (2020).
5. G. Farias, M. Santos and V. López, Making decisions on brain tumor diagnosis by soft computing techniques, *Soft Computing* **14**, 1287 (2010).
6. P. Llamocca, A. Junestrand, M. Cukic, D. U. Puértolas and V. López, Data source analysis in mood disorder research *Proceedings CAEPIA-18* 2018.
7. P. Llamocca, V. López and M. Čukić, The proposition for bipolar depression forecasting based on wearable data collection, *Frontiers in Physiology* **29**, p. doi: 10.3389/fphys.2021.777137. (2022).
8. P. Llamocca, D. Urgelés, M. Cukic and V. Lopez, Bip4cast: Some advances in mood disorders data analysis, in *Proceedings of the 1st International Alan Turing Conference on Decision Support and Recommender Systems, London*, 2019.
9. Berlin database of emotional speech (2005), http://emodb.bilderbar.info/start.html.
10. Emotional speech synthesis database (2012), https://catalog.elra.info/en-us/repository/browse/ELRA-S0329/.
11. P. Boersma and D. Weenink, Praat: doing phonetics by computer [Computer program] Version 6.1.38, retrieved 2 January 2021 http://www.praat.org/, (2021).
12. Y. Jadoul, B. Thompson and B. de Boer, Introducing Parselmouth: A Python interface to Praat, *Journal of Phonetics* **71**, 1 (2018).
13. I. Setti and J. M. Pinto, Análisis de datos en predicción de trastornos emocionales (2019), https://hdl.handle.net/20.500.14352/15227.
14. S. N. Mohammad Shahrul Izham Sharifuddin and A. M. Ali, Comparison of cnns and svm for voice control wheelchair, *IAES International Journal of Artificial Intelligence (IJ-AI)* **9**, p. 387 (2020).
15. C. Henry, K. M'Baïlara, A. Desage, S. Gard, D. Misdrahi and E. Vieta, Towards a reconceptualization of mixed states, based on an emotional-reactivity dimensional model, *Journal of Affective Disorders* **101**, 35 (2007).

Electroencephalography Emotion Recognition by FFT based on CBiGRU

Shunying Dai* and Hongjun Wang†

School of Computing and Artificial Intelligence, Southwest Jiaotong University,
Chengdu, China 610000, China
** dsy960709@126.com*
† wanghongjun@swjtu.edu.cn

Utilizing electroencephalography (EEG) for emotion recognition enables the direct collection of physiological signals, thereby circumventing potential deception associated with facial expressions and language cues. Concurrently, there has been a widespread adoption of deep learning techniques in this domain, aimed at achieving enhanced results through model and parameter adjustments. An innovative approach is introduced by incorporating the Fast Fourier Transform (FFT) for data preprocessing in conjunction with the CBiGRU model, merging the Convolutional Neural Network (CNN) and the Bidirectional Gated Recurrent Unit (BiGRU). Comparative analysis underscores the significance of exploring the intrinsic characteristics of the data itself: the application of FFT not only substantially reduces processing time but also enhances model performance. The integration of CBiGRU and FFT leverages the strengths of both CNN and BiGRU methods, resulting in increased accuracy and rapid convergence.

Keywords: Bidirectional gate recurrent unit; convolutional neural network; EEG; fast Fourier transform; emotion recognition.

1. Introduction

Electroencephalogram (EEG) brain signals record changes in electrical potentials resulting from neuronal discharge activity in the brain. Being directly induced by physiological phenomena, EEG signals are not susceptible to concealment or deception.[1]

In the context of sentiment classification on EEG, temporal data mining has demonstrated its crucial role, with the individual effectiveness of LSTM and GRU notably surpassing that of the CNN model. It was only with the application of the Fast Fourier Transform (FFT), transforming

*Corresponding author.

time-domain signals into frequency-domain signals for easier analysis, that CNN achieved comparable performance. FFT application reduced processing time to 1/4 of the original model's time consumption, leading to both improved efficiency and an accuracy increase exceeding 15% across different deep learning models. Furthermore, results from utilizing the Support Vector Machine (SVM) model surpassed those achieved by pure deep learning approaches.

This chapter proposes a new model, CBiGRU, which is a type of neural network architecture that combines CNN with BiGRU in deep learning and incorporates FFT into the CBiGRU model. Another innovative aspect involves the use of four labels – Arousal, Valence, Dominance, and Liking – to explain people's emotions. Expanding classification beyond common arousal and valence dimensions to include dominance and liking adds significant value to emotion classification in various domains.

The structure of this chapter is outlined as follows: Section 2 provides a summary of related works. Section 3 elucidates the details of the model and its integration. Subsequently, Section 4 delineates the datasets and experiment settings, comparing them with prior studies. Finally, Section 5 engages in a discussion of the implications and significance of our results.

2. Related Work

The shift in EEG-based emotion recognition has evolved from machine learning classifiers before 2019[2] towards deep learning. There are various deep learning frameworks used in EEG emotion recognition, such as Convolutional Neural Network (CNN), Long Short-Term Memory (LSTM), Bidirectional Gated Recurrent Unit (BiGRU), Graph Convolutional Networks (GCN), etc. These machine learning and deep learning extract EEG features to explain time and frequency sequences without manual process or combine with them to enhance the accuracies, which have been proven work in many papers.[3–5]

When using a single model, the effect of LSTM is significantly better than using CNN alone. EEG data exhibits asymmetry due to different cognitive activities activating distinct brain lobes with varying degrees of activation.[6] Additionally, as an improvement on LSTM, BiGRU captures forward and backward dependencies in time series data, adaptive learning of input data importance, and forgetting irrelevant information through the gate mechanism.[7]

Ozdemir introduced a proficient model, named the CNN-LSTM model, for emotion classification, with CNN for electrode location and LSTM for temporal features. This approach yielded notable results, achieving an accuracy of 86.13% for Arousal and 90.62% for Valence.[8]

3. CBiGRU Model

3.1. *Preprocess data: Fast Fourier transform*

FFT is used to derive the statistical features from the five bands wave of EEG signals, capable of reducing Time Complexity to $O(n \log_2 n)$. It transforms signals from the time domain to the frequency domain, allowing the analysis of the spectral information to extract more components. This approach contributes to the extraction of temporal features in EEG, aiding in a more comprehensive understanding of the EEG signal. The Fast Fourier Transform (FFT) is an algorithm for efficiently computing the Discrete Fourier Transform (DFT). The DFT of a discrete signal x(f) is given by the formula:

$$X(f) = \int_{-\infty}^{\infty} x(t) \cdot e^{-j2\pi ft} dt \tag{1}$$

where, X(f) is the Fourier transform of the signal x(t), x(t) is a non-periodic continuous-time signal, f is the frequency, j is the imaginary unit.

The FFT exploits the inherent symmetry and periodicity properties of the complex exponentials to efficiently compute the DFT. It recursively breaks down the DFT calculation into smaller sub-problems, leading to a divide-and-conquer approach.

3.2. *CNN and BiGRU*

This article employs BiGRU and CNN models to investigate information in the time domain, spatial domain, and frequency domain, respectively. CNN leverages the correlation among channels for deep data information mining[9] and proficiently interprets frequency domain information. The following graph Fig.1 describes the data augmentation process and its framework.

CNN is an effective way to structure data and achieved many good scores in papers. CNN contains four layers: convolutional layer, activation function, pooling layer, and fully connected layer. In discrete form, for a two-dimensional image, the convolution operation is expressed as:

$$(I * K)(i,j) = \sum_m \sum_n I(i-m, j-n)K(m,n) \tag{2}$$

where I represents the input image, K is the convolutional kernel.

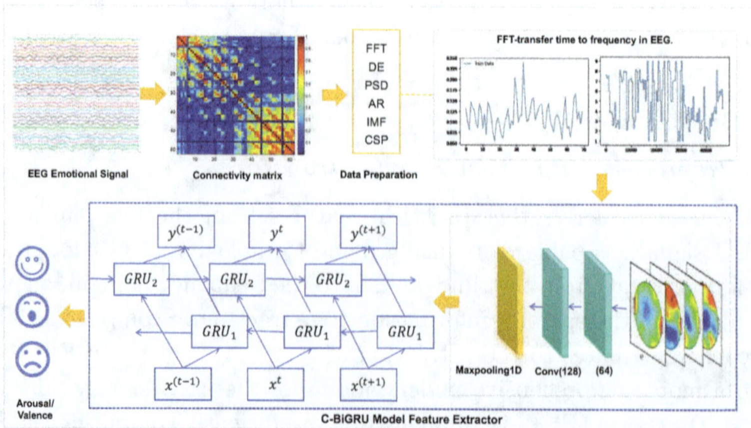

Fig. 1. The proposed integrating method based on CBiGRU

The mathematical representation of max pooling is given by:

$$\text{MaxPooling}(x)(i,j) = \max_{m,n} x(i \times s + m, j \times s + n) \tag{3}$$

BiGRU is a neural network model composed of two unidirectional GRUs with opposite directions, and its output is jointly determined by the states of these two GRUs, forward one and backward one, enhancing its ability to understand sequential dependencies. Meanwhile, BiGRU serve as a derived model of LSTM with fewer parameters and faster convergence, has demonstrated performance that merits further exploration.

$$z_t = \sigma\left(W_z \cdot [h_{t-1}, x_t]\right) \tag{4}$$

$$r_t = \sigma\left(W_r \cdot [h_{t-1}, x_t]\right) \tag{5}$$

$$\tilde{h}_t = \tanh\left(W \cdot [r_t \odot h_{t-1}, x_t]\right) \tag{6}$$

$$h_t = (1 - z_t) \odot h_{t-1} + z_t \odot \tilde{h}_t \tag{7}$$

where represent the update gate, reset gate, the candidate hidden state, and the hidden state at the time step, respectively. In both these two models, we choose the ReLU activation function.

4. Experiments

EEG data is structured into a large matrix (time $\times$ electrodes) through recording devices. An electrode channel typically corresponds to a triggering channel used to synchronize participant responses or stimuli with the EEG signals, segmenting them into multiple time periods. We choose 14 channels that have been demonstrated to show a high correlation with emotion recognition, according to Xiaopeng Si's paper.[10] Then we input the DEAP dataset to the CNN-BiGRU Model to extract features and classify them by Valence and Arousal, then add FFT to improve the result.

4.1. *Dataset*

The DEAP (Database for Emotion Analysis using Physiological Signals) dataset is a commonly used dataset in emotion recognition research. It consists of data collected from 32 subjects (16 males and 16 females), who watched 40 music videos while their physiological signals were recorded through 14 electrodes attached to their scalp. The dataset of DEAP is presented in Table 1:

Table 1: DEAP dataset description

Array	**Shape**	**Array Content**
Data	40×32×8064	video/trial×channels×data
Labels	40×4	video/trial×label

4.2. *Emotional recognition*

According to the labels (Arousal, Valence, Dominance, Liking), it is easy to divide into four sections. Arousal characterizes the physiological excitement level or activity level of emotions or moods. Valence describes the positive or negative nature of emotions or moods. Dominance refers to an individual's perceived level of control or dominance in a specific emotional state. Liking can be used to describe personal preferences and tastes towards the clip.

4.3. *Evaluation metrics*

In the context of statistics and machine learning, TP (True Positives), FP (False Positives), TN (True Negatives), and FN (False Negatives) are

abbreviations for terms used in confusion matrix analysis to describe the results of binary classification:
We will use accuracy evaluate the models:

$$accuracy = \frac{TP + TN}{TP + TN + FP + FN} \tag{8}$$

4.4. *Results*

Compared with deep learning solely, the effective application of FFT to convert time-domain data into frequency-domain data significantly improved the performance of all models by approximately 15%-20%. This process also led to a 75% reduction in running time and decreased the data for one epoch to just 1/10 of the original. This remarkable performance underscores the enduring vitality of traditional feature extraction methods, even in the prevailing popularity of deep learning. In Table 2, 9 algorithms were applied to DEAP for four varieties and showed the results of accuracy.

Table 2: Quantitative results

Methods	**Arousal**	**Valence**	**Dominance**	**Liking**
FFT-SVM	85.49	86.33	84.37	86.14
LSTM	75.12	77.11	77.37	75.43
CNN	64.82	67.36	67.95	66.37
BiGRU	71.44	74.64	45.70	54.33
CNN-GRU	70.23	70.17	72.80	73.31
FFT-BiGRU	90.08	89.64	89.97	92.32
FFT-BiLSTM	91.08	90.65	92.50	93.23
FFT-CNN	90.06	89.69	91.24	92.86
FFT-CBiGRU	**92.09**	**92.23**	**92.51**	**92.61**

It is obvious that FFT can handle the EEG signal properly, with different machine learning or deep learning developing higher accuracies. Subsequently, the two superior performance outcomes with a minimal difference are evaluated by comparing accuracy and loss visualizations to investigate their convergence rates. The results further indicate that, after transforming the data to the frequency domain using FFT, CNN maintains competitive accuracy without substantial lag. Similarly, BiLSTM exhibits a considerable level of accuracy, albeit with slower convergence. Combining both approaches not only marginally improves accuracy but also addresses the convergence issue. This method aids in the selection of a more appropriate model in Fig.2.

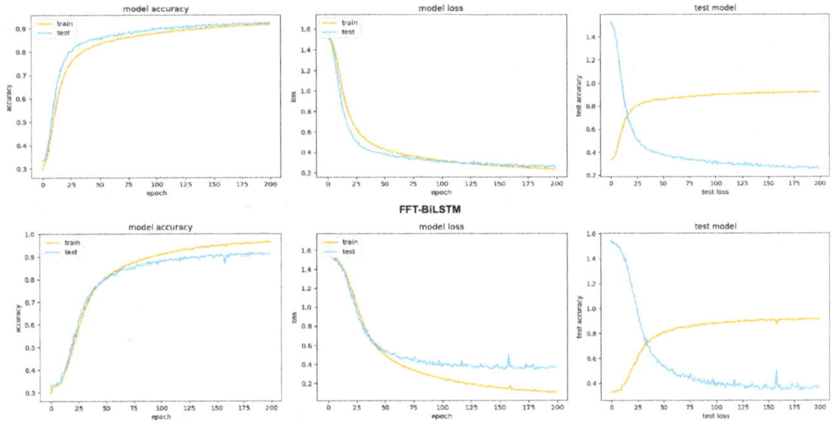

Fig. 2. FFT-CBiGRU and FFT-BiLSTM

5. Conclusion

In this chapter, the CBiGRU model was employed for EEG emotion classification to enhance precision, and the inclusion of FFT as input underscores the necessity of hand-crafted features. Throughout the experiments, we discovered intriguing patterns, and the results demonstrate that optimal performance is achieved within the range of 1 to 3 layers, with accuracy diminishing beyond this threshold.

Based on experiments and past observations, the combination of CNN and BiGRU has significantly improved the performance of single models, yielding better results by accuracy. However, these improvements still significantly lag behind those achieved using manual feature extraction techniques, such as power spectral density (PSD), differential entropy (DE), and FFT, which can easily achieve accuracies exceeding 90%. The experiment demonstrates that manual feature extraction, particularly FFT, is a crucial step for achieving excellent results. Relying solely on deep learning for feature extraction falls short of expectations, primarily due to the insufficient volume of data and the inadequacy in capturing enough features from higher-level temporal and frequency domain information. In the

EEG research, mining information inherent in the data remains crucial, and feature extraction emerges as a key aspect of this exploration.

References

1. Q. Wu, B. Yan, Y. Zeng, C. Zhang and L. Tong, Anti-deception: reliable eeg-based biometrics with real-time capability from the neural response of face rapid serial visual presentation, *Biomedical engineering online* **17**, 1 (2018).
2. N. S. Suhaimi, J. Mountstephens, J. Teo *et al.*, Eeg-based emotion recognition: A state-of-the-art review of current trends and opportunities, *Computational intelligence and neuroscience* **2020** (2020).
3. Y. Yang, Q. Wu, M. Qiu, Y. Wang and X. Chen, Emotion recognition from multi-channel eeg through parallel convolutional recurrent neural network, in *2018 International Joint Conference on Neural Networks (IJCNN)*, 2018.
4. D. Dadebayev, W. W. Goh and E. X. Tan, Eeg-based emotion recognition: Review of commercial eeg devices and machine learning techniques, *Journal of King Saud University - Computer and Information Sciences* **34**, 4385 (2022).
5. M. Jafari, A. Shoeibi, M. Khodatars, S. Bagherzadeh, A. Shalbaf, D. L. García, J. M. Gorriz and U. R. Acharya, Emotion recognition in eeg signals using deep learning methods: A review, *Computers in Biology and Medicine* **165**, p. 107450 (2023).
6. J. A. Coan and J. J. Allen, Frontal eeg asymmetry as a moderator and mediator of emotion, *Biological psychology* **67**, 7 (2004).
7. H. Shi, H. Xie and M. Wu, Emotion recognition based on multi-channel eeg signals, *Journal of Physics: Conference Series* **2078**, p. 012028 (nov 2021).
8. M. A. Ozdemir, M. Degirmenci, E. Izci and A. Akan, Eeg-based emotion recognition with deep convolutional neural networks, *Biomedical Engineering / Biomedizinische Technik* **66**, 43 (2021).
9. X. Chen, Y. Zheng, Y. Niu and C. Li, Epilepsy classification for mining deeper relationships between eeg channels based on gcn, in *2020 International Conference on Computer Vision, Image and Deep Learning (CVIDL)*, 2020.
10. X. Si, D. Huang, Y. Sun and D. Ming, Temporal aware mixed attention-based convolution and transformer network (mactn) for eeg emotion recognition (2023).

Intelligent Transaction Algorithm of Auto Service Value Chain

Li Lu* and Linfu Sun†

School of Computing and Artificial Intelligence, Southwest Jiaotong University, Chengdu, China

Manufacturing Industry Chains Collaboration and Information Support Technology Key Laboratory of Sichuan Province, Southwest Jiaotong University, Chengdu, China

**819692697@qq.com*

†sunlf@vip.sina.com

Traditional transaction systems within auto service chains exhibit several shortcomings, which are untraceable, opaque, and inefficient. To mitigate these challenges, a transaction system based on a consortium blockchain for the auto service value chain is devised to guarantee transaction authenticity and reliability. Initially, a smart contract is a computer protocol that is automatically executed. It integrates into the Byzantine fault-tolerant IBFT consensus algorithm to accomplish transactions in real time and enhance a higher degree of security. Subsequently, the IBFT is fine-tuned to enhance the consortium blockchain performance. In addition, multiple primary nodes execute the IBFT algorithm concurrently via multithreading to mitigate latency and improve throughput. Lastly, experimental results demonstrate that IBFT notably surpasses other consensus algorithms in both throughput and latency.

Keywords: Transaction; smart contract; consensus algorithm; blockchain, value chain.

1. Introduction

The service value chain involves suppliers, manufacturers, and consumers collaborating to create value 0, facilitating parts transactions. Compared with traditional centralized transactions, distributed transactions offer providers and services on cloud platforms a fast, convenient, and secure alternative 0. However, these transactions are untraceable, opaque, and inefficient. To address these challenges, PBFT 0 is implemented on the cloud platform, offering

* Corresponding author.

significant advantages over Paxos 0 and Raft 0, reducing the algorithmic complexity from exponential to polynomial. CDBFT 0 is an improved algorithm designed to improve the performance of the communication process. Additionally, a mechanism for voting rewards and punishments, along with a formulated reputation evaluation mechanism, is proposed to dynamically update the nodes participating in the consensus process. However, the communication complexity of the reputation mechanism is excessive, requiring $O(n^2)$ communication complexity even under optimal conditions. To improve throughput and reduce delay, IPBFT 0 uses multiple leaders to balance the load and quickly eliminate Byzantine nodes in the network through a removal mechanism. However, IPBFT is only suitable for small systems due to its high communication complexity. The HotStuff 0 algorithm offers more advantages compared to the aforementioned consensus algorithms due to its lighter linear complexity. Consequently, HotStuff is widely deployed on cloud platforms. Currently, the HotStuff consensus algorithm achieves better performance with an increased number of network nodes. However, currently, there exists a deficiency in a reliable trading system capable of facilitating trusted transactions between providers and services.

This chapter proposes a transaction system based on a consortium blockchain for auto service chains. A smart contract is intricately designed to facilitate automatic transaction execution. Moreover, the smart contract integrates into the Byzantine fault tolerance IBFT algorithm to enhance the performance. Multiple primary nodes execute the IBFT algorithm concurrently via multithreading to reduce latency and enhance throughput, thereby improving the algorithm's efficiency. finally, experimental results confirm the effectiveness of IBFT.

2. Transaction System

The blockchain system is illustrated in Fig. 1. On the cloud platform, providers and services constitute $n = 3 \times f + 1$ nodes, with a maximum of f Byzantine nodes that the primary node do not generate a new block within the bounded time Δ, all of which engage in the consortium blockchain 0 via an authorization mechanism. Trusted transactions 0 are facilitated through collaboration between providers and services (nodes) within the service value chain. Services release demanded parts, whereas providers share unsalable parts from the shared inventory to reduce inventory backlogs. The cloud platform automatically matches eligible parts within the shared inventory, promptly notifies providers and services of the results, after which the service verifies the results. Trusted transactions are automatically executed using the designed smart contract, integrating an enhanced IBFT algorithm to generate PartsSale orders.

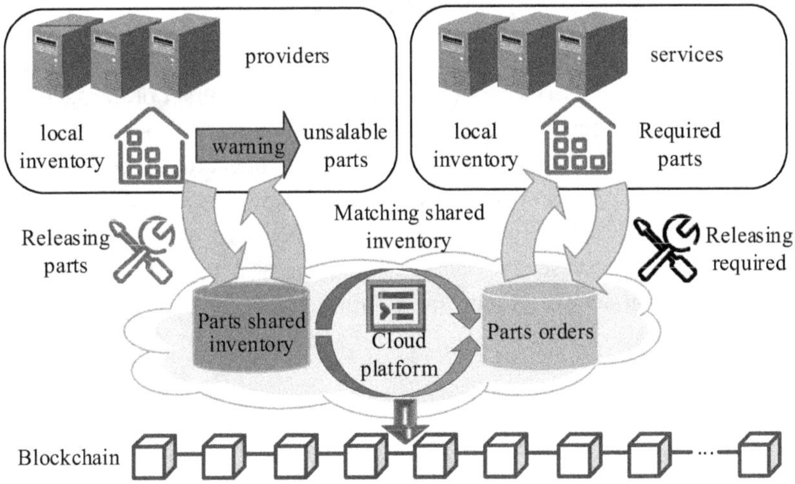

Fig. 1. Intelligent transaction system of providers and services.

3. Model Design

3.1. *Smart contract design*

Algorithm 1: The smart contract of transactions

Input: *amount* and *price*

Output: true or false

```
contract transactionContract {
    function purchase amount () {
        if the transaction is a success then
            Deduct amount × price from the service' accounts
            Generate a PartsSale order
            return true
        else
            return false
    }
    function withdraw () {
        if amount > 0 then
            The account of the service provider is not debited
            return true
        else
            return false
    }
}
```

A smart contract, as described in a computer protocol 0, functions to execute contracts, thus establishing a secure and distributed ledger for transactions between providers and services. In this study, a smart contract of transactions is devised and stored within blockchain nodes, incorporating purchase and withdraw functions. Remarkably, this smart contract not only safeguards the confidentiality of part prices but also optimizes complex workflows, consequently diminishing operational costs. It enables execution by any authorized provider and service, thereby fostering consensus on transaction *amount* and *price*. The execution of such a smart contract (illustrated in Algorithm 1) is pivotal in guaranteeing transaction reliability.

3.2. *IBFT consensus algorithm*

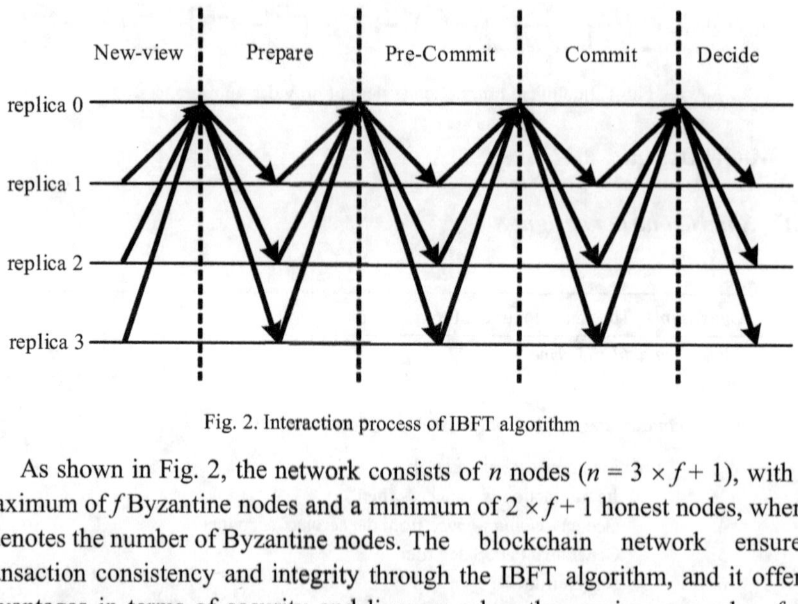

Fig. 2. Interaction process of IBFT algorithm

As shown in Fig. 2, the network consists of n nodes ($n = 3 \times f + 1$), with a maximum of f Byzantine nodes and a minimum of $2 \times f + 1$ honest nodes, where f denotes the number of Byzantine nodes. The blockchain network ensures transaction consistency and integrity through the IBFT algorithm, and it offers advantages in terms of security and liveness when the maximum number f of Byzantine nodes is less than $n/3$. Moreover, the network achieves linear communication complexity, and the network structure remains stable, with n nodes stably connected through a reliable network. The primary node i is labeled as replica 0 in view v, while the other primary nodes are labeled as replica 1, replica 2, and replica 3. These nodes are authorized to perform read, write, and modify operations when the number of Byzantine nodes f is less than $n/3$, while multiple primary nodes execute the IBFT algorithm to achieve consensus in the distributed network.

A block is considered valid if the primary node presents proof that the block's signatures, sourced from a minimum of (n - f) vote messages (partial signatures) spanning the prepare, pre-commit, commit, and decide phases, are consolidated into a vote set *qc*. This consolidated vote set is then relayed to the primary node, reducing communication complexity as opposed to transmitting individual ($n - f$) vote messages.

A timestamp signifies the duration for transactions to reach consensus among these nodes, indicating whether the message was included in the block before a timeout. If the next valid block is not generated before the timeout Δ, the IBFT consensus algorithm for the following view will be activated.

4. Experiments

The experiments used Hyperledger Fabric 2.2 and Ubuntu 20.10 software. An intelligent model and 100 nodes were deployed on Ubuntu for 20 seconds. The performance of IBFT was evaluated based on throughput and average delay 0, and compared with PBFT 0, CDBFT 0, and HotStuff 0 using 60,000 sales data points under identical conditions.

4.1. *Performance analysis*

Table 1 presents a comparison of consensus algorithms. The communication complexity of IBFT is O(n), lower than those PBFT and CDBFT algorithms.

Table 1. The comparison of consensus algorithms

Algorithm	Communication Complexity	Throughput	Confirmation Delay
PBFT 0	O(n^2)	low	high
CDBFT 0	O(n^2)	low	high
Hotstuff 0	O(n)	low	high
IBFT (ours)	O(n)	high	Low

4.2. *Performance evaluation*

The performance evaluations of PBFT 0, CDBFT 0, Hotstuff 0, and IBFT were evaluated from 4 to 100 nodes. As shown in Fig. 3, the horizontal axis represents the number of nodes, and the vertical axis represents throughput. PBFT and CDBFT exhibit lower throughput in the range of 4 to 100 nodes due to increased communication costs, and IBFT optimizes the HotStuff algorithm through multiple primary nodes improve the throughput. the IBFT algorithm demonstrates superior performance than other algorisms.

Fig. 4 presents the average delay of the four algorithms. Although the delay of PBFT 0, CDBFT 0, and Hotstuff 0, and IBFT algorithms increases with the number of nodes, IBFT exhibits the lowest delay. Consequently, IBFT shows superior performance attributed to its higher throughput and diminishing delay with the same number of nodes.

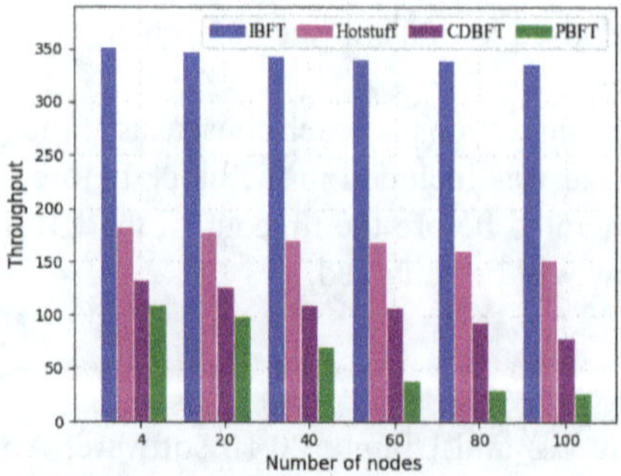

Fig. 3. Throughput comparison of different algorithms.

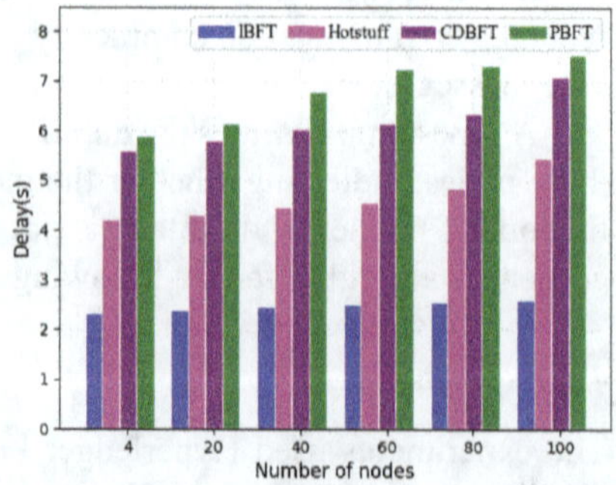

Fig. 4. Delay comparison of different algorithms.

5. Conclusions and Future Works

This chapter introduces a transaction system designed for auto service chains, presenting multiple primary nodes equipped with the IBFT algorithm to autonomously execute transactions via a pre-defined smart contract. Experimental findings illustrate the throughput of IBFT superior relative to other algorithms, sustaining heightened performance with the same number of nodes. Future initiatives will leverage Boneh–Lynn–Shacham for batch verification to mitigate communication complexity.

Acknowledgments

This work is supported by the National Key R&D Program of China (No. 2023YFB3308600).

References

[1] X. WU and C. LI, Multi-Value Chain Auto Parts Demand Prediction Based on Dynamic Heterogeneous Graph Convolution, in *2022 IEEE 24th Int Conf on High Performance Computing & Communications*, (Hainan, China, 2022).

[2] Chen T. Research on Trust Mechanisms of Cloud Manufacturing Service Transaction Based on Blockchain. In: Wuhan University of Technology. 2020.

[3] J. Yang, Z. Jia, R. Su, X. Wu and J. Qin, Improved Fault-Tolerant Consensus Based on the PBFT Algorithm, *IEEE Access*. 10, 30274 (2022).

[4] D. D. Amarasekara, and D. N. Ranasinghe, ZPaxos: An Asynchronous BFT Paxos with a Leaderless Synchronous Group, in *IEEE*, (Basel, Switzerland, 2022).

[5] Z. Ye, X. Tong, W. Fan, Z. Zhang, and C. Jin, A Raft Variant for Permissioned Blockchain, *DASFAA*. (4) 685-689 (2023).

[6] Y. Wang, S. Cai, C. Lin, Z. Chen, T. Wang, Z. Gao and C. Zhou, Study of Blockchains's Consensus Mechanism Based on Credit, *IEEE Access*. 7, 10224 (2019).

[7] L. LU, L. SUN and Y. Zhou, Improved PBFT Algorithm of Multi-primary-node based on Ring of Consistent Hash, *Computer Integrated Manufacturing System*. 29, 25-35 (2023).

[8] M. Yin, D. Malkhi, M. K. Reiter, G. G. Gueta and I. Abraham, HotStuff: BFT Consensus with Linearity and Responsiveness, in *Principles of Distributed Computing*, (Toronto, Canada, 2019).

[9] M. Hao, B. Tan, S. Wang, R. Yu, R. W. Liu and L. Yu, Exploiting blockchain for dependable services in zero-trust vehicular networks, *Frontiers Comput. Sci.* 18, 182805 (2024).

[10] D. Li, W. Liu, L. Deng and B. Qin, Design of multimedia blockchain privacy protection system based on distributed trusted communication, *Trans. Emerg. Telecommun*. Technol. 34, 1 (2023).

[11] B. K. Mohanta, S. S. Panda and D. Jena, An Overview of Smart Contract and Use Cases in Blockchain Technology, in *2018 9th International Conference on Computing, Communication and Networking Technologies (ICCCNT)*, (Bengaluru, India, 2018).

[12] P. Chen, D. Han ,T. H. Weng, K. C. Li, and A. Castiglione, A novel Byzantine fault tolerance consensus for Green IoT with intelligence based on reinforcement, J. Inf. Secur. Appl. 59, 102821 (2021).

Enhanced Detection of Silkworm Cocoon Defects Utilizing the K-YOLOX Algorithm

Bo Xing[*,†,§], Qingqing Shao[*], Lei Xing[‡] and Xianyi Zeng[†,¶]

[*] *College of Textiles, Donghua University, Shanghai 201620, China*
[†] *Univ.Lille, ENSAIT, GEMTEX-Génie et Matériaux Textiles, Lille 59000, France*
[‡] *Henan Qingxinran Environmental Technology Company, Jiaozuo 454000, China*
[§] *bo.xing@ensait.fr*
[¶] *xianyi.zeng@ensait.fr*

In order to enhance the automatic inspection efficacy for cocoons, this chapter introduces a novel algorithm, named K-YOLOX, which is an augmentation of the extant YOLOX object detection framework. This innovation incorporates the distinctive features of cocoons into its operational methodology. By integrating the K-means clustering approach, the YOLOX architecture is optimized, facilitating the derivation of anchors commensurate with the cocoon dimensions. Furthermore, the algorithm's loss function has been recalibrated, with the inclusion of the Generalized Intersection over Union (GIoU) as an additional boundary loss component, thereby significantly refining the precision of cocoon classification and localization. Empirical assessments demonstrate that the K-YOLOX algorithm achieves an average detection accuracy of 94.2% on cocoon imagery, with the comprehensive F1 score reaching 92.5%. These figures represent an enhancement of 10.3% in detection accuracy and 14.2% in the F1 metric when contrasted with the baseline YOLOX model.

Keywords: Defect detection; silkworm cocoon; YOLO.

1. Introduction

With a lengthy history of the silk industry, China is the world's origin of the silk industry and the first country to find and use silk.[1] Following the collection of fresh cocoons, silk reeling enterprises often dry them first before converting them to raw silk through a series of processes, such as mixing, stripping, filtering, boiling, and spinning. Cocoons can be separated into normal and defective cocoons throughout the selection procedure based on their use value in silk spinning manufacture. Normal cocoons are those that are rid of flaws in terms of color, thickness, and fold. Defective cocoons are

[§]Corresponding author.

those that have major issues that prevent or make it impossible to produce regular raw silk cocoons.

Currently, cocoon sorting relies on the cocoon selector to identify the completion of manual sorting. However, because manual sorting is time-consuming and labor-intensive, and the quality of identification is heavily reliant on the cocoon selector's subjective consciousness, researchers at domestic and overseas have begun to investigate how to solve this problem using image recognition technology in recent years. Chen Hao[2] *et al.* proposed an auxiliary detection method for cocoon selection, using MATLAB to graphically process and numerically calculate cocoon images and accurately calibrate them with a camera to finally derive the surface defect area of cocoons. Prasobhkumar[3] *et al.* proposed a cocoon excellent quality rating system by statistically measuring the RGB color threshold of yellow spot cocoons with the specific parameters of normal cocoon ellipse aspect ratio. Yang Guangyou[4] *et al.* proposed the idea of a total neural network application for cocoon detection, integrating BP and RBF neural networks, and the accuracy of cocoon species detection reached 87.8%.

The consistency of domestic and foreign scholars' research on normal and defective cocoons is lesser, which cannot meet the requirements of cocoon selection accuracy in the scheduled production process; the existing methods in the literature for cocoon detection cannot meet the needs of simultaneous multiple targets and cannot be applied to the aforementioned research on cocoon defect detection algorithm.

The target detection algorithms based on deep learning that have emerged in recent years provide new ideas for cocoon defect detection, of which the representative algorithms are R-CNN[5] series and YOLO[6] series. The R-CNN series algorithm is a target detection algorithm based on region generation, although the detection accuracy is high, but the detection time is long and does not meet the criteria for real-time cocoon detection in the production line. The YOLO series algorithm is a target detection algorithm based on the regression method, with a multi-scale detection mechanism, fast recognition speed, and accuracy can also basically reach the level of the R-CNN series algorithm.

In this chapter, we use defective and normal cocoons as research objects, and we use the K-means clustering algorithm[7] to obtain the anchor size suitable for cocoons; furthermore, we use the GIoU function[8] in the loss function of YOLOX and adjust the classification ratio to build the K-YOLOX (Kmeans-YOLOX) network structure based on the basic shape of cocoons and recognition needs. Experiments demonstrate that the new algorithm not only achieves high accuracy in cocoon defect detection but also satisfies the need for multi-target identification.

2. Experiments

2.1. *Construction of cocoon image dataset*

The image collection equipment was used to acquire 500 photographs of cocoons, and the cocoon data was augmented by scaling, flipping, and panning for cocoons in various postures and forms, resulting in a total of 1,0000 images. In batches, all photos were randomly separated into training and test sets.

The data set was labeled using Labelimg software, which automatically generates the corresponding x, y, w, and h values based on the location of the labeled cocoons in each image, records the specific cocoon type information, generates the cocoon image labeling file in XML format, and stores it in a standard VOC data set format after participating in the cocoon image acquisition.

2.2. *K-mean clustering anchor*

Because there are so many distinct things to identify, varied sizes of anchors must be built ahead of time to optimize prediction matching. The standard YOLOX method is pre-trained using 20 items of various sizes from the VOC dataset as examples, resulting in a variety of anchor sizes. However, because cocoon defect detection does not need a substantial size difference between the anchors, the cocoon defect detection anchor must be reset.

The K-mean algorithm clusters the dataset by selecting K objects at random as cluster centers and dividing each of them to the nearest cluster center by calculating the distance between the rest of the objects and the cluster center, and repeating this process until the cluster center no longer changes after the next division.

2.3. *K-YOLOX detection network*

This research builds the K-YOLOX detection network in terms of resetting the loss function for the cocoon's own features to increase the accuracy of cocoon defect identification based on the YOLOX network.

The conventional YOLOX algorithm's typical loss function consists of three-loss functions: coordinate, confidence, and classification. In YOLOX, We primarily increase cocoon flaw detection accuracy by substituting the coordinate loss function and altering the overall percentage of classification loss in the standard loss function. Fig.1 shows the improved network structure.

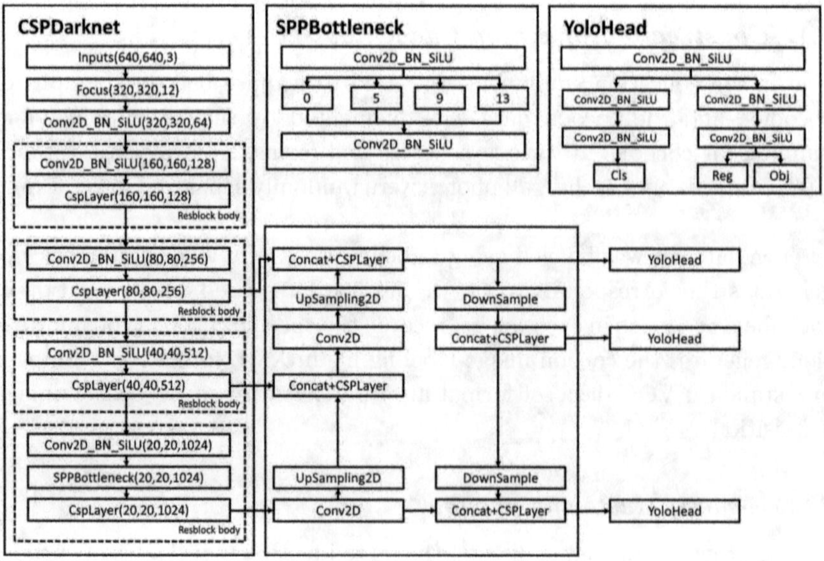

Fig. 1. K-YOLOX detection network

The mean square error function is utilized in the classic YOLOX technique for border coordinate regression (MSE). The MSE function has the drawback of treating the center point coordinates, width, and height of cocoon defect detection as distinct variables when there is a relationship between them, and using it will enhance the instability of the border regression coordinates. As a result, this paper introduces the GIoU function to replace the mean square error function. The GIoU function is an optimization function of IoU, and its duty is to precisely determine the degree of overlap between the predicted frame of cocoon flaws and the real frame. LGIoU loss function specific formula derived as follows:

$$\text{IoU} = \frac{A \cap B}{A \cup B} \tag{1}$$

$$\text{GIoU} = \text{IoU} - \frac{C - (A \cup B)}{C} \tag{2}$$

$$L_{GIoU} = 1 - GIoU \tag{3}$$

For the non-overlapping section between A and B, the LGIoU loss function is likewise included in the attention range. As a result, the LGIoU loss function is chosen as the standard loss function in order to improve the accuracy of the coordinates of the faulty cocoon prediction framework.

When the recognition object is an object with low differentiation of surface features, such as the defect region on the cocoon, the standard loss function is ineffective, so this paper proposes the idea of strengthening the classification weight parameter to increase the total percentage of classification loss in the standard loss function for cocoon defect detection. The redesigned loss function formula is as follows:

$$\begin{aligned}\text{Loss} &= \varphi_{co}\sum_{i=0}^{s^2}\sum_{j=0}^{B} I_{ij}^{obj} L_{GIoU} + \sum_{i=0}^{s^2}\sum_{j=0}^{B} I_{ij}^{obj}\left(C_i^j - \hat{C}_i^j\right)^2 \\ &\quad + \varphi_{no}\sum_{i=0}^{s^2}\sum_{j=0}^{B} I_{ij}^{obj}\left(C_i^j - \hat{C}_i^j\right)^2 \\ &\quad + 10\sum_{i=0}^{s^2} I_i^{no} \sum_{c\in\text{ classes}}^{s^2}\left[P_i^j(c) - \hat{P}_i^j(c)\right]^2\end{aligned} \tag{4}$$

where s is the number of grids in the output layer of the network, B is the number of cocoon identification a priory frames, ϕ_{co} is the coordinate loss weight of the cocoon prediction frame, ϕ_{no} is the confidence loss function for not containing cocoons in the prediction frame, I_{ij}^{obj} is the specific number of cocoons in the ith grid in the cocoon image, P_i^j is the prediction probability when the jth The predicted probability that when the jth prior frame of the ith grid identifies the defective cocoons, $\hat{P}_i^j$ is the true probability that the jth prior frame of the ith grid identifies the specific class of cocoon.

2.4. *Evaluation indicators*

To evaluate the quality of cocoon defect identification, precision P, recall R, mean average accuracy MAP, and F1 score were used as evaluation indexes. The calculation equations are as follows:

$$P = \frac{T_p}{(T_p + F_p)} \tag{5}$$

$$R = \frac{T_p}{(T_p + T_n)} \tag{6}$$

$$F1 = \frac{2PR}{(P + R)} \tag{7}$$

$$MAP = \frac{1}{C}\sum_{k=1}^{C} J(P, R, K) \tag{8}$$

where T_p is the number of correct judgments in the prediction, F_p is the number of incorrect judgments, F_n is the number of missed detections, C is the number of cocoon defective species, K is the species serial number, and J is the area function of the P-R curve and the coordinate axis.

3. Results and Discussions

The improvement effect of K-YOLOX cocoon detection algorithm is tested by extracting images of cocoons with different species distribution and different number distribution in the defective cocoon dataset. The performance tests were performed by constructing three cocoon defect detection algorithms: the original YOLOX algorithm (method 1), the YOLOX algorithm based on K-means clustering (method 2), and the YOLOX algorithm based on K-means clustering with adjusted loss function (method 3). The evaluation indexes mainly use the accuracy rate P, MAP and F1 score, and the test results are shown in Table 1.

Table 1. Improvement effect test

Method	P (%)	MAP (%)	F1 (%)
Method 1	86.8	85.4	80.9
Method 2	90.6	91.9	93.4
Method 3	95.3	95.7	95.1

According to the data in Table 1, it can be obtained that the K-means clustering method improves the mean accuracy of cocoon defect detection by 6.5% and increases the F1 score. By comparing the F1 scores and MAP between method 2 and method 3, it can be concluded that resetting the loss function can effectively improve the recognition accuracy of cocoon defect detection.

The visual inspection results are shown in Fig.2. As can be seen in the figure, the K-YOLOX algorithm can precisely locate and mark defective cocoons with defective markings.

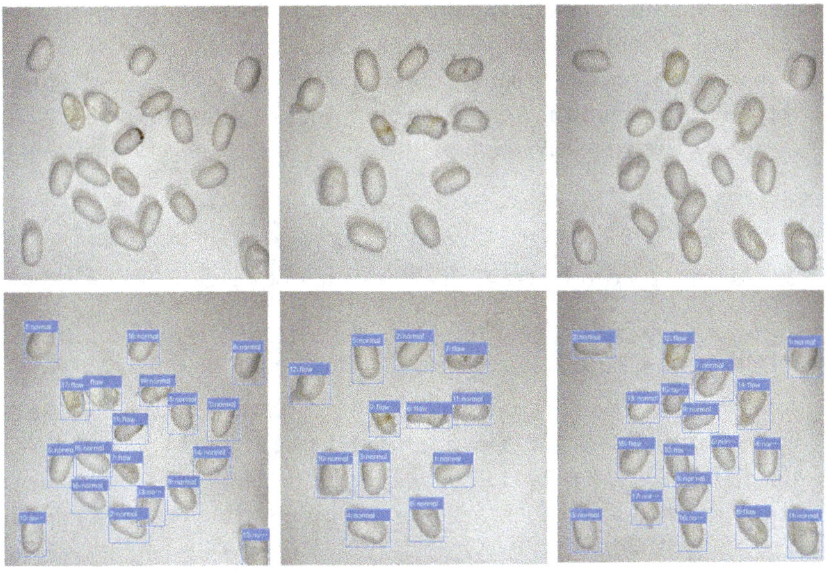

Fig. 2. Visual inspection effect

The K-YOLOX cocoon defect detection algorithm proposed in this paper further adds K-means clustering and resets the loss function on the basis of YOLOX, and according to the test results, the F1 and MAP of K-YOLOX reach 95.1% and 95.7%, respectively. The K-YOLOX algorithm can basically meet the technical requirements for intelligent detection of cocoon selection for the simultaneous identification of multiple samples in the cocoon selection process.

4. Conclusions

To solve the technical problem of fast and accurate identification of defective cocoons in the cocoon selection process, this chapter proposes a cocoon defect detection algorithm (K-YOLOX). Taking defective cocoons and normal cocoons as the research object, the K-means method is used to cluster the cocoon dataset, so as to obtain *a priori* frames suitable for cocoons, which provides technical support for cocoon detection and localization; the classification accuracy of cocoons is improved by adjusting the classification ratio in the loss function and replacing the localization loss function. Comparative experiments are conducted to verify the correctness of the improved behavior by comparing the algorithm before and after the improvement, and the evaluation indexes such as the average recognition accuracy and F1 score are used to compare YOLOX with K-YOLOX. From the experimental results, it can be obtained that the K-YOLOX algorithm has high accuracy, and the average accuracy and F1 score reach 95.7% and 95.1%, respectively, which can meet the cocoon defect detection needs.

Acknowledgments

This study was funded by the China Scholarship Council Fund.

References

1. John Thorley. The silk trade between china and the roman empire at its height, circaa. d. 90–130. *Greece & Rome*, 18(1):71–80, 1971.
2. Chen Hao, Yang Zheng, Liu Xia, and Shao Ling. Study on auxiliary testing method for mulberry silkworm cocoon sorting based on matlab. *Silk*, 53(3):32–36, 2016.
3. Prasobhkumar PP, Francis CR, and Sai Siva Gorthi. Automated quality assessment of cocoons using a smart camera based system. *Engineering in agriculture, environment and food*, 11(4):202–210, 2018.
4. Yang Guangyou, Jiang Jinyu, and Zhou Guozhu. Application of neural network ensemble on the non-destructive testing of pod. *Silk*, 3(05):28–30, 2008.
5. Ren Shaoqing, He Kaiming, Ross Girshick, and Sun Jian. Faster r-cnn: Towards real-time object detection with region proposal networks. *Advances in neural information processing systems*, 28, 2015.
6. Yong-Hwan Lee and Youngseop Kim. Comparison of cnn and yolo for object detection. *Journal of the semiconductor & display technology*, 19(1):85–92, 2020.
7. John A Hartigan and Manchek A Wong. Algorithm as 136: A k-means clustering algorithm. *Journal of the royal statistical society. series c (applied statistics)*, 28(1):100–108, 1979.
8. Zou Chengming and Xue Ronggang. Yolov3 target detection algorithm combining giou and focal loss. *Computer Engineering and Applications*, 56(24):214–222, 2020.

Multi-Objective Cooperative Trajectory Planning for High-speed Gliding Aircraft Cluster

Jiaming Yu, Qinglin Sun* and Hao Sun

College of Artificial Intelligence, Nankai University, Tianjin 300350, China

**sunql@nankai.edu.cn*

The new high-speed gliding aircraft combat strategy aims to adopt end-stage cooperative clusters of high-speed gliding aircraft, capable of coordinated strikes against multiple targets. Real-time planning of feasible trajectories for High speed gliding aircraft clusters is essential for synchronized control and accurate strikes. Traditional trajectory planning methods, when applied to the task of trajectory planning for large clusters of high-speed gliding aircraft, exhibit drawbacks such as lengthy computation times and difficulties in adapting to dynamic environmental changes. To address this issue, this chapter introduces a multi-objective cooperative trajectory planning method for high-speed gliding aircraft clusters based on multi-agent deep deterministic policy gradients. A pre-trained multi-agent trajectory planning model is used, which does not require recalculating trajectories for different starting positions. Detailed simulation results indicate that the proposed high-speed gliding aircraft cluster multi-objective cooperative trajectory planning method offers strong real-time planning performance and high precision in strikes.

Keywords: High-speed gliding aircraft cluster; trajectory planning; multi-agent; deep deterministic policy gradient.

1. Introduction

High-speed gliding aircraft are launched by rocket propulsion or elevated to a certain altitude or operational area by other carrying vehicles. They then utilize aerodynamics to commence unpowered gliding outside the Earth's atmosphere, ultimately engaging in target strikes or landing. High-speed gliding aircraft not only have advantages in speed and maneuverability but also have the ability to perform long-range missions and penetrate air defense systems. The traditional "one-on-one" combat method of high-speed gliding aircraft is becoming increasingly difficult to cope with more challenging tasks, and the new combat method adopts a high-speed gliding aircraft cluster with end-to-end coordination.

Unlike traditional powered aircraft, high-speed gliding aircraft are inherently unpowered and characterized by high cruising speeds, making timely adjustments to their flight trajectories challenging. Therefore, real-time planning of feasible trajectories for clusters of high-speed gliding aircraft becomes an essential prerequisite for coordinated control and precise striking. In recent years, some scholars have conducted research on trajectory planning methods for swarms or formations.[1–3] It is evident that most of these studies focus on unmanned aerial vehicles, with few dedicated to trajectory planning for high-speed glider swarms based on their specific characteristics. Moreover, traditional trajectory planning methods often entail long computation times and pre-set trajectories that struggle to adapt to dynamic environments, leading to delayed responses and low damage efficiency in high-speed gliding aircraft clusters. Deep reinforcement learning offers an innovative solution for multi-objective cooperative trajectory planning for high-speed gliding aircraft clusters, enhancing the coordination and combat capabilities of aircraft clusters.

There is already a considerable amount of research applying deep reinforcement learning to trajectory planning tasks for individual UAV,[4–6] but this chapter shifts the focus to the application of deep reinforcement learning in multi-agent trajectory planning.[7–9]

Accordingly, this chapter proposes a multi-objective cooperative trajectory planning method for high-speed gliding aircraft clusters based on Multi-Agent Deep Deterministic Policy Gradients (MADDPG), utilizing the three degrees of freedom of high-speed gliding aircraft and designing a reward function with initial constraints, terminal constraints, real-time path constraints, and collision avoidance. This method can plan feasible flight trajectories in real-time for each unit in the high-speed gliding aircraft cluster, achieving coordinated multi-aircraft strikes.

2. High Speed Gliding Aircraft Model

The operation of high-speed gliding aircraft is divided into three phases: boost phase, gliding phase, and dive phase. This paper primarily focuses on the trajectory planning during the gliding phase. In the design and trajectory planning of high-speed gliding aircraft, real-time responsiveness is a key consideration. Due to the high-speed characteristics of the aircraft, their motion states may undergo significant changes within an extremely short period. Hence, a model capable of rapid adaptation and computation is necessary to ensure the accuracy of trajectory planning. We have employed a 3-DOF model for the trajectory planning of high-speed gliding aircraft. This choice is primarily based on the high cruising speed characteristics of the high-speed gliding aircraft and the high demand for real-time computation of trajectories. Compared to more complex models with higher degrees of freedom, the 3-DOF model reduces computational load, allowing for trajectory planning to be completed in a shorter time frame, thus meeting the real-time response requirements of high-speed gliding aircraft in dynamic environments. Each aircraft within the high-speed gliding aircraft cluster is established based on the following 3DOF model:

$$\begin{cases} \dot{x} = v_{xy}\cos\varphi + v_{wx} \\ \dot{y} = v_{xy}\sin\varphi + v_{wy} \\ \dot{z} = v_z \\ \ddot{\varphi} = u \end{cases} \tag{1}$$

where [x,y,z] represents the position of each high-speed gliding aircraft in the cluster; $[w_x, w_y, w_z]$ represent the wind speed components along the x, y, and z axes, respectively; v_{xy} denotes the horizontal velocity of each high-speed gliding aircraft; v_z represents the vertical velocity of each high-speed gliding aircraft, which in this study is assumed to be constant as the aircraft are unpowered; φ indicates the yaw angle of each high-speed gliding aircraft; $\ddot{\varphi}$ denotes the angular acceleration of each high-speed gliding aircraft; and u represents the control input during flight, which in this study is set as angular acceleration.

3. Method

3.1. *Deep deterministic policy gradient*

The Deep Deterministic Policy Gradient (DDPG) algorithm is a deep reinforcement learning method specifically designed for complex problems with continuous action spaces. DDPG combines the advantages of policy-based and value-based approaches, utilizing an Actor-Critic framework. In this framework, the Actor is responsible for generating the policy, directly mapping states to actions, while the Critic evaluates the value of actions under the given policy.

The DDPG algorithm employs a deterministic policy μ. Under this deterministic policy, the algorithm outputs the optimal action for a given state:

$$a_t = \mu\left(S_t \mid \theta^{\mu}\right) \tag{2}$$

where $a_t$is the deterministic action value obtained directly under each state through the deterministic policy function μ , and θ^μ represents the parameters of the Actor network used to generate deterministic actions.

Another significant characteristic of DDPG is the introduction of target networks and soft update mechanisms. DDPG creates separate target networks for both the Actor and the Critic, with their parameters gradually approximated through soft updates, a process of slowly tracking the parameters of the learning networks.

The actions determined by the target network of the Actor, coupled with the observed values of the environmental state, are utilized as inputs to the target network of the Critic. This process dictates the direction in which the target network parameters of the Critic are updated. The formula for the parameter update within the Critic network is as follows:

$$y_i = r_i + \gamma Q'\left(S_{i+1}, \mu'\left(S_{i+1} \mid \theta^{\mu'}\right)\right) \mid \theta^{Q'}\right)) \tag{3}$$

$$L = \frac{1}{n}\sum_i^n \left(y_i - Q\left(S_i, a_i \mid \theta^Q\right)\right)^2 \tag{4}$$

where y_i is the actual evaluation value, computed using the target networ, r_i refers to the real reward received γ represents the reward decay rate. The loss function, denoted as L, is defined as the sum of squared deviations between the actual y_i and the estimated values.

The update of the Actor network parameters adheres to a deterministic policy:

$$\nabla_{\theta^\mu} J = \frac{1}{n}\sum_i^n \nabla_a Q(S, a|\theta^Q)|_{S=S_i, a=\mu(S_i)} \nabla_{\theta^\mu}\mu\left(S \mid \theta^\mu\right)|_{S_i} \tag{5}$$

where ∇Q, sourced from the critic, guides the direction for updating the parameters of the actor's network. where ∇_{θ^μ}, sourced from the actor, guides the Actor network to be more likely to choose the above actions

3.2. *Multi-agent deep deterministic policy gradient*

The coordinated strike strategy of high-speed gliding aircraft clusters involves forming a group of multiple aircraft to jointly attack various targets, enhancing the effectiveness of strikes and breach capabilities, and increasing the likelihood of mission completion. Each aircraft in the glider cluster possesses autonomous decision-making and execution capabilities, representing a multi-agent problem where each agent has its unique observations and actions.

Multi-Agent Deep Reinforcement Learning (MADRL) is a branch of reinforcement learning that specifically studies how multiple agents in the same environment can collaborate and compete to achieve their individual goals. In the task of cooperative high-speed gliding aircraft cluster missions, each aircraft is considered an agent that must work in concert within the battlefield environment to effectively strike its respective targets.

This paper employs the Multi-Agent Deep Deterministic Policy Gradient (MADDPG) algorithm, specifically designed for multi-agent problems in continuous spaces, to train a cooperative trajectory planning model for high-speed gliding aircraft. The MADDPG algorithm employs a centralized training approach during training and a decentralized testing method during testing. During the training phase of the MADDPG algorithm, the Critic of each aircraft utilizes the observations and policies of all aircraft, thus serving as a centralized evaluation function. Each aircraft has its own Actor, which, during testing, can select actions based solely on its own observations, making the Actor a decentralized policy.

3.3. *Reward function*

The reward function is a crucial component in the cooperative trajectory planning for high-speed gliding aircraft clusters. Its primary purpose is to provide a quantified evaluation for the start, end, and every step of action during the mission execution, guiding the high-speed gliding aircraft to make optimal decisions.

The real-time path reward focuses on guiding high-speed gliding aircraft towards the target point at each step and ensuring that they precisely reach the target at the stipulated moment:

$$D_t = \sqrt{(x_t - x_f)^2 + (y_t - y_f)^2} \tag{6}$$

$$d = |\frac{z_t}{v_z} \times v_{xy} - D_t| \tag{7}$$

where D_t represents the horizontal distance between the parafoil system in S_t state and the target point $[x_f, y_f]$; d represents real-time path constraint, which is the absolute error value between the remaining flight distance of the parafoil system in the S_t state and the horizontal distance from the target point. v_z is the vertical velocity of a high-speed glideing aircraft, and v_{xy} is the horizontal velocity.

The terminal reward value focuses on the accuracy with which the aircraft reach the predetermined target point at the end of the trajectory planning task. In this paper, this reward is calculated based on the final distance between the aircraft and their target point. Higher rewards are given when the aircraft accurately reach or approach the target poin:

$$D_f = \sqrt{\left(x_{t_f} - x_f\right)^2 + \left(y_{t_f} - y_f\right)^2} \tag{8}$$

where $[x_{tf}, y_{tf}]$ represents the final landing position of the high-speed gliding aircraft.

The collision avoidance reward aims to promote maintaining a safe distance between the aircraft during the mission to prevent collisions and ensure they do not exceed the mission area.

4. Simulation

To verify the correctness of the algorithm proposed in this study, PyCharm was used as the development and simulation environment in Python language version 3.9. The task space size is $105km \times 75km \times 30km$. The initial altitude of the high-speed glider cluster is set to 15km. At the beginning of each round, the high-speed glider cluster will randomly initialize its position in the task area. After 100 rounds of testing, the average error between the high-speed glideing cluster and the target point was 2.1km, with a minimum error of 0.06km and a maximum error of 6.3km. This study assumes that an error of more than 5 kilometers is considered a miss, and the hit rate reaches 96.6%. Fig.1 and Fig.2 display different trajectory planning cases.

The above simulation results demonstrate that the multi-objective trajectory planning method for high-speed gliding aircraft clusters proposed in this paper successfully plans feasible trajectories for each of the 12 aircraft in the trajectory planning task, enabling them to concurrently reach their respective target points at the designated times.

5. Conclusions

This chapter addresses the shortcomings of traditional trajectory planning methods when applied to large-scale high-speed glider groups, such as lengthy computation times and difficulties in adapting to dynamic environmental changes. We propose a multi-objective trajectory planning method

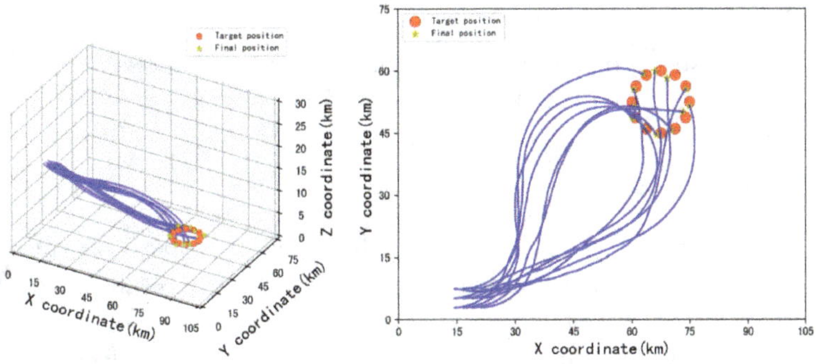

Fig. 1. Simulation result I

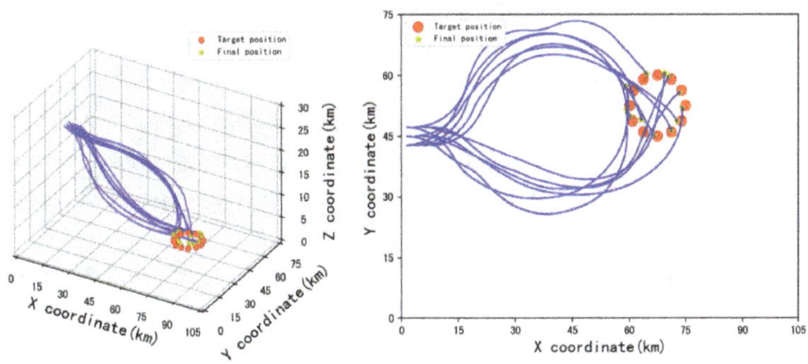

Fig. 2. Simulation result II

for high-speed gliding aircraft clusters based on the MADDPG algorithm. Additionally, a reward function tailored for multi-objective cluster tasks is designed for each aircraft within the cluster. Simulation results demonstrate that the trajectory planning method proposed in this chapter can plan optimal trajectories in real-time for each high-speed gliding aircraft at various positions.

Acknowledgments

This work was supported by the National Natural Science Foundation of China (Grant No. 62103204).

References

1. J. Liu, W. Han, X. Wang and J. Li, Research on cooperative trajectory planning and tracking problem for multiple carrier aircraft on the deck, *IEEE Systems Journal* **14**, 3027 (2019).
2. Y. Chen, J. Yu, X. Su and G. Luo, Path planning for multi-uav formation, *Journal of Intelligent & Robotic Systems* **77**, 229 (2015).
3. C. Xu, M. Xu and C. Yin, Optimized multi-uav cooperative path planning under the complex confrontation environment, *Computer Communications* **162**, 196 (2020).
4. R. Xie, Z. Meng, L. Wang, H. Li, K. Wang and Z. Wu, Unmanned aerial vehicle path planning algorithm based on deep reinforcement learning in large-scale and dynamic environments, *IEEE Access* **9**, 24884 (2021).
5. Q. Liu, L. Shi, L. Sun, J. Li, M. Ding and F. Shu, Path planning for uav-mounted mobile edge computing with deep reinforcement learning, *IEEE Transactions on Vehicular Technology* **69**, 5723 (2020).
6. L. He, N. Aouf and B. Song, Explainable deep reinforcement learning for uav autonomous path planning, *Aerospace science and technology* **118**, p. 107052 (2021).
7. H. Qie, D. Shi, T. Shen, X. Xu, Y. Li and L. Wang, Joint optimization of multi-uav target assignment and path planning based on multi-agent reinforcement learning, *IEEE access* **7**, 146264 (2019).
8. Z. Cui and Y. Wang, Uav path planning based on multi-layer reinforcement learning technique, *Ieee Access* **9**, 59486 (2021).
9. H. Bayerlein, M. Theile, M. Caccamo and D. Gesbert, Multi-uav path planning for wireless data harvesting with deep reinforcement learning, *IEEE Open Journal of the Communications Society* **2**, 1171 (2021).

Time Series: Summarizing Information and Forecasting in Presence of Structural Breaks

Vilém Novák* and Thi Thanh Phuong Truong†

Institute for Research and Applications of Fuzzy Modeling, University of Ostrava, 30. dubna 22, 701 03 Ostrava 1, Czech Republic

* *vilem.novak@osu.cz*

† *Thi.Truong.s01@osu.cz*

In this chapter, we provide the application of selected methods of fuzzy modeling to the identification of structural breaks in time series, forecasting, and automatic summarization of knowledge about time series expressed in natural language.

Keywords: Fuzzy natural logic; fuzzy transform; time series; mining information; evaluative linguistic expressions; intermediate quantifiers.

1. Introduction

In this chapter, we focus on two tasks: First, we present our method for the detection of structural breaks in time series and show that our forecasting method can also forecast them. We statistically demonstrate that indeed the out-samples contain the structural break. Second, we demonstrate examples of how our theory of intermediate quantifiers (e.g., "most, many, a lot of, and a few") can be applied to automatic summarization of knowledge about time series expressed in natural language. The knowledge concerns either single time series (detection and characterization of monotonous segments, future trends, and summarizing information on them), or we can also generate summarizing information on a multitude of time series. The theoretical method used in our applications is *Fuzzy Transform* (F-transform) completed by selected methods of *Fuzzy Natural Logic* (FNL).

This chapter is structured as follows. In Section 2, we introduce the basic concepts of fuzzy transform and fuzzy natural logic. In Section 3, we introduce time series. Section 4 is the main part of this chapter where we present a summarization of knowledge from time series using the theory of intermediate quantifiers, detection, and forecasting of structural breaks.

*Corresponding author.

2. Preliminaries — Fuzzy Transform and Fuzzy Natural Logic

A detailed explanation of basic fuzzy modeling methods is presented in the book[1]. The fundamental idea of the *fuzzy transform* introduced by I. Perfilieva ([2,3]) is to transform a real bounded continuous function $f : [a, b] \longrightarrow [c, d]$, where $[a, b], [c, d] \subset \mathbb{R}$, to a finite vector of components and then transform it back. The principle concept is that of a *fuzzy partition* of $[a, b]$ which consists of a (finite) set of fuzzy sets $\mathscr{A}_h = \{A_0, \ldots, A_n\}$, $n \geq 2$, defined over the set of the nodes $a = c_0, \ldots, c_n = b$ such that $c_{k+1} = c_k + h$ where $h = (b - a)/n$. Based on the given fuzzy partition $\mathscr{A}_h$, the $(n + 1)$-tuple $\mathbf{F}^1[f] = (F_0^1[f], \ldots, F_n^1[f])$ is called first degree *direct fuzzy transform* of f where

$$F_k^1[f](x) = \beta^0[f] + \beta_k^1[f](x - c), \tag{1}$$

where $c \in [a, b]$ is a node.

The *inverse* F-transform is $\hat{f}_h^1(x) = \sum_{k=0}^n F_k^1[f] \cdot A_k(x)$, $x \in [a, b]$. Many properties of the fuzzy transform can be found in the cited literature. The F-transform provides estimation of the absolute values and the first derivative over the area characterized by the fuzzy set $A \in \mathscr{A}_h$ (see[3]), namely

$$\beta^0[f] = f(c) + O(h^2) \quad \text{and} \quad \beta^1[f] = f'(c) + O(h^2).$$

Fuzzy natural logic is a class of special formal theories of mathematical fuzzy logic whose goal is to model the reasoning of people based on using natural language. Its main parts are the theory of *evaluative linguistic expressions* (see[1,4]), for example *small, medium, big, very short, more or less deep, quite roughly strong, extremely high*, and the theory of *intermediate quantifiers*[5–7], for example *All, Most, Almost all, A few, Many, Some, Several, A lot of, A little.* They specify quantity of specimens in the domain of discourse having a certain property:

$$\mathbf{Q\ B\ are\ A} \tag{2}$$

where A, B are properties of the considered specimens and B is the domain of discourse. Example of (2) is "Most (Q) time series in the given set (B) are increasing (A)". FNL contains a formal mathematical theory of the semantics of intermediate quantifiers that are construed by special formulas.

The *Ev* is a metavariable for a formula representing some evaluative linguistic expression: *BiEx* (*extremely big*), *Bi Ve* (*very big*), $\neg$ *Sm* $\bar{v}$ (*not small*), *Sm Ve* (*very small*), *Sm Si* (*significantly small*). Furthermore, let

$z \in Form_{o\alpha}$, $x \in Form_{\alpha}$ be variables and $A, B \in Form_{o\alpha}$ be formulas of the corresponding types. An intermediate quantifier construing the sentence (2) is one of the following formulas:

$$(Q^{\forall}_{Ev}\, x)(B, A) \equiv (\exists z)[(\forall x)((B|z)\, x \Rightarrow Ax) \wedge Ev((\mu B)(B|z))], \tag{3}$$

$$(Q^{\exists}_{Ev}\, x)(B, A) \equiv (\exists z)[(\exists x)((B|z)x \wedge Ax) \wedge Ev((\mu B)(B|z))]. \tag{4}$$

Formula (3) is semantically interpreted by

$$(Q^{\forall}_{Ev} x)(B, A) = \bigvee \left\{ \bigwedge_{m \in N} ((B|Z)(m) \to A(m)) \wedge Ev(\mu(B, B|Z)) \,\middle|\, Z \underset{\sim}{\subseteq} N \right\} \tag{5}$$

where $B, Z, A \underset{\sim}{\subseteq} N$ are fuzzy sets and $\mu(B, B|Z)$ is a *measure of the size* of the Z-part of B.

3. Time Series

A time series is a set $X = \{X(t, \omega) = Tr(t) + C(t) + S(t) + R(t, \omega) \mid t \in \mathbb{T}, \omega \in \Omega\}$ where $\mathbb{T} = \{1, \ldots, p\}$ is a set of natural numbers interpreted as time moment, Tr is a *trend* and $C(t)$ a *cycle* that are often joined into *trend-cycle* $TC(t) = Tr(t) + C(t)$, $S(t)$ is a *seasonal* component that is a mixture of periodic functions and R is a random *noise* with the mean $\mathbf{E}(R(t, \omega)) = 0$ and variance $\mathbf{Var}(R(t, \omega)) < \sigma$, $t \in \mathbb{T}$. For a fixed $\omega \in \Omega$ the time series becomes a real (or complex) valued function $X = \{X(t) \mid t \in \mathbb{T}\}$.

Given $h > 0$ and a fuzzy partition $\mathscr{A}_h$, the inverse F-transform of X w.r.t. $\mathscr{A}_h$ is

$$\hat{X}_h(t) = \hat{TC}_h(t) + \hat{S}_h(t) + \hat{R}_h(t), \qquad t \in \mathbb{T}. \tag{6}$$

It can be proved that setting $h = dT$ for some d and a proper periodicity T, the F-transform removes part or the whole of the seasonal component S and reduces the noise R.

Forecasting of time series is accomplished in two phases: first we forecast future components $F^m_{n+1}[X], \ldots$ and from them using the inverse F-transform we obtain forecasted trend or trend cycle. The forecast of the components is realized by inference on the basis of the learned linguistic description using the PbLD method developed in FLN (for the details see[1]). The latter also explains in natural language, how the forecast has been obtained. The other components (cyclic or seasonal) are forecasted using other methods, for example classical ARIMA or neural networks.

4. Mining Information from Time Series

Mining information from time series is a result of a class of various kinds of methods (cf.[8]) that quite often have statistical nature. Our methods, based on the theory of fuzzy natural logic, have potential to characterize various features occurring in time series using expressions of natural language (cf.[9,10]).

4.1. *Summarization*

One of the possible problems in mining information from time series is identification of intervals with specific monotonous trend, for example *stagnating, sharply increasing/decreasing, slightly increasing/decreasing,* etc. In[9], a simple procedure has been suggested whose result is decomposition of the time domain $\mathbb{T}$ into a set of adjacent intervals $\mathscr{T} = \{\tau_i \mid i = 1, \dots, s\}$, $\bigcup \mathscr{T} = \mathbb{T}$ and consider a tangent $\beta^1[X|\tau]$ in an interval $\tau \in \mathscr{T}$ of the time series X computed using the F-transform. Each τ with the length $|\tau|$ is the *largest interval with the slope evaluated by one evaluative expression.* This has good sense when dealing with long time series whose length is several thousand or more time moments.

A special position among methods for mining information from time series have methods of automatic summarization (see, e.g.,[11–14]). Let us demonstrate both problems on an example. Consider a time series of the length 2000 time points depicted in Figure 1. Using our method, we iden-

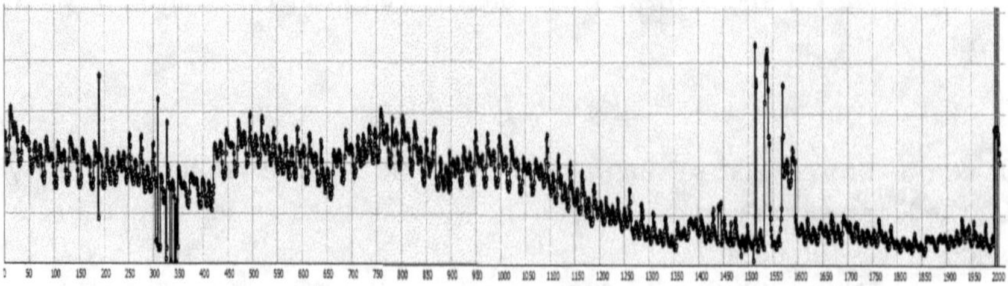

Fig. 1. Time series of the length $|\mathbb{T}| = 2000$.

tified 360 maximal intervals of monotonous trend in it. The quantifier summarizing the information contained in this time series is the following:

In Q maximal intervals, trend of time series is $\mathscr{A}$

where $\mathscr{A}$ is an evaluative expression characterizing the slope $\beta^1[X|\tau]$. In our case, $B = \mathscr{T}$ and so, formula (3) construing this quantifier reduces to

$$(Q^{\forall}_{Ev}\, x)(B, A) \equiv (\forall x)(Bx \Rightarrow Ax) \wedge Ev((\mu B)A)],$$

which, using (5), is semantically equal to

$$(Q^{\forall}_{Ev} x)(A) = \bigwedge_{\tau \in \mathrm{Supp}(A)} A(\tau) \wedge Ev(\mu(\mathscr{T}, A)) \tag{7}$$

where $A \subseteq \mathscr{T}$ is a set of maximal intervals having the trend named by $\mathscr{A}$[a]. Note that $\mu(\mathscr{T}, A)$ is a relative size of A w.r.t. the universe $\mathscr{T}$. In the following table are summarized found maximal intervals. The bracket contains truth value of the respective evaluative expression Ev in (7).

$\mathscr{A}$	**Q**
stagnating:	several (*Sm Ve*: 0.95), a few (*SmSi*: 0.81)
a little increasing:	many ($\neg$ *Sm*: 0.54)
a little decreasing:	many ($\neg$ *Sm*: 0.74)
clearly increasing:	many ($\neg$ *Sm*: 1)
clearly decreasing:	many ($\neg$ *Sm*: 1)
hugely increasing:	several (*Sm Ve*: 1), a few (*SmSi*: 1)
hugely decreasing:	several (*Sm Ve*: 1), a few (*SmSi*: 1)

Clearly, we are not able to identify these intervals manually. Note that they have been obtained by a slight aggregation of 18 intervals with a more detailed evaluation (for example, "a very little increasing" is also "a little increasing").

Other examples of mined summarizing information. In the following examples, we apply intermediate quantifiers (3), (4) to time series.

- In *almost all* (*most, many, few*) time intervals, *increasing* (decreasing) trend is followed by *sharply* (*very slightly, clearly*) *decreasing* (increasing) one.
  Formally: $(Q^{\forall}_{Ev}\tau)(INC(X|\tau), ShINC(X|\tau+1))$ where $\tau + 1$ is a successor of τ.
- *Short* periods of *sharp increase* of trend are *often* followed by *very long periods* of *stagnation.*

[a]In fact, the meaning of $\mathscr{A}$ is an *intension*, i.e., a function from the set of contexts into a set of fuzzy sets. We simplify the explanation in this paper due to the lack of space.

Formally: $(Q^{\forall}_{Bi\ Ve}\tau)(B\tau, C(\tau+1))$ where
$B\tau \equiv (Sm(\mu\mathbb{T})(|\tau|) \wedge (Bi\ Ve)w_{tg}(\beta^1[X|\tau])$ and
$C\tau \equiv (Bi\ Ve(\mu\mathbb{T})(|\tau|) \wedge STAG(\beta^1[X|\tau+1])$

- There are many time series with *stagnating* (increasing, very little decreasing) trend.
  Formally: $(Q^{\forall}_{\neg\ Sm}X)(STAG(\beta^1[X]))$.
- There is period of a *huge* (*slight, clear*) decrease in trend of *almost all* time series.
  Formally: $(Q^{\forall}_{Sm\ Ex}X)(\exists\tau)(HuDEC(\beta^1[X|\tau]))$.
- *Most* (many, few) analyzed time series *stagnated* recently but their expected trend is *slightly increasing.*
  Formally: $(Q^{\forall}_{Bi\ Ve}X)(STAG(X|\tau), SlINC(X|\tau+1))$.

4.2. *Structural breaks and forecast of them*

In [15] we introduced the method for identification of structural breaks in time series. If we want to forecast them, however, structural breaks should occur in the learning set. In Figure 2 we present two time series with apparent structural breaks and their forecast which was obtained using combination

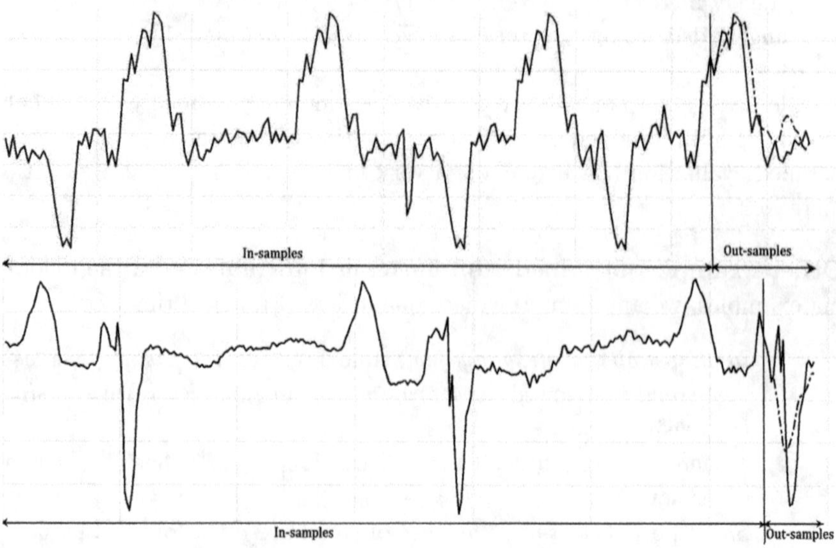

Fig. 2. Two time series TS1 and TS2 containing structural breaks. Their forecast is depicted by dashed line in the out-samples part.

of the F-transform and FNL. First we detected the structural breaks, proved their statistical significance and then we applied the method for learning linguistic description and PbLD (Perception-based Logical Deduction) inference based on it. The learned linguistic description for time series TS1 has the rules of the form

$dX(t)$ &	$d^2X(t)$ &	$d^2X(t-1)$	$\Rightarrow$	$X(t+1)$
-ve sm &	-ex sm &	-sm	$\Rightarrow$	ml me
vr bi &	ro sm &	me	$\Rightarrow$	bi
$\vdots$ &	$\vdots$ &	$\vdots$	$\Rightarrow$	$\vdots$

where $dX(t) = X(t) - X(t-1)$ is the difference (speed of the change), $d^2X(t) = dX(t) - dX(t-1)$ (acceleration of the change). The learned linguistic description for the time series TS2 has the rules of the form $X(t)$ & $X(t-1)$ & $d2X(t)$ & $d2X(t-1) \Rightarrow X(t+1)$.

The basic functions forming the fuzzy partition have width $2h = 6$ for TS1 and $2h = 12$ for TS2 where h is the distance between the nodes. The forecast of both series contains structural break that has been detected by our method and statistically verified by the fuzzy Chow test (the null hypothesis H_0 means that $\beta_k^0 = \beta_{k+1}^0$ and $\beta_k^1 = \beta_{k+1}^1$; cf.[15]):

t	**Fuzzy sets** $[A_k, A_{k+1}]$	F_{test}	F_{crit} $(\alpha = 0.025)$	**Decision**
[186,192]	$[A_{186}, A_{192}]$	229	4.12	Reject H_0
[324,336]	$[A_{324}, A_{336}]$	5.00	3.93	Reject H_0

We see that the detection of the forecasted structural breaks in both time series TS1 and TS2 is statistically significant.

5. Conclusion

In this chapter, we present several ways how information from time series in linguistic form can be mined using our theory of intermediate quantifiers. We also presented forecasting of time series that contain structural breaks and demonstrated that the forecasted structural breaks are statistically significant.

Acknowledgment

This article has been produced with the financial support of the European Union under the REFRESH Research Excellence For REgion Sustainability and High-tech Industries project number CZ.10.03.01/00/22-003/0000048 via the Operational Programme Just Transition.

References

1. V. Novák, I. Perfilieva and A. Dvořák, *Insight into Fuzzy Modeling* (Wiley & Sons, Hoboken, New Jersey, 2016).
2. I. Perfilieva, Fuzzy transforms: theory and applications, *Fuzzy Sets and Systems* **157**, 993 (2006).
3. I. Perfilieva, M. Daňková and B. Bede, Towards a higher degree F-transform, *Fuzzy Sets and Systems* **180**, 3 (2011).
4. V. Novák, A comprehensive theory of trichotomous evaluative linguistic expressions, *Fuzzy Sets and Systems* **159**, 2939 (2008).
5. V. Novák, A formal theory of intermediate quantifiers, *Fuzzy Sets and Systems* **159**, 1229 (2008).
6. P. Murinová and V. Novák, A formal theory of generalized intermediate syllogisms, *Fuzzy Sets and Systems* **186**, 47 (2012).
7. P. Murinová and V. Novák, The theory of intermediate quantifiers in fuzzy natural logic revisited and the model of "Many", *Fuzzy Sets and Systems* **388**, 56 (2020).
8. T.-C. Fu, A review on time series data mining, *Engineering Applications of Artificial Intelligence* **24**, 164 (2011).
9. V. Novák, Linguistic characterization of time series, *Fuzzy Sets and Systems* **285**, 52 (2016).
10. V. Novák, Mining information from time series in the form of sentences of natural language, *International Journal of Approximate Reasoning* **78**, 192 (2016).
11. R. Castillo-Ortega, N. Marín and D. Sánchez, A fuzzy approach to the linguistic summarization of time series, *Multiple-Valued Logic and Soft Computing* **17**, 157 (2011).
12. J. Kacprzyk, A. Wilbik and S. Zadrożny, An approach to the linguistic summarization of time series using a fuzzy quantifier driven aggregation, *International Journal of Intelligent Systems* **25**, 411 (2010).
13. G. Moyse and M. Lesot, Linguistic summaries of locally periodic time series, *Fuzzy Sets and Systems* **285**, 94 (2016).
14. A. Wilbik and R. M. Dijkman, On the generation of useful linguistic summaries of sequences, *2016 IEEEInternationla Conference on Fuzzy Systems* , 555 (2016).
15. V. Novák and P. Truong, A combination of fuzzy techniques and Chow test to detect structural breaks in time series, *MDPI Axioms* **12**, p. 103 (2023).

Formalization of Environment and Concepts

Long Hong* and Ningning Zhou†

School of Computer, Nanjing University of Posts and Telecommunications, Nanjing, Jiangsu, China

**hongl@njupt.edu.cn*

†zhounn@njupt.edu.cn

The environment is the physical and social condition of human live, in which concepts play an important role in scientific discovery, technological innovation, social activities, and daily life. It is extremely important, therefore, to accurately describe the environment and a concept. This chapter aims to study environment formally, to trace the origin of a concept, and to govern the concept based on the formal environment in order to broaden the basis of the existing formal concept. To make the name of specific concept, the formal definition was briefly introduced.

Keywords: Environment; concepts; formalization; intension; extension.

1. Introduction

What is the concept? It is an important fundamental question. In order to formalize the term concept, at the end of the 1970s, the mathematicians at the Technical University of Darmstadt proposed formal concept analysis (FCA) [1]. In 1982, R Wille published the theory of restructuring lattice based on hierarchies of concepts [2]. This theory is now fully developed, which is well recounted in Ref. [3]. And it has been applied to information technology [4], artificial intelligence [5], optimization algorithms [6], chemical engineering [7], and other fields. A formal context is the basis of a formal concept, which is a triplet composed by the set of formal objects, the set of formal attributes, and the relationship between these two sets. To generate formal concepts, Wille corresponds formal attributes and formal objects to intension and extension, respectively [1].

From biological survival, the environment refers to the nature conditions, such as air, water, land, and space [8]. From the scientific research, the

*Corresponding author.

environment is what is omitted by a particular abstraction, and it requires an infinite understanding of things, which is therefore impossible[9]. As a conceptual basis, clearly, the "context" mentioned above reflected only certain aspects of the environment.

Although it is impossible to understand infinity, in practice the meaning of the concept must be clear, otherwise their application to theory research, technique innovation, engineering realization, and social life will produce unimaginable mistakes and even lead to disasters. Therefore, formally describing environment and a concept based on environment are a meaningful work with a foundation. This chapter formally describes the environment, traces the origin of a concept, and governs the concept based on the formal environment in order to broaden the basis of the existing formal concept.

2. Formalization of Environment

The environment surpasses us in every physical dimension [9]. However, we are able to construct a formal environment that contains both the known world $\mathcal{K}$ and the unknown world $\mathcal{U}$, in which the former includes unomitted abstract things and figurative one; the latter includes omitted abstract things at the time.

Let O_k and A_k be the object set and the attribute set of $\mathcal{K}$, respectively; O_u and A_u be the object set and the attribute set of $\mathcal{U}$ respectively, then the description of formal environment is as follows.

Description 1. *The formal environment is the triplet* $\mathcal{E}= (O, A, v_A)$*, where* O *is the set of formal objects and* $O= O_k \cup O_u$*;* A *is the set of formal attributes and* $A=A_k \cup A_u$*; the mapping related to attribute* v_A*:* $O \times A \to \{1, 0, -1\}$*, and*

$$v_A(o, a) = \begin{cases} 1 & o \text{ and } a \text{ interrelated but } \mathcal{U} \\ 0 & \text{no } o \text{ related to } a \text{ and } \mathcal{U} \\ -1 & o \text{ and } a \text{ related to } \mathcal{U} \end{cases} . \quad (1)$$

Theorem 1.

(1) $O_k \cap O_u = \varnothing$.

(2) $A_k \cap A_u = \varnothing$.

Theorem 2.

(1) *Let* $o \in O_k \wedge a \in A_u$*, then* $v_A(o, a)=-1$.

(2) *Let* $o \in O_u \wedge a \in A_k$*, then* $v_A(o, a)= -1$.

(3) *Let* $o \in O_u \wedge a \in A_u$*, then* $v_A(o, a)= -1$.

Theorem 3.

(1) *Let* $o \in O_k \wedge a \in A$*, then* $min(v_A(o, a))=-1$.

(2) *Let* $a \in A_k \wedge o \in O$, *then* $min(v_A(o, a))=-1$.

Theorem 3 shows that it is possible that the objects in the known world have unknown attributes and the objects in the unknown world have known attributes.

The formal environment is shown in Table 1. Here, the line on which the members of formal attribute set, a_1, a_2, …, a_n, a_{u1}, a_{u2}, …, etc. is called the attribute row; the line on which the members of the formal object set, o_1, o_2, …, o_m, a_{u1}, a_{u2}, etc. is called the object column; the part that is on the right side of the object column and below the attribute row is called the value area, which is used to represent the value of $v_A(o, a)$, and the "0/1" in the table is represented as 0 or 1.

Table 1. The formal environment represented by the two-dimensional table.

	a_1	a_2	…	a_n	a_{u1}	a_{u2}	…
o_1	0/1	0/1	…	0/1	-1	-1	-1
o_2	0/1	0/1	…	0/1	-1	-1	-1
⋮	0/1	0/1	…	0/1	-1	-1	-1
o_m	0/1	0/1	…	0/1	-1	-1	-1
o_{u1}	-1	-1	…	-1	-1	-1	-1
o_{u2}	-1	-1	…	-1	-1	-1	-1
⋮	-1	-1		-1	-1	-1	-1

The O_k, A_k and value area represented by 0/1 in Table 1 form the square embedded in the formal environment, which is also, by Theorem 1, the independent square representing the known world. Therefore, the known world can be expressed by triplet $\mathscr{K}$= (O_k, A_k, v_A). Obviously, $\mathscr{K}$ is the sub-environment of $\mathscr{E}$.

3. Formalization of Concepts

From monism, a concept is logical thing [10]. However, a concept is often complexes involving various recognitions, which needs to be broken down into more simple parts, so that its logical structure is displayed [11].

3.1. *Origin of concepts*

Concepts are from sensations, perceptions, and impressions, which are all related to attributes.

3.1.1. *Attribute*

Attribute refers to the nature and characteristics of things [12]. For example, eggs are oval in shape and contain lecithin, which are the properties of eggs.

Attributes and things are inseparable, and the similarities and differences between things are the same as attributes of them. Apples, for example, contain vitamins and are round in shape, which are the attributes of apples and by which we can distinguish apples from eggs.

3.1.2. *Sensation and perception*

Sensation refers to any specific conscious experience caused by irritating of sense organs, or sensory nerves, or sensory areas of the brain [13], and is a simply property of a certain aspect of something that persons recognize [14].

For example, these, the color seen by the eyes, the smell sniffed by the nose and the temperature touched by the skin, are sensations.

Perception refers to the process of mental experiences transformed by sensory irritation, and is an organized whole [13].

In general, persons consider that perception is complex and sensation is simple. The existence of sensory is onefold but perception not. Perception is related to existing experience and the various cognized attributes of a thing.

3.1.3. *Impression*

Impression refers to the signs left by objective things in a person's mind [12].

Because of memory, impressions are multiple sensations and perceptions that leave an imprint in the brain, which is an unclear or inaccurate thinking, feeling, or outlook [15]. Obviously, impression depends on the degree of clarity and certainty of the something's signs left in one's mind.

Here is a Song poem "Plum" written by Anshi Wang:

A few plums in the corner
They unfold alone facing chill
Knew them not snow
Coming for fresh fragrance in bursts

With the help of the wafting aroma, the author deduced that the white object in the corner of the wall in the distance was not snow, and when getting closer, he saw a few golden plum blooming in the cold, and thus felt the cold-resistant quality of the plum. Therefore, the author's impression of plum blossoms comes from the perception formed by sight, smell, and skin senses, as well as the author's own experience.

Sensation and perception are obtained through intuition, while impression is at the junction of perceptual and rational cognition. With the help of a language, impressions enter the realm of rational cognition and form a preliminary concept reflecting things. By refining the initial concept, we get a profound and general concept.

3.2. *Formal concept*

Description 2. *Given $\mathscr{K}$=(O_k, A_k, v_A); let $I \in \mathscr{P}(A_k) \wedge E \in \mathscr{P}(O_k)$. If $\forall o \forall a$ (($o \in E$) $\wedge$($a \in I$) $\wedge$ $v_A(o, a)$=1), then the 2-tuple $\mathscr{C}$=(E, I) is called the formal concept based on $\mathscr{K}$ Where E is the extension of $\mathscr{C}$ and I is the intension of $\mathscr{C}$*

In general, the specific studied problem is a part of the known world, that is, $O_c \subseteq O_k$, $A_c \subseteq A_k$, and $\mathscr{K}_c$=(O_c, A_c, v_A). Obviously, $\mathscr{K}_c$ is also the sub-environment of $\mathscr{K}$ Sub-environments. All examples mentioned below will refer to $\mathscr{K}_c$ unless otherwise specified.

Example 1. The sub-environment is shown in Table 2. Let I_1= {life form}, E_1={dog, frog, fish, eagle, cypress, peach tree, lotus, phragmites}. Since $v_A(E_1, I_1)$=1, the concept $\mathscr{C}_1$=(E_1, I_1)[†] is generated.

Example 2. Let I_2= {lifeform, autonomy moving}. By Table 2, E_2= {dog, frog, fish, eagle}, so $\mathscr{C}_2$= (E_2, I_2).

Table 2. The sub-environment with life-form attributes.

	life form	autonomy moving	Living in water	Living on land	fly	leaves fall	leaves after flowers
dog	1	1	0	1	0	0	0
frog	1	1	1	1	0	0	0
fish	1	1	1	0	0	0	0
eagle	1	1	0	1	1	0	0
cypress	1	0	0	1	0	0	0
peach tree	1	0	0	1	0	1	1
lotus	1	0	1	0	0	1	0
phragmites	1	0	1	1	0	1	0

Theorem 4. *(E, I) → (I→E).*

Theorem 4 states that under a given concept, an intension determines an extension. When the extension is fixed, however, the intension may be different and can generate new concepts. By Example 2, in addition to the two

† "$v_A(E, I)$=1" is equivalent to "if $\forall o(o \in E) \wedge \forall a(a \in I)$, then $v_A(o, a)$=1".

attributes—lifeform and autonomy moving, a frog in $\mathscr{C}_2$ has both attributes of living on land and living in water, so it can also generate new concepts $\mathscr{C}_3=(E_3, I_3)$, where $I_3=$\{lifeform, autonomy moving, living on land, living in water\}, $E_3=$\{frog\}.

Theorem 5. *Let* $I_1, I_2, \ldots, I_n \in \mathscr{P}(A_k)$ *and*

$$I_N = \bigcap_{i=1}^{n} I_i \neq \varnothing. \tag{2}$$

If $E \neq \varnothing \wedge E \in \mathscr{P}(O_k) \wedge v_A(E, I_N)=1$, *then* $\mathscr{C}_N=(E, I_N)$.

Example3. According to Table 2, $I_1=$\{lifeform, autonomy moving, living on land, living in water\}, $I_2=$\{lifeform, autonomy moving, living in water\}. If $I_N=I_1 \cap I_2=$\{lifeform, autonomy moving, living in water\} and when $\mathscr{C}_{12}=(E, I_1 \cap I_2)$, $E=$\{frog, fish\}.

In general, I_N is believed an essential attribute of $\mathscr{C}_N$.

Corollary 6. *A concept has at least one essential attribute.*

3.3. *Discussion*

According to the known world in Description 1, if 1 and 0 in value area are taken as logical true and logical false, respectively, and let $\neg a$ be the attribute that is in contradiction to a, then o does not have a when $v_A(o, a)=0$, which means that o has $\neg a$.

Theorem 7.

(1) $\neg v_A(o, a) \Leftrightarrow v_A(o, \neg a)$;

(2) $v_A(o, a) \Leftrightarrow \neg v_A(o, \neg a)$.

Corollary 8. Given $\mathscr{K}=(O_k, A_k, v_A)$, and let $a \in \mathscr{P}(A_k) \wedge o \in \mathscr{P}(O_k)$. If $\neg v_A(o, a)=1$, then $\mathscr{C}=(o, \neg a)$.

Corollary 9. Given $\mathscr{K}=(O_k, A_k, v_A)$, and let $a_1, a_2 \in \mathscr{P}(A_k) \wedge o \in \mathscr{P}(O_k) \wedge v_A(o, a_1 \cup \neg a_2)=1$, then $\mathscr{C}=(o, a_1 \cup \neg a_2)$.

Given the sub-environment $\mathscr{K}=(O_k, A_k, v_A)$. Let $I \in \mathscr{P}(A_k)$ and $\neg I$ be the negative attribute of I, then ${}^*A_k=A_k \cup \neg I$ is called the set of virtual formal attribute of $\mathscr{K}$; ${}^*\mathscr{K}=(O_k, {}^*A_k, v_A)$ is called the virtual sub-environment of $\mathscr{K}$.

Example 4. The sub-environment is shown in Table 2. Let $a_1=$\{lifeform\} and $a_2=$\{autonomy moving\}, then $\mathscr{C}=(o, a_1 \cup \neg a_2)$ represents the concept of having two attributes of lifeform and un-autonomy moving, which is generated in the virtual sub-environment in Table 2. So, $o=$\{cypress, peach tree, lotus, phragmites\}.

Theorem 10. Let $S_{\mathscr{C}}$ be the concept set generated by $\mathscr{K}$, and ${}^*S_{\mathscr{C}}$ be the concept set generated by ${}^*\mathscr{K}$, then $|S_{\mathscr{C}}| \leq |{}^*S_{\mathscr{C}}|$.

4. Conclusion Remarks

This chapter proposed for the first time formal environment and studied formal concept based on the formal environment. However, the names of specific concepts in the examples mentioned above were all not given. These names should be given by the definition since a definition is a logical way to reveal a concept [16] for which, we extract some contents from another paper written by us as following.

Description 3. *Given* $\mathscr{E}$= *(E, I). Let W be the set of symbol strings and* $d\in W$; *D be the set of existing terms,* $D\cap I=\varnothing$, $m\in\mathscr{P}(I\cup D)$, $\{d\}\cap\mathscr{P}(I\cup D)=\varnothing$. *The two place predicate P(x, y) means that x has the attributes of all members in y, and Q(x, y) means that x is called y. Let t be a thing, if there exists the condition statement*‡

$$P(t, m)\rightarrow Q(t, d),$$

then the quadruple $\mathscr{D}$=*(E, I, D, W) is called formal definition based on* $\mathscr{E}$.

Example 5. According to Description 3 and Table 2, let *m*={lifeform, autonomy moving}, then we have the definition: the thing that has both attributes of lifeform and autonomy moving is called an animal. Here “animal” is from *W*, and it is the will of the person who gives this definition.

Acknowledgments

Authors would like to thank Xia He for valuable suggestions and all kind help.

This work is supported by the Key Program of National Science Foundation of China under Grant No. 18ZDA031 and the National Natural Science Foundation of China under Grant No. 61170322.

References

1. B. Ganter and R. Wille, *Formal concept analysis: mathematical foundations*. (Springer, Heidelberg, 1999).
2. R. Wille, *Restructuring lattice theory: An approach based on hierarchies of concepts,* in *Ordered Sets*, eds. I. Rival, (Boston-Dordrecht: Reidel,1982), pp: 445–470.
3. L. Wei, L. Cao, J. Qi, W. Zhang, Concept reduction and concept characteristics in formal concept analysis, *Sci China. F*, **50**, 1817 (2020).

‡ It is specified that the antecedent of the condition statement is true.

4. V. Snasel, Z. Horak, J. Kocibova, et al., *Analyzing social networks using FCA: Complexity aspects*, in *Proc. IEEE/WIC/ACM Int. Conference on Web Intelligence and Intelligent Agent Technology*, (Milan, Italy, 2009).
5. S. Kuznetsov and S. Obiedkov, Comparing Performance of Algorithms for Generating Concept Lattices, *J. Experimental & Theoretical Artificial Intelligence*, **14**, 189 (2002).
6. L. Ma, F. Chang, H. Zhang, et al., Formal Concept Analysis based Grouping Co-Evolutionary Optimization Algorithms for Large-Scale Global Optimization, *Chinese Journal of Computer*, **44**, 1310 (2021).
7. P. Tang, S. Hui, A. C. M. Fong, A lattice-based approach for chemical structural retrieval, *Engineering Applications of Artificial Intelligence*, **39**: 215 (2015).
8. Longman, *Longman Dictionary of Contemporary English*, New edition. (Longman World Publishing Corporation, 1987).
9. A. N. Whitehead. *Modes of thought*. (Combridge at the University Press, 1956.)
10. G. Ryle, *The Concept of Mind*. (Hutchinson House, 1949).
11. S. Blackhurn, *Oxford dictionary of philosophy*. (Oxford University Press, 1996).
12. Dictionary Editing Room, Institute of Linguistics, Chinese Academy of Social Science, *Contemporary Chinese Dictionary,* 5th edition. (The Commercial Press, 2005).
13. Britannica, The Editors of Encyclopedia, *Concise Encyclopædia Britannica, 15th edition*. (Encyclopedia Britannica Inc., 1985).
14. Q. Yang, *Concise Dictionary of Psychology*. (Jilin People's Publishing House, 1985).
15. A. S. Hornby, *Oxford advanced learner's English-Chinese dictionary*,14 edition. (Oxford University Press, 1997).
16. Y. Jin, *Formal logic,* reprint edition. (People's Publishing House, Beijing, 2006).

Inverse Dynamic Deduction Algorithm of Standard Contradiction Separation Rule Based on Parallel Mechanism

Guoyan Zeng[*,†,§], Guanfeng Wu[*,†], Shuwei Chen[*,†,¶], Yang Xu[*,†] and Peiyao Liu[†,‡]
[*]*School of Mathematics, Southwest Jiaotong University, Chengdu 611756, China*
[†]*National-Local Joint Engineering Laboratory of System Credibility Automatic Verification, Southwest Jiaotong University, Chengdu 611756, China*
[‡]*School of Computer and Software Engineering, XiHua University, Chengdu 610039, China*
[§]*guoyanzeng_math@163.com*
[¶]*swchen@swjtu.edu.cn*

Standard contradiction separation (S-CS) rule is a novel method for automated reasoning. Subsequently, a series of algorithms have been proposed for the implementation of this automated deduction rule. In order to better realize the features of the S-CS rule, such as objective guidance and controllability, and further improve the inference performance of the S-CS algorithm, this chapter proposes a new inverse parallel deduction algorithm based on the S-CS rule. We then embed this new algorithm into Vampire, the leading automated theorem prover, forming a new parallel system called V_IP. V_IP is then evaluated using an international standard test database (e.g., the FOF theorem in CASC-J11). Experimental results indicate that the inverse parallel algorithm effectively improves the performance of Vampire; V_IP is more powerful compared to Vampire.

Keywords: Automatic reasoning; first-order logic; parallel mechanism; inverse deduction; S-CS rule.

1. Introduction

Automated theorem proving (ATP), which is a subfield of mathematical logic and computer science, has become a core part of AI (artificial intelligence). With the development of AI, ATP systems have received extensive attention, and automated theorem provers can solve many practical problems [1]. In automatic reasoning, inference rule is the key to automated reasoning. Resolution principle [2] became the most widely used inference rule due to its efficiency and simplicity. With the widespread use of resolution principle, a large number of ATP systems (e.g., Vampire [3] and Prover 9 [4]) based on binary resolution method have demonstrated strong theorem proving capabilities since the CASC (CADE ATP System Competition) [5] was launched in 1996.

Recently, a new standard contradiction separation (S-CS) inference rule [6] was proposed by Xu *et al.*, which is an extension of the binary resolution method, and some CS-based systems [7, 8] have been designed. The S-CS rule has many advantages [9], such as dynamic and flexible deduction, multi-clause synergized deduction, guided deduction, controllable deduction, deriving a clause with fewer literals, and so on.

Many scholars have conducted extensive research in forward reasoning on the above two inference rules, however, not much research has been done in inverse reasoning. The inverse method [10, 11] is a generic resolution method that uses non-classical logic and classical logic, which was first proposed by Maslov [10] for classical logic. As the name suggests, the inverse method is a bottom-up approach, adding the results of inference rules related to goal sequent, and working forward until the goal formula is derived. The inverse method was then used for non-classical logic in [12] and [13]. In 2014, V. Pavlov [14] used the inverse method in first-order logic theorem proving. However, as discussed in [11], designing systems that are bidirectional in this domain is a challenge, and few ATP systems use the inverse method. Recently, Ref. [15] summarizes the main work on parallel theorem proving. Vampire is also working on parallelism [16] but does not achieve parallelism in the reasoning rules and remains parallel in the strategies.

Aiming at tackling the above challenge, we attempt to further improve the capability and efficiency of automated theorem proving systems based on the S-CS rule, both in inverse and parallel. In Ref. [17], we made a preliminary attempt and there is still a large amount of work to be further studied. Therefore, we design in this chapter a new inverse and parallel algorithm based on the S-CS rule and embed the

[§]Corresponding author.

proposed algorithm into Vampire 4.7 to form a new ATP system called V_IP, which is a parallel bidirectional system.

The remainder of this chapter is structured as follows. We review the S-CS rule and an important strategy in Section 2. Section 3 describes the steps of the S-CS algorithm. In Section 4, we present a novel inverse and parallel dynamic automated deduction algorithm based on the S-CS algorithm. The experimental evaluation is provided in Section 5. In the final section, we demonstrate conclusions and future works.

2. Preliminaries and Motivation

In this section, we introduce the knowledge about S-CS rule and then discuss the motivation of this paper.

2.1. *Basic knowledge of S-CS rule*

Definition 2.1. [6] Suppose a clause set $S = \{C_1, C_2, \ldots, C_m\}$ in first-order logic. Without loss of generality, assume that there does not exist the same variables among $C_1, C_2, \ldots, C_m$(if there exist the same variables, there exists a rename substitution can then be applied to make them different). The following inference rule that produces a new clause from S is called a *standard contradiction separation rule*, in short, an S-CS rule:

For each C_i $(i = 1,2, \ldots, m)$, firstly apply a substitution σ_i to C_i (σ_i could be an empty substitution but not necessary the most general unifier), denoted as $C_i^{\sigma_i}$; then separate $C_i^{\sigma_i}$ into two sub-clauses $C_i^{\sigma_i -}$ and $C_i^{\sigma_i +}$ such that

(1) $C_i^{\sigma_i} = C_i^{\sigma_i -} \vee C_i^{\sigma_i +}$, where $C_i^{\sigma_i -}$ and $C_i^{\sigma_i +}$ have no common literal;
(2) $C_i^{\sigma_i +}$ can be an empty clause itself, but $C_i^{\sigma_i -}$ cannot be an empty clause;
(3) $\bigwedge_{i=1}^{m} C_i^{\sigma_i -}$ is a standard contradiction, that is $\forall (l_1, \ldots, l_m) \in \prod_{i=1}^{m} C_i^{\sigma_i -}$, there exists at least one complementary pair among $\{l_1, \ldots, l_m\}$.

The resulting clause $\bigvee_{i=1}^{m} C_i^{\sigma_i +}$, denoted as $C_m^{s\sigma}(C_1, \ldots, C_m)$ (here "s" means "standard", $\sigma = \bigcup_{i=1}^{m} \sigma_i$, σ_i is a substitution to C_i, $i = 1, \ldots, m$), is called a *standard contradiction separation clause* (S-CSC) of $C_1, \ldots, C_m$, and $\bigwedge_{i=1}^{m} C_i^{\sigma_i -}$ is called a *separated standard contradiction* (S-SC).

In Definition 2.1, if S-CSC is an empty clause, then the clause set S is unsatisfiable, which is actually the Soundness Theorem of S-CS rule as detailed in Theorem 2.1. Our goal is therefore to make the number of literals in S-CSC as small as possible, and preferably 0 when S-CSC is an empty clause, obtaining the conclusion that S is unsatisfiable.

Definition 2.2. [6] Suppose a clause set $S = \{C_1, C_2, \ldots, C_m\}$ in first-order logic. $\Phi_1, \Phi_2, \ldots, \Phi_t$ is called a standard contradiction separation based dynamic deduction sequence from S to a clause Φ_t, denoted as D^s, if

1) $\Phi_i \in S$, $i \in \{1,2, \ldots ,t\}$; or
2) there exist $r_1, r_2, \ldots, r_{k_i} < i$, $\Phi_i = C_{r_{k_i}}^{s\theta_i}(\Phi_{r_1}, \Phi_{r_2}, \ldots, \Phi_{r_{k_i}})$, where $\theta_i = \bigcup_{j=1}^{k_i} \sigma_j$, σ_j is a substitution to Φ_{r_j}, $j = 1,2, \ldots ,k_i$.

This definition reflects the multiple dynamic characteristics of the S-CS rule overall. In addition, the S-CS rule satisfies the following soundness and completeness.

Theorem 2.1. [6] **(Soundness Theorem of the S-CS Based Dynamic Deduction in First-Order Logic)** Suppose a clause set $S = \{C_1, C_2, \ldots, C_m\}$ in first-order logic. $\Phi_1, \Phi_2, \cdots, \Phi_t$ is an S-CS based dynamic deduction from S to a clause Φ_t. If Φ_t is an empty clause, then S is unsatisfiable.

Theorem 2.2. [6] **(Completeness of the S-CS Based Dynamic Deduction in First-Order Logic)** Suppose a clause set $S = \{C_1, C_2, \ldots, C_m\}$ in first-order logic. If S is unsatisfiable, then there exists an S-CS based dynamic deduction from S to an empty clause.

Decision Literal Strategy. [18] There is one literal from each selected clause in the separated standard contradiction that plays an important role on determining the separated standard contradiction, and therefore is called as decision literal. Denote the set of decision literals as D_l in the following. There is usually an order on the decision literals when using them to construct the separated standard contradiction, although the unsatisfiability of the formed separated standard contradiction has no relation with this order. The literals in the subsequent clauses after certain substitutions that are complementary to the previous decision literals will be "pulled" into the separated standard contradiction (S-SC), and the remaining literals, if there is any, will contribute to the S-CSC.

Therefore, the set $\boldsymbol{D_l} = \{l_{d1}, l_{d2}, \cdots, l_{dm}\}$ is a decision literal set in S-SC, where the corresponding order indicates the order in which the literal participates in the S-CS deduction, it also represents the order in which the clauses containing the decision literal participate in the S-CS deduction.

2.2. *Motivation*

Based on Definition 2.1, we have the following idea, which is also the motivation for this article.

1) On the one hand, assuming we have already implemented certain S-CS deduction, we can then obtain a new unit clause from the sub-clause $C_{i1}^{\sigma_{i1}-}$ of the first clause $C_{i1}^{\sigma_{i1}}$ where $C_{i1}^{\sigma_{i1}+}$ is not empty, and this new clause is called an *inverse goal clause*. We replace the original $C_{i1}^{\sigma_{i1}}$ with the inverse goal clause, such that $C_{i1}^{\sigma_{i1}+}$ has no literal and S-CSC is an empty clause. This idea reflects the thought of inverse, because we are trying to find the conditions (inverse goal clauses) that make S-CSC become an empty clause in an inverse way. When the inverse goal clause is deduced, then S-CSC is empty.

2) In order to deduce the above introduced inverse goal clause, we add the negation of the inverse goal clause to the clause set S to form S'. If we obtain an empty clause by S-CS deduction from S', then S can infer the goal clause.

3) On the other hand, we can usually extract more than one inverse goal clause that are independent of each other, so we can process the deduction to infer the inverse goal clause in parallel.

Based on the above motivation analysis, for the inverse goal clause, we need to select ground literals that are inside the contradiction, i.e., they are not among the decision literal set $\boldsymbol{D_l}$. The following example illustrates how to find inverse goal clauses and then perform parallel deduction.

Example 2.1. Let clause set $S=\{C_1, C_2, C_3, C_4, C_5, C_6, C_7, C_8\}$, $C_1=\sim P_4(x_{11}) \vee P_3(f(x_{11})) \vee P_2(x_{11})$, $C_2=\sim P_3(x_{21}) \vee P_4(x_{21}) \vee P_5(x_{21}) \vee \sim P_1(x_{21})$, $C_3=P_1(a)$, $C_4=\sim P_5(x_{41})$, $C_5=\sim P_3(f(x_{51}))$, $C_6=P_5(x_{61}) \vee \sim P_2(x_{61})$, $C_7=P_5(x_{71}) \vee P_3(x_{71})$, $C_8=\sim P_1(x_{81}) \vee \sim P_4(x_{81})$. Here a is a constant, $x_{11}, x_{21}, x_{41}, x_{51}, x_{61}, x_{71}, x_{81}$ are variables, f is a function symbol, P_1, P_2, P_3, P_4, P_5 are predicate symbols. According to the S-CS rule, the S-CS deduction process is shown in Table 1.

Table 1. The S-CS deduction of Example 2.1.

i	$C_i^{\sigma_i-}$	$C_i^{\sigma_i+}$	Decision literal	σ_i
3	$P_1(a)$	$\emptyset$	$P_1(a)$	$\emptyset$
4	$\sim P_5(a/x_{41})$	$\emptyset$	$\sim P_5(a)$	$\{a/x_{41}\}$
5	$\sim P_3(f(a/x_{51}))$	$\emptyset$	$\sim P_3(f(a))$	$\{a/x_{51}\}$
2	$P_4(a/x_{21}) \vee P_5(a/x_{21}) \vee \sim P_1(a/x_{21})$	$\sim P_3(a/x_{21})$	$P_4(a)$	$\{a/x_{21}\}$
1	$\sim P_4(a/x_{11}) \vee P_3(f(a/x_{11}))$	$P_2(a/x_{11})$	$\emptyset$	$\{a/x_{11}\}$

From Table 1, we obtain a S-CSC involving 5 clauses: $C_9=P_2(a) \vee \sim P_3(a)$ and the corresponding S-SC is: $C_3^{\sigma_3-} \wedge C_4^{\sigma_4-} \wedge C_5^{\sigma_5-} \wedge C_2^{\sigma_2-} \wedge C_1^{\sigma_1-}$. We obtain the first clause in the S-CS deduction whose $C_2^{\sigma_2+}$ is not an empty clause, $\boldsymbol{D_l}=\{P_1(a), \sim P_5(a), \sim P_3(f(a)), P_4(a)\}$. Therefore, there are the following inverse goal clauses: $C_{2\_1}=\sim P_1(a)$, $C_{2\_2}=P_5(a)$. Then we apply the S-CS rule in parallel to deduce C_{2_1} and C_{2_2}, so that S can be concluded as unsatisfiable if either one of the clauses in $\{C_{2\_1}, C_{2\_2}\}$ is inferred.

The negation of the inverse goal clauses C_{2_1} and C_{2_2} are denoted as $C_{10}=P_1(a)$ and $C_{11}=\sim P_5(a)$. We add C_{10} and C_{11} to the clause set S, respectively. Then the S-CS deductions as in Tables 2 and 3 are constructed in parallel according to the S-CS rule.

Table 2. The S-CS deduction of $S \cup \{C_{10}\}$.

i	$C_i^{\sigma_i -}$	$C_i^{\sigma_i +}$	Decision literal	σ_i
10	$P_1(a)$	$\emptyset$	$P_1(a)$	$\emptyset$
4	$\sim P_5(a/x_{41})$	$\emptyset$	$\sim P_5(a)$	$\{a/x_{41}\}$
7	$P_5(a/x_{71}) \vee P_3(a/x_{71})$	$\emptyset$	$P_3(a)$	$\{a/x_{71}\}$
2	$P_4(a/x_{21}) \vee P_5(a/x_{21}) \vee \sim P_1(a/x_{21}) \vee \sim P_3(a/x_{21})$	$\emptyset$	$P_4(a)$	$\{a/x_{21}\}$
8	$\sim P_4(a/x_{81}) \vee \sim P_1(a/x_{81})$	$\emptyset$	$\emptyset$	$\{a/x_{81}\}$

Table 3. The S-CS deduction of $S \cup \{C_{11}\}$.

i	$C_i^{\sigma_i -}$	$C_i^{\sigma_i +}$	Decision literal	σ_i
11	$\sim P_5(a)$	$\emptyset$	$\sim P_5(a)$	$\emptyset$
7	$P_5(a/x_{71}) \vee P_3(a/x_{71})$	$\emptyset$	$P_3(a)$	$\{a/x_{71}\}$
9	$P_2(a) \vee \sim P_3(a)$	$\emptyset$	$P_2(a)$	$\emptyset$
6	$P_5(a/x_{61}) \vee \sim P_2(a/x_{61})$	$\emptyset$	$\emptyset$	$\{a/x_{61}\}$

From Table 2, we apply the S-CS rule to the clause set $S \cup \{C_{10}\}$ to get an empty clause, so S can infer C_{2_1}, replacing $C_2^{\sigma_2}$ with C_{2_1} in Table 1, then the clauses C_3, C_4, C_5, C_{2_1} deduce an empty clause, so S is unsatisfiable. Similarly, through Table 3, S can infer C_{2_2}, similar to the operation of C_{2_1}, and ultimately can also obtain that S is unsatisfiable. The S-CS deductions in Tables 2 and 3 are parallel, and if any of these deduction paths is found, then we can conclude that S is unsatisfiable and the whole deduction can stop.

3. Dynamic Deduction Framework of S-CS Rule

Based on the above motivation, we first introduce an S-CS algorithm to implement the S-CS rule, and the S-CS algorithm plays a key role as an important part of the inverse parallel deduction algorithm in the next section. Algorithm 1 describes the details of the pseudo-code and the corresponding steps are as follows.

Step 1. According to the clause selection strategy [7], a clause is selected as the first input clause for S-CS deduction.

Step 2. Select decision literal set D_l according to the decision literal strategy, and we prioritize ground literal as decision literal.

Step 3. Select candidate clause set S_2 according to complementary ratio algorithm [19], where the candidate clause needs to include at least a literal complementary to some decision literal.

Step 4. Traverse each clause C in the candidate clause set S_2, and apply S-CS rule to construct contradiction and generate S-CSC r.

Step 5. If r is empty, then the clause set S is unsatisfiable; otherwise, check whether r is redundant, if r is redundant, go to Step 4; otherwise, put r into R.

Step 6. Check whether the stop conditions (e.g., time and memory thresholds) are reached, if it is, go to Step 7; otherwise, go to Step 4.

Step 7. Apply forward and backward simplification rule, then output R.

Algorithm 1. S-CS algorithm

Input:
a given first-order logic clause set S, selected clauses are denoted as g, C,
a decision literal set D_l, two clause sets are denoted as S_1, S_2(initially empty)
Output: a generated S-CSC set R
1: g=getFirstInputClause(S)
2: $S_1 = S_1 \cup g$
3: D_{l_m}=selectDecisionLiteralSet(g)
4: S_2= CandidateClauseSet (D_l, S)
5: for each $C \in S_2$
6: while (C is a non-unit clause) ≠ null begin
7: r=extendedSplitContradiction(C, S_1)
8: if (r==∅)
9: return "unsatisfiable"
10: else if (chenck_Redundant(r, S))
11: deduction_Backtracking(C)
12: goto 5
13: else
14: $R = R \cup r$
15: $S_1 = S_1 \cup C$
16: if checkStopConditions()
17: goto 21
18: else
19: goto 5
20: end
21: end
22: inner_Simplification(R)
23: for each $c' \in R$
24: backward_Simplification(S, c')
25: $S = S \cup R$

4. Inverse and Parallel Dynamic Deduction Algorithm

In order to further enhance the proof search capacity and efficiency of the S-CS deduction, based on Algorithm 1, we propose a new inverse and parallel algorithm (in short IP algorithm). We extract new inverse deduction goal clauses and then use S-CS algorithm to deduce these goal clauses in parallel. Algorithm 2 describes the details of the pseudo-code and the corresponding steps are as follows.

Step 1. Using S-CS algorithm to construct contradiction and generate S-CSC set R.

Step 2. If R is empty, then the clause set S is unsatisfiable.

Step 3. The clauses whose sub-clause part inside S-CSC is not empty are pulled into clause set Q.

Step 4. Extract the corresponding sub-clause of the first clause in Q inside contradiction, the extracted ground literals as the inverse deduction goals, then take negations of the ground literals and put them into clause set S_3.

Step 5. $|S_3|$ sub-processes are created, where $|S_3|$ counts the number of ground clauses in S_3, and then the function *startFork(i, S,* $S_3[i]$, *S-CS algorithm)* calls the S-CS algorithm to realize the theorem proving task decomposition.

Step 6. The *EndFork* function determines whether the *i*-th child process has ended or not. If it is, analyze the result and output the corresponding deduction message. If one of the *ResultFork*(*i*) is an empty clause, then *S* is unsatisfiable. Otherwise, clear inverse deduction paths and put *R* into *S*.

Step 7. Check whether the stop conditions (e.g., time and memory thresholds) are satisfied, if it is, output R of each child process (i.e., output resultFork()); otherwise, go to Step 1.

Algorithm 2. Inverse Parallel algorithm (IP algorithm)

Input:

a given first-order logic clause set S, selected clauses are denoted as Q,

a clause set S_3 (initially empty)

Output: resultFork()

$R \leftarrow$ S-CS algorithm
if(checkInverse(R))
Q= selectInverseClause (R)
S_3=getNegationInverseGoalClause(Q)
for (i=0;i<$|S_3|$;i++) //$|S_3|$denotes the number of clauses contained in the clause set S_3
startFork(i, S, S_3[i],S-CS algorithm)
while (true)
For (i=0;i <$|S_3|$;i++)
If EndFork(i)
GetresultFork(i)
if resultFork(i)==∅ exit
else ClearingInverseDeductionPaths
S=$S \cup R$
if checkStopConditions() output resultFork(i)
else goto 1

In the inverse parallel deduction algorithm, we first use the S-CS algorithm to generate S-CSC clause, then check whether inverse deduction is necessary based on S-CSC. If necessary, the ground literals inside the contradiction are extracted to construct the inverse deduction goal clauses, and the S-CS algorithm is applied again, but to different clause sets, in parallel. If one of the goal clauses is deduced, we conclude that S is unsatisfiable. This algorithm is bidirectional, i.e., it is possible to alternate between forward (without inverse goal clauses) S-CS algorithm and inverse S-CS algorithm with inverse goal clauses.

5. Experimental Analysis

Vampire is the top international automatic theorem prover for first-order logic, and it has been the CASC champion for decade of years. In order to further improve the performance of Vampire, we embed IP algorithm into Vampire 4.7[1] to form a new ATP system called V_IP. The operation mode of V_IP is in the following setting. First, solve the problem in the benchmark library with Vampire and exit if it is successful, otherwise solve it using the IP algorithm. If the IP algorithm fails to solve it as well, then the proof tasks assigned by the IP algorithm are formed into new problems for Vampire to solve. Multiple Vampires then solve these different new problems in parallel, if one of the new problems is solved by Vampire, the original clause set is unsatisfiable. Then we tested V_IP using the FOF (First Order Form) division theorems from the CASC competition[2] (CASC-J11), totaling 500 theorems. The experiments are performed on a PC with a 3.6 GHz Inter(R) Core (TM) i7-7700 processor and 32GB of RAM, OS Ubuntu 20.04 64-bit, with the standard default CPU time limit of 300 s.

Figure 1 shows the result of the CASC-J11 competition for V_IP and Vampire 4.7. The number of theorems proved for V_IP is 469 with 13 more problems than that proved by Vampire 4.7 solely. The average time for proving the 469 theorems is 16.06 s for V_IP, and the average time for proving the 456 theorems is 16.09 s for Vampire 4.7. For efficiency, the average solving time of V_IP is less than that of

1 https://vprover.github.io/download.html.

2 https://www.tptp.org/CASC/.

Vampire 4.7. The experimental results demonstrate that the inverse parallel inference mechanism based on S-CS rule can improve the proving ability of Vampire to large extent.

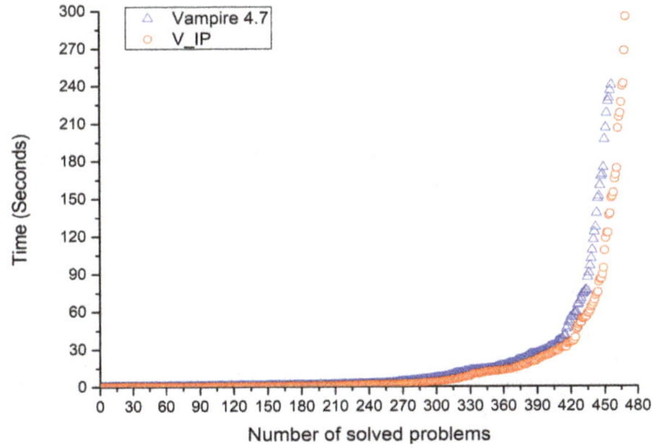

Fig. 1. Performance comparison between V_IP and Vampire 4.7 on CASC-J11.

In Table 4, Rating denotes the difficulty coefficient of the competition theorem and is a real number taking value in [0.0, 1.0]. The difficulty of proving the theorem is proportional to the rating value. Table 4 shows that V_IP can solve some problems that Vampire 4.7 cannot, especially those with higher values of rating, and even theorems with large ratings that contain large number of formulae, such as CSR177+1 and HWV107+1. It shows that the IP algorithm can improve the ability of Vampire. Analyzing the experimental results further, we analyze the proof path and find that multiple Vampires prove different goal clauses in parallel. Not only the S-CS algorithm can be executed in parallel, but also Vampire can be invoked in parallel. Therefore, it is the main reason why V_IP is superior to Vampire. In summary, V_IP has a stronger performance compared to Vampire 4.7, which means that the IP algorithm effectively enhances the ability of Vampire.

Table 4. List of 13 problems solved by V_IP but unsolved by Vampire 4.7 on CASC-J11 competition

Theorem	Rating	Number of formulae	Theorem	Rating	Number of formulae
BOO109+1	0.69	3	NUM694+4	0.86	573
CSR036+3	0.44	8006	NUN056+1	0.69	19
CSR177+1	0.92	7433	NUN075+1	0.94	19
GRP746+1	0.5	9	PUZ077+1	0.75	7
HWV107+1	0.92	9614	REL029+4	0.6	17
ITP009+2	0.92	37	SWB092+1	0.97	560
LAT372+1	0.92	103			

6. Conclusions and Future Work

In this chapter, we first extend the S-CS algorithm to form a new inverse parallel deduction algorithm (IP algorithm); the IP algorithm is then embedded into Vampire 4.7 to form a parallel prover V_IP. V_IP and Vampire 4.7 are tested with CASC-J11 in the same environment, and the comparison experiments have shown that V_IP has better performance than Vampire 4.7. Based on the specific analysis of the deduction paths, it can be seen that the inverse parallel approach is goal-oriented, and the parallel algorithm is efficient, which is an excellent complement to the resolution method. In conclusion, V_IP has the advantages of the S-CS rule and can also utilize the capabilities of the Vampire, reflecting goal-oriented and efficient parallelism. In the future, the inverse parallel deduction algorithm still needs to be improved, e.g., effective heuristic strategies not only enhance the efficiency of deduction but also better control the direction of deduction.

Acknowledgments

This work has been partially supported by the National Natural Science Foundation of China (Grant No.62106206, 62206227, 62366017).

References

1. J. P. P. Flor, W. Swierstra, Y. Sijsling. Pi-Ware: hardware description and verification in Agda, in 21st International Conference on Types for Proofs and Programs, 2015, pp. 1–27, https://doi.org/10.4230/LIPIcs.TYPES.2015.9.
2. J. A. Robinson. A machine-oriented logic based on the resolution principle, Journal of the ACM (JACM), 1965, 12(1): 23-41.
3. L. Kovács, A. Voronkov. First-order theorem proving and Vampire, International Conference on Computer Aided Verification. Berlin, Heidelberg: Springer Berlin Heidelberg, 2013, pp. 1-35. https://doi.org/10.1007/978-3-642-39799-8_1.
4. W. McCune. Prover9 and Mace4 (2005–2010). http://www.cs.unm.edu/~mccune/prover9/.
5. G. Sutcliffe. The CADE ATP system competition—CASC. AI Magazine, 2016, 37(2): 99-101.
6. Y. Xu, J. Liu, S. Chen, et al. Contradiction separation based dynamic multi-clause synergized automated deduction, Information Sciences, 2018, 462: 93-113.
7. F. Cao, Y. Xu, J. Liu, S. Chen, J. Yi, A multi-clause dynamic deduction algorithm based on standard contradiction separation rule, Information Sciences. 2021, 566: 281–299, https://doi.org/10.1016/j.ins.2021.03.015.
8. P. Liu, S. Chen, J. Liu, Y. Xu, F. Cao, An efficient contradiction separation based automated deduction algorithm for enhancing reasoning capability, Knowledge-Based Systems. 2023, 261: 110217.
9. Y. Xu, S. Chen, J. Liu, et al. Distinctive features of the contradiction separation based dynamic automated deduction, Data Science and Knowledge Engineering for Sensing Decision Support: Proceedings of the 13th International FLINS Conference (FLINS 2018), 2018, pp. 725-732.
10. S. Y. Maslov. An inverse method of establishing deducibility in the classical predicate calculus, Doklady Akademii Nauk. Russian Academy of Sciences, 1964, 159(1): 17-20.
11. J. A. Robinson and A. Voronkov, Handbook of automated reasoning, Vol. 1 and 2, the MIT Press and North Holland, 2001.
12. V. Pavlov, V. Pak. The inverse method application for non-classical logics. Nonlinear Phenomena in Complex Systems, 2015, 18(2): 181-190.
13. A. Voronkov. Theorem proving in non-standard logics based on the inverse method. In D. Kapur, editor, 11th International Conference on Automated Deduction, volume 607 of Lecture Notes in Artificial Intelligence, Springer Verlag, 1992, pp. 648-662.
14. V. Pavlov, V. Pak. The inverse method and first-order logic theorem proving, Nonlinear Dynamics and Applications. 2014, 20: 127-135.
15. M. P. Bonacina. Parallel theorem proving. Handbook of Parallel Constraint Reasoning, 201 8: 179-235.
16. M. Rawson, G. Reger. A Multithreaded Vampire with Shared Persistent Grounding, in: F-MCAD. 2021, pp. 280-284.
17. G. Wu. Study on parallel sat solver and parallel automated deduction reasoning systems, Southwest Jiaotong University, 2019 (in Chinese). https://doi.org/10.27414/d.cnki.gxnju.2019.002484.
18. S. Chen, Y. Xu, Y. Jiang, J. Liu, X. He. Some synergized clause selection strategies for contradiction separation based automated deduction, in: 2017 12th International Conference on Intelligent Systems and Knowledge Engineering, ISKE, 2017, pp. 1–6, https://doi.org/10.1109/ISKE.2017.8258741.
19. G. Zeng, S. Chen, J. Liu, et al. A complementary ratio based clause selection method for contradiction separation dynamic deduction. Knowledge-Based Systems, 2023: 111238. https://doi.org/10.1016/j.knosys.2023.111238.

Quantifier's Semantics Discrimination and Logical Cube*

Xurong Li*,‡, Guohua Jiang*,§ and Long Hong†,¶

*School of Computer Science, Nanjing University of Aeronautics and Astronautics
Nanjing, Jiangsu 211106, China

†School of Computer Science, Nanjing University of Posts
and Telecommunications
Nanjing, Jiangsu, China
‡hollylxr@aliyun.com
§jianggh@nuaa.edu.cn
¶hongl@njupt.edu.cn

Quantifiers are the logical symbols in first-order languages. The basic explanation for them is as follows: universal quantifier (∀) means "for all" or "everything"; existing quantifier (∃) means "for some" or "at least one". However, there are semantic differences in the no unique interpretation for one quantifier. After introducing the interesting semantics of quantifiers in a statement form with 1-predicate, this chapter analyzes the semantics of statement forms with two predicates, proposes ∀-conjunction and ∃-implication (∀c∃i) and ∀-implication ∃-conjunction (∀i∃c) in definition and discusses their properties by theorems, and describes the relationship between these theorems. According to the characteristics of **A**^ and **E**^ that are expansion forms of **A** and **E** in traditional logic, two trees were constructed separately; based on the chain theorem and the main relationship between ∀c∃i and ∀i∃c, it was diagrammed as a novel logical cube that is compatible with the traditional square of opposition. These works could open up a new research direction for logic and provide new tools for logic applications.

Keywords: Quantifiers semantics; c∀i∃; i∀c∃; chain theorem; logic cube.

1. Introduction

In the classical logical deduction, there are four categorical propositions [1], [2] about the relationship between classes (sets): universal affirmative proposition written as "All S is P" is expressed by A, universal negative proposition written as "No S is P" is expressed by E, particular affirmative proposition written as "Some S is P" is expressed by I and particular negative proposition written as

---

§Corresponding author.

"Some S is not P" is expressed by O. The letters S and P above represent subject and predicate terms, respectively. In addition, there are affirmation propositions and negative propositions about individuals.

B. Russell pointed out that in addition to logic and pure mathematics, no word possesses accurate meaning [3]. Therefore, modern logic uses symbols for description and calculation to strive to attain universality and the greatest possible exactness.

In the first-order languages, a predicate is used to characterize the nature of variables or the relationship between variables and often written in capital English letters. Quantifiers in predicate calculus refer to universal quantifier $\forall$ and existential quantifier $\exists$, $\forall$ represents "for all" or "everything" or "for any one", $\exists$ represents "exists for some" or "at least one" or "for some", and their relationships are as follows:

$$\forall x\mathrm{A}(x) = \neg\exists x\neg\mathrm{A}(x) \tag{q1}$$

$$\exists x\mathrm{A}(x) = \neg\forall x\neg\mathrm{A}(x) \tag{q2}$$

Based on the above brief introduction, the four categorical propositions in classical logic such as universal affirmation proposition can be translated into the corresponding statement forms:

$$\mathbf{A} =_{\mathrm{df}} \forall x(\mathrm{S}(x)\wedge(\mathrm{S}(x)\rightarrow\mathrm{P}(x))) \tag{1}$$

$$\mathbf{E} =_{\mathrm{df}} \forall x(\mathrm{S}(x)\wedge(\mathrm{S}(x)\rightarrow\neg\mathrm{P}(x))) \tag{2}$$

$$\mathbf{I} =_{\mathrm{df}} \exists x(\mathrm{S}(x)\wedge\mathrm{P}(x)) \tag{3}$$

$$\mathbf{O} =_{\mathrm{df}} \exists x(\mathrm{S}(x)\wedge\neg\mathrm{P}(x)) \tag{4}$$

And for constant a, there are affirmative proposition $\mathrm{S}(a)$ and negative proposition $\neg\mathrm{S}(a)$.

The relationships between **A**, **E**, **I** and **O** are analyzed briefly in the following:

Theorem 1. $A \Leftrightarrow \neg O$.

Corollary 2. $O \Leftrightarrow \neg A$.

Theorem3. $I \Leftrightarrow \neg E$.

Corollary 4. $E \Leftrightarrow \neg I$.

Theorem5. $A \Rightarrow I$.

Theorem6. $E \Rightarrow O$.

The logical square representing the relationship between **A**, **E**, **I** and **O** is shown in Figure 1.

This chapter describes quantifier discrimination and analyzes the semantics of statement forms with two predicates. It uses the method of modern logic to establish the expansion forms of **A**, **E**, **I** and **O** in classical logic and studies the relationships in these forms to try to establish a new theory of relationship between classes with compatibility.

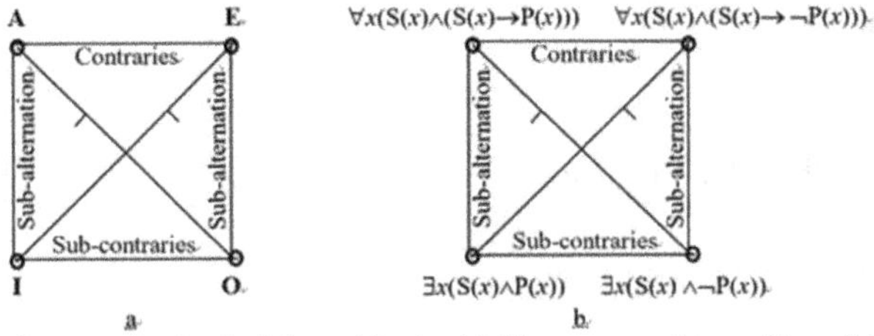

Fig 1. A logical square in classical formal logic. (a) The square of opposition; (b) the square of opposition with formulas descripting **A**, **E**, **I**, **O**. The symbol ⊣ expresses contradictory relationship.

2. On Quantifiers in Statements with 1-Predicate

Since G. Frege's publication of the *Concept Writing*, quantifiers have become one of the main hallmarks of modern logic, in the study of logical foundations and logic applications, quantifying variables has become an interest among scholars..

2.1. *General description of quantifiers*

To achieve a general description of the quantifiers, the n-level universal quantifier $\forall_n x\mathrm{A}(x)$, n-level existential quantifier $\exists_n x\mathrm{A}(x)$ and n-level exact quantifier $\Lambda_n x\mathrm{A}(x)$ were given in [4]; their semantics are below.

$\forall_n x\mathrm{A}(x)$: Except for n-1 xs, all others make A(x) established;

$\exists_n x\mathrm{A}(x)$: No more than n xs make A(x) valid;

$\Lambda_n x\mathrm{A}(x)$: Just n-1 xs make A(x) work.

Obviously, $\forall_1 x\mathrm{A}(x) = \forall x\mathrm{A}(x)$; $\Lambda_2 x\mathrm{A}(x) = \exists! x\mathrm{A}(x)$.

2.2. *Discrimination for quantifier's semantics*

In general, universal quantifier is interpreted indiscriminately as "for all" or "everything". When studying the problem of infinity, Wujia Zhu strictly distinguished between "for all" and "everything", and created the enumerate quantifier E [5], [6], [7]. The potential infinity and actual infinity in mathematics correspond to negative infinite and the positive infinite in philosophy. Taking natural numbers as examples, the actual infinite theory holds that there exists $\aleph_0$ as the cardinal number of nature number set. The potential infinite theory thinks that there are infinite natural numbers, but its set has no cardinality, in which if using one by one and when counting n, next is n+1, thus counting is never finished. Zhu expressed the potential infinity as *f-going,* in which the progressive tense means forever counting; and he used *gone* to express the actual infinity that must be perfect tense, which means that all nature numbers have been counted, hence there is $\aleph_0$. Zhu also gave two quantifiers that have difference semantics as follow.

$\forall =_{df}$ for all,

$E =_{df}$ everything.

3. ON Semantics of 2-Predicates Statements

3.1. *Discrimination of quantifier's semantics*

There is a common patterm in the statement form with two unary predicates, in which universal quantifier is followed by the implication and the existential quantifier is followed by the conjunction. For instance, $\forall x(S(x)\rightarrow P(x))$ means that for all x, if they have property S, then also have property P. And $\exists x(S(x)\wedge P(x))$ means that there exist some x with both property S and property P. However, this patterm is not eternal.We can interpret the statement form $\forall x(S(x)\wedge P(x))$ as " for every x, it has both property S and property P". For example, let x= watermelon, the predicate S is "big" and P is "sweet". Then $\forall x(S(x)\wedge P(x))$ can be translated to "Every watermelon is both big and sweet".

From the natural language semantics and structure,$\forall x(S(x)\wedge P(x))$ is affirmative for both $S(x)$ and $P(x)$, while $\forall x(S(x)\rightarrow P(x))$ need condition, so the former is a stronger assertion than the latter. From Zhu's research results on the problem of infinite view, $\forall x(S(x)\wedge P(x))$ reflects the features of the enumerated quantifiers also. We can also interpret the statement form $\exists x(S(x)\rightarrow P(x))$ as "existence x, if it has property S, then it also has property P. " For example, let x= city, the predicate S "located in Jiangsu" and P "located in China". Then $\exists x(S(x)\rightarrow P(x))$ can be translated to " These cities are located in China, if they are located in Jiangsu Province." Clearly, in terms of the semantic and natural language structures, $\exists x(S(x)\rightarrow P(x))$ is a weaker assertion than $\exists x(S(x)\wedge P(x))$.

In the following, the pattern that a conjunction is after the universal quantifier and an implication follows an existential quantifier is called $\forall$-conjunction and $\exists$-implication ($\forall$c$\exists$i); the pattern that an implication is after the universal quantifier and a conjunction follows an existential quantifier is called $\forall$-implication and $\exists$-conjunction ($\forall$i$\exists$c).

3.2. *$\forall$-conjunction and $\exists$-implication ($\forall$c$\exists$i)*

Definition1. *$\forall$c$\exists$i is composed of the following formulas:*

$$A^{\wedge} =_{df} \forall x(S(x)\wedge P(x)) \tag{5}$$

$$E^{\wedge} =_{df} \forall x(S(x)\wedge \neg P(x)) \tag{6}$$

$$I^{\rightarrow} =_{df} \exists x(S(x)\rightarrow P(x)) \tag{7}$$

$$O^{\rightarrow} =_{df} \exists x(S(x)\rightarrow \neg P(x)) \tag{8}$$

Theorem7. $A^{\wedge} \Leftrightarrow \neg O^{\rightarrow}$.

Corollary 8. $O^{\rightarrow} \Leftrightarrow \neg A^{\wedge}$.
Theorem 9. $E^{\wedge} \Leftrightarrow \neg I^{\rightarrow}$.
Corollary 10. $I^{\rightarrow} \Leftrightarrow \neg E^{\wedge}$.
Theorem 11. $A^{\wedge} \Rightarrow I^{\rightarrow}$.
Theorem12. $E^{\wedge} \Rightarrow O^{\rightarrow}$.
Theorem13. $\neg (A^{\wedge} \wedge E^{\wedge})$.

Theorem 13 indicates there is no true relationship between **A^** and **E^**.

Theorem14. $S(x)=T,\ P(x)=T \Rightarrow I^{\rightarrow} \Rightarrow \neg O^{\rightarrow}$.
Theorem15. $S(x)=F \Rightarrow I^{\rightarrow} \Rightarrow O^{\rightarrow}$.
Theorem16. $S(x)=T,\ P(x)=F \Rightarrow O^{\rightarrow} \Rightarrow \neg I^{\rightarrow}$.
Theorem17. $S(x)=F \Rightarrow O^{\rightarrow} \Rightarrow I^{\rightarrow}$.

Theorem14 to Theorem 17 indicate the relationship between $\mathbf{I}^{\rightarrow}$ and $\mathbf{O}^{\rightarrow}$ is uncertain. According to Theorem 7 to Theorem 17, logical square matrix is obtained shown in Figure 2.

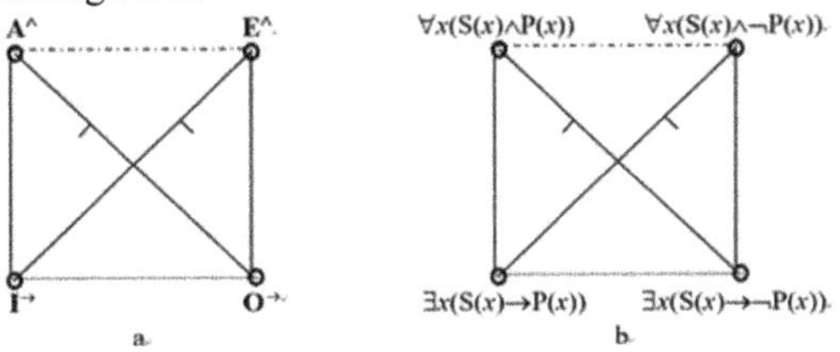

Fig 2. Logical square of ∀c∃i. Here a: square with symbols; b: square with formula. Solid lines represent the certain relationship; dashed lines indicate the uncertain one; dot dash lines indicate opposing one.

3.3. *∀-implication and ∃-conjunction (∀i∃c)*

Definition2. *∀i∃c is composed of the following formula:*

$$\mathbf{A}^{\rightarrow} =_{\mathrm{df}} \forall x(\mathrm{S}(x) \rightarrow \mathrm{P}(x)) \tag{9}$$

$$\mathbf{E}^{\rightarrow} =_{\mathrm{df}} \forall x(\mathrm{S}(x) \rightarrow \neg \mathrm{P}(x)) \tag{10}$$

$$\mathbf{I}^{\wedge} =_{\mathrm{df}} \exists x(\mathrm{S}(x) \wedge \mathrm{P}(x)) \tag{11}$$

$$\mathbf{O}^{\wedge} =_{\mathrm{df}} \exists x(\mathrm{S}(x) \wedge \neg \mathrm{P}(x)) \tag{12}$$

Theorem18. $A^{\rightarrow} \Leftrightarrow \neg O^{\wedge}$.
Corollary 19. $O\wedge \Leftrightarrow \neg A^{\rightarrow}$.
Theorem 20. $I^{\wedge} \Leftrightarrow \neg E^{\rightarrow}$.
Corollary 21. $E^{\rightarrow} \Leftrightarrow \neg I^{\wedge}$.
Theorem 22. $S(x)=T \Rightarrow A^{\rightarrow} \Rightarrow I^{\wedge}$.
Theorem 23. $S(x)=F \Rightarrow A^{\rightarrow} \Rightarrow \neg I^{\wedge}$.

Theorem 22 and Theorem 23 indicate the relationship between $\mathbf{A}^{\rightarrow}$ and **I^** is uncertain, that is, **I^** cannot be necessarily inferred from $\mathbf{A}^{\rightarrow}$.

Theorem 24. $S(x)=T,\ P(x)=T \Rightarrow A^{\rightarrow} \Rightarrow \neg E^{\rightarrow}$.

Theorem 25. $S(x)=F \Rightarrow \boldsymbol{A}^{\rightarrow} \Rightarrow \boldsymbol{E}^{\rightarrow}$.
Theorem 26. $S(x)=T, P(x)=F \Rightarrow \boldsymbol{E}^{\rightarrow} \Rightarrow \neg \boldsymbol{A}^{\rightarrow}$.
Theorem 27. $S(x)=F \Rightarrow \boldsymbol{E}^{\rightarrow} \Rightarrow \boldsymbol{A}^{\rightarrow}$.

Theorem 24 to Theorem27 indicate the relationship between **A**$^{\rightarrow}$and **E**$^{\rightarrow}$ is uncertain.

Theorem 28. $S(x)=T, \neg P(x)= T \Rightarrow \boldsymbol{E}^{\rightarrow} \Rightarrow \boldsymbol{O}^{\wedge}$.
Theorem 29. $S(x) =F, \neg P(x)= T \Rightarrow \boldsymbol{E}^{\rightarrow} \Rightarrow \neg \boldsymbol{O}^{\wedge}$.

Theorem 28 and Theorem 29 indicate the relationship between **E**$^{\rightarrow}$ and **O**^ is uncertain.

Theorem 30. $\exists x(S(x) \wedge P(x)), \exists y(S(x) \wedge \neg P(x)), P(x)=S(x) \Rightarrow \boldsymbol{I}^{\wedge} \Rightarrow \neg \boldsymbol{O}^{\wedge}$.
Theorem 31. $\exists x(S(x) \wedge P(x)), \exists y(S(x) \wedge \neg P(x)), P(x) \subset S(x) \Rightarrow \boldsymbol{I}^{\wedge} \Rightarrow \boldsymbol{O}^{\wedge}$.

Theorem 30 and Theorem31 indicate the relationship between **I**^ and **O**^ is uncertain, and can be represented by Venn diagram in set theory [8]. Logical square obtained by Theorem 18 to Theorem 31 is shown in Figure 3, and it is called modern square of opposition [9].

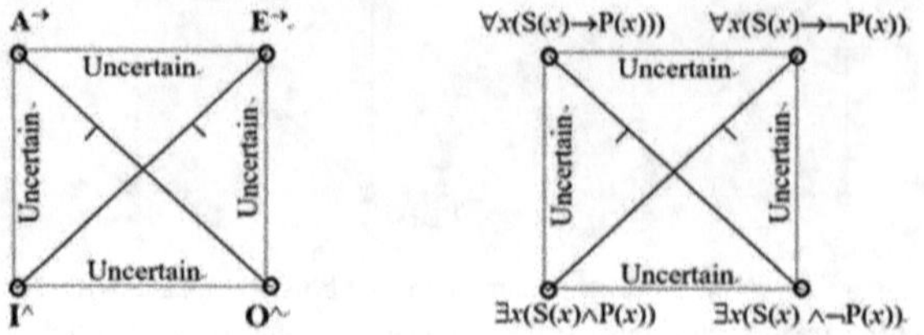

a b

Fig 3. logical square of i∀c∃. Here a: the square of i∀c∃; b: the square descriped by formulas.

3.4. *Logic cube*

Theorem 32. $A^{\wedge} \Rightarrow \boldsymbol{I}^{\wedge}$.
Theorem 33. $E^{\wedge} \Rightarrow \boldsymbol{O}^{\wedge}$.
Theorem 34. (Chain theorem 1). $A^{\wedge} \Rightarrow A^{\rightarrow} \Rightarrow I^{\wedge} \Rightarrow I^{\rightarrow}$.

Proof. ‡

1. **A**^	premiss
2. $\forall x(S(x) \wedge P(x))$	1, Eq. (5)
3. $S(a) \wedge P(a)$	2, ∀-
4. $S(a)$	3, ∧-
5. $P(a)$	3, ∧-
6. $S(a) \rightarrow P(a)$	4, 5, →+
7. $\forall x(S(x) \rightarrow P(x))$	6, ∀+
8. **A**→	7, Eq. (9); **A**^ ⇒**A**→

‡ The partial notations in basis of this proof are from natural reasoning system [10][11].

9. $S(a) \to P(a)$ 7, $\forall$-
10. $P(a)$ 4, 9, $\to$-
11. $S(a) \wedge P(a)$ 4, 10, $\wedge$+
12. $\exists x(S(x) \wedge P(x))$ 11, $\exists$+
13. $\mathbf{I}^\wedge$ 12, Eq. (11); $\mathbf{A}^{\to} \Rightarrow \mathbf{I}^\wedge$
14. $S(a) \wedge P(a)$ 12, $\exists$-
15. $S(a)$ 14, $\wedge$-
16. $P(a)$ 14, $\wedge$-
17. $S(a) \to P(a)$ 15, 16, $\to$+
18. $\exists x(S(x) \to P(x))$ 17, $\exists$+
19. $\mathbf{I}^{\to}$ 18, Eq. (7); $\mathbf{I}^\wedge \Rightarrow \mathbf{I}^{\to}$ □

Theorem 35（Chain theorem 2). $E^\wedge \Rightarrow E^{\to} \Rightarrow O^\wedge \Rightarrow O^{\to}$.

'Theorem 11, Theorem32, Theorem34' and 'Theorem12, Theorem 33, Theorem35' respectively show that they are the conclusions deducted alone from **A^** and **E^**, and their tree structure is shown in Figure 4.

a   b.

Fig4. Tree structure. Here a: the tree with **A^** as the root; b: the tree with **E^** as the root.

Combining two chain theorems and related theorems and referring to the form of Figures 2 and 3, the main relationships among **A^**, **E^**, **I→**, **O→**, **A→**,**E→**, **I^** and **O^** shown in Figure 5.

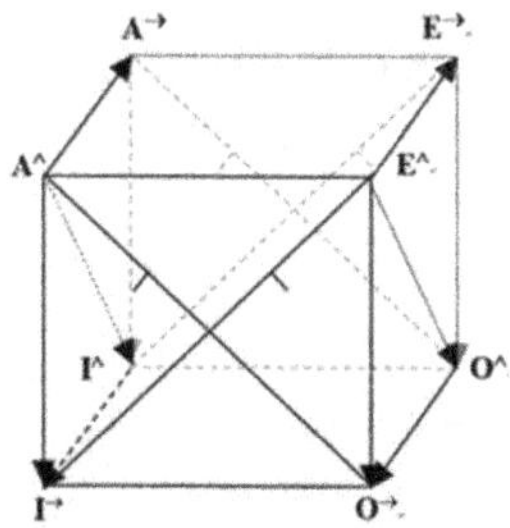

Fig 5. Logic cube. This square based on ∀c∃i, ∀i∃c and tree structure of **A^** and **E^** describe the main relationships among **A^**, **E^**, **I→**, **O→**, **A→**, **E→**, **I^** and **O^**.

4. Conclusion

This chapter belongs to basic research on logic. It analyzed the different semantics of quantifiers in statements containing two predicates and gave the corresponding statement form according to their differences, presented ∀c∃i and ∀i∃c in definition and described the relationships between them, established the chain theorem and diagrammed logic cube describing main relationships among **A^**, **E^**,

I$^{\rightarrow}$, **O**$^{\rightarrow}$, **A**$^{\rightarrow}$, **E**$^{\rightarrow}$, **I**^ and **O**^. These works could open up a new direction for logic research and provided new tools for logic applications hoping to make the historic contribution to the development of logic.

Acknowledgments

This work is supported by the Key Program of National Science Foundation of China under Grant No. 18ZDA031 and the National Natural Science Foundation of China under Grant No. 61170322.

References

1. Y. Jin, *Formal logic, republication,* (People's Publishing House, Beijing, 2006).
2. Irving M. Copi, Carl Cohen, Kenneth McMahon, *Introduction to Logic,* 14th edition, (Pearson Education Limited, 2014).
3. B. Russell, *Human knowledge, its scope and limits*, (Simon and Schuster, 1948).
4. S. Mo. Enumeration of the quantifier and predicate calculus. *J. math.***14**, 218 (1964).
5. W. Zhu, *The logical foundation of mathematical infinity and mediation,* (Science Press, Beijing, 2012).
6. W. Zhu, Y. Lin, G. Du, N. Gong, The inconsistency of countable infinite sets, *Kybernetes: Int. J. of systems & cybernetics*, **37**, 446 (2008).
7. W. Zhu, Y. Lin, G. Du, N. Gong, Inconsistency of uncountable infinite sets under ZFC framework, *Kybernetes: Int. J. of systems & cybernetics*, **37**, 453 (2008).
8. X. He, G. Du, L. Hong, Graphic Deduction Based on Set, *J. Univers. Comput. Sci.* **26**: 1331 (2020).
9. P. Hurley, *A concise introduction to logic*, 10th edition, (Wadsworth Publishing, 2011).
10. S. Hu, Z. Lu, *Fundamentals of mathematical logic,* Vol.1, (Science Press, Beijing, 1981).
11. W. Zhu, X. Xiao, *Introduction to mathematical logic*. (Nanjing University Press, Nanjing, 1995).

Actualization Probabilities of Idempotent Nullnorms and Uninorms

J. Baz[*,‡], S. Montes[*] and I. Díaz[†]

[*] *Department of Statistics and OR and DM, University of Oviedo, Oviedo 33007, Spain*

[†] *Department of Computer Science, University of Oviedo, Gijón 33204, Spain*

[‡] *bazjuan@uniovi.es*

Nullnorms and uninorms are used, in some contexts, to recursively aggregate random data. However, there is a possibility that, given the current value of the aggregation, an addition of a new value does not change the output. In this chapter, we focus on the probability of having an actualization when aggregating independent and identically distributed continuous random variables with idempotent nullnorms and uninorms. For idempotent nullnorms, a closed expression for the actualization probability is given and, for idempotent uninorms, an upper and a lower bound of such probability are derived. The expected number of actualizations when aggregating an infinite amount of values is also computed.

Keywords: Aggregation; Markov chains; nullnorms; uninorms.

1. Introduction

Recursivity[1] is a desirable property for aggregation functions, since it has computational benefits.[2] Two of the most remarkable families of recursive aggregations are nullnorms and uninorms,[3] which are associative, symmetric and with, respectively, an annihilator or a neutral element. Idempotent nullnorms and uninorms have been studied in detail in the literature[4,5] and generalize maximum and minimum operators. They are widely used in fuzzy neural networks to process, in some cases, a large amount of data.[6,7]

However, when working with idempotent nullnorms and uninorms, there is a possibility that, given the current value of the aggregation, an addition of a new value does not change the output. The aim of this chapter is to quantify this phenomena, by computing the probability of having an actualization in each step of the recursive process. In particular, we are going to suppose that the inputs are independent continuous random variables with the same distribution. This idea has been studied in detail in probability theory for the maximum, in the so-called record statistics.[8]

The rest of this chapter is structured as follows. In Section 2, the basic notions of aggregation functions and probability theory that we need for the development of this chapter are presented. Then, we link our problem with Markov chains in Section 3. Sections 4 and 5 are devoted to the computation of actualization probabilities of, respectively, idempotent nullnorms and uninorms. Finally, the conclusions are discussed in Section 6.

2. Preliminaries

2.1. *Aggregation functions*

Aggregation functions are, intuitively, functions that fuse some values on the same real interval and return another value in the same interval, fulfilling some boundary conditions and increasing monotonicity. In particular, in this paper we will focus on extended aggregation functions.

Definition 2.1.[3] Let I be a real interval. A function $A : \cup_{n\in\mathbb{N}} I^n \to I$ is said to be an (extended) aggregation function if for any $n \in \mathbb{N}$, the function $A^{(n)} : I^n \to I$ defined as $A^{(n)}(\vec{x}) = A(\vec{x})$ fulfills

- $A^{(n)}$ is increasing in each component
- $\inf_{\vec{x}\in I^n} A^{(n)}(\vec{x}) = \inf I$ and $\sup_{\vec{x}\in I^n} A^{(n)}(\vec{x}) = \sup I$

In the rest of the paper, we will use the term aggregation function to refer to extended aggregation functions and denote the components of a vector as $\vec{x} = (x_1, \dots, x_n)$. There exist a huge amount of properties defined for aggregation functions. We will focus on the following ones.

Definition 2.2. [3] Let $A : \cup_{n\in\mathbb{N}} I^n \to I$ be an aggregation function. Then,

- if $A^{(2)}\left(A^{(n)}(\vec{x}), A^{(m)}(\vec{y})\right) = A^{(n+m)}(\vec{x}, \vec{y})$ for any $\vec{x} \in I^n$, $\vec{y} \in I^m$ and $n, m \in \mathbb{N}$, A is said to be associative,
- if $A(\vec{x}) = A(\vec{y})$ for any $\vec{x}, \vec{y} \in \cup_{n\in\mathbb{N}} I^n$ such that $\vec{y}$ is a permutation of the elements of $\vec{x}$, A is said to be symmetric,
- if $A(x, \dots, x) = x$ for any $x \in I$, A is said to be idempotent,
- if $\min(\vec{x}) \le A(\vec{x}) \le \max(\vec{x})$ for any $\vec{x} \in \cup_{n\in\mathbb{N}} I^n$, A is said to be internal,
- if $A(\vec{x}) = a$ for any $\vec{x} \in \cup_{n\in\mathbb{N}} I^n$ such that $a \in \{x_1, \dots, x_n\}$, then a is said to be an annihilator element of A,
- if $A(x_1, \dots, x_{i-1}, e, x_{i+1}, \dots, x_n) = A(x_1, \dots, x_{i-1}, x_{i+1}, \dots, x_n)$ for any $\vec{x} \in \cup_{n\in\mathbb{N}} I^n$ and $n, i \in \mathbb{N}$ such that $i \le n$, then e is said to be a neutral element of A.

A well-known result states that an aggregation function is idempotent if and only if is internal[3]. Each aggregation function has at most one annihilator element and a neutral element. We end this section by introducing nullnorms and uninorms.

Definition 2.3. [3] An aggregation function $A : \cup_{n \in \mathbb{N}}[0,1]^n \to [0,1]$ is said to be a nullnorm if is symmetric, associative and has an annihilator element.

Definition 2.4. [3] An aggregation function $A : \cup_{n \in \mathbb{N}}[0,1]^n \to [0,1]$ is said to be an uninorm if is symmetric, associative and has a neutral element.

2.2. *Probability notions*

Let us briefly introduce the concepts of Probability theory needed along this paper. Let us first recall the notion of conditional independence.

Definition 2.5. [9] Let X, Y, Z be three random variables with joint density function $f : \mathbb{R}^3 \to \mathbb{R}^+$. Then, it is said that X and Y are conditionally independent given Z if there exist $g, h : \mathbb{R}^2 \to \mathbb{R}^+$ such that $h(x, y, z) = f(x, z)g(y, z)$ for all $x, y, z \in \mathbb{R}$.

A Markov chain is a sequence of random variables in which, given a value of a variable, the previous and posterior random variables are independent. This is usually interpreted as a random variable evolving in time for which the future and the past are conditionally independent given the present.

Definition 2.6. [10]A sequence of random variables $(X_n)_{n \in \mathbb{N}}$ is said to be a Markov chain (fulfills the Markov property) if, given X_j, X_{j+1} is independent from X_i for any $i, j \in \mathbb{N}$ with $i < j$.

The different values that a Markov chain can take are known as states. If a state $a \in \mathbb{R}$ fulfills that, if $X_i = a$, then $X_k = a$ for any $k \geq i$, it is said that a is an absorbent state. We end this section by stating some well-known results related to the maximum and minimum of random variables.

Proposition 2.1.[11] *Let $X_1, \ldots, X_n$ be n independent continuous random variables with the same distribution. Then,*

- $P(\max(X_1, \ldots, X_n) \leq t) = P(X_1 \leq t)^n$ *for any* $t \in \mathbb{R}$.
- $P(X_n = \max(X_1, \ldots, X_n)) = P(X_n = \min(X_1, \ldots, X_n)) = \frac{1}{n}$

3. Associative Aggregation as Markov chains

In the following, we are going to consider $(X_n)_{n\in\mathbb{N}}$ a sequence of random variables with the same distribution and independent. Then, given an associative aggregation function $A : \cup_{n\in\mathbb{N}} I^n \to I$, we are going to study the sequence $(A^{(n)}(X_1, \ldots, X_n))_{n\in\mathbb{N}}$, which fulfills the Markov property.

Proposition 3.1. *Let $(X_n)_{n\in\mathbb{N}}$ be a sequence of independent random variables and let $A : \cup_{n\in\mathbb{N}} I^n \to I$ be an associative aggregation function. Then, the sequence $(A^{(n)}(X_1, \ldots, X_n))_{n\in\mathbb{N}}$ is a Markov chain.*

Proof. Denote $Z_n = A^{(n)}(X_1, \ldots, X_n)$. Then, applying the associativity, one has $Z_{n+1} = A^{(2)}(Z_n, X_{n+1})$. Given the value of Z_n, Z_{n+1} is a function of X_{n+1}. For any $k \leq n$, Z_k is a function of $X_1, \ldots, X_k$. Since $X_1, \ldots, X_k$ are independent of X_{n+1} and transformations of independent variables are also independent[9], it is concluded that Z_{n+1} and Z_k with $k \in \{1, \ldots, n-1\}$ are conditionally independent given Z_n, that is, the Markov property. □

Latter result also holds if we consider recursivity[1] instead of associativity. With this approach, annihilator elements are absorbent states and neutral elements can be seen as states that restart the Markov chain.

Proposition 3.2. *Let $(X_n)_{n\in\mathbb{N}}$ be a sequence of independent random variables and let $A : \cup_{n\in\mathbb{N}} I^n \to I$ be an associative aggregation function. Then, if a is an annihilator element of A, then a is an absorbent state of $(A^{(n)}(X_1, \ldots, X_n))_{n\in\mathbb{N}}$. In addition, if e is a neutral element of A and $(X_n)_{n\in\mathbb{N}}$ is a sequence of random variables with the same distribution, then, given $A^{(k)}(X_1, \ldots, X_k) = e$, $(A^{(k+n)}(X_{k+1}, \ldots, X_{k+n}))_{n\in\mathbb{N}}$ has the same distribution as $(A^{(n)}(X_1, \ldots, X_n))_{n\in\mathbb{N}}$.*

Proof. For the first item, since $A^{(2)}(a, x) = a$, one has that the probability of $A^{(n)}(X_1, \ldots, X_n) = a$ given $A^{(n-1)}(X_1, \ldots, X_{n-1}) = a$ is 1. Therefore, a is an absorbent state of the Markov chain.

For the second one, since $A^{(2)}(e, x) = x$, given $A^{(k)}(X_1, \ldots, X_k) = e$, one has $A^{(k+n)}(X_1, \ldots, X_{k+n}) = A^{(n)}(X_{k+1}, \ldots, X_{k+n})$. Since the random variables in $(X_n)_{n\in\mathbb{N}}$ have the same distribution, then $A^{(n)}(X_{k+1}, \ldots, X_{k+n})$ has the same distribution as $A^{(n)}(X_1, \ldots, X_n)$. □

In the next sections, we will study actualizations of random sequences associated with idempotent nullnorms and uninorms, defined as follows.

Definition 3.1. Let $(X_n)_{n\in\mathbb{N}}$ be a random sequence. Then, it is said that there is an actualization in the i-th value if $X_i \neq X_{i-1}$ with $i > 1$. In addition, the first value will be considered to be always an actualization.

4. Actualization Probabilities of Idempotent Nullnorms

In this section, we will consider idempotent nullnorms. They have the following characterization in terms of the maximum, the minimum and the annihilator element.

Lemma 4.1.[12] *Let $A : \cup_{n=1}^{\infty}[0,1]^n \to [0,1]$ be an idempotent nullnorm with annihilator element $a \in [0,1]$. Then,*

$$A^{(2)}(x,y) = \begin{cases} \max(x,y) & \text{if } x,y < a \\ \min(x,y) & \text{if } x,y > a \\ a & \text{elsewhere} \end{cases}$$

Using the latter result, the actualization probability can be computed.

Theorem 4.1. *Let $(X_n)_{n\in\mathbb{N}}$ be a sequence of independent random variables sharing the same continuous distribution over $[0,1]$ and let $A : \cup_{n=1}^{\infty}[0,1]^n \to [0,1]$ be an idempotent nullnorm with annihilator element $a \in [0,1]$. Then,*

$$P\left(A^n(X_1,\dots,X_n) \neq A^{n-1}(X_1,\dots,X_{n-1})\right) = \frac{p_a^{n-1} + (1-p_a)^{n-1}}{n}$$

for any $n \in \mathbb{N}$ with $n \geq 2$ where $p_a = P(X_1 \leq a)$.

Proof. $(A^{(n)}(X_1,\dots,X_n))_{n\in\mathbb{N}}$ is a discrete-time continuous-state Markov chain with absorbent state a by applying Propositions 3.1 and 3.2.

Suppose that $X_1 \leq a$. If there exists $i \in \{2,\dots,n-1\}$ such that $X_i \geq a$, then $X_j = a$ for any $j \geq i$ and there is not an actualization. The other case happens with probability p_a^{n-2} and implies $A^n(X_1,\dots,X_n) = \min(X_1,\dots,X_n)$, which has an actualization probability of $\frac{1}{n}$. Multiplying all the probabilities the result is $\frac{p_a^{n-1}}{n}$. Working with the case $X_1 > a$, we obtain a probability of $\frac{(1-p_a)^{n-1}}{n}$. The result holds by summing both probabilities. □

We can derive the expected number of actualizations just by computing $1+\sum_{n=2}^{\infty}\frac{p_a^{n-1}+(1-p_a)^{n-1}}{n}$. The computation can be reached by using usual tools, thus we are going to omit the steps.

Proposition 4.1. *Let $(X_n)_{n\in\mathbb{N}}$ be a sequence of independent random variables with the same continuous distribution over $[0,1]$ and let A :*

$\cup_{n=1}^{\infty}[0,1]^n \to [0,1]$ *be an idempotent nullnorm with annihilator element* $a \in [0,1]$. *Then, the expected number of actualizations is* $-\frac{\ln(p_a)}{p_a} - \frac{\ln(p_a)}{1-p_a} - 1$ *with* $p_a = P(X_1 \leq a) \in (0,1)$.

5. Actualization Probabilities of Idempotent Uninorms

Idempotent uninorms have a more difficult characterization than idempotent nullnorms, given in the following result.

Lemma 5.1.[12] *Let* $A : \cup_{n=1}^{\infty} \to [0,1]$ *be an idempotent uninorm with neutral element* $e \in [0,1]$. *Then, there exists a decreasing function* $g : [0,1] \to [0,1]$ *with* $g(e) = e$ *such that:*

$$A(x,y) = \begin{cases} \min(x,y) & \text{if } y < g(x) \\ \min(x,y) \text{ or } \max(x,y) & \text{if } y = g(x) \\ \max(x,y) & \text{if } y > g(x) \end{cases}$$

In this case, we give just bounds for the actualization probability.

Theorem 5.1. *Let* $(X_n)_{n\in\mathbb{N}}$ *be a sequence of independent random variables sharing the same continuous distribution over* $[0,1]$ *and* $A : [0,1]^2 \to [0,1]$ *an idempotent uninorm with neutral element* $e \in [0,1]$. *Then,*

$$\frac{p_e^n + (1-p_e)^n}{n} \leq P\left(A^n(X_1,\ldots,X_n) \neq A^{n-1}(X_1,\ldots,X_{n-1})\right)$$
$$\leq 1 - \left(p_e^{n-1} + (1-p_e)^{n-1}\right)\left(1 - \frac{1}{n}\right)$$

for any $n \in \mathbb{N}$ *with* $n \geq 2$ *where* $p_e = P(X_1 \leq e)$.

Proof. Using Lemma 5.1, consider the sets of points over the unit square in which there is actualization $\{x, y \in [0,1] \mid x > y, y < g(x)\}$ and $\{x, y \in [0,1] \mid x < y, y > g(x)\}$. The remaining points in which there is actualization have measure 0, thus we can dismiss them since we are considering continuous random variables.

For the lower bound, if we want to minimize the sets of points in which are actualization, we have to choose the function g in Lemma 5.1 as $g(x) = 1$ if $x \in [0,e)$, $g(e) = e$ and $g(x) = 0$ if $x \in (e,1]$. Suppose $A^{(n-1)}(X_1,\ldots,X_{n-1}) < e$. Since A is internal, the probability of $A^{(n-1)}(X_1,\ldots,X_{n-1}) < e$ is greater than the probability of the maximum, which is p_e^{n-1} by Proposition 2.1. In that case, there is an actualization if $X_n < A^{(n-1)}(X_1,\ldots,X_{n-1})$. First, we need $X_n < e$, which has probability

p_e and then having, given that $A^{(n-1)}(X_1, \ldots, X_{n-1}) < E$ and $X_n < e$, that $X_n < A^{(n-1)}(X_1, \ldots, X_{n-1})$. This probability is greater than the one of $X_n < \min(X_1, \ldots, X_{n-1})$, which is $\frac{1}{n}$. Multiplying the terms, we obtain a probability of $\frac{p_e^n}{n}$. For the case $A^{(n-1)}(X_1, \ldots, X_{n-1}) > e$, similarly we obtain $\frac{(1-p_e)^n}{n}$. Summing both terms the lower bound is reached.

For the upper bound, if we want to maximize the sets of points in which are actualization, we have to choose the function g to be $g(x) = e$ for all $x \in [0, 1]$. Suppose $X_n < e$, which happens with probability p_e. Then, there is an actualization if $X_n < A^{(n-1)}(X_1, \ldots, X_{n-1})$. That always happens if $A^{(n-1)}(X_1, \ldots, X_{n-1}) > e$, which has a probability upper bounded by $P(\max(X_1, \ldots, X_{n-1}) > e) = 1 - p_e^{n-1}$. For the other case, if $A^{(n-1)}(X_1, \ldots, X_{n-1}) < e$, we can bound again the probability by considering the one of $X_n < A^{(n-1)}(X_1, \ldots, X_{n-1})$ when $X_n < e$ and $\max(X_1, \ldots, X_n) < e$, which is $\frac{1}{n}$. Summarizing all the probabilities, we obtain $p_e(1 - p_e^{n-1} + p_e^{n-1}\frac{1}{n}) = p_e - p_e^n\left(1 - \frac{1}{n}\right)$. If we suppose that $X_n > e$ we can similarly obtain $1 - p_e - \frac{(1-p_e)^n}{n}$. Summing both terms we reach the upper bound. □

If $e \in \{0, 1\}$, we recover the actualization probabilities of the minimum and maximum. We also provide a lower bound for the expected number of actualizations. For the upper bound, we cannot obtain a similar result since it does not converge.

Corollary 5.1. *Let $(X_n)_{n\in\mathbb{N}}$ be a sequence of independent random variables with the same continuous distribution over $[0, 1]$ and let $A : \cup_{n=1}^{\infty}[0, 1]^n \to [0, 1]$ be an idempotent uninorm with neutral element $e \in [0, 1]$. Then, the expected number of actualizations has a lower bound of $-\ln(p_e) - \ln(1 - p_e)$ with $p_e = P(X_1 \leq e) \in (0, 1)$.*

6. Conclusions

The probability of actualization for idempotent nullnorms and upper and lower bounds for idempotent uninorms are presented. We want to remark that these results quantify the sensibility of these aggregations to the addition of new data, which is relevant, for instance, in fuzzy neural networks. In particular, when defining the architecture for them, it is relevant to consider that, if the probability of actualization is small, having layers with a big number of neurons could lead to computational costs without an impact in its performance.

The main open problem is to derive the exact probability of actualization of idempotent uninorms in terms of the function g in Lemma 5.1. In addition, we wonder if similar results can be done for dependant random variables.

Funding Details

The authors have been supported by Programa Severo Ochoa of Principality of Asturias (BP21042) and the Ministry of Science and Innovation (PID2022-139886NB-I00).

References

1. A. del Amo, J. Montero and E. Molina, Representation of consistent recursive rules, *European Journal of Operational Research* **130**, 29 (2001).
2. L. Magdalena, L. Garmendia, D. Gómez, R. G. del Campo, J. T. Rodríguez and J. Montero, Types of recursive computable aggregations, in *2019 IEEE International Conference on Fuzzy Systems (FUZZ-IEEE)*, 2019.
3. M. Grabisch, *Aggregation functions* (Cambridge University Press, 2009).
4. G. D. Çaylı and F. Karaçal, Idempotent nullnorms on bounded lattices, *Information Sciences* **425**, 154 (2018).
5. B. De Baets, Idempotent uninorms, *European Journal of Operational Research* **118**, 631 (1999).
6. M. Hell, F. Gomide and P. Costa, Neurons and neural fuzzy networks based on nullnorms, in *2008 10th Brazilian symposium on neural networks*, 2008.
7. A. Lemos, W. Caminhas and F. Gomide, New uninorm-based neuron model and fuzzy neural networks, in *2010 Annual Meeting of the North American Fuzzy Information Processing Society*, 2010.
8. M. Ahsanullah, *Record statistics* (Nova Science Publishers, 1995).
9. V. K. Rohatgi and A. M. E. Saleh, *An introduction to probability and statistics* (John Wiley & Sons, 2015).
10. J. R. Norris, *Markov chains*, no. 2 (Cambridge university press, 1998).
11. H. A. David and H. N. Nagaraja, *Order statistics* (John Wiley & Sons, 2004).
12. E. Czogała and J. Drewniak, Associative monotonic operations in fuzzy set theory, *Fuzzy sets and systems* **12**, 249 (1984).

Synthetic Data Set Generation Method for Object Detection

Ningning Zhou*, Tong Li, and Jie Yang

School of Computer, Nanjing University of Posts and Telecommunications, Nanjing, Jiangsu 200023, China

*zhounn@njupt.edu.cn

Aiming at the problem of high construction cost of data set for object detection, especially in industrial application scenarios where sufficient sample images cannot be obtained from the Internet, this chapter proposes a method to automatically generate a synthetic data set .First, create and render the 3D models for the target devices to ensure that the synthetic images have a realistic patch-level feeling. Second, construct a simulation environment and then integrate the 3D models into the simulated environment based on global domain randomization. Finally, automatically annotate the label of the target objects in this synthetic image by computer graphics methods. It effectively reduces the cost of data acquisition while ensuring the detection accuracy of the model. Some current mainstream object detection models are trained with the synthetic data set. Experiments on real images show that the models trained with the synthetic images have a relatively high accuracy.

Keywords: Object detection; synthetic data set; global domain randomization; automatic label annotation.

1. Introduction

Object detection is an important and well-developed branch in the field of artificial intelligence. The first step of deep learning-based object detection is the need to collect and label a large number of sample images. The larger the amount of data, the more general it is, and the more the derived law conforms to the real one, which leads to a higher performance and generalization of the final model.

The traditional method of acquiring data using a camera to photograph the target is inefficient and costly for complex tasks with dozens or hundreds of detection objects. Image annotation work is done using marker tools, such as Labeling and VOTT. Even with the help of auxiliary tools[1]-[4], human-based manual annotation is still costly and subjective differences are inevitable.

In addition, for some specific target detection tasks, especially the detection of equipment in industrial application scenarios, the special type and diversity of the equipment, the unavailability of sufficient sample images from the Internet, coupled with the diversity of the target equipment working environment, further exacerbate the difficulty of data collection and annotation, requires a lot of human effort and is costly.

In order to solve the above problems, a comprehensive and diverse set of training samples is rapidly constructed by computers to improve the target detection accuracy in this chapter. First, create and render the 3D models for the target devices to ensure that the synthetic images have a realistic patch-level feeling. Second, construct a simulation environment and then integrate the 3D models into the simulated environment based on global domain randomization. Finally, automatically annotate the label of the target objects in this synthetic image by computer graphics methods. It effectively reduces the cost of data acquisition while ensuring the detection accuracy of the model. Some current mainstream object detection models[5–7] are trained with the synthetic data set of anti-vibration dampers of power equipment. Experiments on real images show that the models trained with the synthetic images have an accuracy of 0.882.

2. Our Method

In view of the special type and diversity of target devices in some industrial application scenarios, which cannot get enough sample images from the Internet, coupled with the diversity of working environments of some target devices, this paper will first explore the use of 3D models of objects to generate the training data set required for target detection neural networks, and then train the current mainstream object detection neural network models on the obtained synthetic data set. The main idea is shown in Figure 1.

3. Construction of the Synthetic Data Set

The whole process of synthetic data set construction is shown in Figure 2. Firstly, 3D object models with the local realism are established based on the detection targets. Then a simulation environment is constructed using the global randomization method. Finally, the 3D object models are fused into the simulation environment to obtain synthetic images and the labels of the target objects in the images are automatically annotated by graphics technology. This section takes the typical detection object of power anti-vibration damper as an example.

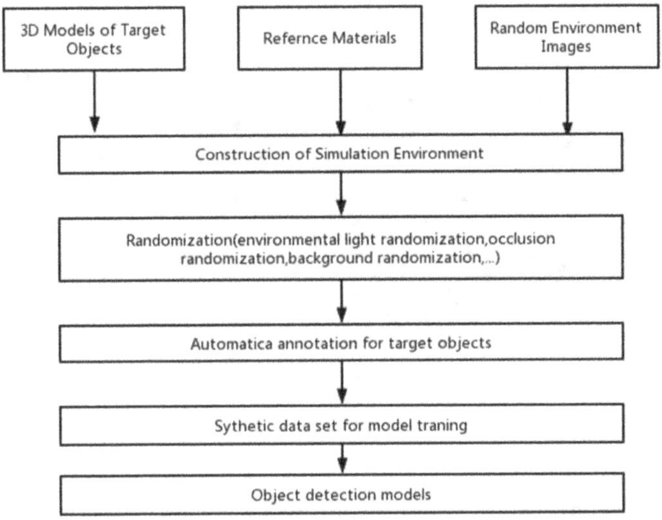

Fig. 1. Training the object detection network with the synthetic data set

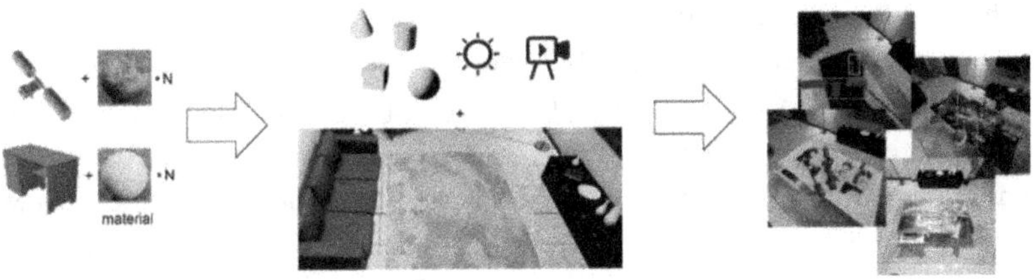

Fig. 2. Construction process of the synthetic data

3.1. *Modeling of target devices based on local realism*

Local realism refers to the realism of the detected object (power anti-vibration damper) in the whole simulation environment, which is required to have the characteristics of the real-world object (power anti-vibration damper), such as contour, material, texture, and so on. Therefore, the first step is to create a 3D model of the target object and the second step is to render the texture .

3.1.1. *3D modeling*

The profile of an anti-vibration damper in a power scenario is usually characterized by a long, curved and streamlined, hollow structure. Its contour and shape characteristics are critical for target detection. As a result,3DMAX and other software are used to establish a three-dimensional model of it. Two kinds of 3D models of anti-vibration dampers created by 3DMax are shown in Figure 3.

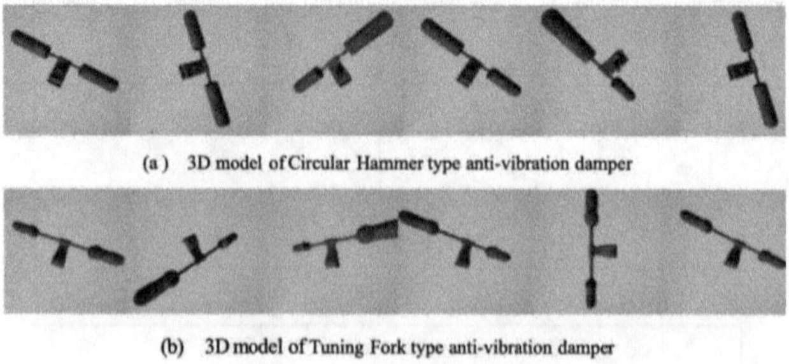

Fig. 3. 3D models of anti-vibration damper

3.1.2. *Texture rendering*

In this paper, we use ExtremePBR as a material library for PBR (Physically Based Rendering)[8], which is a rendering technique based on real physical laws. The real power vibration dampers are shown in Figure 4(a) , and the vibration damper models applying PBR material are shown in Figure 4(b), and the binary images generated from 3D models at similar Euler angles, are shown in Figure 4(c). It can be observed that the synthetic images reproduce the rust and metal texture of the real-world vibration dampers to some extent, while the binary images show the contour features of the real-world vibration dampers almost perfectly.

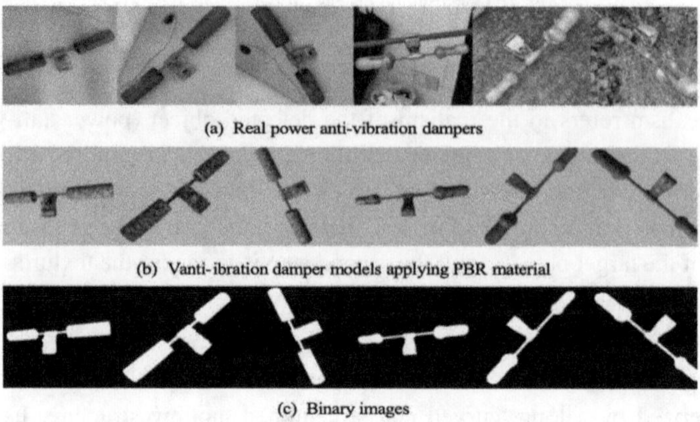

Fig. 4. Real images, Synthetic images and Binary images

3.2. *Fusion the 3D model of target device with simulation environment based on global domain randomization*

3.2.1. *Construction of the simulation environment*

The simulation environment in this paper is built based on the existing open source simulation environment.

Photographing anti-vibration dampers in a university setting, the frequent object is the desk. In order to link the simulation environment with the real world, this paper models the desk and integrates it into the simulation environment. An anti-vibration damper model is placed on a desk model to simulate a real-world anti-vibration damper placed on a real desk. Adding some random materials to the desktop, which could be grass, sky, stencil, etc., allows to simulate real-world anti-vibration dampers in various backgrounds without having to do a lot of random transformations of the whole simulation environment.

3.2.2. *Using random elements to increase sample diversity*

Global domain randomization means adding random elements to the simulation environment to simulate the real environment in a random way. The random elements involved in this paper include ambient light randomization, viewpoint distance randomization, occlusion randomization and background randomization.

Figure 5. shows the synthetic images of two typical devices in the power environment: tuning fork type anti-vibration damper and round hammer type anti-vibration damper obtained by the method of this paper. Figure 5(a) are the real images, and Figure 5(b) are the synthetic images obtained from all random components, which are very close to the real images in both contours and textures.

(a)Real images

(b)Synthetic images

Fig. 5. Real images and Synthetic images

3.3. *Automatic annotation of target object*

After the target object is transformed in the random field, the camera captures the frame containing the target object to generate a number of bounding boxes.

After calculating the 3D bounding box coordinate information, the graphics simulator will calculate the bounding box information in the 2D image based on the spatial information of the camera and the bounding box.

The coordinates of the diagonal vertex can be obtained using a similar calculation method. The position information of the bounding box is determined by the two diagonal vertices, and the graphics simulator outputs an annotation file, as shown in Figure 6(a). Figure 6(b) displays the standard format data used for training neural networks.

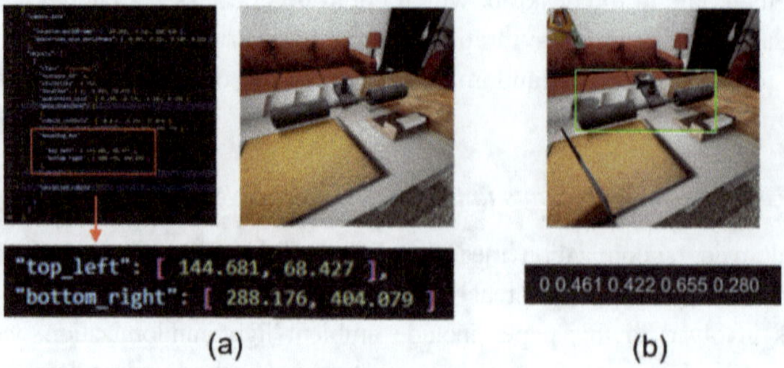

Fig. 6. Standardized annotation

4. Experiments

4.1. *Experimental environmental and assessment indicator*

The experimental environment for training and testing in this paper is a CUDA computing architecture written in Python version 3.7, AMD Ryzen 7 3700X 8-Core Processor 3.59 GHz, NVIDIA GeForce RTX 3060, and windows 11. The algorithm evaluation indicator used in this paper is mAP .

4.2. *Training set and testing set*

The experiment was conducted for the detection of anti-vibration damper in the inspection of power equipment. According to the method of synthesizing images in section 3 of this paper, 10 sets of 512*512 pixel simulation images were generated .The synthetic data set was transformed according to the annotation format of the coco2017 public data set.

The pictures used in the test set are taken from ordinary Android phone. In order to test the the model with robustness,the pictures of the anti-vibration damper were collected under different environments such as different backgrounds, different postures and different light intensities, with a total of 384 pictures and 417 rectangular boxes.

4.3. *Analysis of the experimental results*

The Neural network models were trained on the synthetic data sets with sample number ranging from 2k to 50k, and the network performance was evaluated using cocoM, and the results are shown in Table \ref{tab2}. The accuracy of all models increases first and then decreases along with the number of synthetic images, and basically, all performed the best when using 10k synthetic images . The results show that training models with synthetic images is not the more images the better. When the data scale reaches a certain extent, more data is counterproductive to the model performance. Ideally, more synthetic images mean more semantic variety covered by the data set, and so the higher the accuracy of the training. However, synthetic images inevitably differ from real images, so more synthetic images also mean more noise. When the advantage of multiple semantic varieties from more images is not enough to compensate for the disadvantage of noise, the model accuracy tends to decrease. Therefore, we try to minimize the difference between the synthetic images and the real images as far as possible when synthesizing the data, and select the appropriate quantity when choosing the number of data.

Table 1. Impact of different sized data sets on model performance

	d2k	d5k	d10k	d20k	d35k	d50k
Model 1[1]	0.682	0.737	0.776	0.742	0.735	0.730
Model 2[2]	0.811	0.809	0.838	0.821	0.823	0.811
SSD	0.682	0.701	0.753	0.769	0.732	0.721
YOLOV3	0.742	0.769	0.780	0.742	0.735	0.682
YOLOV7	0.783	0.823	0.882	0.861	0.842	0.821

[1]Model 1 is the version of Faster RCNN that uses Resnet 50.

[2]Model [2] is the version of Faster RCNN that uses Resnet 101.

5. Conclusion

The experimental results in this chapter show that synthetic images are an effective solution to the difficulties of data collection in supervised learning. The model trained with simulated images can approximate the accuracy trained with real data. Using a locally realistic simulation environment differs from a globally realistic simulation environment in that a locally realistic solution is faster, less costly, and more compatible with target detection tasks in industrial scenarios. In the future research, we will focus on using AI's methods to generate and render 3D models, and to improve the naturalness of model-environment integration, which future improves the realism of the synthetic images.

Acknowledgments

This work is supported by Jiangsu Province Graduate Practice Innovation Project No. (SJCX22_0270).

References

1. P. Ammirato, P. Poirson, E. Park, J. Koˇsecká and A. C. Berg, A dataset for developing and benchmarking active vision, in *2017 IEEE International Conference on Robotics and Automation (ICRA),* (Singapore, 2017).
2. Y. Furukawa and J. Ponce, Accurate, dense, and robust multi-view stereopsis (pmvs), in *IEEE Computer Society Conference on Computer Vision and Pattern Recognition,* (USA, 2007).
3. Y. Furukawa, B. Curless, S. M. Seitz and R. Szeliski, Towards internet scale multi-view stereo, in *2010 IEEE computer society conference on computer vision and pattern recognition,* (USA, 2010).
4. C. Vondrick, D. Patterson and D. Ramanan, Efficiently scaling up crowdsourced video annotation: A set of best practices for high quality, economical video labeling, *International journal of computer vision* 101,184 (2013).
5. S. Ren, K. He, R. Girshick and J. Sun, Faster r-cnn: Towards real-time object detection with region proposal networks, *Advances in neural information processing systems* ,28 (2015).
6. W. Liu, D. Anguelov, D. Erhan, C. Szegedy, S. Reed, C.-Y. Fu and A. C. Berg, Ssd: Single shot multibox detector, in *Computer Vision–ECCV 2016: 14th European Conference, Proceedings, Part I 14* , (Netherlands, 2016).
7. C.-Y. Wang, A. Bochkovskiy and H.-Y. M. Liao, Yolov7: Trainable bag-of-freebies sets new state-of-the-art for real-time object detectors, arXivpreprint arXiv:2207.02696 (2022).
8. M. Pharr, W. Jakob and G. Humphreys, *Physically based rendering: From theory to implementation*, (Morgan Kaufmann, 2016).

An Efficient Attribute Reduction Method Using Self-information for Ordered Decision Systems

Bo Zou*, Binbin Sang†,‡, Shaoguo Cui* and Jianhang Yu†

*College of Computer and Information Science, Chongqing Normal University, Chongqing 401331, China

†Chongqing Key Laboratory of Computational Intelligence, Chongqing University of Posts and Telecommunications, Chongqing 400065, China

‡sangbinbin@cqnu.edu.cn

Attribute reduction has the ability to identify essential features, reduce feature dimensions, and improve the classification of learning models. The study of attribute reduction methods for ordered decision systems, which are data with preferred order relations, is a popular research topic. In this chapter, a heuristic attribute reduction method based on self-information is proposed by improving the computation of the approximation set. First, by choosing the triangular matrix form, we cleverly optimize the matrix computation process, which facilitates the storage of relations among multiple objects and saves time effectively at the same time. Second, we introduce a more efficient method to compute the approximation set, which further improves the computational efficiency of the algorithm. The adoption of self-information as the uncertainty measure considers both deterministic information and possible categorization information. Finally, the validation practice on multiple datasets fully proves the effectiveness of the algorithm. From the experimental results, the algorithm in this chapter can effectively remove irrelevant or redundant attributes, and at the same time, it is also more efficient compared to other algorithms due to the improved approximation set calculation method.

Keywords: Ordered decision system; attribute reduction; self-information; dominance-based rough set approach.

1. Introduction

Attribute reduction is a crucial step in data preprocessing, enhancing classification accuracy by selecting information-rich attributes and eliminating redundant ones. This process finds widespread application in intelligent computing, data mining, and machine learning.[1–5]

Rough set theory (RST), proposed by Pawlak in 1982, provides an effective mathematical framework for solving real-world problems characterized by vagueness, imprecision, or uncertainty, is completely data-driven, relies on existing data to solve uncertainty problems, and has received widespread attention in the field of attribute reduction.[6,7] However, RST processes information systems through equivalence relations, which has significant limitations in dealing with real-world problems, so researchers have introduced alternative relations such as similarity, adjacency, and dominance, and well-known models such as Neighborhood Rough Sets (NRS),[8] Dominance Rough Sets (DRS),[9] and Variable Precision Rough Sets (VPRS)[10] have emerged. This chapter is particularly concerned with datasets involving dominance relations.

In attribute reduction methods, assessing attribute importance and quantifying data inconsistency usually relies on uncertainty measures, and a series of extension studies[11,12] have been proposed since Shannon proposed information entropy.[13] In order to consider both deterministic and possibly categorical information, this chapter uses self-information as the uncertainty measure.[14] In addition, an upper triangular matrix-based form is used to compute the approximation set to simplify the computational process.

The rest of this chapter is structured as follows. Section 2 gives a background on the basics of rough sets; Section 3 details our proposed algorithm, Section 4 experimentally verifies the effectiveness and efficiency of our algorithm, and, finally, the conclusion will emphasize the efficiency and effectiveness of our algorithm.

2. Preliminary

This section provides a concise overview of DRSA.

A decision system is typically denoted by a quaternion as $S = \langle U, AT \cup D, V, f \rangle$, where U represents a universe, AT is the set of conditional attributes, and D is the set of decisions. V is defined as the union of all V_{a_k} where $a_k \in V \cup D$. Here, V_{a_k} represents the domain of attribute a_k, and the function $f : U \times C \to V$ is defined as an information function, and $f(x_i, a_k)$ denotes the value under attribute a_k. For each $a \in AT \cup D$, if V_a is increasing or decreasing in partial order, the decision system is said to be an ordered decision system, denoted $S^{\succeq} = \langle U, AT \cup D, V, f \rangle$.

Definition 2.1.[15] Given an ODS $S^{\succeq} = \langle U, AT \cup D, V, f \rangle$, for any $P \subseteq AT$, the dominance-based relation is defined as

$$\mathcal{D}_P^{\succeq} = \{(x_i, x_j) \in U \times U \mid f(x_i, a) \nless f(x_j, a), \forall a \in P\}. \qquad (1)$$

Definition 2.2.[15] Given an ODS $S^{\succeq} = \langle U, AT \cup D, V, f \rangle$, as the decision set D encompasses d distinct values, where $d = |D|$, U can be partitioned into d equivalence classes based on the preference-order relation determined by D. Consequently, this collection of equivalence classes is termed the decision class and can be expressed as $Cl = \{Cl_n \mid n \in D, D = \{D_1, \cdots, D_d\}\}$. Two comprehensive sets of classes necessitating approximation exist: the upward union set and the downward union set, defined as follows.

$$Cl_{\overline{n}}^{\succeq} = \bigcup_{n \geq n'} Cl_{n'}, Cl_{\overline{n}}^{\preceq} = \bigcup_{n \leq n'} Cl_{n'}, \forall n', n \in D, D = \{D_1, \cdots, D_d\}. \tag{2}$$

3. Proposed Method and Algorithm

Definition 3.1. Given an ODS $S^{\succeq} = \langle U, AT \cup D, V, f \rangle$, for any $P \subseteq AT$, $n \in D$, the approximation of the upward union $Cl_{\overline{n}}^{\succeq}$ of the lower and upper approximations are redefined respectively:

$$\underline{\mathcal{D}_P}(Cl_{\overline{n}}^{\succeq}) = min\{f(x_i, d) \mid \langle x_j, x_i \rangle \in \mathcal{D}_{\overline{P}}^{\succeq}, x_j \in U\}, \tag{3}$$

$$\overline{\mathcal{D}_P}(Cl_{\overline{n}}^{\succeq}) = max\{f(x_i, d) \mid \langle x_i, x_j \rangle \in \mathcal{D}_{\overline{P}}^{\succeq}, x_j \in U\}. \tag{4}$$

In our approach, to save time in matrix computation, we will use triangular matrices to represent the dominance relation(TDRM).

Definition 3.2. Given an ODS $S^{\succeq} = \langle U, AT \cup D, V, f \rangle$, for any $P \subseteq AT$, we define TDRM as

$$\mathrm{TM}_{\overline{U}}^{\succeq P} = \begin{bmatrix} m_{(1,1)}^{\succeq P} & m_{(1,2)}^{\succeq P} & \cdots & m_{(1,n)}^{\succeq P} \\ & m_{(2,2)}^{\succeq P} & \cdots & m_{(2,n)}^{\succeq P} \\ & & \ddots & \vdots \\ & & & m_{(n,n)}^{\succeq P} \end{bmatrix}, \tag{5}$$

where

$$m_{(i,j)}^{\succeq P} = \begin{cases} 0, & \langle x_i, x_j \rangle \notin \mathcal{D}_{\overline{P}}^{\succeq}, \\ 1, & \langle x_i, x_j \rangle \in \mathcal{D}_{\overline{P}}^{\succeq} \\ 2, & \langle x_j, x_i \rangle \in \mathcal{D}_{\overline{P}}^{\succeq}. \\ 3, & f(x_j, P) = f(x_i, P). \end{cases} \tag{6}$$

$$m_{(i,j)}^{\succeq P} = \begin{cases} m_{(i,j)}^{\succeq P}, & \langle x_i, x_j \rangle \notin \mathcal{D}_{\overline{P}}^{\succeq} \ or \ x_j = x_i \\ \sim m_{(i,j)}^{\succeq P}, & \langle x_i, x_j \rangle \in \mathcal{D}_{\overline{P}}^{\succeq} \ or \ \langle x_j, x_i \rangle \in \mathcal{D}_{\overline{P}}^{\succeq} \end{cases} \tag{7}$$

Further explanation of Eq. 7: Assuming the TDRM is complete, there is $m_{(i,j)} =\sim m_{(j,i)}$, and $1(01)_2$ and $2(10)_2$ are inverse by bit when expressed in binary. To find $P_l(x_i)$ and $P_u(x_i)$ in the upper triangular matrix, the relational identity of all samples with respect to the current sample is used. In this case, the first element of the row of the matrix in which the current sample is located, i.e., $m_{(i,i)}$, and all the elements of the column in which it is located that are either 1 or 2 need to be interchanged when comparing them, i.e., 1 converted to 2, and 2 converted to 1.

Definition 3.3. Given an ODS $S^{\succeq} = \langle U, AT \cup D, V, f \rangle$, for any $P \subseteq AT$, due to $Cl^{\succeq} = D$, the matrix-based self-information of $Cl^{\succeq}$ can also be defined as

$$\mathrm{TMI}_P^U(Cl^{\succeq}) = -\sum_{n=Cl_1^{\succeq}}^{Cl_d^{\succeq}} \left(1 - \frac{|\,\mathcal{D}_P(Cl_n^{\succeq})\,|}{|\,\overline{\mathcal{D}_P}(Cl_n^{\succeq})\,|}\right) \ln \frac{|\,\mathcal{D}_P(Cl_n^{\succeq})\,|}{|\,\overline{\mathcal{D}_P}(Cl_n^{\succeq})\,|} \tag{8}$$

Definition 3.4. Given an ODS $S^{\succeq} = \langle U, AT \cup D, V, f \rangle$, for any $P \subseteq AT$ and any element $a \in P$, the representation of the inner significance and the outer significance of a using TMI is formulated as

$$sig_{inner}^{\succeq U}(a, P, D) = \mathrm{TMI}_{P-\{a\}}^U(Cl_n^{\succeq}) - \mathrm{TMI}_P^U(Cl_n^{\succeq}) \tag{9}$$

and

$$sig_{outer}^{\succeq U}(a, P, D) = \mathrm{TMI}_P^U(Cl_n^{\succeq}) - \mathrm{TMI}_{P\cup\{a\}}^U(Cl_n^{\succeq}), \tag{10}$$

respectively.

Next, a attribute reduction algorithm based on matrix dominance self-information is proposed.

4. Experiment and Analysis

In this section, we present a series of comparative experiments to verify the effectiveness of our improved algorithm.

All datasets are public datasets downloaded from the UCI database, and the platform on which all comparison experiments in this section are run is a computer configured with a 64-bit Microsoft Windows 10 system with an Inter(R) Core(TM) i9-10900X CPU at 3.70 GHz and 128 GB of RAM, and all algorithms are implemented by the Matlab R2020a.

In this experiment, we remove some missing attributes from the data. Subsequently, tests were performed using the K-nearest neighbor (KNN) and naive bayes (NB) classifiers and compared with the original data as well as the FS_DRS[16], FS_FDRS[17] and AR_MDSI algorithms. The classification results are shown in Table 1 and 2, with the best performing data

Algorithm 3.1 AR_MDSI algorithm

Input: An ODS $S^{\succeq} = \langle U, AT \cup D, V, f \rangle$.
Output: A reduct $R_{S^{\succeq}}$.
Initialize $R_{S^{\succeq}} \leftarrow \emptyset$;
Calculate $\mathbb{TMI}^{U}_{AT}(Cl^{\succeq}_{n})$ via using Definition 3.3;
for $a \in AT$ **do**
  Calculate $sig^{\succeq U}_{inner}(a, AT, D) = \mathbb{TMI}^{U}_{AT-\{a\}}(Cl^{\succeq}) - \mathbb{TMI}^{U}_{AT}(Cl^{\succeq})$ by Eq. 9;
  if $sig^{\succeq U}_{inner}(a, AT, D) > 0$ **then**
    $R_{S^{\succeq}} \leftarrow R_{S^{\succeq}} \cup \{a\}$;
  end if
end for
Let $B \leftarrow R_{S^{\succeq}}$;
while $\mathbb{TMI}^{U}_{R_{S^{\succeq}}}(Cl^{\succeq}) \neq \mathbb{TMI}^{U}_{AT}(Cl^{\succeq})$ **do**
  for $a \in (AT - B)$ **do**
    Calculate $sig^{\succeq U}_{outer}(a, B, D) = \mathbb{TMI}^{U}_{B}(Cl^{\succeq}) - \mathbb{TMI}^{U}_{B \cup \{a\}}(Cl^{\succeq})$ by Eq. 10;
  end for
  $a_{max} = arg\ max\{sig^{\succeq U}_{outer}(a, B, D), a \in (AT - B)\}$;
  $B \leftarrow B \cup \{a_{max}\}$
end while
for $a \in B$ **do**
  Calculate $\mathbb{TMI}^{U}_{B-\{a\}}(Cl^{\succeq})$ via using Eq. 9;
  if $\mathbb{TMI}^{U}_{B-\{a\}}(Cl^{\succeq}) = \mathbb{TMI}^{U}_{B}(Cl^{\succeq})$ **then**
    $B \leftarrow B - \{a\}$;
  end if
end for
$R_{S^{\succeq}} \leftarrow B$;
return $R_{S^{\succeq}}$;

highlighted in bold, and it can be seen that in the KNN classifier results, our algorithm shows the highest performance in most of the datasets (including the mean).

We also evaluate the efficiency of the algorithm AR_MDSI in terms of runtime, and the results are listed in table 3, where the values represent the time in seconds, and the shortest time has been boldly labeled. Table 3 clearly shows that the improved attribute reduction algorithm in this paper has almost the shortest average running time for all datasets.

Table 1. The comparison of classification accuracies of different algorithms on KNN (%).

datasets	RAW	FS_DRS	FS_FDRS	AR_MDSI
Derm	32.00±0.81	43.07±1.07	43.61±0.94	**43.86±1.37**
WDBC	90.93±0.36	96.19±0.29	**96.29±0.25**	91.12±0.28
Aust	57.61±0.56	75.96±0.70	61.72±0.73	**76.58±0.45**
Germ	65.04±0.78	92.99±0.45	78.56±0.52	**99.32±0.09**
Mice	63.75±0.67	**64.70±0.62**	39.16±0.81	63.57±m0.70
Wred	69.38±0.43	**83.92±0.42**	71.82±0.72	81.71±0.31
Wite	69.16±0.37	**94.47±0.15**	71.15±0.42	88.38±0.29
HVWN	**99.52±0.05**	82.94±0.53	99.03±0.05	**99.52±0.05**
Card	87.33±0.37	88.05±0.40	**88.26±0.17**	87.49±0.31
Sonar	85.05±0.80	69.23±0.91	76.20±0.99	**85.10±0.51**
Average	73.58±0.50	80.61±0.53	74.15±0.55	**82.94±0.44**

Table 2. The comparison of classification accuracies of different algorithms on NB (%).

datasets	RAW	FS_DRS	FS_FDRS	AR_MDSI
Derm	21.80±1.62	47.38±1.03	**47.49±0.72**	31.15±1.66
WDBC	92.30±0.23	94.15±0.17	**96.94±0.24**	92.30±0.16
Aust	89.58±0.57	92.71±0.25	92.12±0.17	**93.65±0.24**
Germ	50.00±0.00	96.68±0.15	96.68±0.36	**98.40±0.13**
Mice	56.34±0.91	48.31±0.51	40.36±0.54	**56.35±0.70**
Wred	80.09±0.35	**85.88±0.23**	81.68±0.31	84.60±0.28
Wite	73.35±0.18	**80.86±0.13**	75.02±0.15	79.82±0.15
HVWN	89.79±0.09	87.13±0.25	99.37±0.20	**99.52±0.05**
Card	**79.64±0.28**	77.92±0.20	79.27±0.12	74.59±1.73
Sonar	80.67±0.71	72.36±0.82	72.69±0.67	**80.91±0.64**
Average	73.28±0.47	79.67±0.37	79.47±0.36	**80.49±0.56**

Table 3. The comparison of time of different algorithm (s).

datasets	FS_DRS	FS_FDRS	AR_MDSI
Derm	1.21	**0.97**	3.22
WDBC	7.53	2.37	**2.27**
Aust	**2.70**	3.71	2.93
Germ	10.85	12.89	**8.11**
Mice	76.06	36.57	**22.29**
Wred	28.07	24.90	**12.95**
Wite	1104.20	885.24	**90.55**
HVWN	**9.32**	10.35	11.23
Card	170.67	134.77	**54.06**
Sonar	1.50	0.84	**0.72**
Average	128.40	101.17	**18.99**

5. Conclusion

In this chapter, we explore the challenges of attribute reduction for ordered decision-making systems, taking into account both identified and potential categorical information, and propose an enhanced attribute reduction method that can efficiently extract key information in this system. Validation on different datasets shows that the algorithm is significantly effective and, at the same time, less time consuming compared to other methods.

Acknowledgment

This work is supported by the Nature Science Foundation Project of Chongqing Science and Technology Bureau (Nos. CSTB2023NSCQ-MSX1010, CSTB2022NSCQ-MSX1206); the National Natural Science Foundation of China (Nos. 62306054); the Science and Technology Research Program of Chongqing Municipal Education Commission (No. KJQN202300549); the Chongqing Normal University Foundation Project (Nos. 22XLB019); the Key Science and Technology Research Program of Chongqing Municipal Education Commission (No. KJZD-K202200510); the Technology Foresight and System Innovation Project of Chongqing Municipal Science and Technology Bureau (No. CSTB2022TFII-OFX0042); the Humanities and Social Sciences Planning Fund of Ministry of Education of China (No. 22YJA870005); the Chongqing Social Science Planning Project(No. 2022NDYB119); the Natural Science Foundation of Chongqing (No. sl202100000312).

References

1. X. Zhang, J. Yang and L. Tang, Three-way class-specific attribute reducts from the information viewpoint, *Information Sciences* **507**, 840 (2020).
2. W. Xu and Y. Yang, Matrix-based feature selection approach using conditional entropy for ordered data set with time-evolving features, *Knowledge-Based Systems* **279**, p. 110947 (2023).
3. O. O. Aremu, R. A. Cody, D. Hyland-Wood and P. R. McAree, A relative entropy based feature selection framework for asset data in predictive maintenance, *Computers & Industrial Engineering* **145**, p. 106536 (2020).
4. Y. Peng and Q. Zhang, Uncertainty measurement for set-valued data and its application in feature selection, *International Journal of Fuzzy Systems* **24**, 1735 (2022).

5. L. Sun, L. Wang, W. Ding, Y. Qian and J. Xu, Feature selection using fuzzy neighborhood entropy-based uncertainty measures for fuzzy neighborhood multigranulation rough sets, *IEEE Transactions on Fuzzy Systems* **29**, 19 (2020).
6. Z. Jiang, K. Liu, X. Yang, H. Yu, H. Fujita and Y. Qian, Accelerator for supervised neighborhood based attribute reduction, *International Journal of Approximate Reasoning* **119**, 122 (2020).
7. L. Sun, L. Wang, W. Ding, Y. Qian and J. Xu, Neighborhood multi-granulation rough sets-based attribute reduction using lebesgue and entropy measures in incomplete neighborhood decision systems, *Knowledge-Based Systems* **192**, p. 105373 (2020).
8. Y. Yao, Relational interpretations of neighborhood operators and rough set approximation operators, *Information sciences* **111**, 239 (1998).
9. S. Greco, B. Matarazzo and R. Slowinski, Rough approximation of a preference relation by dominance relations, *European Journal of operational research* **117**, 63 (1999).
10. W. Ziarko, Variable precision rough set model, *Journal of computer and system sciences* **46**, 39 (1993).
11. Y. Qian and J. Liang, Combination entropy and combination granulation in rough set theory, *International Journal of Uncertainty, Fuzziness and Knowledge-Based Systems* **16**, 179 (2008).
12. Q. Hu, X. Che, L. Zhang, D. Zhang, M. Guo and D. Yu, Rank entropy-based decision trees for monotonic classification, *IEEE Transactions on Knowledge and Data Engineering* **24**, 2052 (2011).
13. C. E. Shannon, A mathematical theory of communication, *The Bell system technical journal* **27**, 379 (1948).
14. L. Yang, K. Qin, B. Sang and C. Fu, A novel incremental attribute reduction by using quantitative dominance-based neighborhood self-information, *Knowledge-Based Systems* **261**, p. 110200 (2023).
15. S. Greco, B. Matarazzo and R. Slowinski, Rough sets theory for multi-criteria decision analysis, *European journal of operational research* **129**, 1 (2001).
16. W. S. Du and B. Q. Hu, A fast heuristic attribute reduction approach to ordered decision systems, *European journal of operational research* **264**, 440 (2018).
17. Q. Hu, D. Yu and M. Guo, Fuzzy preference based rough sets, *Information Sciences* **180**, 2003 (2010).

Present and Future of Computable Aggregations

L. Garmendia[*,¶], J. Montero[†,‖], L. Magdalena[‡,**] and D. Gómez[§,††]

[*] *Departamento de Ingenería del Software e Inteligencia Artificial, Universidad Complutense de Madrid, Madrid, Spain*
[†] *Facultad de Ciencias Matemáticas, Universidad Complutense de Madrid, Madrid, Spain*
[‡] *Departamento de Matemática Aplicada, Universidad Politécnica de Madrid, Madrid, Spain*
[§] *Departamento de Estadistica y Ciencia de los Datos, Universidad Complutense de Madrid, Madrid, Spain*

[¶] *lgarmend@fdi.ucm.es*
[‖] *monty@mat.ucm.es*
[**] *luis.magdalena@upm.es*
[††] *dagomez@estad.ucm.es*

The concept of aggregation within fuzzy sets has been deeply modified from the initial view as mappings where a fixed number of aggregated items, and the aggregation itself, take values in the unit interval to assume a family of mapping and impose certain restrictions to assure consistency and computability, till focus on aggregation as algorithms on lists of objects. This chapter points out how new concepts such as recursivity, map reducibility or hierarchy can be translated into computable aggregations.

Keywords: Computable aggregation; aggregation function; recursivity; non-deterministic; map reducible; hierarchical; prioritized aggregation.

1. Introduction: Aggregation Functions

Definition 1.1. [1,2] An aggregation operator is usually defined as a function $Ag : [0,1]^n \to [0,1]$, where it is assumed to satisfy the following properties:

[¶]Corresponding author.

- Ag is monotonic and non-decreasing
- $Ag(0, \ldots, 0) = 0$
- $Ag(1, \ldots, 1) = 1$

2. Computable Aggregation

Aggregation functions are commonly linked to aggregation processes in the existing literature. Typically, the act of combining information is represented by a function or a set of functions.

However, in the pioner work of Montero et al.[3], it is introduced the notion of computable aggregations. In this context, an aggregation process is associated with a program, which may not necessarily be articulated solely in terms of functions. This perspective emphasizes how each aggregation is derived, focusing on the procedure responsible for generating the aggregated value.

Definition 2.1.[3] Let $L < T >$ be a finite and non-empty list of n elements with type T. A computable aggregation is a program P that transforms the list $L < T >$ into an element of T, where T is a complete lattice.

In the next sections we introduce several specific approached to computable aggregation.

3. Recursive Computable Aggregations

Several recursive computable aggregations are studied, giving a recursive function ϕ (see[4,5] for more details) that operates the aggregation of a list of n values with a new value, without the need to look again to the list when one more value is added to recompute the aggregation. This means that an aggregation can be computed by running n times the recursive function.

Definition 3.1.[5] A computable aggregation P is recursive if and only if there exists a sequence of binary programs defined with lists of dimension two $\{\phi_n\}_{n>1}$ verifying:

$$P(L) = \begin{cases} 0 & \text{If } |L| = 0 \\ x_1 & \text{If } L = \{x_1\} \\ \phi_2(x_1, x_2) & \text{If } L = \{x_1, x_2\} \\ \vdots & \\ \phi_n(P(L \setminus \{x_n\}), x_n) & \text{If } L = \{x_1, \ldots, x_n\} \end{cases}$$

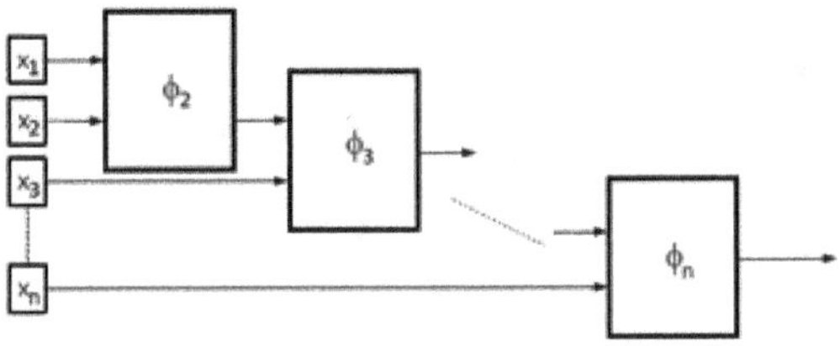

Fig. 1. Recursive (Hierarchical) Computable Aggregation

4. Parallel Computable Aggregations

Given a hierarchical structure such as a tree and some computable aggregations for its levels, the concept of tree hierarchical computable aggregation is defined depending on the depth of the tree.

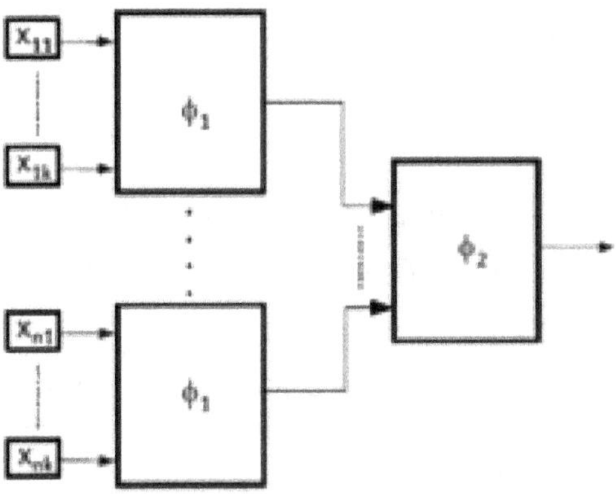

Fig. 2. 2.3-level tree (Hierarchical) Computable Aggregation

5. Prioritized Computable Aggregations

The concept of prioritized aggregation operators is presented by Yager in[6], where he considers the existence of many situations in which a decision process involving the aggregation of several criteria, requires those criteria to be ordered and grouped to establish a prioritization in the aggregation process. Here it is important to note that if we have a motorbike with low value in safety, most probably it is not even necessary to evaluate the criteria related with the cost, since the global value of the aggregation should be low.

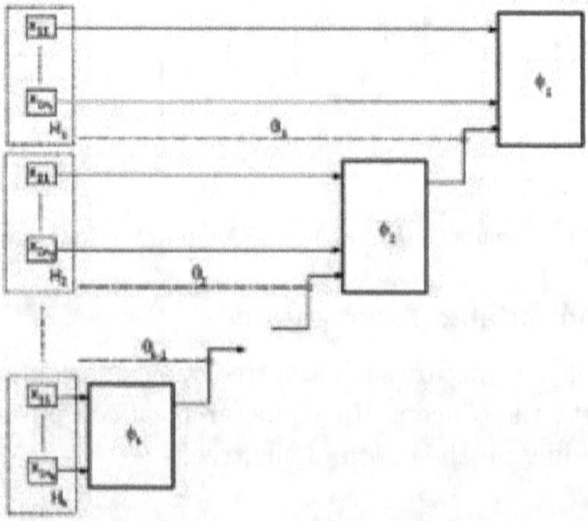

Fig. 3. Prioritized (Hierarchical) Computable Aggregation

6. Non-deterministic Computable Aggregations

A non-deterministic program does not always give the same outputs when the same inputs are entered.

Definition 6.1.[7] A computable aggregation P over the set T is a non-deterministic computable aggregation if and only if the program implementing it is non-deterministic.

Definition 6.2.[7] (Empirical distribution of a computable aggregation). Given a computable aggregation P and given a list $L_1 \in L < T >$, the distribution of results obtained after m executions of P over L_1 is represented by $LP_m(L_1)$.

Definition 6.3.[7] **Strong $\leq$-monotonicity of non-deterministic computable aggregations.** A non-deterministic computable aggregation P is strong $\leq$-monotone if and only if, when $L_1 \leq L_2$, then $DPL_{1m} \leq DPL_{2m}$ for any $m \geq 1$.

Definition 6.4.[7] **Max-Min preorder.** A list L_1 in L is Max-Min lower or equal than a list L_2 in L if and only if the highest degree in L_1 is lower or equal than the minimum value of L_2.

Proposition 6.1.[7] *The Probability sampled and k-sampled computable aggregations are Strong $\leq$MaxMin - Monotonic.*

Definition 6.5. (Theoretical distribution of a computable aggregation). Given a computable aggregation P that aggregates a list $L_1 \in L < T >$, let us denote by $P(L_1)$ the theoretical distribution after all possible realizations of P over the fixed list L_1.

Proposition 6.2. *P is population asymptotic monotone if P is $\leq S1to1$-asymptotic monotone.*

Proposition 6.3. *PAg, k computable aggregation is population monotone.*

7. Computable Aggregations of Random Variables

The concept of aggregation of random variables was defined in[8], extending the boundary conditions and the monotonicity using stochastic orders.

Definition 7.1. Given an aggregation function $Ag : \mathbb{R}^n \longrightarrow \mathbb{R}$ and a list of random variables $X_1, \ldots, X_n$, the random variable X_{Ag} associated with the aggregation function $Ag(u, v) = (u + v)/2$ is defined as $X_{Ag} = Ag(X_1, \ldots, X_n)$.

Example 7.1. Taking into account previous considerations, the random variable X_{Ag} associated with the aggregation function $Ag(u, v) = (u+v)/2$ for the Bernoulli random variables X_1, X_2 could be denoted as

$$X_{Ag} = Ag(X_1, X_2) = \frac{X_1 + X_2}{2}$$

formally defined as $X_{Ag} : \Omega^2 \longrightarrow \mathbb{R}$.

$$X_{\mathrm{Ag}}(c, c) = \frac{X_1(c) + X_2(c)}{2} = 1,$$

$$X_{\text{Ag}}(c,x) = \frac{X_1(c) + X_2(x)}{2} = 0.5,$$

$$X_{\text{Ag}}(x,c) = \frac{X_1(x) + X_2(c)}{2} = 0.5,$$

$$X_{\text{Ag}}(x,x) = \frac{X_1(x) + X_2(x)}{2} = 0.$$

So, the support of this random variable is $\mathcal{D}_{X_{\text{Ag}}} = \{0, 0.5, 1\}$ and the mass function can be computed assuming the independence among X_1 and X_2 as:

$$f_{X_{\text{Ag}}}(0) = P(X_1 = 0, X_2 = 0) = P(X_1 = 0)P(X_2 = 0) = (1 - p_1)(1 - p_2),$$

$$f_{X_{\text{Ag}}}(0.5) = P(X_1 = 1, X_2 = 0) + P(X_1 = 0, X_2 = 1) = p_1(1-p_2) + (1-p_1)p_2$$

and

$$f_{X_{\text{Ag}}}(1) = P(X_1 = 1, X_2 = 1) = p_1 p_2.$$

8. Conclusions

In this chapter, we provide a comprehensive overview of the concept of computable aggregation introduced in Ref. 3 which differs significantly from the traditional understanding of aggregation operators. We delve into several properties that have been defined specifically for function aggregation operators, highlighting the need for a redefinition when it comes to programs or algorithms. These new definitions and concepts range from recursive, parallel, to prioritized and non-deterministic computable aggregations. Each of these approaches addresses different aspects, such as hierarchy, structure and decision prioritization, improving the versatility and adaptability of the aggregation process. This shift in focus from classical aggregation operators to computable aggregations broadens the understanding of aggregation processes, paving the way for a more nuanced exploration of their underlying procedures and functionalities.

It is also important to mention that not only is it necessary to redefine classic concepts for aggregation operators in a computing context, but that within the framework of computable aggregators, other ways of aggregation appear (such as non-deterministic computable aggregators) that were not contemplated within the conventional aggregation operators. These new aggregation processes contemplate situations as real as statistical estimation and sampling processes that cannot be address in a conventional approach or processes where the object is to aggregate random variables as is done in many machine learning problems.

Acknowledgment

This research has been partially supported by the Government of Spain, grant PID2021-122905NB-C21.

References

1. G. Beliakov, A. Pradera, T. Calvo *et al.*, *Aggregation functions: A guide for practitioners* (Springer, 2007).
2. T. Calvo, G. Mayor and R. Mesiar, *Aggregation operators: new trends and applications* (Springer Science & Business Media, 2002).
3. J. Montero, R. González-del Campo, L. Garmendia, D. Gómez and J. T. Rodríguez, Computable aggregations, *Information Sciences* **460**, 439 (2018).
4. R. González-del Campo, L. Garmendia and J. Montero, Complexity of increasing phi-recursive computable aggregations, in *XIX Congreso Español sobre Tecnologías y Lógica Fuzzy (CAEPIA-ESTYLF)*, 2018.
5. L. Magdalena, L. Garmendia, D. Gómez, R. G. del Campo, J. T. Rodríguez and J. Montero, Types of recursive computable aggregations, in *2019 IEEE International Conference on Fuzzy Systems (FUZZ-IEEE)*, 2019.
6. R. R. Yager, Prioritized aggregation operators, *International Journal of Approximate Reasoning* **48**, 263 (2008).
7. L. Magdalena, D. Gómez, L. Garmendia and J. Montero, Analysing monotonicity in non-deterministic computable aggregations: The probabilistic case, *Information Sciences* **583**, 288 (2022).
8. J. Baz, I. Díaz, L. Garmendia, D. Gómez, L. Magdalena and S. Montes, Computable aggregations of random variables, *Information Sciences* **654**, p. 119842 (2024).

On Generalized Overlap Functions and Overlap Indices

T. C. Asmus*,¶, H. S. Santos*,||, G. P. Dimuro*,**, H. Camargo†,††,
C. Marco-Detchart‡,‡‡ and H. Bustince§,§§

* *Universidade Federal do Rio Grande, Campus Carreiros, 96201-900 Rio Grande, Brazil*
† *Federal University of São Carlos, Campus São Carlos, 13565-905 São Carlos, Brazil*
‡ *Universitat Politècnica de València, Camino de Vera s/n, 46022 Valencia, Spain*
§ *Universidad Publica de Navarra, Campus Arrosadia, 31006 Pamplona, Spain*

¶ *tiagoasmus@furg.br*
|| *helida@furg.br*
** *gracalizdimuro@furg.br*
†† *heloisa@dc.ufscar.br*
‡‡ *cedmarde@upv.es*
§§ *bustince@unavarra.es*

Overlap indices are functions measuring the fuzzy intersection between two fuzzy sets and can be built employing overlap functions, which are aggregation functions with features that provide good results when applied to practical problems, such as image processing and classification. Many studies have generalized the concept of overlap functions, such as n-dimensional overlap functions, general overlap functions, and quasi-overlap functions, along with their application in practical problems. However, the concept of overlap index has not been nearly as extended or generalized as the one of overlap function. So, the objective of this chapter is to introduce the concepts of n-dimensional overlap indices and general overlap indices, with special attention to the development of their construction methods based on generalized overlap functions.

Keywords: Overlap functions; overlap indices; n-dimensional overlap functions; general overlap functions.

1. Introduction

Overlap indices are functions that measure the fuzzy intersection between two fuzzy sets. They were inspired by Zadeh's consistency index [1] and initially developed to be applied in interpolative fuzzy systems [2]. They have also been applied in the construction of rule weights in Fuzzy Rule-Based

---

¶Corresponding author.

Classification Systems (FRBCSs) [3]. One important aspect of overlap indices is that they can be constructed by an aggregation of overlap functions [4], which are a special type of aggregation function with properties that provide good results when applied to practical problems, such as image processing [5] and classification [6].

Since its introduction, many studies have been dedicated to the generalization of the concept of overlap function, also obtaining successful results in practical applications. For example, n-dimensional overlap functions were developed to measure the overlapping of more than two classes [7], general overlap functions have less strict boundary conditions [8], quasi-overlap functions disregard the continuity property [9], and pseudo-overlap functions do not consider the commutativity property [10].

However, the concept of overlap index has not been nearly as extended or generalized as the one of overlap function. Since they are related concepts, with somewhat similar properties, there is promise in generalizing overlap indices, which may be constructed via generalized overlap functions, and observing their performance when applied to practical problems.

With that in mind, the objective of this chapter is to introduce the concepts of n-dimensional overlap indices and general overlap indices, with special attention to the development of their construction methods based on generalized overlap functions.

This chapter is organized as follows. In Section 2 we recall some necessary preliminary concepts. Section 3 introduces n-dimensional overlap indices and some construction methods. In Section 4, general overlap indices are defined, also with some construction methods. Finally, Section 5 is the conclusion.

2. Preliminary Concepts

Let us denote $\vec{x} = (x_1, \ldots, x_n) \in [0,1]^n$.

Definition 2.1. [11] An aggregation function is any function $M\colon [0,1]^n \to [0,1]$ that respects the following conditions: **(M1)** M is increasing in each variable; **(M2)** $M(0,\ldots,0) = 0$ and $M(1,\ldots,1) = 1$.

Definition 2.2. [7] A function $On\colon [0,1]^n \to [0,1]$ is called an n-dimensional overlap function if the following conditions hold, for all $\vec{x} \in [0,1]^n$: **(On1)** On is commutative; **(On2)** $On(\vec{x}) = 0$ if and only if $\prod_{i=1}^{n} x_i = 0$; **(On3)** $On(\vec{x}) = 1$ if and only if $\prod_{i=1}^{n} x_i = 1$; **(On4)** On is increasing; **(On5)** On is continuous.

A 2-dimensional overlap function is just called an overlap function. Considering an n-dimensional overlap function On, one can loosen conditions **(On2)** to: **(On2')** $\prod_{i=1}^{n} x_i = 0 \Rightarrow On(\vec{x}) = 0$, obtaining an n-dimensional 0-overlap function. Analogously, a function On is considered an n-dimensional 1-overlap function if the condition **(On3)** is loosened to: **(On3')** $\prod_{i=1}^{n} x_i = 1 \Rightarrow On(\vec{x}) = 1$. By combining both concepts, we have the following definition:

Definition 2.3. [8] A function $GO\colon [0,1]^n \to [0,1]$ is called a general overlap function (GOF, for short) if the following conditions hold, for all $\vec{x} \in [0,1]^n$: **(GO1)** GO is commutative; **(GO2)** If $\prod_{i=1}^{n} x_i = 0$ then $GO(\vec{x}) = 0$; **(GO3)** If $\prod_{i=1}^{n} x_i = 1$ then $GO(\vec{x}) = 1$; **(GO4)** GO is increasing; **(GO5)** GO is continuous.

A quasi-overlap function disregards continuity [9], a pseudo-overlap function does not need to be commutative [10] and by disregarding both properties in generalized overlap functions, we have the notion of general pseudo quasi-overlap function [12].

One may obtain GOF by truncated versions of n-dimensional overlap functions [8]. If continuity is not necessary in the aggregation process, one may construct general quasi-overlap functions as follows:

Proposition 2.1. *Take $\alpha, \beta \in [0,1]$, such that $\alpha \leq \beta$, and let $On\colon [0,1]^n \to [0,1]$ be an n-dimensional overlap function. Then, the function $QTOn_{\alpha,\beta} : [0,1]^n \to [0,1]$, defined, for all $\vec{x} \in [0,1]^n$, by*

$$QTOn_{\alpha,\beta}(\vec{x}) = \begin{cases} 0, & \text{if } On(\vec{x}) \leq \alpha \\ On(\vec{x}), & \text{if } \alpha < On(\vec{x}) < \beta \\ 1, & \text{if } On(\vec{x}) \geq \beta, \end{cases} \tag{1}$$

is a general quasi-overlap function.

Proof. Since On is an n-dimensional overlap function, properties **(On1)** and **(On4)** are immediately satisfied. Also, by **(On2)**, it holds that $\prod_{i=1}^{n} x_i = 0 \Leftrightarrow On(\vec{x}) = 0 \Rightarrow On(\vec{x}) \leq \alpha \Rightarrow QTOn_{\alpha,\beta}(\vec{x}) = 0$, meaning that condition **(On2')** is satisfied. Finally, by **(On3)**, one has that $\prod_{i=1}^{n} x_i = 1 \Leftrightarrow On(\vec{x}) = 1 \Rightarrow On(\vec{x}) \geq \beta \Rightarrow QTOn_{\alpha,\beta}(\vec{x}) = 1$, showing that condition **(On3')** is also satisfied. □

We denote by $FS(U)$ the space of all fuzzy sets defined over an universe $U = \{u_1, \ldots, u_m\}$.

Definition 2.4. [2] A mapping $\mathcal{O}\colon FS(U) \times FS(U) \to [0,1]$ is said to be an overlap index if it satisfies the following conditions, for all $A, B, C \in FS(U)$: **($\mathcal{O}$1)** $\mathcal{O}(A,B) = 0$ if and only if, for all $u \in U, A(u) \cdot B(u) = 0$; **($\mathcal{O}$2)**

$\mathcal{O}$ is commutative: $\mathcal{O}(A,B) = \mathcal{O}(B,A)$; ($\mathcal{O}$**3**) If $B \leq C$, meaning that $B(u) \leq C(u)$ for all $u \in U$, then $\mathcal{O}(A,B) \leq \mathcal{O}(A,C)$. For an overlap index to be called normal, it also has to satisfy the following condition: ($\mathcal{O}$**4**) If there exists $u \in U$ such that $A(u) \cdot B(u) = 1$, then $\mathcal{O}(A,B) = 1$.

The following theorem shows a construction method for overlap indices:

Theorem 2.1. *[2] Consider an aggregation function $M\colon [0,1]^m \to [0,1]$ with the condition "**(M1)**: $M(x_1,\ldots,x_m) = 0 \Leftrightarrow x_1 = \ldots = x_m = 0$" and an overlap function $O\colon [0,1]^2 \to [0,1]$. Then, the function $\mathcal{O}_M^O\colon FS(U) \times FS(U) \to [0,1]$, given, for all $A, B \in FS(U)$, by*

$$\mathcal{O}_M^O(A,B) = M(O(A(u_1),B(u_1)),\ldots,O(A(u_m),B(u_m))),$$

is an overlap index.

3. n-Dimensional Overlap Indices

As stated in the Introduction, overlap indices are generalized measures of fuzzy intersections between two fuzzy sets. In this section, we expand this concept to a function that measures the fuzzy intersection between n fuzzy sets, namely, n-dimensional overlap indices.

Definition 3.1. A mapping $\mathcal{O}n\colon FS(U)^n \to [0,1]$ is said to be an overlap index if it satisfies the following conditions, for all $A_1,\ldots,A_n,B \in FS(U)$: ($\mathcal{O}$**n1**) $\mathcal{O}n(A_1,\ldots,A_n) = 0$ if and only if, for all $u \in U, \prod_{i=1}^n A_i(u) = 0$; ($\mathcal{O}$**n2**) $\mathcal{O}n$ is commutative[a]; ($\mathcal{O}$**n3**) If $A_1 \leq B$, then $\mathcal{O}n(A_1,A_2,\ldots,A_n) \leq \mathcal{O}n(B,A_2,\ldots,A_n)$. For an n-dimensional overlap index to be called normal, it also has to satisfy the following condition: ($\mathcal{O}$**n4**) If there exists $u \in U$ such that $\prod_{i=1}^n A_i = 1$, then $\mathcal{O}n(A_1,\ldots,A_n) = 1$.

A 2-dimensional overlap index is just an overlap index, as per Def. 2.4. Most properties from overlap indices follow naturally for n-dimensional overlap indices. Since our main interest is obtaining construction methods for them, the following result is an adaptation of Theorem 2.1 for n-dimensional overlap indices:

Theorem 3.1. *Consider an aggregation function $M\colon [0,1]^m \to [0,1]$ that respects condition **(M1)** (see Theorem 2.1), and an n-dimensional quasi-1-overlap function $QOn^1\colon [0,1]^n \to [0,1]$ (an n-dimensional overlap function*

[a]Here, we use the term "commutative" to signify an n-ary function in which a permutation of the inputs does not impact the output. We do not use the term "symmetric" to avoid confusion since, in [2], the authors have introduced the concept of *symmetric overlap index*, which has a different meaning.

*without condition **(On5)** and with condition **(On3')** instead of **(On3)**). Then, the function $\mathcal{O}n_M^{QOn^1}: FS(U) \times FS(U) \to [0,1]$, given, for all $A_1, \ldots, A_n \in FS(U)$, by*

$$\mathcal{O}n_M^{QOn^1}(A_1, \ldots, A_n) = M(QOn^1(A_1(u_1), \ldots, A_n(u_1)), \ldots, QOn^1(A_1(u_m), \ldots, A_n(u_m))),$$

is an n-dimensional overlap index.

Proof. Let us check if $\mathcal{O}n_M^{QOn^1}$ respects all conditions from Definition 3.1:

($\mathcal{O}$**n1**) Suppose that $\mathcal{O}n(A_1, \ldots, A_n) = 0$. Then, it holds that:

$$\begin{aligned}
\mathcal{O}n(A_1, \ldots, A_n) = 0 \\
&\Leftrightarrow M(QOn^1(A_1(u_1), \ldots, A_n(u_1)), \ldots, QOn^1(A_1(u_m), \ldots, A_n(u_m))) = 0 \\
&\Leftrightarrow QOn^1(A_1(u_1), \ldots, A_n(u_1)) = \ldots = QOn^1(A_1(u_m), \ldots, A_n(u_m)) = 0 \\
&\Leftrightarrow \prod_{i=1}^{n}(A_i(u)) = 0 \text{ for all } u \in U.
\end{aligned}$$

($\mathcal{O}$**n2**) It follows from **(On1)**.

($\mathcal{O}$**n3**) Immediate, since M and QOn^1 are increasing in each variable. □

Since every n-dimensional-1-overlap function, quasi-overlap function, and overlap function are also an n-dimensional quasi-1-overlap function, the following results are immediate.

Corollary 3.1. *Consider an aggregation function $M: [0,1]^m \to [0,1]$ that respects condition **(M1)** and an n-dimensional-1-overlap function $On^1: [0,1]^n \to [0,1]$. Then, the function $\mathcal{O}n_M^{On^1}: FS(U) \times FS(U) \to [0,1]$, given, for all $A_1, \ldots, A_n \in FS(U)$, by*

$$\mathcal{O}n_M^{On^1}(A_1, \ldots, A_n) = M(On^1(A_1(u_1), \ldots, A_n(u_1)), \ldots, On^1(A_1(u_m), \ldots, A_n(u_m))),$$

is an n-dimensional overlap index.

Corollary 3.2. *Consider an aggregation function $M: [0,1]^m \to [0,1]$ that respects condition **(M1)** and an n-dimensional quasi-overlap function $QOn: [0,1]^n \to [0,1]$. Then, the function $\mathcal{O}n_M^{QOn}: FS(U) \times FS(U) \to [0,1]$, given, for all $A_1, \ldots, A_n \in FS(U)$, by*

$$\mathcal{O}n_M^{QOn}(A_1, \ldots, A_n) = M(QOn(A_1(u_1), \ldots, A_n(u_1)), \ldots, QOn(A_1(u_m), \ldots, A_n(u_m))),$$

is an n-dimensional overlap index.

Corollary 3.3. *Consider an aggregation function $M: [0,1]^m \to [0,1]$ that respects condition **(M1)** and an n-dimensional overlap function $On: [0,1]^n \to [0,1]$. Then, the function $\mathcal{O}n_M^{On}: FS(U) \times FS(U) \to [0,1]$, given, for all $A_1, \ldots, A_n \in FS(U)$, by*

$$\mathcal{O}n_M^{On}(A_1, \ldots, A_n) = M(On(A_1(u_1), \ldots, A_n(u_1)), \ldots, On(A_1(u_m), \ldots, A_n(u_m))),$$

is an n-dimensional overlap index.

Remark 3.1. Since one can apply the constructions methods presented in Theorem 3.1 and Corollaries 3.1 and 3.2 with $n = 2$, these are viable methods to construct overlap indices (Definition 2.4) based on quasi-1-overlap functions, 1-overlap functions, and quasi-overlap functions, respectively. Also, when $n = 2$, Corollary 3.1 coincides with Theorem 2.1.

4. General Overlap Indices

In this section, we generalize the notion of n-dimensional overlap index by loosening condition ($\mathcal{O}$**n1**), in the same manner as n-dimensional 0-overlap functions and general overlap functions (GOF) replace condition **(On2)** by **(On2')** and **(GO2)** in their respective definitions.

Definition 4.1. A mapping $\mathcal{GO}\colon FS(U)^n \to [0,1]$ is said to be a general overlap index if it satisfies the following conditions, for all $A_1, \ldots, A_n, B \in FS(U)$: ($\mathcal{GO}$**1**) If, for all $u \in U, \prod_{i=1}^n A_i = 0$, then $\mathcal{GO}(A_1, \ldots, A_n) = 0$; ($\mathcal{GO}$**2**) $\mathcal{GO}$ is commutative; ($\mathcal{GO}$**3**) If $A_1 \leq B$, then $\mathcal{GO}(A_1, A_2, \ldots, A_n) \leq \mathcal{GO}(B, A_2, \ldots, A_n)$. For an n-dimensional overlap index to be called normal, it also has to satisfy the following condition: ($\mathcal{GO}$**4**) If there exists $u \in U$ such that $\prod_{i=1}^n A_i = 1$, then $\mathcal{GO}(A_1, \ldots, A_n) = 1$.

Immediately, any n-dimensional overlap index is also a general overlap index, but the converse does not hold. As follows, we present a construction method inspired in the one from Theorem 3.1 for general overlap indices.

Theorem 4.1. *Consider an aggregation function $M\colon [0,1]^m \to [0,1]$ that respects condition* **(M1)** *(see Theorem 2.1) and a general quasi-overlap function $QGO\colon [0,1]^n \to [0,1]$. Then, the function $\mathcal{GO}_M^{QGO}\colon FS(U) \times FS(U) \to [0,1]$, given, for all $A_1, \ldots, A_n \in FS(U)$, by*

$$\mathcal{GO}_M^{QGO}(A_1, \ldots, A_n) = M(QGO(A_1(u_1), \ldots, A_n(u_1)), \ldots, QGO(A_1(u_m), \ldots, A_n(u_m))),$$

is a general overlap index.

Proof. Let us check if $\mathcal{GO}_M^{QGO}$ respects all conditions from Definition 4.1:
($\mathcal{GO}$1) Suppose that $\prod_{i=1}^n (A_i(u)) = 0$, for all $u \in U$. Then, it holds:

$$\begin{aligned}
&\prod_{i=1}^n (A_i(u)) = 0, \text{ for all } u \in U \\
&\quad \Rightarrow QGO(A_1(u_1), \ldots, A_n(u_1)) = \ldots = QGO(A_1(u_m), \ldots, A_n(u_m)) = 0 \\
&\quad \Leftrightarrow M(QGO(A_1(u_1), \ldots, A_n(u_1)), \ldots, QGO(A_1(u_m), \ldots, A_n(u_m))) = 0 \\
&\quad \Leftrightarrow \mathcal{GO}(A_1, \ldots, A_n) = 0.
\end{aligned}$$

($\mathcal{GO}$**2**) It follows from **(GO1)**.

($\mathcal{GO}$**3**) Immediate, since M and QGO are increasing in each variable. □

Corollary 4.1. *Consider an aggregation function $M\colon [0,1]^m \to [0,1]$ that respects condition **(M1)** and a general overlap function $GO\colon [0,1]^n \to [0,1]$. Then, the function $\mathcal{GO}_M^{GO}\colon FS(U) \times FS(U) \to [0,1]$, given, for all $A_1, \ldots, A_n \in FS(U)$, by*

$$\mathcal{GO}_M^{GO}(A_1, \ldots, A_n) = M(GO(A_1(u_1), \ldots, A_n(u_1)), \ldots, GO(A_1(u_m), \ldots, A_n(u_m))),$$

is a general overlap index.

Proof. Straight, since every GOF is a general quasi-overlap function. □

Remark 4.1. From Theorem 4.1 and Proposition 2.1, one can obtain general overlap indices based on choices of an aggregation function M that respects condition **(M1)** and general quasi-overlap function QGO, which, in turn, can be constructed via an n-dimensional overlap function On that is truncated by chosen values of α and β. This allows us to create very flexible overlapping indices, enable to adapt to different practical applications, each one with its own set of constraints.

Example 4.1. Let us construct a general quasi-overlap function $QTO_{\alpha,\beta}$ based on Proposition 2.1, with arbitrary values $\alpha, \beta \in [0,1]$ such that $\alpha \leq \beta$, and a chosen n-dimensional overlap function O. By taking $On = O_p$ (the product), $\alpha = 0.2$ and $\beta = 1$, we have the following expression for $QT_{0.2,1}^{O_p}$:

$$QT_{0.2,1}^{O_p}(\vec{x}) = \begin{cases} 0, & \text{if } \prod_{i=1}^{n} x_i \leq 0.2 \\ \prod_{i=1}^{n} x_i, & \text{if otherwise.} \end{cases} \tag{2}$$

Now, we construct a general overlap index through Theorem 4.1, based on the general quasi-overlap function $QT_{0.2,1}^{O_p}$ and an aggregation function M that respects condition **(M1)**. By taking $M = AM$ (the arithmetic mean), the constructed general overlap index $\mathcal{GO}_{AM}^{QT_{0.2,1}^{O_p}}$ is expressed by:

$$\mathcal{GO}_{AM}^{QT_{0.2,1}^{O_p}}(A, B) = \frac{1}{m} \sum_{i=1}^{m} QT_{0.2,1}^{O_p}(A_1(u_i), \ldots, A_n(u_i)). \tag{3}$$

Observe that $\mathcal{GO}_{AM}^{QT_{0.2,1}^{O_p}}$ is a general overlap index, but it is neither an overlap index nor an n-dimensional overlap index.

5. Conclusion

This chapter developed some generalizations on the concept of overlap indices, either by considering the overlapping of n fuzzy sets or by relaxing the condition in which an overlap index may return the value zero. Furthermore, we explored different ways of constructing such indices employing generalized overlap functions.

Since overlap indices can be applied to obtain rule weights in FRBCSs [3], one immediate possibility of future work is to generate several rule weights based on different general overlap indices and compare their effect in the classification process.

References

[1] L. Zadeh, Fuzzy sets as a basis for a theory of possibility, *Fuzzy Sets and Systems* **1**, 3 (1978).

[2] S. Garcia-Jimenez, H. Bustince, E. Hüllermeier, R. Mesiar, N. R. Pal and A. Pradera, Overlap indices: Construction of and application to interpolative fuzzy systems, *IEEE Transactions on Fuzzy Systems* **23**, 1259 (Aug 2015).

[3] M. Elkano, M. Galar, J. A. Sanz, P. F. Schiavo, S. Pereira, G. P. Dimuro, E. N. Borges and H. Bustince, Consensus via penalty functions for decision making in ensembles in fuzzy rule-based classification systems, *Applied Soft Computing* **67**, 728 (2018).

[4] H. Bustince, J. Fernandez, R. Mesiar, J. Montero and R. Orduna, Overlap functions, *Nonlinear Analysis: Theory, Methods & Applications* **72**, 1488 (2010).

[5] A. Jurio, H. Bustince, M. Pagola, A. Pradera and R. Yager, Some properties of overlap and grouping functions and their application to image thresholding, *Fuzzy Sets and Systems* **229**, 69 (2013).

[6] G. Lucca, J. A. Sanz, G. P. Dimuro, B. Bedregal, H. Bustince and R. Mesiar, CF-integrals: A new family of pre-aggregation functions with application to fuzzy rule-based classification systems, *Information Sciences* **435**, 94 (2018).

[7] D. Gómez, J. T. Rodríguez, J. Montero, H. Bustince and E. Barrenechea, n-dimensional overlap functions, *Fuzzy Sets and Systems* **287**, 57 (2016), Theme: Aggregation Operations.

[8] L. De Miguel, D. Gómez, J. T. Rodríguez, J. Montero, H. Bustince, G. P. Dimuro and J. A. Sanz, General overlap functions, *Fuzzy Sets and Systems* **372**, 81 (2019).

[9] R. Paiva, R. Santiago, B. Bedregal and E. Palmeira, Lattice-valued overlap and quasi-overlap functions, *Information Sciences* **562**, 180 (2021).

[10] D. García-Zamora, R. Paiva, A. Cruz, J. Fernandez and H. Bustince, Some construction methods for pseudo-overlaps and pseudo-groupings and their application in group decision making, *Axioms* **12** (2023).

[11] G. Beliakov, H. Bustince and T. Calvo, *A Practical Guide to Averaging Functions* (Springer, Berlin, New York, 2016).

[12] R. E. B. Paiva and B. R. C. Bedregal, General pseudo quasi-overlap functions on lattices, *Axioms* **11** (2022).

Exploring Large Language Models for Automated Review Notes Distribution in Animation Production

Diego Garcés[*,¶], Matilde Santos[†,‖] and David Fernández-Llorca[‡,§,**]

[*] *Computer Science Faculty, Complutense University of Madrid, Spain*
Skydance Animation Madrid, Spain
[†] *Institute of Knowledge Technology, Complutense University of Madrid, Spain*
[‡] *European Commission, Joint Research Centre, Seville, Spain*
[§] *Computer Engineering Department, University of Alcalá, Madrid, Spain*

[¶] *digarces@ucm.es*
[‖] *msantos@ucm.es*
[**] *david.fernandez-llorca@ec.europa.eu*

Supervisors and directors meet daily to review in-progress material during the production of an animated movie. These movies take years to produce, and thousands of text notes about necessary fixes are generated. These notes are manually distributed to different departments for correction. As with any manual process, many notes are not properly assigned and are never addressed, affecting the final quality of the movie. This chapter introduces a method for automatically distributing these notes using multi-class text classification with Large Language Models (LLMs). Starting with the Llama 2 7B model, we apply one-shot learning, as well as fine-tuning with QLoRA. The best results are obtained with the adapted model, achieving a mean accuracy of 0.853 in just 30 minutes of training on an RTX-3090.

Keywords: Movie production; review notes; text classification; large language models (LLMs); Llama 2; fine-tuning; zero- and one-shot learning.

1. Introduction

Iteration is a crucial aspect of movie production. Daily review meetings are held to examine material in progress, where directors and supervisors offer feedback on how to improve or fix the presented material. These comments are transcribed into text notes and uploaded into a production database, before being manually distributed by the producers to the different artistic

[¶]Corresponding author.

departments responsible for addressing them. Despite animation production being a highly technical field, this part of the process remains entirely manual. Combined with the fast-paced nature of movie production, many notes are not correctly allocated to the appropriate departments, leaving certain improvements unattended, which can lower the quality of the final movie.

Assigning these notes to the appropriate department is a text classification problem. The notes are transcriptions, written in free text, of the director's comments during review meetings and are drafted by the producer present at the meeting.

There are numerous research about text classification and how it is applied in different fields.[1,2] These techniques have been applied to ticket assignment,[3] news categorization,[4] topic classification,[5] or sentiment analysis.[6] Nevertheless, the application of automatic text classification with Natural Language Processing (NLP) techniques in certain areas has not been fully explored, making it a novel area of research.

Producing an animated movie requires highly technical and specialized knowledge. Artists are grouped into different departments depending on the task they perform. Modeling, Surfacing, Rigging, CFX, Animation, Layout, Final Layout (FL), FX, Compo, or Lighting are examples of typical departments found in any animation studio.

A review session usually touches on several aspects of the material. Due to that, an individual note might comment on the animation and placement of props, which needs to be assigned to two different departments (e.g., to Animation and Final Layout). Therefore, the problem is a multi-label text classification task.[7] However, as a preliminary step in the use of LLMs in this context, we have limited our scope to notes assigned exclusively to one department, thereby transforming it into a multi-class text classification problem.[8]

The notes in this chapter bear significant differences from those found in publicly available datasets related to critical reviews of released movies. The latter are often used for sentiment analysis tasks,[6] wherein movie reviews are classified as positive or negative. In contrast, our dataset contains comments that focus on specific elements of a movie, such as a shot or a character, and provides instructions on what to do and how to approach it. A few examples from our dataset are presented in Table 1. For the sake of brevity, only a portion of each note is displayed in the table.

To the best of our knowledge, there are no recent publications about the application of text classification techniques using LLMs for review notes in

Table 1. Examples of review notes and target departments.

Note text	Destination Departments
When she opens the eyes widely, the eyelashes are penetrating the eyelid.	CFX
Have the girl's expression with a warmer smile, brighter expression.	Animation
Move the stickers down on the corner of the building so it feels a human person placed them.	Final Layout

movie production. The content of these notes is highly confidential since it pertains to significant intellectual property. As a result, the dataset used in this research cannot be made publicly available.

This proposal builds on top of our recent previous work,[9,10] where we presented a comparison of several transformer-based language models and methods for their application in classifying movie review notes. In this chapter, we extend our work by exploring the use of LLMs (Llama 2 7B model) through both zero- and one-shot learning and by applying fine-tuning using QLoRA. We assess the performance of these approaches in the task of movie review notes classification.

2. Proposed Method Overview

Fig. 1 depicts a complete overview of the primary processes created for the assignment of movie review notes among departments within an animation studio. The diagram shows the data transformations and processes executed to achieve the final outcome. More detailed information about the Movie Production Tracking, Data Acquisition, CSV Documents, Label Estimation and Tokenization can be found in our previous work[9,10].

3. Data Filtering

As mentioned in Sec. 1, during a review meeting many comments are not confined to a single topic or discipline and can encompass various subjects. As a result, the notes often need to be distributed to several departments. This scenario renders the label estimation stage a multi-label task. However, as an initial step towards solving the complete task, this research focuses on a subset of notes - those directed solely to one department. This approach transforms the issue into a multi-class task. This process is carried out in the data filtering stage. After filtering, the dataset comprises a total of 27,884 samples. The distribution of labels is presented in Table 2.

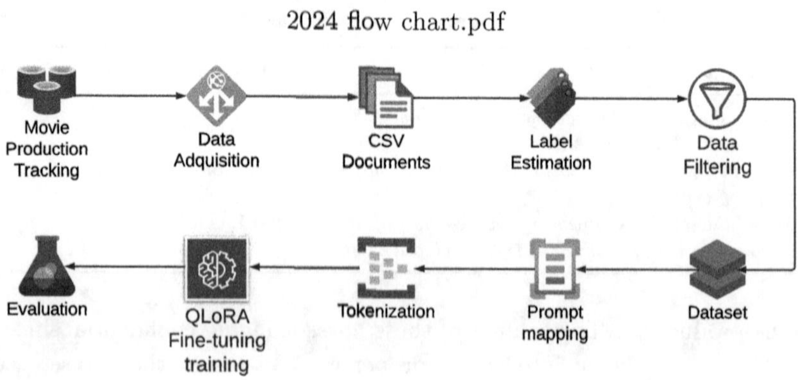

Fig. 1. Flow chart of the proposed approach for automatic distribution of review notes in movie production.

Table 2. Label distribution of filtered movie review notes multi-class dataset.

CFX	Final Layout	Modelling	Layout	Fx	Lighting	Compo	Animation
29.81%	11.42%	5.20%	0.42%	3.04%	15.43%	0.55%	30.24%

4. Prompt Mapping

LLMs like Llama 2[11] are trained and evaluated via prompts that specify the task at hand. Given that the model will be used with prompts, its training should also be structured similarly. The prompts used for training share the same structure as those that will be used upon its deployment, ensuring optimized results. At this stage, the dataset is transformed to integrate the note text into a prompt template, containing the description of the task and the list of departments required for classification. As the Llama 2 model is being trained, the prompt template should conform to the Llama 2 prompt format as outlined in the model's documentation[11].

5. QLoRA Fine-Tuning Training

Fine-tuning a language model is computationally cheap compared to performing training from scratch. However, it can still be prohibitively expensive for current Large Language Models, like Llama 2, used in this research.

LoRA[12] has been used to be able to train the model on a RTX 3090 GPU. The idea behind LoRA is to approximate the fine-tuned weight update with a Rank Decomposition Matrix, decomposed as the product of two smaller matrices. This reduces the number of trainable parameters significantly as LoRA only optimizes the Rank Decomposition Matrix instead of all the parameters in the LLM. Hugging Face Parameter Efficient Fine-Tuning (PEFT) library was used in this research to train the model with LoRA, with the following weight updating:

$$\mathrm{W_{ft}} = \mathrm{W_{pt}} + \Delta\mathrm{W} = \mathrm{W_{pt}} + \frac{\alpha}{\mathrm{r}}\mathrm{AB} \tag{1}$$

where W_{pt} are the weights of the pre-trained model, ΔW is the update to the model weights, AB is the Rank Decomposition Matrix, r is the dimension of the smaller matrices and α is a scaling factor used to balance the importance of the pre-trained weights and the updated ones.

To improve even more the performance, QLoRA [13] was used to train the model. QLoRA is a version of LoRA where the pretrained model weights are loaded as quantized 4-bit weights, instead of the original 8-bits format. QLoRA was incorporated by using the bitsandbytes library with PEFT.

6. Experiments and Results

To evaluate the effectiveness of the proposed method, the fine-tuned LLM model was compared with the model based on one-shot learning. We used the variant of the Llama 2 model with 7 billion parameters for this purpose.

To classify the review notes using the pre-trained model, in-context learning was applied. The prompt begins with some explanatory text outlining the task, followed by a list of departments for text classification. Zero-shot inference yielded very low accuracy, so one-shot inference was applied, adding one example of note text for each of the departments, in compliance with the Llama 2 prompt format. Few-shot learning was not considered, as the number of departments and the length of the notes would result in a prompt exceeding the model's context length limit, which stands at 4k for Llama 2. The template employed for one-shot learning is depicted in Fig. 2, where {Example Note text} is replaced by the text of one of the notes from the train dataset, {department} is replaced by the label of that note, and {Input Note text} is replaced by the text of the note from the test dataset under evaluation. An example note text is added for each department to the prompt.

The fine-tuning was performed with an initial learning rate of 10^{-4}. The LoRA r and α hyperparameters were respectively set to 16 and 64. The model comprised a total of 3,540,389,888 parameters, of which 39,976,960

```
Categorize the movie review note text into one of the 8 categories:
CFX, Final Layout, Modelling, Layout, FX, Lighting, Compo, Animation

Examples:
Input:
{Example Note text}
### Response: {department}
...
Input:
{Input Note text}
### Response:
```

Fig. 2. Template used to implement one-shot learning, as recommended in [11].

were trainable. A paged optimizer was employed during the training process to avoid memory spikes and out-of-memory errors. The training was performed on a local computer equipped with an RTX 3090 GPU with 24GB of VRAM, completing in 30 minutes. The SciKit-learn library was used to generate the metrics (accuracy and F1-score). The prompt used for fine-tuning is the same used in one-shot learning, depicted in Fig. 2, removing the examples.

To ensure consistency in the results, both approaches were tested using the same dataset. The filtered dataset, containing 27884 notes, was divided such that 5% was used for testing, with the remainder allocated for training and validation. The graph depicting the accuracy and F1-score values obtained from evaluating both the one-shot model and the fine-tuned model per label is presented in Fig. 3.

From the data, it is evident that fine-tuning the model significantly enhances performance when compared to the use of the pre-trained model with one-shot inference. The average F1-score, which weights all labels with the quantity of notes assigned to each label, increases from 0.45 with the pre-trained model to 0.85 with the fine-tuned model.

The per-label metrics indicate that one-shot learning performs best with widely recognized departments (e.g., Animation or Lighting), about which the model may contain prior knowledge from the pre-training datasets. However, for more specialized departments such as Final Layout, the results are considerably inferior compared to those obtained with the fine-tuned model. In any case, the fine-tuned model yields inferior results for departments with a limited number of samples, though still superior to the one-shot alternative.

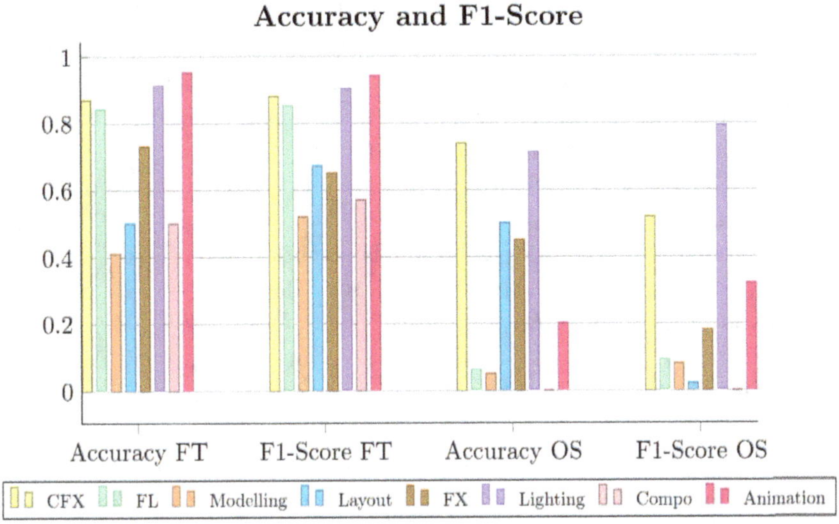

Fig. 3. Accuracy and F1-Score per label for fine-tuned (FT) and one-shot (OS) models.

7. Conclusion and Future Work

This chapter concludes that movie review notes can be automatically assigned using the proposed approach of fine-tuning an LLM, such as Llama 2, on the specific review notes dataset. Furthermore, it demonstrates that the fine-tuned model significantly outperforms the pre-trained model with its default weights, even when one example per label is included within the supplied prompt.

As future works, we want to augment this approach with the complete dataset containing the multi-label review notes. Also, the dataset can be improved by balancing it and improving the label estimation process. Besides, hyper-parameters could be optimized to improve the accuracy. Comparing the fine-tuned model against a few-shot method with the pre-trained model could also give more information about the training needs of the model with this type of dataset. Exploring other LLMs like Falcon,[14] GPT-4,[15] or Mistral[16] and comparing their results and resource consumption are future lines of research.

References

1. A. Gasparetto, M. Marcuzzo, A. Zangari and A. Albarelli, A Survey on Text Classification Algorithms: From Text to Predictions, *Information* **13** (2022).

2. S. Minaee *et al.*, Deep Learning–Based Text Classification: A Comprehensive Review, *ACM Computing Surveys* **54** (2021).
3. L. Feng, J. Senapati and B. Liu, TaDaa: real time Ticket Assignment Deep learning Auto Advisor for customer support, help desk, and issue ticketing systems, *arXiv:2207.11187* (2022).
4. T. Martin, The Reuters Dataset https://martin-thoma.com/nlp-reuters, (2017).
5. M. Rahman and Y. Akter, Topic Classification from Text Using Decision Tree, K-NN and Multinomial Naïve Bayes, in *1st Int. Conf. on Advances in Science, Eng. and Rob. Tech. (ICASERT)*, 2019.
6. L. Zhang, S. Wang and B. Liu, Deep Learning for Sentiment Analysis : A Survey, *WIREs Data Mining and Knowledge Discovery* **8**, p. e1253 (2018).
7. M.-L. Zhang and Z.-H. Zhou, A Review on Multi-Label Learning Algorithms, *IEEE Transactions on Knowledge and Data Engineering* **26**, 1819 (2014).
8. E. Lee, C. Lee and S. Ahn, Comparative study of multiclass text classification in research proposals using pretrained language models, *Applied Sciences* **12** (2022).
9. D. Garcés, M. Santos and D. Fernández-Llorca, Language models for automatic distribution of review notes in movie production, in *Intelligent Data Engineering and Automated Learning (IDEAL)*, 2023.
10. D. Garcés, M. Santos and D. Fernández-Llorca, Text classification for automatic distribution of review notes in movie production, in *18th Int. Conf. Soft Comp. Models Ind. Env. Apps. (SOCO)*, 2023.
11. H. Touvron *et al.*, Llama 2: Open foundation and fine-tuned chat models, *arXiv 2307.09288* (2023).
12. E. J. Hu *et al.*, LoRA: Low-rank adaptation of large language models, in *International Conference on Learning Representations*, 2022.
13. T. Dettmers, A. Pagnoni, A. Holtzman and L. Zettlemoyer, QLoRA: Efficient Finetuning of Quantized LLMs, *arXiv preprint arXiv:2305.14314* (2023).
14. G. Penedo *et al.*, The RefinedWeb Dataset for Falcon LLM: Outperforming Curated Corpora with Web Data, and Web Data Only, *arXiv:2306.01116* (2023).
15. Y. Liu *et al.*, Summary of chatgpt/gpt-4 research and perspective towards the future of large language models, *arXiv:2304.01852* (2023).
16. A. Q. Jiang *et al.*, Mistral 7B, *arXiv 2310.06825* (2023).

Analyzing Dispersion Measures for Ordinal Variables

N. Martínez*,‡, J. Montero*,§ and D. Gómez†,¶

*Facultad de Ciencias Matemáticas, Universidad Complutense de Madrid
Madrid 28040, Spain

†Facultad de Estudios Estadísticos, Universidad Complutense de Madrid
Madrid 28040, Spain

‡nuriamm77@hotmail.com

§monty@mat.ucm.es

¶dagomez@estad.ucm.es

One of the most relevant concepts in statistics is data variability or dispersion. We can find a wide number of studies related to the measurement of dispersion for quantitative data, however, the measurement of dispersion for qualitative data is being poorly developed despite the increasing weight in the Science of linguistic terms to manage information. In fact, only a few measures can be found within a qualitative framework, and their properties have not received much attention. In this chapter, we stress this terrible theoretical lack, exploring a desirable set of properties to be accomplished by an ordinal dispersion measure that allows a structured view of most existing ordinal dispersion measures and a better understanding of ordinal dispersion.

Keywords: Analytical statistics; computing for statistics; decision procedures and theory; data analysis; fuzzy sets.

1. Introduction

Variability, dispersion and variance are key concepts in statistics. For continuous statistical variables, there are clear dispersion measures, such as standard deviation, variance or quasi-variance, which have been defined and studied in-depth over the years. Although there are many disciplines which work with ordinal variables as a basis such as psychology, economics or marketing research, the situation is very different when willing to measure the dispersion of these ordinal variables. Even today, there is no standardized metric to calculate the dispersion of ordinal variables (like, for example, a Likert scale). One way to handle the dispersion of ordinal variables is often through the entropy.[1] We include it in the analysis carried out

‡Corresponding author.

in this chapter to show how entropy is an appropriate measure for nominal variables but not for ordinal ones. Another common approach is to assign numbers to each ordinal category and convert the ordinal estimation into a numerical one. Classical dispersion measures, are then applied. Some authors[2,3] have stated that this procedure is undesirable because it can lead to misunderstandings. Therefore, they defined or reviewed ordinal dispersion measures[2,3] that deal with ordinal statistical variables better than other classical measures, such as entropy, standard deviation, variance or quasivariance. This chapter is organized as follows: the following section is dedicated to remind the notion of ordinal statistical variables and to list the most relevant ordinal dispersion measures found over the state of the art. In Section 3, we propose the set of properties to be accomplished by an ordinal dispersion measure and we analyze these properties on the key ordinal measures. We conclude this chapter with some conclusions and final remarks.

2. Preliminaries

In this section we recall the concept of ordinal statistical variables and list the most relevant dispersion measures for this kind of variables:

2.1. *Ordinal statistical variable*

An ordinal statistical variable is a qualitative statistical variable X which takes non-numeric values amongst which an ordering criteria can be applied. Note that when there is no order, these are called nominal statistical variables.

Given a statistical ordinal variable X with k categories, we define:

- $A_x = \{a_i, i \in \{1, \ldots, k\}\}$ as the scale or set of labels that X can take.
- $L_x = \{i, i \in \{1, \ldots, k\}\}$ as the set of rankings associated with the labels in the scale.

It is important to note that the numerical assignment applied to the labels refers exclusively to the position of these labels in the scale, but there is no relationship between these numbers and the distance defined amongst categories.

In this paper, for the sake of simplicity, we will work with an ordinal variable X with k categories ($k \in \mathbb{N}$ and $k > 1$) labeled as $\{1, 2, \ldots, k\}$

defined through a vector of relative frequencies $f = (f_1, \dots, f_k)$ or $F = (F_1, F_2, \dots, 1)$ the vector of accumulated frequencies.

2.2. *Ordinal dispersion measure*

There are several authors[4–6] who have analyzed the possible ways to approach dispersion measures on ordinal variables trying to give a formal definition. However, as it is pointed out in Lehman[7], after Leik's dispersion measure[8] in 1966, no other ordinal dispersion measures can be found in the literature until 1992 when a new one appeared[9]. And even since then, there are not many proposals and there is no single definition that has been universally accepted. Some authors[3,10] show extensive summaries of the existing measures up to 2015. We found few other authors (see García-Lapresta & Borge[11] amongst others), who have worked on this topic after that date.

Regarding the set of desirable properties that a dispersion measure should accomplish, although some authors mention the basic property of scale invariance[2,12], we have found little literature[3,13] where the properties have been studied and even then, the analysis carried out are vague and not rigorously defined.

In a general and mathematical way, given an ordinal statistical variable X, an ordinal dispersion measure D can be viewed as:

$$\begin{aligned} D : \mathcal{X} &\longrightarrow \mathbb{R}^+ \\ f &\mapsto D(f) \end{aligned} \tag{1}$$

where $\mathcal{X} = \{f = (f_1, \dots, f_k) \in [0,1]^k \text{ with } \sum_{l=1}^{k} f_l = 1\}$.

Remark 2.1. There are authors[8,9] that define dispersion measures in terms of absolute frequencies. For these cases, we have rewritten those expressions in terms of relative frequencies as per the previous definition.

Also, let us note that there are some ordinal dispersion measures which provide a qualitative output[5,11] and obviously do not match the previous definition. In this paper however we have focused only on dispersion measures with quantitative outputs.

2.3. *Most relevant ordinal dispersion measures*

We will list here the most relevant ordinal dispersion measures found in the state-of-the-art. An extensive analysis was carried out in Ref.[14]. In Figure 1 we show some of the key ordinal dispersion measures:

Short Name	Name, Author & Year	Formula
LOV	Leik's Ordinal Variation Leik, 1966	$LOV(f) = \frac{\sum_{i=1}^{k} d_i(f)}{\underset{g\in[0,1]^k}{Max} \sum_{i=1}^{k} d_i(g)} = \frac{2}{(k-1)}\sum_{i=1}^{k} d_i(f)$ con $d_i(f) = \begin{cases} F_i = \sum_{i=1}^{i} f_i & \text{si } F_i \le 1/2, \\ 1-F_i = 1-\sum_{i=1}^{i} f_i & \text{en otro caso.} \end{cases}$
$IOV_{\_N}$	Index of Ordinal Variation[a] Berry & Mielke, 1992	$IOV_{\_N}(f) = \frac{4}{k-1}\sum_{i<j} f_i f_j (j-i)$ con $1 \le i, j \le k$
COV	Coefficient of Ordinal Variation Kvalseth, 1995	$COV(f) = 1-\sqrt{(1-\frac{4}{(k-1)}\sum_{i<j} f_i f_j \lvert i-j \rvert)}$
Disensus	Tastle & Wierman, 2007	$Dsn_{L,X}(f) = -\sum_{i=1}^{k} f_i \log_2(1-\frac{\lvert i-\mu_X \rvert}{k-1})$, con $\mu_X(f) = \sum_{i=1}^{k} f_i\, i$
Entropy	Shannon, 1948	$H_{X,L}(f) = -\sum_{i=1}^{k} f_i \log_2(f_i)$

Fig. 1. Ordinal Dispersion Measures [1,8,9,12,13]

3. Properties and Analysis of Ordinal Dispersion Measures

Given X an ordinal statistical variable with k categories $k > 1$ and an ordinal dispersion measure D defined as (1), we define the following properties:

Definition 3.1. (Property 1) We will say that a dispersion measure D is an **Index** if D takes values on the bounded interval $[0, 1] \in \mathbb{R}$.

Researchers agree on the fact that the minimal dispersion takes place when all the sample is concentrated on a single category, and in this case, the dispersion measure takes the value 0. However, it is less clear when getting into the definition of maximal dispersion. In this section, we will define 2 different classes of boundary conditions that differentiate nominal from ordinal variables.

Definition 3.2. (Property 2.1) We will say that a dispersion measure D satisfies the **uniform or spread boundary conditions** when the following holds:

- $D(f_1, f_2, \ldots, f_k) = 0 \iff \exists i \in \{1, \ldots, k\}$ so that $f_i = 1$ and $f_j = 0\, \forall j \in \{1, \ldots, k\}, j \neq i$.
- $D(f_1, f_2, \ldots, f_k) = 1 \iff f = \left(\frac{1}{k}, \frac{1}{k}, \ldots, \frac{1}{k}\right)$ (uniform vector).

Definition 3.3. (Property 2.2) We will say that a dispersion measure D satisfies the **polar boundary conditions** when the following holds:

- $D(f_1, f_2, \ldots, f_k) = 0 \iff \exists i \in \{1, \ldots, k\}$ so that $f_i = 1$ and $f_j = 0\,\forall j \in \{1, \ldots, k\}$, $j \neq i$.
- $D(f_1, f_2, \ldots, f_k) = 1 \iff f = \left(\frac{1}{2}, 0, \ldots, 0, \frac{1}{2}\right)$ (polar vector).

Definition 3.4. (Property 3) A dispersion measure D satisfies the property of **permutation invariance** if and only if $D(f) = D(\Pi(f))$ for all Π permutations of k elements, $f \in \mathcal{X}$ vector of relative frequencies.

Note that the dispersion measures that satisfy this property are valid for qualitative nominal variables but not for qualitative ordinal variables, so it is not a desirable property for ordinal variables.

Kvalseth[12] proposed a new dispersion measure (denoted as COV) because he detected that the previously defined IOV dispersion measure[9] was biased when evaluating intermediate values. In order to differentiate these two measures, he proposed the study of monotony in $k = 2$ that can be formalized as follows:

Given an ordinal variable X with $k = 2$ categories, the vector of relative frequencies can be expressed in terms of a single parameter α, $\alpha \in [0, \frac{1}{2}]$ with $f = (\alpha, 1 - \alpha)$ and $F = (\alpha, 1)$. Let us note that dispersion measures can be expressed for this situation as unidimensional functions defined in $\alpha \in [0, 1/2]$.

Definition 3.5. (Property 4) Given an ordinal variable X with $k = 2$ categories, we will say that a dispersion measure D is:

- **Linear**, if $D(\alpha)$ is a linear function.
- **Concave**, if $D(\alpha)$ is a concave function.
- **Convex**, if $D(\alpha)$ is a convex function.

The property of invariance to complementary distributions has been introduced in a non-rigorous way[3,15]. In this paper, we formalize this property. In order to do so, we first need to define what complementary distributions are:

Definition 3.6. Given two vectors of relative frequencies f and f' with $f, f' \in \mathcal{X}$, we say that f and f' are **complementary** if $f'_i = f_{k-i+1}$ for all $i \in \{1, \ldots, k\}$. Note that this means that $f' = (f'_1, f'_2, \ldots, f'_k) = (f_k, f_{k-1}, \ldots, f_1)$.

Definition 3.7. (Property 5) A dispersion measure D satisfies the property of being **invariant to complementary distributions** when $D(f) = D(f')$ for all f, f' complementary vectors.

The following property captures the intuitive idea that any ordinal dispersion measure should satisfy: when the frequency of a certain category (or part of it) is evenly split into adjacent categories, the dispersion of the new vector is higher. For example, suppose that category 3 has 20% of the population and 10% is moved to category 2 and the remaining 10% is moved to category 4. Then the dispersion of the new distribution should be higher. This can be formalized as follows:

Definition 3.8. (Property 6) Given $f \in \mathcal{X}$ a vector of relative frequencies, $r \in \mathbb{N}$ with $1 < r < k$, and $\alpha \in \mathbb{R}$ with $0 \leq \alpha \leq f_r$, let $f' \in \mathcal{X}$ be defined as:

$$f'_i = \begin{cases} f_i + \frac{\alpha}{2} & \text{if } i = r-1 \text{ or } i = r+1, \\ f_i - \alpha & \text{if } i = r, \\ f_i & \text{otherwise.} \end{cases}$$

A dispersion measure D satisfies the **mitosis** property if $D(f) \leq D(f')$.

The **translations invariance** property is a very common and standard one on dispersion measures of quantitative variables, however, translating this property to the context of qualitative variables is not a trivial task. In the following definition we formalize the idea presented vaguely in Ref.[3].

Definition 3.9. Given a parameter $a \in \mathbb{N}$ and given $f \in \mathcal{X}$ a vector of relative frequencies with $f_j = 0$ for all $j \in \{k-a+1, \ldots, k\}$, we define $f(a)$ as the result of the translation of a positions in f when $f(a)_j = 0$ for all $j \in \{1, \ldots, a\}$ and $f(a)_i = f_{i-a}$ for all $i \in \{a+1, \ldots, k\}$.

Definition 3.10. (Property 7) A dispersion measure D satisfies the property of **translations invariance** when $D(f) = D(f(a))$ for all $f \in \mathcal{X}$ and $f(a)$ the resulting vector of the translation of a positions in f.

There are several research studies[16,17] that focus on the values that extreme categories take, therefore it seems interesting to analyze how dispersion measures take into account these values. The proposed property analyzes the impact of extreme values on dispersion measures when working with symmetric distributions, i.e. distributions whose measure of central tendency lies on the central category of the scale.

Definition 3.11. The **unique median of** X is defined as $Med_u(X) = min\{j, j \in \{1, \ldots, k\}$ so that $F_j > \frac{1}{2}$ and $F_i \leq \frac{1}{2}$ for all $i < j\}$.

Definition 3.12. Let $f \in \mathcal{X}$ be a vector of relative distributions of X. We will say that f **is a symmetric distribution** if k is odd and $Med_u(f) = j = \frac{k+1}{2}$.

Definition 3.13. (Property 8) Let $f \in \mathcal{X}$ vector of relative frequencies be a symmetric distribution. Given $\alpha \in \mathbb{R}$ so that $f^\alpha = (f_1 - \alpha, f_2, \ldots, f_{k-1}, f_k + \alpha) \in [0,1]^k$ is also a symmetric distribution. A dispersion measure D is **independent towards extreme values on symmetric distributions** if $D(f) = D(f^\alpha)$.

To conclude this section we include a table summarizing the analysis of the presented properties on the most relevant dispersion metrics (see 2.3).

PROPERTY	LOV	IOV_N	COV	Disensus	Entropy
1. Index	Yes	Yes	Yes	Yes	**No**
2. Boundary Conditions	Polar	Polar	Polar	Polar	**Spread**
3. Permutation Invariance	No	No	No	No	**Yes**
4. Monotonicity in $k=2$	Linear	Non-linear, increasing and concave	Linear	Non-linear, increasing and concave	Non-linear, increasing and concave
5. Invariant to complementary distributions	Yes	Yes	Yes	**No**	Yes
6. Mitosis	Yes	Yes	Yes	No	No
7. Translations Invariance	Yes	Yes	Yes	**No**	Yes
8. Independency of extreme values on symmetric dist.	**Yes**	No	No	No	No

4. Conclusions and Final Remarks

Many authors[3,5,11] expressed the need to identify some desirable properties that any dispersion measure should satisfy. So, with the main objective of analyzing the most well-known ordinal dispersion measures, in this chapter, we have tried to establish a series of properties that have allowed us to identify differences among them. It is important to emphasize that some of the properties defined in this chapter, such as properties 4, 5 or 7, are based on existing vague concepts that were not rigorously formalized. Other properties, such as properties 6 or 8, are new concepts that allow us to understand better the behavior of ordinal dispersion measures.

References

1. C. E. Shannon, A mathematical theory of communication, *The Bell System Technical Journal* **27**, 379 (1948).
2. J. Blair and M. G. Lacy, Measures of variation for ordinal data as functions of the cumulative distribution, *Perceptual & Motor Skills* **82**, 411 (1996).
3. T. Gadrich and E. Bashkansky, Ordanova: analysis of ordinal variation, *Journal of Statistical Planning and Inference* **142**, 3174 (2012).

4. J. Blair and M. G. Lacy, Statistics of ordinal variation, *Sociological Methods & Research* **28**, 251 (2000).
5. F. Franceschini, M. Galetto and M. Varetto, Qualitative ordinal scales: the concept of ordinal range, *Quality Engineering* **16**, 515 (2004).
6. E. Bashkansky and T. Gadrich, Evaluating quality measured on a ternary ordinal scale, *Quality and Reliability Engineering International* **24**, 957 (2008).
7. R. S. Lehman, *Statistics and Research Design in the Behavioral Sciences.* (Wadsworth/Thomson Learning, 1991).
8. R. K. Leik, A measure of ordinal consensus, *Pacific Sociological Review* **9**, 85 (1966).
9. K. J. Berry and P. W. Mielke Jr, Assessment of variation in ordinal data, *Perceptual and Motor Skills* **74**, 63 (1992).
10. T. Gadrich, E. Bashkansky and R. Zitikis, Assessing variation: a unifying approach for all scales of measurement, *Quality & Quantity* **49**, 1145 (2015).
11. J. L. García-Lapresta and L. Borge, Measuring dispersion in the context of ordered qualitative scales, *The Mathematics of the Uncertain: A Tribute to Pedro Gil* , 605 (2018).
12. T. O. Kvålseth, Coefficients of variation for nominal and ordinal categorical data, *Perceptual and Motor Skills* **80**, 843 (1995).
13. W. J. Tastle and M. J. Wierman, Consensus and dissention: A measure of ordinal dispersion, *International Journal of Approximate Reasoning* **45**, 531 (2007).
14. N. Martínez, D. Gómez, P. Olaso, K. Rojas and J. Montero, A novel ordered weighted averaging weight determination based on ordinal dispersion, *International Journal of Intelligent Systems* **34**, 2291 (2019).
15. T. Gadrich and Y. N. Marmor, Two-way ordanova: analyzing ordinal variation in a cross-balanced design, *Journal of Statistical Planning and Inference* **215**, 330 (2021).
16. G. Albaum and B. Murphy, Extreme response on a likert scale, *Psychological Reports* **63**, 501 (1988).
17. I. Clarke, Extreme response style in cross-cultural research: An empirical investigation, *Journal of Social Behavior and Personality* **15**, p. 137 (2000).

Anomaly Detection in the Uniaxial Vibration in Wind Turbines Applying Linear Regression and Z-score

D. Coronel[*,§], C. Guevara[†,¶] and M. Santos[‡,||]

[*]*Computer Science Faculty, Complutense University of Madrid*
Madrid 2840, Spain

[†]*Quantitative Methods Department, CUNEF University*
Madrid 28040, Spain

[‡]*Institute of Knowledge Technology, Complutense University of Madrid*
Madrid 2840, Spain

[§]*dcoron01@ucm.es*
[¶]*cesar.guevara@cunef.edu*
[||]*msantos@ucm.es*

Currently, condition monitoring and anomaly detection in wind turbines show great promise for research. Wind energy offers notable benefits, such as sustainability and a low environmental impact. However, wind turbines encounter significant challenges, including operational issues and high maintenance costs. Therefore, this chapter focuses on anomaly detection in wind turbines using Supervisory Control and Data Acquisition (SCADA) vibration data. The proposed approach involves an iterative algorithm that enhances robustness and adaptability by continually adjusting thresholds through a combination of linear regression and z-score. The proposed algorithm detected 246 anomalies with a precision rate of 71.13% and an error rate of 28.87%.

Keywords: Wind turbine; Z-score; linear regression; anomalies detection; neural networks.

1. Introduction

Condition monitoring and anomaly detection in wind turbines have emerged as highly promising research areas [1]. Between 2000 and 2018, wind power accounted for 4.8% of total electricity production and is projected to reach 12% by 2040. However, wind energy faces several challenges, particularly in operation and maintenance (O&M) costs [2]. These maintenance expenses represent

[§]Corresponding author.

between 20% and 30% of the total electricity generated over the lifetime of the project [3],[4].

Critical failures in wind turbines result from their prolonged exposure to harsh environments [5],[6]. Examples of O&M research on wind turbines can be found in Phong [7], who introduced an approach for monitoring and diagnosing wind turbine (WT) generator gearbox and bearing failures. This method uses the Cumulative Sum-centered Multiple Linear Regression (CUSUM) model with Supervisory Control and Data Acquisition (SCADA) data of wind turbine temperature and wind speed, collected from a wind farm in Poland. The proposed model achieved an accuracy of 95.5% in fault detection. The article by Amin [8] developed a predictive model for detecting anomalies in the gondola of wind turbines. It is based on the application of an MLP neural network with two hidden layers and multilayer linear regression method. The data used come from a SCADA system, primarily wind speed, power, and temperature. These data were collected from the Westmill wind farm in Swindon, UK. The results obtained with this model demonstrate a remarkable 99.8% accuracy in predicting anomalies.

So, research in this field is of vital importance as it contributes significantly to reduce economic losses and fosters the development of the wind industry. Thus, the main objective of this chapter is to develop an algorithm for efficiently identifying anomalies in wind turbines by using SCADA data related to vibration signals. The implementation of this iterative approach will establish a robust foundation for consistently fine-tuning the model, enabling it to accommodate fluctuations inherent in the data. This chapter uses historical blade vibration data, utilizing five-data windows. This approach works by continually adjusting thresholds through a combination of linear regression and z-score. Although initially it only detects failures, working with the models will allow us forecasting.

This document is structured as follows. Section 2 presents the methods and materials used in the research. Section 4 elaborates on the development of the proposed algorithm. Section 5 presents the results obtained and compares them with other relevant works. In Section 6, conclusions are presented and potential directions for future work are discussed.

2. Materials and Methods

2.1. *Wind turbine vibration database*

The database used for this study comprises uniaxial vibration data collected from wind turbines. This dataset, compiled by Mustansiriyah University [9], includes multiple sets of vibration data corresponding to WT in both healthy and faulty conditions. The faulty conditions in the datasets include blade cracks, blade surface degradation, and blade imbalance. In total, the database has 35 datasets collected at various wind speeds, with each dataset containing 500 records.

In this study, only blade surface cracking are used. The variables considered include time, in seconds (s), and vibration, measured in acceleration units of gravity (g), with 1g equivalent to 9.80665 m/s². Table I presents the variables extracted from the wind turbine vibration database. In the database, vibrations are given with six decimal digits although in our study we use three decimal in accordance with actual sensor precision.

The data are divided into 5-log data windows, enabling the implementation of an iterative approach to analyze time segments and variations in blade vibration.

Tabla I Description of the wind turbine vibration database.

ATTRIBUTE NAME	DATA TYPE	DECIMALS	EXAMPLE
Vibration (g)	Numeric	6	0.007209
Time (s)	Numeric	3	0.100

Figure 1 displays a scatter plot containing 500 data points within a time window ranging from 0 to 0.499 seconds. The plot shows the vibration of the wind turbine blade. Each black dot represents an individual measurement, with a maximum vibration of 0.018g, a minimum of -0.012g, and a mean of 0.0005g.

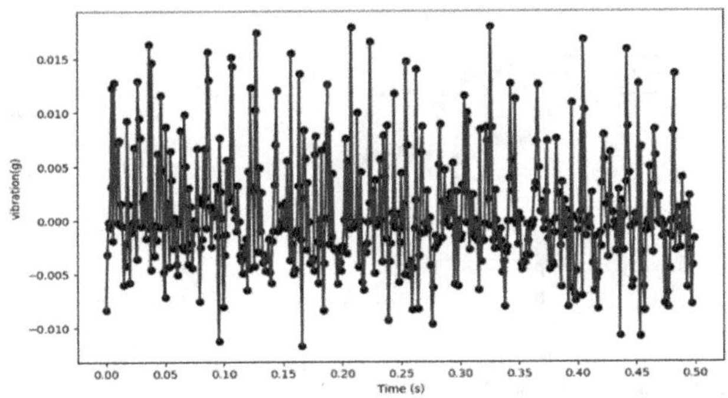

Fig. 1 Wind turbine rotor vibration dispersion diagram.

2.2. *Linear regression*

Linear regression will be used. It is a statistical technique that examines the correlation between an independent and a dependent variable. Typically it is used for predicting or forecasting trends within a dataset. Eq. (1) represents the mathematical expression of linear regression.

$$Y = B_0 + \sum B_I X_i + \epsilon_i \quad (1)$$

Where *Y* denotes the dependent variable, *X_i* represents the independent variables, ∈ represents the error values, and *B* is the linear parameter estimates that need to be calculated.

2.3. *Z-score*

Z-score is a statistical measure that quantifies the dispersion or variability of a data set Eq. (2). In general, this method indicates how far individual values are on average from the mean of the data set. Any point outside the specific threshold is considered an outlier.

$$Z = \left[\frac{x - \mu}{\sigma}\right] \quad (2)$$

Were x represents the sample dataset, σ is the standard deviation of the samples, and μ denotes the mean of the dataset [10].

3. Proposed Anomaly Detection Algorithm

An anomaly detection algorithm has been designed. This algorithm relies on the combined use of linear regression and z-score techniques to detect and identify potential anomalous data points. It follows an iterative approach, where each phase feeds and establishes a new threshold for the subsequent data. The steps of the algorithm which pseudo code is presented in Algorithm 1 are as follows.

Algorithm 1: Anomaly Detector Pseudo Code	
Function DetectAnomalies(Data)	# Function Initialization
Read(Data)	# Data Reading
AnomalyList = []	# Initialize List to Store Anomalies
sizeGroup = 5	# Initialize size of group
for i in range(0, len(datos), sizeGroup):	# Iteration over the Data
DataGrup = datos.iloc[i:i + sizeGroup]	# Extract data for the current group
if (i == 0):	# Check if it is the first group
FutureValue = linearRegression(DataGrup)	# Calculate future value using linear regression

```
    Threshold = CalculateThreshold(DataGrup, FutureValue)       # Calculate Threshold
  else:                                                         # For subsequent groups
    μ = Mean(DataGrup)                                          # Calculate the mean
    σ = StandardDeviation(DataGrup)                             # Calculate the standard
                                                                deviation
    z_score = (DataGrup - μ) / σ                                # Calculate Z-Score
    IsAnomaly = (z_score > Threshold) | (z_score < -Threshold)  # Identify Outlying Values
    AnomalyList.append(IsAnomaly)                               # Store Anomaly Status in the
                                                                List
    FutureValue = linearRegression(DataGrup)                    # Calculate future value using
                                                                linear regression
    Threshold = CalculateThreshold(DataGrup, FutureValue)       # Calculate Threshold
  Return AnomalyList                                            # Return List of Anomalies
End Function
```

3.1. *Phase 1: Data division into groups*

In the initial phase, the data are split into groups, each one of 5 records. Each group represents a specific time interval for the analysis. The first group establishes the forward-looking value as a starting point for the next group. This iteration of groups is repeated until all data have been considered.

3.2. *Phase 2: Prediction of future data*

In this phase, the future vibration value (FutureValue) is determined by applying a linear regression model to the current data series (dataActual). This model is trained using time as the independent variable and the vibration of the current dataset as the dependent variable. Once the model is trained, a prediction of the next vibration value is generated. This predicted value serves as an indicator of how the data series is expected to evolve in the next group.

3.3. *Phase 3: Threshold generation*

Once the future value of the vibration (FutureValue) has been obtained as the result of the linear regression prediction, it is incorporated into the current data segment. After grouping these data, the mean (μ) and standard deviation (σ) are obtained Eq. (3).

$$\sigma = \sqrt{\frac{\Sigma(X_i - \mu)^2}{N}} \tag{3}$$

Where X_i represents the individual value, μ is the mean, and N is the total number of values in the set. The threshold (th) for the next dataset is obtained by Eq. (4).

$$th = \frac{(\text{FutureValue}-\mu)}{\sigma} \tag{4}$$

3.4. *Phase 4: Stopping of anomalies*

With the generated threshold, we generate a new dataset (dataGrup) and calculate the mean μ and standard deviation σ using Eq. (3). Then the algorithm calculates the z-score for each data point in the new dataset using Eq. (5) and compared each record using Eq. (6).

$$z\_score = \frac{(\text{dataGrup}_i - \mu)}{\sigma} \tag{5}$$

$$anomalous = (z\_scores_i) > \text{threshold} \tag{6}$$

Any point outside the specified threshold is considered an outlier. Once the outliers in the current dataset have been detected, each of the steps is repeated until all the data records are compared.

4. Results and Discussion

The results obtained through the implementation of the proposed algorithm are the following: it successfully detected a total of 246 anomalies, with a detection of 71.13%, and an error of 28.87%.

Figure 2 shows the anomalous data represented by red circles. The green line represents the high threshold, maximum of 0.016g, while the orange line denotes the low iterative threshold, with a minimum of -0.012g. The red anomalous data falling within the high and low thresholds are considered false positives.

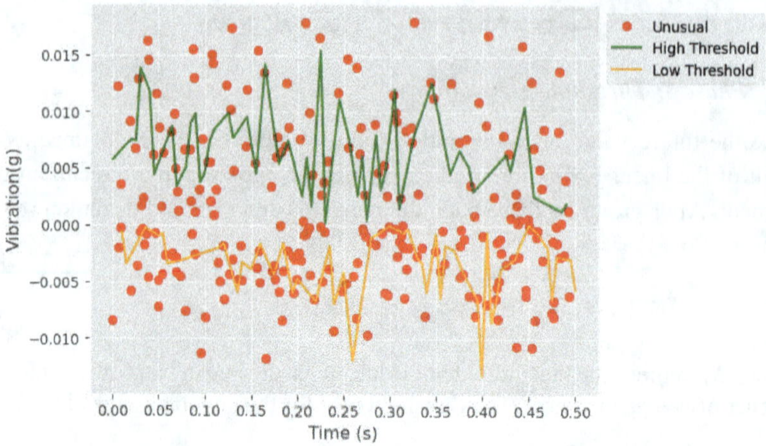

Fig. 2 Scatter plot depicting the results of the proposed algorithm.

In Figure 3, normal data are shown by blue circles. With the same thresholds. blue dots lying outside the high and low thresholds are considered false positives.

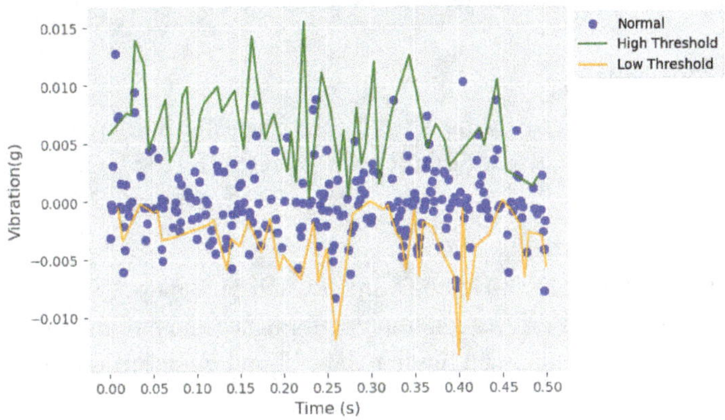

Fig. 3 Scatter plot representing the normal results of the proposed algorithm.

When comparing our results with those presented in the papers by Phong [7] and Amin [8], similarities and differences can be discussed. Both studies, the same as ours, are grounded on the use of real SCADA data, that are more challenging because of the noise and potential measurement errors. However, the mentioned papers achieve an accuracy exceeding 95% and even reaching 99.8%. This may be due to the combination of various methods, including cumulative summation, neural networks, and linear regression, that enables it to effectively capture the inherent complexity of the data. Those researches use a database with a substantial volume of records that is not publicly available. This aspect is crucial as it allows the models to learn more complex patterns and adapts to the data. Our model gives promising results, a 71,13% accuracy with a much smaller dataset. It can be better refined by incorporating new techniques and large datasets.

5. Conclusions and Future Work

In this chapter, an algorithm for anomaly detection in wind turbines has been proposed. The algorithm combines linear regression and z-score techniques, using SCADA data of vibration signals of WT. It has been designed iteratively, involving the segmentation of data into groups, generation of thresholds, and detection of anomalies at each stage.

The integration of these two methods gives good results and proves the feasibility of this algorithm for O&M of WT. Furthermore, the interactivity of the algorithm enhances the adaptability of the model to changes in wind turbine behavior by continuously adjusting thresholds based on predicted values and current conditions.

For future research, we plan the inclusion of additional variables and expanding the dataset to enrich the analysis. Additionally, the integration of different artificial intelligence techniques should be explored.

Acknowledgments

This work has been partially financed by the Spanish Ministry of Science and Innovation with the MCI/AEI/FEDER project PID2021-123543OB-C21.

References

1. R. Pandit, D. Astolfi, J. Hong, D. Infield, and M. Santos, "SCADA data for wind turbine data-driven condition/performance monitoring: A review on state-of-art, challenges and future trends," Wind Engineering, vol. 47, no. 2, pp. 422–441, 2023.
2. R. Pandit, D. Infield, and M. Santos, "Accounting for Environmental Conditions in Data-Driven Wind Turbine Power Models," IEEE Trans Sustain Energy, vol. 14, no. 1, pp. 168–177, 2023.
3. L. Xiang, P. Wang, X. Yang, A. Hu, and H. Su, "Fault detection of wind turbine based on SCADA data analysis using CNN and LSTM with attention mechanism," Measurement, vol. 175, p. 109094, 2021.
4. B. Zhou et al., "Review of Key Technologies for Offshore Floating Wind Power Generation," Energies, vol. 16, no. 2, p. 710, 2023.
5. M.-H. ; Wang et al., "Fault Detection of Wind Turbine Blades Using Multi-Channel CNN," Sustainability, vol. 14, no. 3, p. 1781, Feb. 2022.
6. J.E. Sierra-Garcia et al., "Identification of Vibration Modes in Floating Offshore Wind Turbines," Journal of Marine Science and Engineering, vol. 11, no. 10, p. 1893, Sep. 2023.
7. P. B. Dao, B. Stephen, and M. Benbouzid, "A CUSUM-Based Approach for Condition Monitoring and Fault Diagnosis of Wind Turbines," Energies, vol. 14, no. 11, p. 3236, Jun. 2021.
8. J. Amini, Kanfoud, and T.H. Gan, "An Artificial Intelligence Neural Network Predictive Model for Anomaly Detection and Monitoring of Wind Turbines Using SCADA Data," Applied Artificial Intelligence, vol. 36, no. 1, 2022.
9. F. Ogaili, A. AbdulhadyJaber, and M. N. Hamzah, "Wind Turbine Blades Fault Diagnosis based on Vibration Dataset Analysis," vol. 4, 2023.
10. M. Zou and S. Z. Djokic, "A Review of Approaches for the Detection and Treatment of Outliers in Processing Wind Turbine and Wind Farm Measurements," Energies, vol. 13, no. 16, p. 4228, Aug. 2020.

Research on Mining Negative Online Reviews on E-commerce Platforms Based on Social Network Analysis and LDA Model

Yanrong Huang[*,†,‡], Zhiyi He[*], Han Lv[†] and Jian Min[†,§]

[*]*College of Economics & Management, Zhejiang University of Water Resources and Electric Power, Hangzhou 310018, China*

[†]*School of Management, Wuhan University of Technology Wuhan 430070, China*
[‡]*hyanrong@whu.edu.cn*
[§]*mjsw@whut.edu.cn*

Negative online reviews have become essential decision-making information for businesses. By conducting text mining on negative online reviews of e-commerce platforms to accurately identify problems in online platform transactions, using social network analysis to clarify the correlation between critical factors in negative reviews, and applying the LDA topic model to mine eight significant themes of negative reviews, namely platform rider disputes, education refund difficulties, difficulty in canceling or changing reservations, damage or loss of goods, taxi disputes, payment harassment complaints, platform member disputes, and slow response on customer service. This chapter is of great significance for improving the quality of products and services, enhancing customer satisfaction, and effectively regulating e-commerce platforms by the government.

Keywords: Negative online reviews; text mining; e-commerce platforms; SNA; LDA.

1. Introduction

Online shopping has become a consumption trend, and more and more consumers post text reviews on online platforms to provide feedback on their shopping experiences or service evaluations. The product or service issues consumers face during shopping are often reflected in negative online reviews, which often contain their true feelings and problems encountered. Negative online review text on e-commerce platforms are essential data for providing consumer experiences and feelings feedback, with rich semantic content, such as

[‡]Corresponding author.

people's knowledge, perspectives, and preferences [2,3]. By mining helpful information from negative comments, we can understand the shortcomings and improvement directions of the product, helping merchants improve service quality and product sales.

With the rapid growth of unstructured text data online, analyzing consumer attitudes or opinions toward goods or services has become an essential field of social network text research. Applying text mining to the study of negative reviews on e-commerce platforms, quickly screening and identifying problems with goods and services, has become a meaningful way to improve consumer satisfaction and loyalty. Mining negative reviews on e-commerce platforms mainly requires solving two significant problems. First, how to extract keywords from negative online reviews through text comment mining analysis and clarify the interrelationships between keywords in negative online reviews? Second, how to identify the main product or service issues faced by e-commerce platforms and make targeted improvements?

You can delete our sample text and replace it with the text of your own contribution to the proceedings. However, we recommend that you keep an initial version of this file for reference.

2. Related Work

The academic community has conducted relevant research on negative online comments on e-commerce platforms, mainly focusing on two aspects: first, exploring the impact of negative online comments on consumer behavior. The impact of conflicting online comments [4], responding to negative online comments [5], and the proportion and quality of comments [7] on consumer shopping decisions. The boundary conditions of the relationship between negative online evaluations and online sales [6], the impact on customer perception of hotel image and behavioral intentions [8,9], the impact on brand assets, and the mediating role of emotional contagion [10], etc.

Secondly, analyze the online comment corpus through text mining methods. Using text mining methods and negative binomial regression methods, explore how embedded emotional expressions in online hotel reviews affect consumer perception [11], identify food safety issues on online ordering platforms based on machine learning algorithms [12], and so on. Currently, the Latent Dirichlet Allocation (LDA), as a classic method for text clustering, can overcome label limitations, semantic ambiguity, and multi-dimensional and sparse data in links [14] and is widely used in comment text mining [3,13]. Scholars have identified topics discussed in Twitter tweets by modeling potential Dirichlet assignments, and used social network analysis to model user interactions as weighted directed networks [14]. These studies provide new methodological ideas.

Secondly, analyze the online comment corpus through text mining methods. Using text mining methods and negative binomial regression methods, explore how embedded emotional expressions in online hotel reviews affect consumer perception [12]; identify food safety issues on online ordering platforms based on machine learning algorithms [13]; use text data mining techniques to analyze how the three factors of "content theme, proportion, and consistency" affect online sales [14]; Using support vector machine (SVM) methods for reverse classification to determine the number of critical features related to positive/negative overall attitudes [15], and so on.

In summary, current research on negative reviews on e-commerce platforms mainly focuses on the impact of negative reviews on sales performance. It is worth exploring how to comprehensively use methods such as text feature extraction, text clustering, and social network analysis to extract keywords from negative online comments on e-commerce platforms, clarify the product and service problems faced by e-commerce platforms, analyze the interrelationships between keywords, and propose effective improvement measures.

3. Research Steps and Methods

3.1. *Research steps*

This article intends to comprehensively use the LDA method and social network analysis to analyze negative comments on e-commerce platforms, with four main steps: collecting negative comments on e-commerce platforms, text data preprocessing, keyword social network analysis, and LDA topic model for keyword topic clustering.

3.2. *Research steps and methods*

3.2.1. *Latent dirichlet allocation (LDA)*

LDA (Latent Dirichlet Allocation) is a probabilistic statistical model widely used in natural language processing and text mining. It aims to identify implicit topic structures in large-scale document collections and discover potential topic patterns in document collections. In the LDA topic model, confusion is used as an indicator to evaluate the model. The degree of confusion measures the quality of the topic model's prediction of the distribution of words in unseen documents. The lower the level of confusion, the better the model fitting effect, and the higher the differentiation between topics. The calculation formula is as follows:

$$Perplexity=exp\left\{-\left(\sum_{m-1}^{M}\sum_{n-1}^{N_m}\log\left(\sum_{k-1}^{K}p(w_n|z_k)p(z_k|d_m)\right)\right)/\left(\sum_{m-1}^{M}N_m\right)\right\} \quad (1)$$

In Eq. (1), where M is the number of texts in the data set, N_m is the total number of words in the Mth text, and K is the number of topics, $p(w_n|z_k)$ indicates the probability of the words w_n under the topic z_k, and $p(z_k|d_m)$ is the probability of the test text d_m under the topic z_k.

3.2.2. *Social network analysis (SNA)*

Social Network Analysis (SNA) is a methodology for studying social structure, which uses quantitative methods to study the relationships (or edges) between individuals (or "nodes"), revealing key individuals (or "central nodes"), group structure, overall characteristics of the network (such as density, agglomeration coefficient), and the flow of information, influence, or resources in the network.

4. Data Sources and Preprocessing

4.1. *Data sources*

This article collects negative review text of e-commerce on the platform of black cat complaint platforms（ https://tousu.sina.com.cn ）. A total of 10123 negative reviews text data were collected from the Black Cat Complaint Platform. After removing comments unrelated to e-commerce platforms, deleting duplicate data, and invalid and low information text comments, a total of 9041 negative comment texts were obtained, covering multiple consumer fields, including shopping platforms, financial payments, entertainment life, freight logistics, education and training, takeout catering, tourism and accommodation, beauty and clothing, real estate and home furnishings, etc.

4.2. *Data preprocessing*

The raw text data obtained has noise and cannot be directly analyzed for mining, so preprocessing is required, including text cleaning, text segmentation, removal of stop words, and part of speech tagging. This article uses Python 3.12.1 software for text preprocessing and uses the Jieba Chinese word segmentation tool to segment comment text data. Using the Harbin Institute of Technology's stop word list, remove words from the text that have no significant contribution to sentiment analysis or task objectives, including conjunctions, interjections, pronouns, etc., and annotate the part of speech of the text data.

5. Experimental Results

Using negative texts from e-commerce platforms obtained from the Black Cat platform as a case set, the correlation between keywords was discovered through keyword co-occurrence and social network analysis. The TF-IDF algorithm extracted critical information about comment text features, and the LDA topic model was used for clustering analysis.

5.1. *Text feature extraction*

The high-frequency text features frequently appearing in comments can reveal the core issues the complainant is concerned about. By analyzing high-frequency feature keywords, it was found that content such as "refund," "customer service," "contact," and "transaction" received significant attention. The word cloud of high-frequency words is shown in Figure 1.

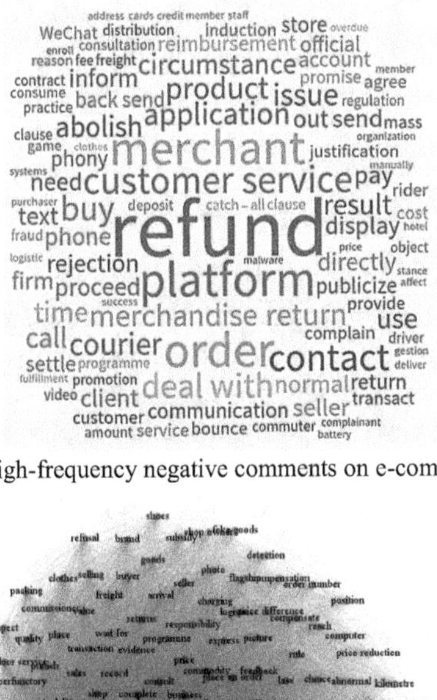

Fig. 1 Cloud diagram of high-frequency negative comments on e-commerce platforms

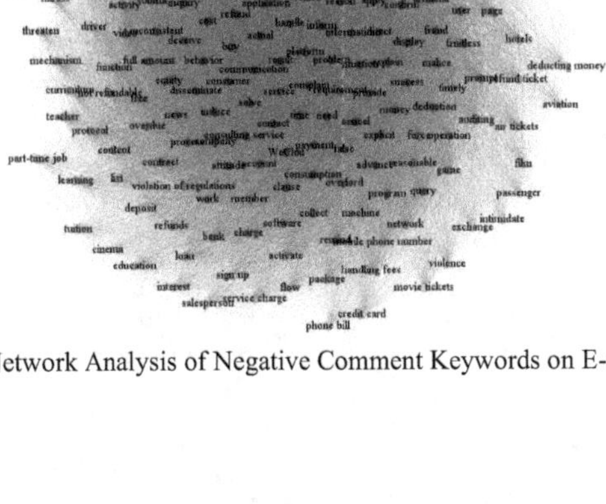

Fig. 2. Social Network Analysis of Negative Comment Keywords on E-commerce Platforms

5.2. *Feature correlation analysis based on social networks*

By analyzing the relationships between different influencing factors through social networks, to discover the correlations between the main features. In addition, the importance of comment text features can be further explored by analyzing the central nodes in social networks. Visualize the keyword co-occurrence network using Gephi 0.9.3 software and calculate the network metrics, such as average degree, average weighting degree, network diameter, graph density, and average path length, as shown in Figure 2.

The average degree of this network is 174.239, which means that each keyword co-occurs with an average of 174 other keywords, indicating a high degree of interconnectivity between nodes in the network. The average weighted degree is 5551.043, and the graph density is 0.834, indicating that the nodes in the network are highly interconnected and there is a co-occurrence relationship between most keywords. The average path length is 1.172, indicating that any two keywords in the network can reach each other through very few intermediate steps, reflecting the compactness of the network and the efficiency of information dissemination.

5.3. *Text topic analysis based on LDA topic model*

In this study, word frequency-inverse document frequency (TF-IDF) was used to extract text features, and the confusion evaluation method was used to determine the optimal number of topics. We calculated the perplexity of LDA models for 1 to 12 topics. The results show that when the number of topics was set to 8, the perplexity of the model reached a local minimum, showing an evident "turning point" phenomenon. This study determined eight as the optimal number of topics, which can cover a wide range of text information while avoiding vocabulary overlap caused by too many topics, thus ensuring the interpretability of the topics.

Table 1 list of theme- feature words of negative reviews text topic on e-commerce platforms

Theme number	Feature words 1	Feature words 2	Feature words 3	Feature words 4	Feature words 5	Feature words 6	Feature words 7	Feature words 8
Theme1	rider	fined	part-time	meal delivery	illegal deduction	clause	exceeding time limit	route
Theme2	education	refund	drawback	learning	course	fee	registration	agreement
Theme3	cancel	the order	ticket	refund	plane	airline's flight	airline company	change of ticket
Theme4	express	account	contact	customer service	logistics	complaints	request	information
Theme5	driver	order	cancel	customer	defect	complaint	vehicle	system
Theme6	phone calls	urgent collection	contact	harassmen t	platform	repayment	time limit exceeded	negotiation
Theme7	customer service	membership	refunds	renewals	recharges	requests	scenarios	automated
Theme8	refunds	merchants	problems	purchases	customer service	contact	processing	aftermarket

Using the Latent Dirichlet Allocation (LDA) model for text clustering. Due to LDA being an unsupervised distributed model, the feature vocabulary extracted by the model may contain vocabulary that is not significantly related to the topic. In order to more effectively reveal the core content of each topic, we selected the top 8 words with the highest TF-IDF weight from the vocabulary extracted from each topic as keywords. These keywords best represent the core concepts and characteristics of their respective themes. Finally, the subject feature word extraction results of negative comment texts on e-commerce platforms are shown in Table 1.

By using the LDA model to cluster and analyze negative review texts on e-commerce platforms, eight main themes of consumer complaints were identified, namely platform rider disputes, education refund difficulties, difficulty in canceling or changing reservations, damage or loss of goods, taxi disputes, payment harassment complaints, platform member disputes and slow response on customer service.

The evolution of e-commerce platform relationships is manifested as a shift from a single merchant consumer model to a diversified network involving complex interactions among platforms, labor, merchants, and consumers. The "platform rider dispute" theme extracted in this article indicates that the relationship between platforms and labor has become a new focus.

The forms of e-commerce disputes are diverse, and service quality has become a focus. With the rapid development and diversification of e-commerce, the types of disputes involved are also increasing. The theme of "payment harassment complaints" indicates that payment harassment is an operational issue of online financial platforms and a complex issue involving legal, ethical, and social responsibility. From the perspective of comment themes, e-commerce platforms have transformed from simple buying and selling disputes between merchants and customers to "education refund difficulties," "difficulty in canceling or changing reservations," "damage or loss of goods," "taxi disputes," "platform membership disputes," and "slow customer service response." The theme of "Payment Harassment Complaints" indicates that payment harassment is not only an operational issue of online financial platforms but also a complex issue involving legal, ethical, and social responsibility. Resolving these disputes requires the platform to improve service quality, dispute resolution mechanisms, transparency, and user experience. At the same time, strengthening industry supervision, establishing a sound legal and regulatory system for e-commerce disputes, and ensuring adequate protection of consumer rights are the keys to the healthy development of e-commerce.

Acknowledgments

This work is supported by the National Natural Science Foundation of China under grant 72101235. Work partially supported by the State Scholarship Fund under grant 202108330330.

References

1. Q. H. Pang, X. W. Dong, B Zhou, etc., Negative online comment keyword extraction based on sentiment analysis and TextRank, *Information Science*(China),40, 111-117(2022).
2. W. J. Duan, G. Bin, and Andrew B, "Do online reviews matter?—An empirical investigation of panel data." *Decision support systems* 45, 1007-1016 (2008).
3. Y. R. Huang, R. Wang, B. Huang, B. Wei,S. L. Zheng, and M. Chen, Sentiment classification of crowdsourcing participants' reviews text based on LDA topic model. *IEEE Access*, 9, 108131-108143(2021).
4. B. Le, Y. Chen, and X. L. Yang. "The impact of contradictory online reviews on consumer online purchase decision: experimental evidence from China." *SAGE Open* 13, 21582440231180107 (2023).
5. J. H. Wang, Genre change in the online context: Responding to negative online reviews and redefining an effective genre construct on Amazon. Com. *Journal of Business and Technical Communication,* 35, 297-332 (2021).
6. A. Shen, P. Wang, and Y. Ma. When crowding-in and when crowding-out? The boundary conditions on the relationship between negative online reviews and online sales. *Managerial and Decision Economics* 43, 2016-2032 (2022).
7. M. R. Shihab and A. P. Putri, Negative online reviews of popular products: Understanding the effects of review proportion and quality on consumers' attitude and intention to buy, *Electronic Commerce Research*, 19, 159-187(2019).
8. F. Meng, R. B. Dipietro, J. H. Gerdes, et al. How hotel responses to negative online reviews affect customers' perception of hotel image and behavioral intent: An exploratory investigation, *Tourism Review International*, 22, 23-39(2018).
9. H. Wan, M. Q. Mei, J. Yan, J. Xiong, and L. Wang, How does apology matter? Responding to negative customer reviews on online-to-offline platforms, *Electronic Commerce Research and Applications*, 61, 101291 (2023).
10. S. Y. Kwak, M. Shin, M. Lee, K. J. Back, Integrating the reviewers' and readers' perceptions of negative online reviews for customer decision-making: a mixed-method approach. *International Journal of Contemporary Hospitality Management*, (2023).
11. M. Lee, M. Jeong, J. Lee, Roles of negative emotions in customers' perceived helpfulness of hotel reviews on a user-generated review website:

A text mining approach, *International Journal of Contemporary Hospitality Management*, 29, 762-783(2017).

12. Y. R. Huang, X. Wang, R. Wang, J. Min, Analysis and recognition of food safety problems in online ordering based on reviews text mining, *Wireless Communications and Mobile Computing*, 1-15 (2022).
13. H. Y. Gao, J. W. Liu, S. X. Yang, Topic mining of online medical reviews based on improved LDA, *Journal of Beijing University of Technology*, 39, 427-434(2019).
14. X. K. Li, Z. Yao, and L. Y. Mei, Analyzing online reviews of foreign tourists to destination attractions in China: a novel text mining approach, *Asia Pacific Journal of Tourism Research* , 28, 647-666(2023).

Reciprocal Trust Region Policy Optimization

Haotian Xu*, Junyu Xuan†, Guangquan Zhang‡ and Jie Lu§

Australia Artificial Intelligence Institute, University of Technology Sydney, New South Wales, Australian

* *Haotian.Xu@student.uts.edu.au*

† *Junyu.Xuan@uts.edu.au*

‡ *Guangquan.Zhang@uts.edu.au*

§ *Jie.Lu@uts.edu.au*

Trust region policy optimization (TRPO) is one of the most representative and successful policy iteration optimization algorithms, which maximizes a surrogate objective based on a generalized advantage function and importance sampling, subject to a Kullback–Leibler (KL) divergence constraint between two consecutive policies. Theoretically, it is desired to improve the surrogate objective at each iteration. However, without a positive objective limit, this theoretical requirement is not always satisfied. In this chapter, we aim to find out a new TRPO formulation. According to the reciprocal optimization technique, we exchange the objective and the constraint in TRPO and then formulate a reciprocal TRPO (rTRPO). Our rTRPO is to minimize the KL divergence, subject to a minimal positive surrogate return, which could guarantee a positive improvement of the return. Further, an approximate solution is designed via a linear approximation for the return and a quadratic approximation for the KL divergence, which induces a convex quadratic programming. Our experiments on three different benchmark environments show that our proposed rTRPO is superior to three existing techniques (i.e., actor–critic algorithm, the original TRPO, and entropy regularized TRPO).

Keywords: Reinforcement learning; reciprocal optimization; trust region policy optimization; Kullback–Leibler divergence.

1. Introduction

In machine learning community, reinforcement learning (RL) [1–3] deals with a special problem of how to act optimally in an unknown dynamic environment. The agent is allowed to execute a sequence of actions on the environment to manipulate its corresponding states. Then, according to the rewards that it accumulates, the agent learns how to act more optimally. A policy implies the learning agent's way of behaving at a given time, which is defined as a mapping from states to probabilities of selecting

*Corresponding author.

possible action.[1] The agent aims to improve its policy to obtain higher rewards, which is named policy optimization.[1]

Policy optimization results in two kinds of RL methods: value-based and policy-based optimizations,[1] which are usually implemented by critic and actor architectures. Currently, combining these two kinds of architectures builds widely used actor–critic architecture, whose approaches employ the optimization of value function as guidance of policy improvements.[1] Several frequently mentioned methods include actor–critic (AC) algorithm[4] and its variants,[1] and trust region policy optimization (TRPO)[5] and its various variants, for example, entropy regularized TRPO,[6] and Proximal Policy Optimization (PPO).[7]

The TRPO is formulated to maximize a surrogate objective based on generalized advantage function and importance sampling, subject to a limited KL divergence constraint between two adjacent policies.[5] Such a TRPO approach has gained many successful applications, due to its scalability and efficiency. Theoretically, it is desired to improve its surrogate objective at each update. However, without a positive objective constraint, this theoretical requirement is not always satisfied. It is still an opening issue to how to reformulate TRPO, which is our aim in this chapter.

According to reciprocal optimization technique,[8] we exchange the objective and constraint to minimize the KL divergence, subject to the constraint that the surrogate return is greater than a positive pre-defined threshold. This new TRPO formulation derives a reciprocal TRPO. Moreover, according to a linear approximation for the surrogate return and a quadratic approximation for KL divergence, we build an approximate solution based on convex quadratic programming. Finally, we selected three different benchmark environments: Cart Pole, continuous Mountain Car, and Reacher, to validate the effectiveness of our rTRPO, by comparing it to three previous methods (i.e., AC,[4] TRPO,[5] and EnTRPO[6]).

2. Backgrounds

The reinforcement learning comprised of a discrete time discounted Markov decision process (MDP) defined by a tuple $\{\mathcal{S}, \mathcal{A}, P, r, \gamma, \rho_0\}$. Here, $\mathcal{S}$ and $\mathcal{A}$ denote the state and action spaces of the agent, respectively; $P(s'|s,a) : \mathcal{S} \times \mathcal{A} \times \mathcal{S} \mapsto \mathbb{R}$ (one-dimensional real space) is the transition probability distribution from the state $s \in \mathcal{S}$ to a next state $s' \in \mathcal{S}$ under executing an action $a \in \mathcal{A}$; $r(s,a) : \mathcal{S} \times \mathcal{A} \mapsto \mathbb{R}$ is the reward function of s and a; $\gamma \in (0,1)$ represents the discount factor; and $\rho_0 = P(s_0)$ denotes the initial state distribution[1–3].

Let $\pi : \mathcal{S} \times \mathcal{A} \mapsto [0,1]$ be a stochastic policy. For the current state s_t, the agent executes an action $a_t \in \mathcal{A}$ following a conditional probability distribution $\pi(a_t|s_t)$, and the the agent receives a reward $r_t = r(s_t, a_t)$. At each time step t, with its state-action value function $Q^\pi(s_t, a_t)$ and state value function $V^\pi(s_t)$, we define an advantage function:

$$A^\pi(s_t, a_t) = Q^\pi(s_t, a_t) - V^\pi(s_t). \tag{1}$$

Using this stochastic policy π, the agent would collect a trajectory: $\tau = \{s_t, a_t\}_{t=1}^{\infty}$, which will be used to estimate the expected discounted reward $J(\pi)$ or the return:

$$J(\pi) = \mathbb{E}_{s_0,a_0,\ldots\sim\pi}\left[\sum_{t=0}^{\infty}\gamma^t r_t\right] = \mathbb{E}_{\tau\sim\pi}\left[\sum_{t=0}^{\infty}\gamma^t r_t\right] = \mathbb{E}_{s_0\sim\rho_0}\left[V^\pi(s_0)\right]. \tag{2}$$

For another policy $\hat{\pi}$, its expected return could be expressed as the following identity, according to the advantage function over π[3,5]:

$$J(\hat{\pi}) = J(\pi) + \mathbb{E}_{\tau\sim\hat{\pi}}\left[\sum_{t=0}^{\infty}\gamma^t A^\pi(s_t, a_t)\right] \tag{3}$$

which indicates that if the second term is positive, any policy update could guarantee the improvement of policy performance[5]. However, it is difficult to optimize this equation (3), since it depends on $\hat{\pi}$. To this end, such a relation is locally approximated as

$$J(\hat{\pi}) = J(\pi) + \mathbb{E}_{\tau\sim\pi}\left[\sum_{t=0}^{\infty}\gamma^t A^\pi(s_t, a_t)\right] \tag{4}$$

which is based on π, rather than $\hat{\pi}$, via ignoring the changes in the state visitations from the policy change. Further, via importance sampling, the second term in (4) could be reformulated as

$$\bar{R}(\hat{\pi}) = \mathbb{E}_{\tau\sim\pi}\left[\sum_{t=0}^{\infty}\gamma^t \frac{\hat{\pi}(a_t|s_t)}{\pi(a_t|s_t)} A^\pi(s_t, a_t)\right] = \mathbb{E}_{\tau\sim\pi}\left[\sum_{t=0}^{\infty}\frac{\hat{\pi}(a_t|s_t)}{\pi(a_t|s_t)}\hat{A}^\pi(s_t, a_t)\right] \tag{5}$$

where $\hat{A}(s_t, a_t) = \sum_{i=0}^{\infty}(\gamma\lambda)^i\delta_{t+i}$ is a generalized advantage estimation (GAE), in which $\lambda \in [0,1]$ is a new added control parameter and δ_t is the temporal difference error: $\delta_t = r_t + \gamma V^\pi(s_{t+1}) - V^\pi(s_t)$. It is noted that this error is also used to train the critic network in many actor-critic methods.

3. Reciprocal Trust Region Policy Optimization

In this section, we will introduce reciprocal optimization technique[8] at first and propose a reciprocal trust region policy optimization (rTRPO) after reviewing the original TRPO[5].

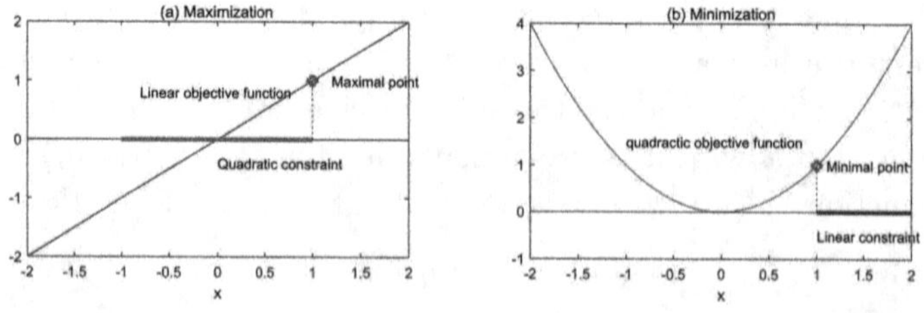

Fig. 1. (a) Primary maximization problem and (b) its reciprocal minimization one.

3.1. *Reciprocal optimization technique*

In[8], the following two optimization problems are investigated. The primary problem is formulated as

$$\min f(\boldsymbol{x}),\ \text{s.t. } g(\boldsymbol{x}) \geq \alpha,\ \boldsymbol{x} \in \mathcal{X} \tag{6}$$

where $f : \mathcal{X} \to \mathcal{R}$, $g : \mathcal{X} \to \mathcal{R}$, $\alpha \in \mathcal{R}$, and $\mathcal{X} \subseteq \mathcal{R}^n$ (n-dimensional real space) is open. Given $\beta \in \mathcal{R}$, the reciprocal problem associated to (6) is

$$\max g(\boldsymbol{x}),\ \text{s.t. } f(\boldsymbol{x}) \leq \beta, \boldsymbol{x} \in \mathcal{X}. \tag{7}$$

Sufficient conditions are established[8] to guarantee that an optimal solution of (6) is also optimal for (7), and vice-versa, which is stated as follows: if $\boldsymbol{x}^*$ is the optimal solution of (6) and (7), then $\alpha = g(\boldsymbol{x}^*)$ and $\beta = f(\boldsymbol{x}^*)$.

Here we give a simple one-dimensional example, as shown in Fig. 1. The maximization task is

$$\max g(x) = x,\ \text{s.t. } f(x) = x^2 \leq 1. \tag{8}$$

whose optimal solution is $x = 1$ and its objective $g(1) = 1$. Its corresponding minimization one is

$$\max f(x) = x^2,\ \text{s.t. } g(x) = x \geq 1. \tag{9}$$

3.2. *Reciprocal trust region policy optimization*

Before introducing our reciprocal trust region policy optimization, we review the original TRPO. Now, assume that a policy π is parameterized by a parameter vector $\boldsymbol{\theta}$, which is indicated by $\pi_{\boldsymbol{\theta}}$. With the previous policy $\pi_{\boldsymbol{\theta}_{old}}$, the agent interacts with the environment to collect some trajectory data, and then the TRPO constructs a constrained maximization problem:

$$\begin{aligned} \max\ & \hat{R}(\boldsymbol{\theta}) = \hat{\mathbb{E}}_t \left[\frac{\pi_{\boldsymbol{\theta}}(a_t|s_t)}{\pi_{\boldsymbol{\theta}_{old}}(a_t|s_t)} \hat{A}(a_t, s_t) \right] \\ \text{s.t.}\ & \hat{D}_{KL}(\boldsymbol{\theta}) = \hat{E}_t[D_{KL}(\pi_{\boldsymbol{\theta}}(a_t|s_t)||\pi_{\boldsymbol{\theta}_{old}}(s_t|a_t))] \leq \delta \end{aligned} \tag{10}$$

where $\boldsymbol{\theta}$ and $\boldsymbol{\theta}_{old}$ are two parameter vectors corresponding to the current and previous policies, respectively; the $D_{KL}(\cdot||\cdot)$ indicates the KL divergence (or relative entropy)[9]; and δ is the threshold of KL divergence. This constrained optimization (10) is used to learn the policy π or actor network in TRPO. Theoretically, it is desired to achieve a positive $\hat{R}(\boldsymbol{\theta})$, which is not guaranteed in (10), without a constraint $\hat{R}(\boldsymbol{\theta}) > 0$

According to the above reciprocal optimization technique, we reformulate TRPO (10) into a constrained minimization problem:

$$\min \hat{D}_{KL}(\boldsymbol{\theta}),\ \text{s.t.}\ \hat{R}(\boldsymbol{\theta}) \geq \eta_0 \tag{11}$$

where $\eta_0 > 0$ is a surrogate return threshold. This formulation means that we should find out a minimal policy update to achieve the least positive surrogate return η_0. Further, we referred to this version as reciprocal trust region policy optimization or simply rTRPO.

3.3. *Solution method for rTRPO*

In this sub-section, we design an effective solution procedure for rTRPO. At first, the $\hat{R}(\boldsymbol{\theta})$ is approximated linearly around $\boldsymbol{\theta}_{old}$, i.e.,

$$\hat{R}(\boldsymbol{\theta}) \approx \hat{R}(\boldsymbol{\theta}_{old}) + (\boldsymbol{\theta} - \boldsymbol{\theta}_{old})^T \nabla \hat{R}(\boldsymbol{\theta}_{old}) = \hat{R}(\boldsymbol{\theta}_{old}) + \mathbf{g}^T \mathbf{d} \tag{12}$$

where the descent direction $\boldsymbol{d} = \boldsymbol{\theta} - \boldsymbol{\theta}_{old}$ and and the gradient of $\hat{R}(\boldsymbol{\theta})$ with respect to $\boldsymbol{\theta}$ is $\boldsymbol{g} = \nabla \hat{R}(\boldsymbol{\theta}_{old})$.

The constraint $\hat{D}_{KL}(\boldsymbol{\theta})$ is approximated quadratically

$$\begin{aligned} \hat{D}_{KL}(\boldsymbol{\theta}) &\approx \hat{D}_{KL}(\boldsymbol{\theta}_{old}) + (\boldsymbol{\theta} - \boldsymbol{\theta}_{old})^T \nabla \hat{D}_{KL}(\boldsymbol{\theta}_{old}) \\ &\quad + \tfrac{1}{2}(\boldsymbol{\theta} - \boldsymbol{\theta}_{old})^T \nabla^2 \hat{D}_{KL}(\boldsymbol{\theta}_{old})(\boldsymbol{\theta} - \boldsymbol{\theta}_{old}) = \tfrac{1}{2}\boldsymbol{d}^T \boldsymbol{H} \boldsymbol{d} \end{aligned} \tag{13}$$

where the Hessian matrix $\boldsymbol{H} = \nabla^2 D_{KL}(\boldsymbol{\theta}_{old})$ is called as Fisher information matrix specially, and $D_{KL}(\boldsymbol{\theta}_{old}) = 0$ and $\nabla D_{KL}(\boldsymbol{\theta}_{old}) = \mathbf{0}$[9].

Now the above constrained problem (11) is approximated as standard quadratic programming (QP):

$$\min \tfrac{1}{2}\boldsymbol{d}^T \boldsymbol{H} \boldsymbol{d},\ \text{s.t.}\ \boldsymbol{g}^T \boldsymbol{d} \geq \eta. \tag{14}$$

It is worth noting that the $\eta > 0$ is a return increment from the previous policy to guarantee the return improvement. In addition, since the Fisher information matrix $\boldsymbol{H}$ is positive definite, such a problem (14) is convex.

Now we define a Lagrangian function to convert the QP problem (14) into an unconstrained one:

$$L(\boldsymbol{d}) = \frac{1}{2}\boldsymbol{d}^T \boldsymbol{H} \boldsymbol{d} - \lambda(\boldsymbol{g}^T \boldsymbol{d} - \eta) \tag{15}$$

where $\lambda > 0$ is a non-negative Lagrangian multiplier. Let the gradient of $L(\boldsymbol{d})$ with respect to $\boldsymbol{d}$ be zero, we derive a set of linear equations:

$$\boldsymbol{H}\boldsymbol{d} = \lambda \boldsymbol{g}. \tag{16}$$

After neglecting λ, we apply conjugate gradient method to solve $\boldsymbol{H}\boldsymbol{u} = \boldsymbol{g}$ to achieve a search direction $\boldsymbol{u}$ and thus have a descent direction $\boldsymbol{d} = \lambda \boldsymbol{u}$, where λ also acts as a step length.

Now, considering the constraint in (14):

$$\boldsymbol{g}^T \boldsymbol{d} = \frac{1}{\lambda} \boldsymbol{d}^T \boldsymbol{H} \boldsymbol{d} = \lambda \boldsymbol{u}^T \boldsymbol{H} \boldsymbol{u} \geq \eta \tag{17}$$

we have a minimal step size:

$$\lambda_{min} = \frac{\eta}{\boldsymbol{u}^T \boldsymbol{H} \boldsymbol{u}} > 0. \tag{18}$$

Since the Fisher information matrix $\boldsymbol{H}$ is positive definite, the denominator is positive. To guarantee a positive λ_{min}, we would set a positive $\eta = \rho|\hat{R}(\boldsymbol{\theta}_{old})|$ and thus we have $\eta_0 = \hat{R}(\boldsymbol{\theta}_{old}) + \rho|\hat{R}(\boldsymbol{\theta}_{old})|$, where $\rho \in \{0, 1\}$ is a factor to control the return improvement.

Further, since the rTRPO (11) is nonlinear, it is desired to find out a larger improvement of the return. At first, we estimate a minimal KL divergence with the minimal step length λ_{min}:

$$D_{KL}^{min} = D_{KL}(\boldsymbol{\theta}_{old} + \lambda_{min} \boldsymbol{u}), \tag{19}$$

and then we apply a linear search to find an optimal step length index i^*:

$$i^* = \underset{i=1,\ldots,n}{\operatorname{argmax}} \ \hat{R}(\boldsymbol{\theta}_{old} + \lambda_0^i \lambda_{min} \boldsymbol{u}), \ \text{s.t.} \ \hat{D}_{KL}(\boldsymbol{\theta}_{old} + \lambda_0^i \lambda_{min} \boldsymbol{u}) \leq D_{KL}^{min} \tag{20}$$

where n is the number of linear searches, e.g., $n = 15$, and $\lambda_0 > 1$ is an expanding factor. Lastly, we update the $\boldsymbol{\theta}$ vector:

$$\boldsymbol{\theta} = \boldsymbol{\theta}_{old} + \lambda_0^{i^*} \frac{\rho|\hat{R}(\boldsymbol{\theta}_{old})|}{\boldsymbol{u}^T \boldsymbol{H} \boldsymbol{u}} \boldsymbol{u} \tag{21}$$

which is used to train the actor in our rTRPO.

4. Experiments

In this section, we will conduct some experiments to validate our proposed method: rTRPO, via comparing to AC[4], TRPO[5], and EnTRPO[6].

We collected three benchmark environments: Cart Pole, Mountain Car and Reacher from [a] to validate and compare three experimented reinforcement learning methods, as shown in Fig. 2 and Table 1.

[a]https://pypi.org/project/gym/

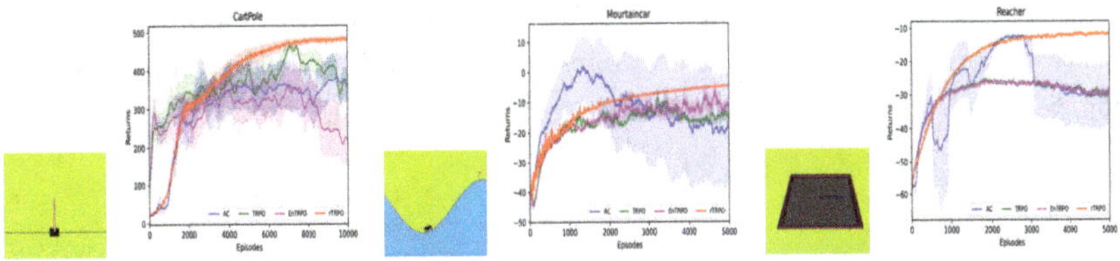

(a) Cart pole (b) Mountain car (c) Reacher

Fig. 2. Three benchmark environments and performance comparison.

Table 1: Three benchmark environments and their performance

Environment	CartPole	MountainCar	Reacher
State space	4	2	11
Action space	2(discrete)	1(continuous)	2(continuous)
AC	374.15±3.87	-18.22±1.21	-30.08±0.36
TRPO	382.79±26.16	-15.49±0.96	-29.18±0.76
EnTRPO	234.93±13.19	-11.12±1.68	-29.25±0.56
rTRPO	**480.77**±1.89	**-5.15**±0.37	**-11.80**±0.26

Then, we conduct 10 trials using 10 random seeds on three benchmark environments for four algorithms and show their experimental results in Fig. 2 and Table 1. Table 1 lists the mean and standard deviation from the last 1000 episodes of each method, where the best results are highlighted in **bold**. We observe that (a) AC is not stable, especially in two continuous environments; (b) EnTRPO performs better and worse than TRPO on CartPole and Mountaincar; (c) our rTRPO improves its performance gradually and is superior to three compared methods significantly. In short, we validate the superiority of our proposed method rTRPO experimentally.

5. Conclusions

Trust region policy optimization is a successful reinforcement learning method. However, its theoretical improvement requirement is not always satisfied, without a positive objective constraint. In this chapter, we apply reciprocal optimization technique and then reformulate TRPO into a reciprocal trust region policy optimization via exchanging the objective and KL divergence constraint, i.e., to minimize the KL divergence, subject to a least return increment, which guarantees a positive improvement of the surrogate return. Our experiments on three environments validate the effectiveness of our proposed method, via comparing it to three existing methods. Our future work will focus on updating our method via entropy trick.[10]

Acknowledgments

This work was supported by the Australian Research Council through the Discovery Project under Grant DP200100700.

References

1. R. S. Sutton and A. G. Barto, *Reinforcement Learning: An Introduction*, 2nd edn. (MIT Press, Cambridge MA, USA, 2018).
2. J. Xuan, J. Lu, Z. Yan and G. Zhang, Bayesian deep reinforcement learning via deep kernel learning, *International Journal of Computational Intelligence Systems* **12**, 164 (2018).
3. T. Wang, S. Xie, M. Gao, X. Chen, Z. Zhang and H. Yu, Offline reinforcement learning via policy regularization and ensemble q-functions, in *The 34th International Conference on Tools with Artificial Intelligence*, (Virtual Event, 2022).
4. V. R. Konda and J. N. Tsitsiklis, Actor-critic algorithms, in *The 14th Annual Conference on Neural Information Processing Systems*, (Vancouver, Canada, 2000).
5. J. Schulman, S. Levine, P. Moritz, M. Jordan and P. Abbeel, Trust region policy optimization, in *The 32nd International Conference on Machine Learning*, (Lille, France, 2015).
6. S. Roosraie and M. M. Ebadzadeh, EnTRPO: trust region policy optimization method with entropy regularization, *arXiv:2110.13373* , 1 (2021).
7. J. Schulman, F. Wolski, P. Dhariwal, A. Radford and O. Klimov, Proximal policy optimization algorithms, *arXiv:1707.06347v2* , 1 (2017).
8. P. Favati and M. Pappalardo, On the reciprocal vector optimization problems, *Journal of Optimization Theory and Applications* **47**, 181 (1985).
9. T. van Erven and P. Harremoes, Renyi divergence and Kullback-Leibler divergence, *IEEE Transactions on Information Theory* **60**, 3797 (2014).
10. J. Xuan, X. Luo, G. Zhang, J. Lu and Z. Xu, Uncertainty analysis for the keyword system of web events, *IEEE Transactions on Systems, Man, and Cybernetics: Systems* **46**, 829 (2016).

A Novel Approach for Measuring Demographic Parity Fairness in Group Recommendation

R. Yera*, M. J. Barranco and L. Martínez

Computer Science Department, University of Jaén, Jaén, Spain

**ryera@ujaen.es*

Fairness is currently becoming a necessary dimension to consider in contemporary artificial intelligence (AI)-based systems, according to the recent Ethics Guidelines for a Trustworthy AI. Being recommender systems popular applications that incorporate AI to a larger or lesser extent, the literature analysis identifies a research gap related to the exploration of demographic parity fairness in the group recommendation scenario. This chapter focuses on this gap, developing a group recommendation framework that has as main novelty the measuring of the consumer fairness taking into account the presence of advantaged and disadvantaged classes of users. Experimental studies are developed for measuring the performance of the proposal in a real recommendations scenario, illustrating that it is able to distinguish different fairness levels across the delivered recommendations.

Keywords: Group recommender systems; demographic parity fairness; experiments.

1. Introduction

The Ethics Guidelines for a Trustworthy AI,[a] published by the European Commission in 2019, identified that the development, deployment, and use of artificial intelligence (AI) systems should adhere to the ethical principles of respect for human autonomy, prevention of harm, explainability, and fairness. Particularly, fairness concerns a commitment to ensure that individuals and groups in the system are free from unfair bias and discrimination in AI, assuring an equal and fair distribution of benefits and costs. Considering its social impact on gender and diversity-related issues, the studies on measuring and improving the systems' fairness are becoming a relevant and necessary topic in AI research.

Recently, the concept of fairness has been brought to the recommender systems (RS) research area. As popular AI-based systems, RSs are focused

[a]https://digital-strategy.ec.europa.eu/en/library/ethics-guidelines-trustworthy-ai

on providing users with items that best fit their preferences and needs, in an overloaded search space of possible options.[1] Since the 1990s, they have been applied in scenarios such as e-commerce, e-learning, e-health, and so on.[1]

Formally, fairness is associated with the recognition of RS as a multi-stakeholder scenario, in which the main stakeholders are consumers (C) and providers (P).[2] Specifically, C-fairness, the focus of this chapter, is formerly centered on the disparate impact of recommendations on protected classes of consumers, associated with sensitive features, e.g., gender, race, and age.

In other direction, recommender systems have been recently brought to scenarios where users consume items in groups and not individually, such as TV programs, restaurants, or tourism packages.[3,4] For these scenarios, the concept of group recommender systems has been developed for suggesting items focused on satisfying the overall preference of the group beyond the particularities of each individual preference of the users.

However, research on fairness has been mainly centered on individual RS,[2,5,6] being on the other hand very limited the research on recommendation fairness in group recommender systems (GRS). Current GRS proposals only consider fairness by optimizing the exposure of items in the final recommendation lists (e.g., the recommendation of items that are preferred by the whole group, as far as possible[7,8]), without taking into account the user's demographic attributes, which are considered as key information for characterizing algorithmic fairness.[9] Here the item exposure makes reference to the position of the item in the list of generated recommendations, having a top position being considered as a higher exposure.

At this stage, we consider that it is necessary to boost the study of fairness in the traditional GRS scenario. With this goal in mind, this chapter extensively brings to the GRS context the **C-fairness** concept proposed by Burke[2] which is supported by the demographic attributes of the consumers.

This chapter is structured as follows. Section 2 presents some antecedents on fairness in group recommendation. Section 3 explores how to measure demographic attributes-related fairness in GRS. Section 4 develops the evaluation of the presented approach. Section 5 concludes the contribution.

2. Fairness in Group Recommender Systems

In contrast to fairness in individual RS[5,6], fairness in GRS has been hardly covered by the specialized literature, and with a much more limited scope. Xiao et al.[10] proposed several definitions of fairness, encouraging the group members to minimize the gap between least and highest recommendation utilities. Serbos et al.[11] explore fairness in the package-to-group recommendation problem, by formalizing the concepts of m-proportional fairness and m-envy-freeness, modelling greedy algorithms to find appropriate recommendations. Sacharidis[7] also focused on fairness in GRS, defining the member utility of a group recommendation list as the similarity value of the list and the member ground truth. The fairness of the list is then defined as the lowest member utility of the group, and is reached through the items probability to be in the top-n pareto-optimal items for the group.

A similar goal is followed by Stratigi et al.[12] focused on an iterative process to find a top-n list for the group that maximizes global utility. Also, Felfernig et al.[13] present FAI, a method where group's individuals incrementally build the recommendation list, in a fair turn-based way.

Finally, Kaya et al.[8] have focused on a rank-sensitive balancing of relevance across the group. Herein, the group recommendation list is built in a greedy way, according to a score that depends on the probability that the item is relevant to each member and that there is at least one relevant item in the list for each group's member.

In short, current proposals for GRSs consider fairness by optimizing user fairness in item exposure at the final recommendation list, a particular C-fairness scenario in individual RS. Furthermore, Pitoura et al.[9] have recently pointed out that in GRS the few fairness-aware approaches have been centered only on individual and consumer fairness. It is then necessary to take into account fairness from a more holistic viewpoint than in individual RS, by evaluating C-fairness from the demographic parity viewpoint (performance in protected classes, e.g. age, gender), in the GRS. This is the goal of the current work.

3. Formulating Consumer Fairness in the GRS Scenario

This section is focused on presenting the framework that will be used in the current work, for exploring fairness in group recommendation. This architecture is composed of three stages (see Figure 1): 1) group formation, 2) group recommendation process, and 3) fairness measuring in group recommendation.

Group formation process. The typical group formation process in GRS, usually predefines the associated groups in a previous stage before the recommendation generation. The current work will consider the selection of groups with users having a large similarity in their rating profiles as group

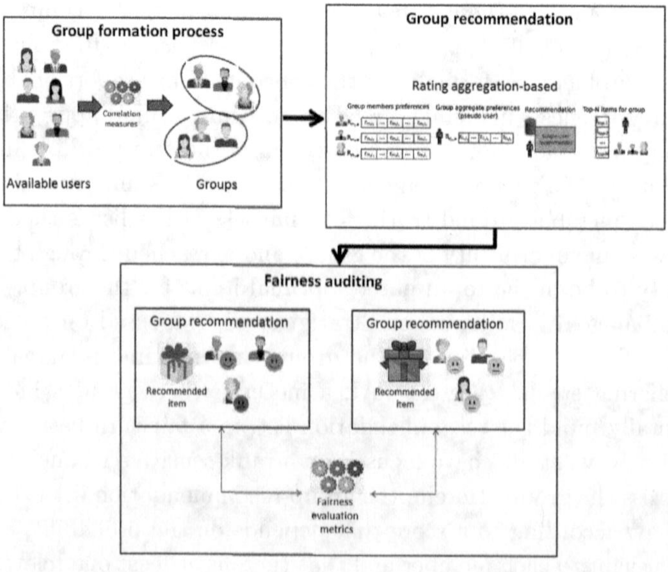

Fig. 1. Overall architecture for measuring fairness in group recommendation

formation strategy. This approach has been explored by several research works focused on this issue[3].

Group recommendation process. In this stage it will be used a rating aggregation-based GRS approach, as one of the basic approaches in group recommendation[4]. It is based on the creation, for each group, of a pseudo-user that contains the preferences of the underlying users. This pseudo-user is then used for generating recommendations as it was an individual, typical user[4]. In this work it will be considered the average approach for aggregating the user's preferences in the composition of the group profile.

Fairness measuring in GRS. Here we will present the novel strategy for characterizing fairness in group recommendation. As novel feature, it is focused on bringing the demographic attributes-related fairness, into the framework of group recommendation. Specifically, it is centered on exploring whether the group members belonging to the disadvantaged classes (regarding users' demographic features), are adversely affected by the generated recommendations.

With this purpose in mind, at first we identify the level of deviation of the actual user preferences r_{ui}, in relation to the predictions r_{Gi} done for the group using a GRS algorithm (Equation 1). In this case we consider a signed deviation regarding we are interested in exploring whether the users' ratings are over or under the predicted group preferences.

$$User\_Group\_Dev(u) = \frac{1}{|R_u|} \sum_{r_{ui} \in R_u} (r_{ui} - r_{Gi}) \tag{1}$$

In a second stage, we characterize the recommendation unfairness $Unfairn$ of the user u in a group recommendation framework, as how close are his/her $User\_Group\_Dev(u)$ value, in relation to the $User\_Group\_Dev(v)$ values of the other users $v \in G$ (Equation 2). If this difference is far enough, we consider that the underlying GRS model performed different for this user, being associated this behavior with a larger unfairness.

$$Unfairn(u) = min(|User\_Group\_Dev(u) - User\_Group\_Dev(v)|); v \in G \tag{2}$$

Finally, the unfairness level associated to a group G (Equation 3), is represented as the absolute difference between the average unfairness of the users respectively belonging to the set A and D of the advantaged and disadvantaged users in the group G.

$$Unfairn(G) = |\frac{1}{|A|} \sum_{u \in A} Unfairn(u) - \frac{1}{|D|} \sum_{v \in D} Unfair(v)| \tag{3}$$

The next section will be focused on exploring this unfairness criteria in a traditional RS dataset.

4. Experiments

In this study we will use the traditional Movielens 100K dataset[14]. It contains the ratings provided by 943 users over a set of 1682 movies, having 100000 ratings values in the range [1, 5], and a sparsity around 94%[14]. In addition to the preference data, there is available some demographic information for each user, specifically age, gender, and occupation.

The Pearson correlation coefficient is used as similarity metric, building groups with users with a pairwise similarity $\alpha \geq 0$. Particularly, we will build groups of size 4; 3 out of these 4 users belonging to the *advantaged* demographic user class, and the fourth user belonging to the *disadvantaged* user class. In this context, the goal of the current experiment is to characterize the Unfairn value for each of the 4 members of the group (Equation 2), as well as the overall Unfairn value for the whole group (Equation 3).

The used GRS approach was the rating aggregation-based GRS approach with the average aggregation function[4].

Table 1 illustrates the unfairness degree associated to the mentioned dataset, according to three possible disadvantaged classes of users, which are $Gender = Woman$, $Age < 25$, and $Age > 55$. The selection of these three classes follows a common sense connected to the user gender (F), to young people ($Age < 25$), and old people ($Age > 55$). In addition, it is also presented the average unfairness associated to the users in the advantaged class (M, $Age \geq 25$, and $Age \leq 55$).

According to Table 1, the larger unfairness values were associated to the gender attribute. In this case, the selected disadvantaged class was associated to the larger unfairness values, lying around 0.2454. However, in the case of the age criteria, it was obtained a higher unfairness value for the advantaged class in relation to the disadvantaged class. Beyond this result, it was obtained a larger unfairness value for the criteria $Age > 55$ in relation to $Age < 25$.

Table 2 illustrates the average unfairness values associated to the groups. Here it is interesting to point out that even though the disadvantaged class of users with the larger unfairness according to Table 1 was $Gender = Woman$, in this second table the larger group unfairness was associated to $Age > 55$. This behavior can be explained by the fact that in the case of gender it was obtained a high individual unfairness according to Equation 2 for both disadvantaged and advantaged classes, leading to a lower unfairness

Table 1. Average unfairness values according to Equation 2, for disadvantaged and advantaged classes in group members

	Gender	Age	Age
Disadvantaged users	$Gender = Woman$ 0.2454±0.1893	$Age < 25$ 0.1572±0.1202	$Age > 55$ 0.1866±0.1476
Advantaged users	$Gender = Man$ 0.2070±0.1469	$Age >= 25$ 0.2007±0.1516	$Age <= 55$ 0.2312±0.1573

Table 2. Average unfairness values for the identified groups and the disadvantaged classes of users, according to Equation 3

	Gender	Age	Age
Group unfairness	$Gender = Woman$ 0.1790±0.1029	$Age < 25$ 0.1455±0.0741	$Age > 55$ 0.2037±0.1291

for the groups according to Equation 3 that is focused on measuring the difference between unfairness at advantaged and disadvantaged classes. In a different direction, the lower unfairness value was obtained for "$Age < 25$".

Overall, the experimental results lead to the following findings:

-The approach exposed at Section 3 for characterizing unfainess in group recommendation, is able to identify different unfairness levels in GRS.

-The difference in the results obtained in Tables 1 and 2 regarding the larger unfairness value ($Gender = Woman$ in Table 1; $Age > 55$ in Table 2), illustrate that it would be interesting to manage two different unfairness notions in this GRS scenario, which are 1) a different prediction nature (over or under the actual preferences) in relation to the other group members, and 2) a different prediction nature in relation to the users belonging to other demographic attributes.

-The criterion $Age < 25$ leads to the lower unfairness value regarding the considered disadvantaged classes, being interesting the exploration of other ages based on the same criteria.

5. Conclusions

This research work has been centered on bringing a novel framework for characterizing demographic parity fairness in group recommendation. The performed experiments illustrate that the proposal is able to distinguish different fairness levels in demographic attributes, such as gender and age. Future works will be focused on studying other kinds of attributes and other fairness dimensions in the GRS scenario.

Acknowledgments

This research is supported by the European Union's Horizon Europe research and innovation program under the Marie Sklodowska-Curie grant agreement number 101106164; and by the Andalusian PAIDI 2020 Program under the project PROYEXCEL_00257.

References

1. J. Lu, D. Wu, M. Mao, W. Wang and G. Zhang, Recommender system application developments: a survey, *Decision Support Systems* **74**, 12 (2015).
2. R. Burke, Multisided fairness for recommendation, in *Workshop on Fairness, Accountability, and Transparency in Machine Learning*, 2017.

3. Y. Pérez-Almaguer, R. Yera, A. A. Alzahrani and L. Martínez, Content-based group recommender systems: a general taxonomy and further improvements, *Expert Systems with Applications* **184**, p. 115444 (2021).
4. T. De Pessemier, S. Dooms and L. Martens, Comparison of group recommendation algorithms, *Multimedia Tools and Applications* **72**, 2497 (2014).
5. J. Bobadilla, R. Lara-Cabrera, A. Gonzalez-Prieto and F. Ortega, Deepfair: Deep learning for improving fairness in recommender systems., *International Journal of Interactive Multimedia and Artificial Intelligence* **6**, 86 (2021).
6. Y. Deldjoo, D. Jannach, A. Bellogin, A. Difonzo and D. Zanzonelli, Fairness in recommender systems: research landscape and future directions, *User Modeling and User-Adapted Interaction* , 1 (2023).
7. D. Sacharidis, Top-n group recommendations with fairness, in *Proceedings of the 34th ACM/SIGAPP Symposium on Applied Computing*, 2019.
8. M. Kaya, D. Bridge and N. Tintarev, Ensuring fairness in group recommendations by rank-sensitive balancing of relevance, in *Proceedings of the 14th ACM Conference on Recommender Systems*, 2020.
9. E. Pitoura, K. Stefanidis and G. Koutrika, Fairness in rankings and recommendations: An overview, *The VLDB Journal* **31**, 431 (2022).
10. L. Xiao, Z. Min, Z. Yongfeng, G. Zhaoquan, L. Yiqun and M. Shaoping, Fairness-aware group recommendation with pareto-efficiency, in *Proceedings of the eleventh ACM conference on recommender systems*, 2017.
11. D. Serbos, S. Qi, N. Mamoulis, E. Pitoura and P. Tsaparas, Fairness in package-to-group recommendations, in *Proceedings of the 26th international conference on world wide web*, 2017.
12. M. Stratigi, J. Nummenmaa, E. Pitoura and K. Stefanidis, Fair sequential group recommendations, in *Proceedings of the 35th Annual ACM Symposium on Applied Computing*, 2020.
13. A. Felfernig, L. Boratto, M. Stettinger, M. Tkalčič *et al.*, *Group recommender systems: An introduction* (Springer, 2018).
14. F. M. Harper and J. A. Konstan, The movielens datasets: History and context, *ACM Transactions on Interactive Intelligent Systems (TIIS)* **5**, 1 (2015).

TOROS: Target Oriented O(n) Recommender System

Seda Polat Erdeniz*,‡, Ilhan Adiyaman*, Tevfik Ince*, Ata Gür* and Alexander Felfernig†,§
*Frizbit S.L., P.O. Box 1212, Barcelona, Spain
†Graz University of Technology, Graz, Austria
‡team@frizbit.com
§felfernig@ist.tugraz.at

Scalability challenges in recommender systems refer to the difficulties that arise when maintaining systems that can handle growing datasets. On the other hand, state-of-the-art recommender systems are focusing only on increasing the number of transactions (by using evaluation metrics based on rating or ranking). However, the success of a recommender system may be reflected in business metrics, such as increased sales, revenue, user retention, or customer satisfaction. In this chapter, we aim to overcome these two challenges together: "how to define own targets (evaluation metrics) on a recommender system?" and in the meanwhile "how to scale it on big data?". We proposed a collaborative filtering method called "TOROS: Target Oriented *O(n)* Recommender System". TOROS reduces the similarity calculation complexity from $O(n^2)$ to $O(n)$ and it has been evaluated on both publicly available datasets and also real-world e-commerce datasets of an e-commerce services provider company Frizbit S.L. We have compared TOROS with state-of-the-art recommender system algorithms and evaluated based on time and space consumption yet. As future work, we also evaluate the efficiency of TOROS in terms of various business-specific targets.

Keywords: KNN; recommender systems; evaluation metrics; big data.

1. Introduction

Recommender systems[1] in e-commerce contribute to improved user experiences, increased sales and revenue, customer engagement and retention, enhanced user satisfaction, and a competitive advantage in the market. In recommender systems, there are two common challenges which we focus in this chapter: scalability and evaluation.

Scalability challenges in recommender systems arise when these systems need to handle large and growing datasets. As the data volume grows, the computational demands for data storage, retrieval, and processing increase significantly. Scalability becomes a concern when dealing with a large user

‡Corresponding author.

base. Maintaining user profiles, preferences, and interaction history for a substantial number of users requires efficient storage and retrieval mechanisms. Similarly, the size of the item catalog can be extensive in e-commerce and content recommendation systems. Efficiently indexing and managing a large number of items can be challenging.

Evaluation of a recommender system is essential to assess its performance and determine how effectively it meets its goals. Several evaluation techniques and metrics can be used to assess different aspects of a recommender system. Recommender systems can be evaluated based on *Rating* (e.g., MAE and RMSE), *Ranking* (e.g., precision and recall), or other *Business-specific* metrics (increased sales, revenue, user retention, or customer satisfaction, etc.). State-of-the-art recommender models are supporting rating and ranking metrics but not business-specific metrics.

To overcome these challenges together, we proposed a novel method "TOROS" which addresses the scalability limitations by scoring users and items. This enables us to scale the implementations of both *item-based* and *user-based* similarity approaches linearly, with the size of the data, which is $O(n)$. On the other hand, its objective function can be defined to maximize business specific goals, such as boosting the revenue.

2. Background

In this section, we provide background information by describing item and user based collaborative filtering, evaluation metrics in recommender systems and two algorithms which are the basis algorithms (K-Nearest Neighbors and Global Baseline Estimator) of our proposed method.

Collaborative Filtering[2] recommends items based on the preferences of similar users or the similarity of items themselves. It can be further divided into two sub-types: user-based and item-based collaborative filtering. *User-Based Collaborative Filtering* identifies users who have similar preferences to the target user and recommends items that those similar users have liked. It assumes that users with similar tastes will have similar preferences for items. In *Item-Based Collaborative Filtering*, the system identifies similar items based on user preferences and recommends items that are similar to the ones the user has already shown interest in. It assumes that if a user likes one item, they are likely to enjoy similar items.

Evaluation metrics of recommender systems are still under research due to the unique nature of these systems and the need to assess their performance accurately. In default evaluation metrics for recommender systems, the emphasis is typically on technical and algorithmic performance rather than business metrics. Default evaluation metrics are designed to assess the accuracy, effectiveness, and quality of recommendations based on historical user-item interactions. State-of-the-art machine learning libraries for recommender systems (e.g., microsoft-recommenders[3], Surprise[4]) are all using *traditional evaluation metrics* : *Rating Metrics* (how accurate a recommender is at predicting ratings that users gave to items. For example: Root Mean Square Error (RMSE), R Squared (R2), Mean Absolute Error (MAE), and Explained Variance), *Ranking Metrics* (how relevant recommendations are for users. For example: Precision, Recall, Normalized Discounted Cumulative Gain (NDCG), and Mean Average Precision (MAP)), *Classification metrics* (binary labels. For example: Area Under Curver (AUC), and Logistic loss (Logloss)).

These *traditional evaluation metrics* are valuable for assessing the technical performance of the recommendation algorithms, they don't always directly translate into business success. However, *business metrics* are directly tied to the strategic objectives of the company. They measure how well the recommender system contributes to achieving these goals, such as increasing revenue, improving customer retention, or boosting user engagement. By focusing on business metrics, organizations can ensure that their recommender system investments align with their broader objectives.

K-Nearest Neighbors (KNN)[5] is a popular machine learning algorithm used for both classification and regression tasks. It has also a wide usage in recommender systems. KNN is relatively simple and intuitive, but it can be computationally expensive, especially when dealing with large datasets or high-dimensional feature spaces. For an offline recommender system, calculating the complete similarity matrix with KNN has $O(n^2)$ time and space complexity where n is the number of users.

There are several approaches applied on KNN to make it scalable such as approximate KNN methods, using dimensionality reduction techniques, clustering techniques, data sampling, indexing structures, using parallel computing frameworks like Sparkor Hadoop, and using incremental learning techniques. However, scalability of KNN is still a research topic in recommender systems domain. Moreover, applying business-specific evaluation metrics in the objective function is also not provided by any KNN library.

Global Baseline Estimator (GBE)[6] is a technique with $O(n)$ complexity which is used in recommender systems to estimate the baseline ratings or preferences of users or items. It provides a simple baseline prediction that is independent of the specific user or item characteristics. The global baseline helps in modeling the overall tendencies or biases in data.

3. Proposed Method: TOROS

Aforementioned, in recommender systems there are open research questions in terms of scalability and evaluation. To solve those scalability and evaluation challenges, our proposed method TOROS combines KNN and GBE to employ the advantages of both methods. We employ GBE to reach $O(n)$ complexity and adapt its scoring function into an objective function for an adaptable target. An objective function with adaptable targets gives us opportunity to also define business-specific metrics (e.g., revenue boost), instead of only using the available metrics (RMSE or precision) in the available models as mentioned above. We combine it with KNN because KNN approach can be applied also for both item based collaborative filtering.

Algorithm 3.1 TOROS

```
Function scoring():
    μ = stdDev(all ratings)
    For each u in n_users:
        score_u = stdDev(ratings of u)
    For each i in m_items:
        score_i = stdDev(ratings of i)
Function TOROS_user_based_ratings():
    For each u in n_users:
        For each i in m_items:
            if rating_ui == 0:
                For each u2 in k_nearest_scored_users:
                    rating_ui += score_u2
                rating_ui = μ + (rating_ui)/k
Function TOROS_item_based_ratings():
    For each u in n_users:
        For each i in m_items:
            if rating_ui == 0:
                For each i2 in k_nearest_scored_items:
                    rating_ui += score_i2
                rating_ui = μ + (rating_ui)/k
```

As shown in Algorithm 3.1, we defined a scoring function which is used to calculate similarity. Instead of computing the complete similarity matrix, we use the scores of users (or items) to find the k nearest neighbors of them. In other words, if score of an item is 0.81, the closest scored three nearest neighbors can have scores 0.78, 0.82, 0.87. Same strategy applies when calculating nearest neighbors of a user. This scoring based similarity calculation decreases the computational dimensions from 2d-array (a complete similarity matrix) to 1d-array (scores array). Thus, TOROS has linear ($O(n)$) complexity. After scoring function calculates scores of all items and users, ratings are calculated based on the nearest score items and users. In order to keep it simple to demonstrate in Algorihm 3.1, we did not put a function for getting the k nearest scored items/users. It is straightforward once scores are calculated and sorted.

In the basic version, we use standard deviation as in GBE. However, according to the targeted evaluation metrics, we can adapt scoring function into many versions. Using GBE helps us to find similarly rating users or similarly rated items. In this scenario, nearest neighbors of a user would be users with similar rating behaviours (highly critical users or users who are giving always positive feedback).

4. Evaluation

In this section, we evaluate our proposed method in terms of time and space complexity by comparing with state of the art recommender system algorithms on the various datasets using Surprise library[4] in Python using Global Baseline Estimator (GBE)[6], KNN item-based[4], KNN user-based[4], Matrix Factorization (MF) with SVD[7].

Table 1. Results on Movielens 100k data with 0.9k users and 1.6k items.

	Movielens Data (100k)		
	Training Time (s)	Testing Time (s)	Memory (GB)
GBE	0.21	0.10	0.012
KNN item-based	0.60	2.77	0.033
KNN user-based	0.61	1.71	0.019
Matrix Fact. SVD	0.68	0.12	0.026
TOROS item-b.	0.20	1.02	0.011
TOROS user-b.	0.21	0.17	0.012

As observed in Table 1, the most efficient methods in terms of runtime are the $O(n)$ methods like GBE and TOROS. However, as discussed earlier in this paper, GBE does not support item based recommendation and business-metrics based training. Thus, TOROS is a more advantageous option than GBE in terms of its business-metrics based training opportunity.

On the other hand, memory consumption of KNN item based is higher than KNN user based since number of items (1.6k) are higher than number of users (0.9k) in this dataset. For user based similarity, a similarity matrix with the shape 1.6kx1.6k is prepared wheras for item based similarity it is 0.9kx0.9k. However, actually in most of the real world scenarios, number of users are much more higher than number of items on an e-commerce system. Therefore, item based collaborative filtering methods become more computationally efficient for such systems. Therefore, we also tested the performance on our real world events data.

For the second test, we used 10 days transactions data of an e-commerce customer of Frizbit. In this small amount of time range, data includes 6.4M events with around 1.5M users and 20k products. For this second test, we have selected algorithms as in Table 2 because of the data size.

Table 2. Results on Frizbit data: 6.4M ratings, 1.4M users, and 20k items.

	Frizbit Data (6.4M)		
	Training Time (s)	Testing Time (s)	Memory (GB)
GBE	63.27	45.35	3.2
KNN user-based	Error	Error	15098.02+
KNN item-based	1440.23	62.17	15.12
MF SVD	3246.28	124.35	12.24
TOROS item-based	78.34	63.87	4.1
TOROS user-based	82.13	56.34	5.2

As also shown in Table 2, KNN training on Frizbit 6.4M dataset needs 15.1 TB memory allocation as provided in the error message from Surprise library, so it could not be trained. We can understand that creating a user similarities matrix for around 1.5M users need 15.1 TB space. However, we used a computer with 60 GB memory. On the same computer, KNN item-based did not create a problem, for around 20k items it used 15 GB memory. Therefore, with this machine's capacity and the needed space allocation by KNN similarity matrix, we can only train KNN item based for maximum around 80k items, but user based similarity is impossible since the number of users in the data is much more than 80k.

GBE has again a good scalability performance on this data. However, its disadvantages remains the same: not available for item-based recommendation and no business-metrics based training opportunity. This means, TOROS brings a solution for collaborative filtering on high amount of users (1M+) and ratings (6M+) data. Using TOROS, both item and user based similarities can be calculated effectively with also training based on business-specific metrics.

5. Future Work

In this paper, we have evaluated the proposed method TOROS in terms of scalability performance (with metrics time and memory consumption). Therefore, as future work we would like to work on defining new scoring functions to increase the recommendation quality in terms of accuracy and business-specific metrics (such as revenue).

For example, if the evaluation metric wanted to be set as *High Revenue*, the scoring function can be adapted to recommend high price products which can be interesting for the user. Let's assume Tom and Bob are buying products on a camping products seller e-commerce company. If the recommender system of this store is using a standard collaborative filtering approach which is trained only according to Ratings, Camping Tent B is recommended to both Tom and Bob, so the revenue of the company will be 600$. However, if the recommender system of this store is adapted a "High Revenue" quality measure using ScoredKNN, Camping Tent A is recommended to Tom, and Camping Tent B is recommended to Bob, so the revenue of the company will be 1200$. Therefore, the *Opportunity Cost* of ignoring "High Revenue" as a quality metric is 600$ in this example.

Table 3. Example user acceptances for two recommendable items

	Camping Tent A (900$, Rating:4.8)	Camping Tent B (300$, Rating:5)
Tom (High Score)	ACCEPT	ACCEPT
Bob (Low Score)	REJECT	ACCEPT

6. Conclusions

Scalability is a common issue of recommender systems which occurs very often with a large number of users. Moreover, state-of-the-art recommender models do not support business-specific evaluation metrics. In this chapter, we proposed a method to overcome these challenges by using a neighborhood-based approach called "TOROS". We have evaluated TOROS by comparing with publicly available collaborative filtering algorithms based on popular Movielens dataset and also our own e-commerce transactions dataset at Frizbit S.L. As a result, we have observed that TOROS makes item and user-based collaborative filtering feasible on the big data since it reduces the complexity of KNN from $O(n^2)$ to $O(n)$. Moreover, its objective function can be easily adapted to any business-specific evaluation metric. This makes it very flexible for directly boosting business-specific targets instead of only focusing on rating or ranking optimization.

This research is funded by ACCIO's Tecniospring INDUSTRY program with the project number "ACE026/21/000108".

References

1. A. Felfernig, S. Polat-Erdeniz, C. Uran, S. Reiterer, M. Atas, T. N. T. Tran, P. Azzoni, C. Kiraly and K. Dolui, An overview of recommender systems in the internet of things, *Journal of Intelligent Information Systems* **52**, 285 (2019).
2. S. P. Erdeniz, A. Felfernig, R. Samer and M. Atas, Matrix factorization based heuristics for constraint-based recommenders, in *Proceedings of the 34th ACM/SIGAPP Symposium on Applied Computing*, 2019.
3. A. Argyriou, M. González-Fierro and L. Zhang, Microsoft recommenders: Best practices for production-ready recommendation systems, in *Companion Proceedings of the Web Conference 2020*, 2020.
4. N. Hug, Surprise: A python library for recommender systems, *Journal of Open Source Software* **5**, p. 2174 (2020).
5. R. H. Singh, S. Maurya, T. Tripathi, T. Narula and G. Srivastav, Movie recommendation system using cosine similarity and knn, *International Journal of Engineering and Advanced Technology* **9**, 556 (2020).
6. Y. Koren, Factor in the neighbors: Scalable and accurate collaborative filtering, *ACM Transactions on Knowledge Discovery from Data (TKDD)* **4**, 1 (2010).
7. A. Mnih and R. R. Salakhutdinov, Probabilistic matrix factorization, *Advances in neural information processing systems* **20** (2007).

An Analysis Comparing the Step-wise Weight Assessment Ratio Analysis (SWARA) and the Cardinal Sum Reciprocal (CSR) Weighting Methods Based on Simulation Techniques

Z. Chergui and A. Jiménez-Martín*

Decision Analysis and Statistics Group, Universidad Politécnica de Madrid, Campus de Montegancedo S/N, Boadilla del Monte, 28660 Madrid, Spain
**antonio.jimenez@upm.es*

Many weighting methods based on ordinal information have been proposed by different authors within multi-attribute utility/value theory. Moreover, the literature has also addressed the possibility of considering additional information regarding the ranking of criteria, mainly focused on the strength of the differences between the weights of consecutive criteria. However, although comparison analyses have been carried out to analyze the performance of weighting methods based on ordinal information, no comparative analyses of these methods have been conducted when additional information is available. In this chapter, we compare the step-wise weight assessment ratio analysis (SWARA) and the cardinal sum reciprocal (CSR) using Monte Carlo simulation techniques on the basis of the hit ratio and the rank order correlation as quality measures.

Keywords: Weighting methods; ordinal information; comparison analyses; SWARA; CSR.

1. Introduction

The additive model is considered as a valid approach for most real decision-making problems within multi-attribute utility/value theory (MAUT/MAVT),[10,12] where the DM's preferences are incorporated into the analyses by means of component utility functions (u_i) and weights (w_i) representing the relative importance of criteria (attributes) as follows:

$$u(A^j) = \sum_{i=1}^{n} w_j \times u_i(x_i^j),$$

where n is the number of attributes and x_i^j is the performance on the ith attribute for alternative A^j.

Different weighting methods have been proposed in the literature. They can be divided into two groups: *subjective* methods, where the DM provides information about his/her own preferences, and *objective* methods, which consider only the mathematical features of the measuring procedures without looking at the details provided by the DM.

In this chapter, we consider subjective methods where the DM provides ordinal information, i.e., a ranking of the attributes in descending order of importance. Moreover, additional information is provided in terms of the strength of preferences between consecutive weights in the ranking. Reference 4 reviews subjective weighting methods, including a ranking of differences, a semantic scale, precise/imprecise cardinal information, ratio-based methods, or preference statements.

However, although comparison analyses have been carried out to assess the performance of weighting methods based on ordinal information, which are also reviewed in Ref. 4, there are no comparative analyses of methods accounting for additional information.

In this chapter, we perform a comparison analysis of two outstanding weighing methods based on ordinal information, which provide additional information: the *step-wise weight assessment ratio analysis* (SWARA)[7] and the *cardinal sum reciprocal* (CSR) method.[5]

In the simulation process, we consider many scenarios accounting for different numbers of criteria and alternatives. Alternative component utilities and attribute weights (TRUE weights) are first randomly generated. Then, the SWARA and SR methods are applied on the basis of the TRUE weights, and the corresponding ranking of alternatives is derived by each method. They are then compared with the TRUE ranking, derived from the TRUE weights. To do this, we use the *hit ratio* and the *rank-order correlation* as quality measures. They are the most commonly used metrics in comparison analysis.

This chapter is organized as follows. In Section 2, we provide further information about the SWARA and CSR methods. The simulation procedure is described and the results are interpreted in Section 3. Finally, some conclusions are provided in Section 4.

2. Weighting Methods based on Ordinal Information with Additional Information

Different weighting methods based on ordinal information are available in the literature. These weighting methods account for the strength of the differences between consecutive criteria in the ranking. They differ as to how this strength is expressed by the DM. In Ref. 4 four possibilities are described. The first consists of providing a ranking of the differences between consecutive criteria[2]. Another possibility is to use a semantic scale[5]. The last two alternatives consist of providing ratios between consecutive and the last criteria or using preference statements, respectively. Several methods were identified within this framework.

In this paper, we focus on two outstanding weighting methods, the cardinal sum reciprocal (CSR) method, in which uses a semantic scale, and the step-wise weight assessment ratio analysis method (SWARA), a ratio based method.

The CSR[5] method can be considered an extension of the sum reciprocal (SR) method, a surrogate weighting method. The following scale is proposed to represent the strength of the differences between consecutive criteria in the ranking: $>_0$, equally important; $>_1$, slightly more important; $>_2$, more important (clearly more important), and $>_3$, much more important.

Then, "An ordinal number is assigned to each importance scale position, starting with the most important position as number 1. Let the total number of importance scale positions be Q. Each criterion i has the position $p(i) \in \{1, ..., Q\}$ on this importance scale, such that for every two adjacent criteria c_i and c_{i+1}, whenever $c_i >_{s_i} c_{i+1}$, $s_i = |p(i+1) - p(i)|$. The position $p(i)$ then denotes the importance as stated by the DM. Thus,

$$Q = \sum_{i=1}^{n-1} s_i + 1\text{''}.$$

Finally, the CSR method allocates the weights according to:

$$w_i = \frac{1/p(i) + \frac{Q+1-p(i)}{Q}}{\sum\limits_{j=1}^{n}\left(1/p(j) + \frac{Q+1-p(j)}{Q}\right)}, i = 1, ..., n \tag{1}$$

The performance of the CSR method has been analyzed using simulation techniques[5,8,9], and it was found to outperform other cardinal and linear programming methods in all cases.

Besides, the SWARA method[7] is a ratio-based method, where, after scoring, the differences of the consecutive criteria are determined to be incorporated in the weight elicitation as follows:

$$Y_i = \frac{Y_{i-1}}{1 + (s_i - s_{i-1})}, \quad i = 2, ..., n, \tag{2}$$

with $Y_1 = 1$ and where s_i is the score provided by the DM for the i-th criterion.

Once the Y_i, $i = 2, ..., n$, are computed, a normalization process is performed to derive the weights: $w_i = \frac{Y_i}{\sum_i^n Y_i}$, $i = 1, ..., n$.

The of SWARA method's main feature is the possibility to estimate the opinion of DMs or interest groups about the significance ratio of the attributes in the process of their weight elicitation. SWARA method has been applied in several real decision-making problems and comparison studies[6,11,13]. However, it has never been compared to a semantic scale-based method.

3. Simulation Process and Results

In the simulation process, we consider many scenarios accounting for different numbers of criteria $n \in \{3, 6, 10, 15, 25\}$ and alternatives $m \in \{3, 7, 10, 15, 20\}$. Ten thousand random instances are generated for each scenario.

Fig. 1 summarizes the simulation process for each instance. First (in black), a component utility matrix, $U_{m\times n}$, is randomly generated using a uniform distribution. This is normalized, leading to $\mathbf{U}'$. Then, weights are randomly generated from a uniform or normal distribution. They are then normalized, leading to the TRUE weight vector, $\mathbf{w} = (w_1, ..., w_n)$. Finally, the additive model is used to evaluate the alternatives under consideration on the basis of $\mathbf{U}'$ and $\mathbf{w}$ to derive a ranking of the alternatives: the TRUE ranking.

Note that, as pointed out above, a normal distribution $N(5, 2.5)$ in addition to a uniform distribution in $[0, 1]$ will be considered to randomly generate weights. Refs. 3 and 1 describe how the weights are generated and normalized, respectively. The results derived for each of the two distributions, which represent different situations, are analyzed.

Secondly (in green), the SWARA method is applied on the basis of the randomly generated weights after normalization, r_j, $j = 1, ..., n$, which are multiplied by 100, applying Eq. (2) and the corresponding normalization process. The additive model is again used on the basis of $\mathbf{U}'$ and the weight vector derived by SWARA to obtain a ranking of the alternatives: the SWARA ranking.

Next (in red), the CSR method is applied. The CSR weight vector is extracted from TRUE weight vector. First, ordinal information is directly obtained from the TRUE weight vector, whose elements are ordered from

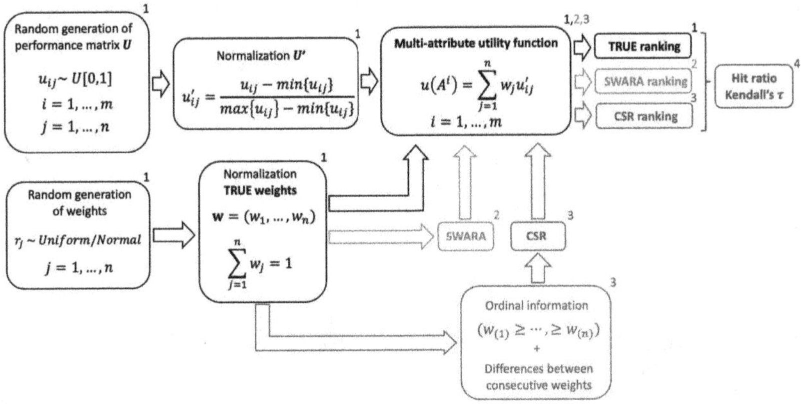

Fig. 1. Simulation process

the highest to the lowest $(w_{(1)} \geq \cdots \geq w_{(n)})$. Then, the difference between each two consecutive criteria $(w_{(i)} - w_{(i+1)})$, $i = 1, \ldots, n-1$, is computed and the value of the best criteria is set $p(1) = 1$. Subsequently, according to the difference values, $p(i)$, $i = 2, \ldots, n$, quantities are defined subject to the conditions below:

- if $(w_i - w_{i+1}) < \frac{1}{4n}$, then $p(i) = p(i-1)$ (equally important),
- if $\frac{1}{4n} < (w_i - w_{i+1}) \leq \frac{1}{2n}$, then $p(i) = p(i-1) + 1$ (slightly more important),
- if $\frac{1}{2n} < (w_i - w_{i+1}) \leq \frac{1}{n}$, then $p(i) = p(i-1) + 2$ (more important),
- if $\frac{1}{n} < (w_i - w_{i+1})$, then $p(i) = p(i-1) + 3$ (much more important).

Then, Eq.(1) is applied to derive the CSR weight vector.

Finally, the TRUE ranking is compared to the CSR and the SWARA rankings across the 10000 instances executed for the scenario under consideration to analyze their performance. To do this, we use the hit ratio and the rank-order correlation as quality measures. They are the most commonly used metrics in comparison analysis.

The hit ratio represents the proportion of the 10000 instances executed in which the corresponding method (CSR or SWARA) outputs the same best alternative as in the TRUE ranking. The rank-order correlation (Kendall's τ) measures how similar the overall rank structures of alternatives are in the TRUE ranking and in the ranking derived from the corresponding method. We take the average of the rank-order correlations across the 10000 instances.

Table 1 shows the hit ratios and rank-order correlation (Kendall's τ) averages for the different scenarios under consideration and for the two distributions used to generate the TRUE weights: the uniform and normal distributions. The bottom row reports the average values for all the scenarios under consideration.

Table 1. Simulation results

		Hit ratio (%)				Kendall's τ			
		Normal distr.		Uniform distr.		Normal distr.		Uniform distr.	
Alt.	Crit.	CSR	SWARA	CSR	SWARA	CSR	SWARA	CSR	SWARA
3	3	94.70	96.02	94.87	82.14	0.9279	0.9481	0.9294	0.7630
	6	93.58	94.45	93.14	75.39	0.9150	0.9250	0.9068	0.6626
	10	93.16	93.76	90.79	71.09	0.9083	0.9172	0.8738	0.6004
	15	93.52	93.85	88.99	69.77	0.9120	0.9170	0.8493	0.5761
	25	93.38	93.67	85.07	66.62	0.9097	0.9132	0.7975	0.5272
7	3	92.66	94.33	91.18	75.11	0.9332	0.9515	0.9262	0.7793
	6	90.60	91.80	88.76	65.02	0.9169	0.9274	0.9041	0.6770
	10	89.84	90.96	85.94	59.18	0.9133	0.9210	0.8758	0.6120
	15	89.64	90.29	82.96	55.78	0.9095	0.9146	0.8499	0.5766
	25	88.94	89.31	78.04	51.73	0.9082	0.9113	0.8013	0.5318
10	3	92.16	93.99	89.42	72.27	0.9331	0.9507	0.9251	0.7778
	6	89.50	90.91	86.18	61.09	0.9189	0.9291	0.9026	0.6728
	10	88.73	89.80	83.72	54.93	0.9124	0.9199	0.8780	0.6145
	15	87.80	88.45	80.84	51.28	0.9091	0.9146	0.8473	0.5779
	25	88.27	88.73	75.06	47.06	0.9083	0.9116	0.8030	0.5354
15	3	92.15	94.01	87.95	69.49	0.9339	0.9514	0.9255	0.7805
	6	88.77	90.12	84.98	57.65	0.9182	0.9290	0.9025	0.6728
	10	87.59	88.61	81.60	50.76	0.9123	0.9194	0.8770	0.6147
	15	86.84	87.32	78.35	46.59	0.9097	0.9148	0.8487	0.5756
	25	86.74	87.18	72.56	43.57	0.9073	0.9109	0.8045	0.5342
20	3	91.39	93.61	87.06	68.09	0.9343	0.9514	0.9260	0.7806
	6	88.23	89.64	84.53	56.76	0.9192	0.9294	0.9032	0.6736
	10	86.36	87.47	80.14	48.60	0.9119	0.9194	0.8769	0.6161
	15	85.90	86.66	76.43	44.05	0.9098	0.9151	0.8470	0.5765
	25	85.84	86.29	70.12	39.50	0.9070	0.9105	0.8044	0.5352
average		90.11	91.10	0.917	0.924	0.915	0.924	0.915	0.925

First, we find that, as expected, the greater the number of alternatives is, the worse the values for the two metrics are with respect to both distributions under consideration. The same applies for the number of criteria for any given number of alternatives.

Besides, the results of both methods are very similar when the normal distribution is used, although SWARA slightly outperforms CSR in all scenarios for both metrics, with the difference becoming smaller and smaller as the number of criteria increases. However, CSR clearly outperforms SWARA when a uniform distribution is used in both the hit ratio and Kendall's τ for all the scenarios under consideration. The difference now increases as the number of criteria grows.

Taking into account the fact that the randomly generated weights from a normal distribution tend to be similar and that the uniform distribution leads to weight vectors accounting for all possibilities (from similar to very different weights), we should attach more importance to the results obtained with the uniform distribution and interpret the results for the normal distribution as referring to of a particular situation.

4. Conclusions

We have compared the performance of two weighting methods based on ordinal information which account for additional information on the strength of preferences between consecutive weights in the rank by means of either a semantic scale (CSR method) or ratios (SWARA). To do this, we have performed a Monte Carlo simulation process considering many scenarios accounting for a different number of alternatives and criteria, together with two alternative random weight generation distributions.

We have found that the performance of both methods is very similar, although SWARA slightly outperforms CSR when the normal distribution is used. In the more general case when a uniform distribution is used, however, CSR clearly outperforms SWARA for all the scenarios under consideration, and the difference increases as the number of attributes grows.

On the basis of the simulation results, therefore, we recommend CSR, especially for a high number of attributes. Moreover, some DMs might find it difficult to provide scores in SWARA, and we consider that it is less stressful for them to provide ordinal information in conjunction with linguistic terms.

Acknowledgments

This work was supported by the Grants PID2021-122209OB-C31 and RED2022-134540-T funded by MICIU/AEI/10.13039/501100011033.

References

1. E.A. Aguayo, A. Mateos and A. Jiménez-Martín, A new dominance intensity method to deal with ordinal information about a DM's preferences within MAVT, *Knowl. Based Syst.* **69**, 159 (2014).

2. B. S. Ahn, Approximate weighting method for multiattribute decision problems with imprecise parameters, *Omega* **72**, 87 (2017).
3. H.K. Alfares and S. Duffuaa, Simulation-based evaluation of criteria rank-weighting methods in multi-criteria decision-making, *Int. J. Inf. Tech. Decis. Making* **15**(1), 43 (2016).
4. Z. Chergui and A. Jiménez-Martín, On ordinal information-based weighting methods and comparison analyses, *Omega*, under review, (2014).
5. M. Danielson and L. Ekenberg, A robustness study of state-of-the-art surrogate weights for MCDM, *Group Decis. Negot.* **26**, 677 (2017).
6. S.J. Ghoushchi, H. Garg, S.R. Bonab and A. Rahimi, An integrated SWARA-CODAS decision-making algorithm with spherical fuzzy information for clean energy barriers evaluation, *Expert Syst. Appl.* **223**, 119884 (2023).
7. V. Kersuliene, E.K. Zavadskas and Z. Turskis, Selection of rational dispute resolution method by applying new step-wise weight assessment ratio analysis (SWARA), *J. Bus. Econ. Manag.* **11**, 243 (2010).
8. S. Lakmayer, M. Danielson and L. Ekenberg, Automatically generated weight methods for human and machine decision-making. In H. Fujita, Y. Wang, Y., Xiao & A. Moonis (eds). *Advances and Trends in Artificial Intelligence. Theory and Applications. LNCS* **13925**, 195 (Springer, 2023).
9. S. Lakmayer, M. Danielson and L. Ekenberg, Aspects of ranking algorithms in multi-criteria decision support systems. In H. Fujita & G. Guizzi (eds). *New Trends in Intelligent Software Methodologies, Tools and Techniques. Frontiers in Artificial Intelligence and Applications* **371**, 63 (IOS Press, 2023).
10. H. Raiffa, *The Art and Science of Negotiation* (Harvard University Press, Cambridge, 1982).
11. T. Sivageerthi, S. Bathrinath, M. Uthayakumar and R.K.A. Bhalaji, A SWARA method to analyze the risks in coal supply chain management, *Materials Today: Proceedings*, **50** (5), 935 (2022).
12. T.J. Stewart, Robustness of additive value function method in MCDM, *J. Multi-Crit. Decis. Anal.* **5**, 301 (1996).
13. N. Tayyar and M. Durmu, Comparison of Max100, SWARA and pairwise weight elicitation methods. *Int. J. Eng. Res. Appl.* **7**, 67 (2017).

Simulation-informed Gaussian Processes for Accelerated Bayesian Optimisation

Santiago Ramos Garces[*,†,§], Ivan De Boi[*], Rudi Penne[*], João Pedro Ramos[†], Marc Dierckx[†] and Stijn Derammelaere[*,‡]

[*] *Department of Electromechanics, CoSysLab/InViLab, University of Antwerp, Groeneborgerlaan 171, Antwerp 2020, Belgium*
[†] *Belgian Nuclear Research Centre SCK CEN, Boeretang 200, Mol 2400, Belgium*
[‡] *AnSyMo/Cosys, Flanders Make*
[§] *santiago.ramosgarces@uantwerpen.be*

Bayesian optimisation, known for its minimal requirement of design parameter evaluations, is vital for global industrial process optimisation. It typically employs Gaussian processes as surrogate models for specific objectives. Traditional Bayesian optimisation is directly applied to the actual system. However, numerous industrial applications rely on simulation models. Although these models fail to represent the real system fully, they offer valuable insights to improve optimisation algorithms. A model that relies on the transfer of physical knowledge from the simulations to a Gaussian process model of the actual system is called a *physics-informed Gaussian process* model. Inspired by this, this chapter proposes a novel approach called *simulation-informed Gaussian process.* This approach constructs a Gaussian process kernel from simulation results to better capture design parameter-objective function correlations. This results in an accelerated Bayesian optimisation convergence of the actual system. We show this by comparing our method to conventional and physics-informed Bayesian optimisation. In addition, we offer insights into the consequences of integrating potentially misleading information into the Gaussian process framework.

Keywords: Gaussian process; Bayesian optimisation; simulation-informed.

1. Introduction

Optimising some industrial applications encounters the challenge of expensive objective function evaluations or large input parameter spaces. This is evident in diverse fields, such as polymer design[1] and accelerator tuning.[2] These fields demand efficient optimisation algorithms capable of identifying optimal solutions with limited evaluations. Simulation models approximating system behaviour are beneficial in such optimisations. However, they often fail to capture complete system dynamics, prompting a shift towards data-driven machine learning methods. One such approach is implementing Gaussian processes[3] (GP) as a surrogate model of the underlying expensive objective function.

The advent of so-called physics-informed machine learning, integrating physical laws or simulations into these methods, represents notable progress. This is particularly significant in solving Partial Differential Equations (PDEs) using Neural networks and GPs.[4–6] Another noteworthy technique is the multi-fidelity GP,[7] merging information from various sources with differing fidelity levels to enhance the surrogate model. However, these methods often demand substantial computational resources, limiting their practical use in optimising expensive functions or scenarios with a single information source. Bayesian optimisation,[8,9] widely used for the global optimisation of expensive objective functions, effectively integrates with GPs. Developing a physics-informed Gaussian process (PI-GP) means integrating simulation data into the GP to enhance the BO convergence. A notable method in the field[2] replaces the conventional GP kernel with one including the simulation model's Hessian, introducing information about the function near the optimum to speed up convergence.

This chapter introduces an alternative PI-GP construction, considering that the initial design parameter setting influences the Hessian and might be based on a local rather than a global optimum. We distinguish between the *simulation model* (used for kernel computation of the PI-GP) and the *application model* (representing the actual system). Our GP has a kernel constructed by learning the correlation between design parameters and the optimal solution directly from the simulation model. After fully precomputing this kernel, it is used in the Bayesian optimisation process for the application model. We dub our method *simulation-informed* Gaussian process (SI-GP). Unlike multi-fidelity GP, which fuses multi-source simulation data, SI-GP transfers knowledge from the simulation model to the GP of the application model. This characteristic sets it apart from other PI approaches, which combine measurements from the actual system with simulated data to construct the kernel.[6] When tested on an application model, our simulation-informed kernel construction outperforms the conventional GP BO and the PI-GP BO incorporating the Hessian. Our findings highlight the effect of incorporating inaccurate information into the GP, which can degrade performance and, in some cases, make a conventional GP more effective.

2. Gaussian Process Formulation

Gaussian process regression aims to predict an unknown function $\mathbf{y} = f(\mathbf{x}) + \epsilon$, where ϵ is zero-mean Gaussian-distributed noise with variance σ_r^2. In this context, $\mathbf{x} = (x_1, \ ..., \ x_d) \in \mathbb{R}^d$ represents an input location, and $\mathcal{F} = \{\mathbf{X}, \mathbf{y}\}$ denotes the set of N observations from the function, where $\mathbf{y} = \{\mathrm{y}_1, \ ..., \ \mathrm{y}_N\}^T \in \mathbb{R}^N$, and $\mathbf{X} = \{\mathbf{x}_1; \ ...; \ \mathbf{x}_N\} \in \mathbb{R}^{N\times d}$. The GP can be interpreted as a distribution over functions[3], expressed as $\mathbf{y} \sim \mathcal{GP}(\boldsymbol{\mu}(\mathbf{x}), k(\mathbf{x}, \mathbf{x}'))$, with mean $\boldsymbol{\mu}(\mathbf{x})$ and kernel function $k(\mathbf{x}, \mathbf{x}')$. In this work, we assume smooth varying functions without discontinuities, a condition met by employing the squared exponential kernel, which inherently yields such functions. It is described as

$$k(\mathbf{x}, \mathbf{x}') = \sigma_f^2 \exp\left[-0.5\,(\mathbf{x} - \mathbf{x}')\,\boldsymbol{\Lambda}\,(\mathbf{x} - \mathbf{x}')^T\right], \tag{1}$$

where $\boldsymbol{\Lambda} = \mathrm{diag}(\mathbf{l})^{-2} \in \mathbb{R}^{d\times d}$ and $\mathbf{l} = (l_1, \ ..., \ l_d)$. The final aim of GP regression is to make M predictions $\widehat{\mathbf{y}} \sim f(\mathbf{X}_*)$ at query points $\mathbf{X}_* = \{\mathbf{x}_{*1}; \ ...; \ \mathbf{x}_{*M}\}^T \in \mathbb{R}^{M\times d}$. This is achieved through the joint posterior distribution, obtained by conditioning the joint prior distribution on the observations. An in-depth explanation of GP can be found in the book[3].

3. Bayesian Optimisation

Bayesian optimisation[8,9] aims to determine the input location $\mathbf{x}^{\text{opt}}$ that minimises an objective function, $f(\mathbf{x}^{\text{opt}}) = \mathrm{argmin}_{\mathbf{x}}\, f(\mathbf{x})$, with as few function evaluations as possible. This paper adopts expectation improvement (EI) as acquisition function α. It balances exploitation, which focuses on promising local regions, and exploration, which explores the entire design space defined by the boundaries of the inputs. EI's closed-form expression is computationally efficient and lightweight, enabling online optimisation. While other acquisition functions are described in the literature[8,9], they are beyond the scope of this work.

4. Simulation-informed Gaussian Process

In the physics-informed Gaussian process[2], the Hessian matrix ($\mathbf{H}_{i,j} = \partial_{x_i}\partial_{x_j} f_{\text{sim}}(\mathbf{x})|_{\mathbf{x}=\mathbf{x}^{\text{opt}}}$) is computed via numerical differentiation around the global optimum ($\mathbf{x}^{\text{opt}}$) of the simulation's model objective function (f_{sim}). This Hessian replaces the matrix $\boldsymbol{\Lambda}$ in Eq. (1), enabling kernel computation based solely on simulations. Subsequently, the precomputed kernel is used during the Bayesian optimisation iterations of the application model. In contrast, this paper introduces an alternative method for constructing the SI-GP kernel to replace $\mathbf{\Lambda}$ in the GP. Our approach does not rely on prior knowledge of the simulation model's optimum. BO is employed for optimising the objective function of the simulation model. Each run of the BO algorithm is referred to as *simulation-informed optimisation* (SI optimisation). The set of optimal input locations obtained from n SI optimisations is denoted as $\mathbf{X}^{\text{opt}}$. Meanwhile, $\mathbf{X}^{\text{sim}}$ gathers the input locations that led to a decrease in f_{sim} during each SI optimisation, and $\mathbf{X}^{\text{SI}}$ compiles these $\mathbf{X}^{\text{sim}}$ sets from all n SI optimisation runs. The pseudo-code is outlined in Algorithm 4.1.

Algorithm 4.1 yields two primary outcomes. First, it identifies the optimal input location from the n SI optimisations. Second, it derives the matrix $\mathbf{K}^{\text{SI}}$ for SI-GP kernel construction, which replaces $\mathbf{\Lambda}$ in Eq. (1). This matrix, $\mathbf{K}^{\text{SI}}$, effectively captures input correlations that reduce the simulation model's objective function, accelerating optimisation convergence in the application model. Central to this method is physics-informed optimisation in the simulation model to pinpoint optimal regions, guiding the algorithm towards these areas during the application model optimisation. Unlike direct Hessian computation in PI-GP, this method gathers more information from the n SI optimisations. However, despite the efficiency of the proposed method, it incurs the computational cost of conducting n SI optimisations. In contrast, only a single optimisation is possibly required in PI-GP. Additionally, the computation of both $\mathbf{H}$ and $\mathbf{K}^{\text{SI}}$ incurs extra computational costs relative to conventional GP.

Algorithm 4.1. Simulation-informed Gaussian process
Input: Initial simulation model observations $\mathcal{F}_0$, query locations $\mathbf{X}_*$, initial GP model $\mathcal{M}_0 = \mathcal{N}(\boldsymbol{\mu}_0, k_0)$ built from initial observations $\mathcal{F}_0$, acquisition function α, noise ϵ, initial best location $\mathbf{x}^+$, and $\mathbf{X}_0^{\text{opt}}, \mathbf{X}_0^{\text{SI}}, \mathbf{X}_0^{\text{sim}}, \mathbf{x}_0^{\text{opt}} = \{\}$.
Output: Optimal input location of the simulation model $\mathbf{x}^{\text{opt}}$, matrix $\mathbf{K}^{\text{SI}}$.

for $s = 1$ to n **do** — Limit of SI simulations
  for $i = 1$ to h **do** — Limit of function evaluations
    $\mathbf{x}_i^* \leftarrow \text{argmax}_{\mathbf{x}\in\mathbf{X}_*}\ \alpha(\mathbf{x}|\mathcal{M}_{i-1})$
    $\mathrm{y}_i \leftarrow f_{\text{sim}}(\mathbf{x}_i^*) + \epsilon$
    $\mathcal{F}_i \leftarrow \mathcal{F}_{i-1} \cup (\mathbf{x}_i^*, \mathrm{y}_i)$ — Update observations
    $\mathcal{M}_i \leftarrow \mathcal{M}_{i-1}|\mathcal{F}_i$ — Update GP model
    if $f_{\text{sim}}(\mathbf{x}_i^*) < f_{\text{sim}}(\mathbf{x}^+)$ **then**
      $\mathbf{X}_i^{\text{sim}} \leftarrow \mathbf{X}_{i-1}^{\text{sim}} \cup (\mathbf{x}_i^*, \mathrm{y}_i)$
      $\mathbf{x}^+ \leftarrow \mathbf{x}_i^*$ — Update best location
  end

$\mathbf{X}_s^{\text{opt}} \leftarrow \mathbf{x}_{s-1}^{\text{opt}} \cup \mathbf{x}^+$ Update sets after each SI optimisation
$\mathbf{X}_s^{\text{PI}} \leftarrow \mathbf{X}_{s-1}^{\text{PI}} \cup \mathbf{X}^{\text{sim}}$
end
$\mathbf{x}^{\text{opt}} = \text{argmin}_{\mathbf{x} \in \mathbf{X}_s^{\text{opt}}} f_{\text{sim}}(\mathbf{x})$
Compute correlation $\mathbf{K}^{\text{SI}}$ from $\mathbf{X}^{\text{SI}}$
return $\mathbf{x}^{\text{opt}}, \mathbf{K}^{\text{SI}}$

5. Results and Discussion

We evaluate our method by performing BO on two benchmark functions: Styblinski-Tang[10] and Hosaki[11]. We compare our proposed SI-GP with conventional GP and PI-GP[2]. These benchmark functions serve as simulation models for computing the GP kernel for the PI-GP and our SI-GP. A modified version of the function is obtained by adding a negative Gaussian tail

$$g(x_1, x_2) = -\frac{A}{2\pi\sigma_{x_1}\sigma_{x_2}} \exp\left[-\left(\frac{(x_1 - \mu_{x_1})^2}{2\sigma_{x_1}^2} + \frac{(x_2 - \mu_{x_2})^2}{2\sigma_{x_2}^2}\right)\right].$$

This modified version represents the application model, i.e. the real-world function for which we want to find a global minimum. By varying parameters $\mu_{x_1}, \mu_{x_2}, \sigma_{x_1}, \sigma_{x_2}$ and A, we effectively alter the exact location of the original global minimum of the benchmark functions. Fifty random tail locations were chosen for each benchmark function, generating fifty distinct application models. Each application model's optimisation was conducted five times, and the number of evaluations was limited to 100. In each instance, we started from the simulation model's global minimum. For PI kernel construction, PI-GP began at the locations denoted with a plus marker in Fig. 1. In contrast, our proposed SI-GP employed two additional nearby starting points for each function. We conducted ten SI optimisations per starting point for both benchmark functions, resulting in the data sets $\mathbf{X}^{\text{SI}}$. These data sets were then used to calculate the $\mathbf{K}^{\text{SI}}$ matrix for the SI-GP kernel corresponding to each function.

In our study, we consider the global optimum of the simulation model as unknown. Thus, the PI-GP must ascertain the optimum before computing the Hessian. We employed the interior-point algorithm for this purpose. However, this algorithm led to the local minimum for both benchmark functions when started from the locations indicated in Fig. 1. Consequently, the resulting Hessian matrix inadequately represents directions to the global minimum. Conversely, SI-GP consistently identified the global optimum,

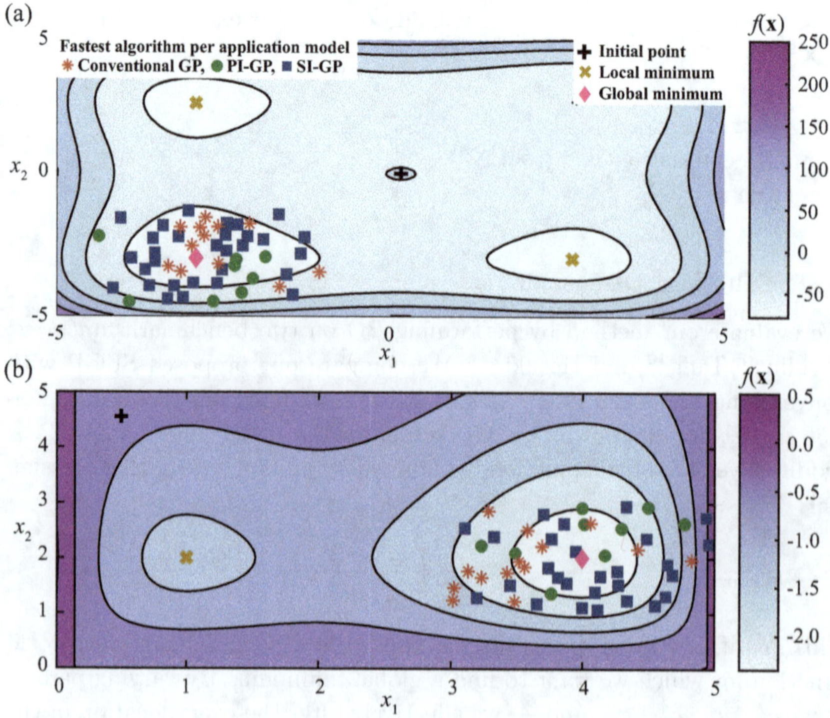

Fig. 1. Contour plot of (a) the Styblinksy-Tang and (b) the Hosaki functions.

enabling its $\mathbf{K}^{\mathrm{SI}}$ matrix to effectively guide the BO process towards the global minimum.

In Fig. 1, the new global minimum locations for the application models are indicated by stars (conventional GP), circles (PI-GP), and squares (SI-GP). These markers represent the GP type that achieved the fastest convergence for each application model. SI-GP outperformed the others in 60% of the Styblinksy-Tang and 50% of the Hosaki cases. It was followed by conventional GP, which accounted for 24% of Styblinski-Tang and 32% of Hosaki cases, and PI-GP, which was effective in 16% of Styblinski-Tang and 18% of Hosaki scenarios. These findings suggest that Hessians based on local minima can potentially misguide BO from the global minimum. Consequently, PI-GP's performance degrades, making conventional GP more effective in certain scenarios.

Furthermore, Fig. 2 presents box plots illustrating the variation in BO convergence for each GP type across all fifty models. Four key observations emerge: (1) The proposed SI-GP enhances BO convergence by 14% in the Styblinski-Tang and 20% in the Hosaki benchmarks, outperforming other methods. (2) PI-GP and SI-GP approaches reduce the variance in convergence speed, as they guide BO in a specific direction rather than extensive exploration. (3) When PI-GP's kernel construction is based on local minima, it fails to improve BO convergence, yielding performance comparable to conventional GP. (4) For both benchmark functions, the 75th percentile for SI-GP is notably lower than that of other methods, suggesting that when other methods converge faster than SI-GP, the margin of advantage is only slight. Finally, this study highlights scenarios in which obtaining an accurate Hessian for global optima is challenging, resulting in SI-GP outperforming PI-GP.

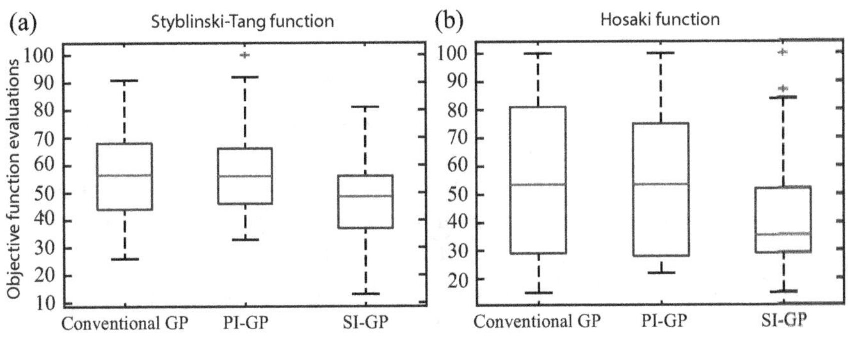

Fig. 2. BO convergence variation for the fifty application models.

6. Conclusions

This chapter presents a simulation-informed Gaussian process implemented in a Bayesian optimisation context. We validate its efficacy in accelerating the convergence of the BO algorithm. This research compares the proposed SI-GP with conventional GP and the PI-GP,[2] employing two benchmark functions. The findings indicate that SI-GP effectively incorporates input correlation into the SI kernel, leveraging the BO algorithm's global convergence properties and noise robustness within the simulation model. SI-GP is particularly advantageous for industrial applications where evaluating the objective function in experimental or real-world settings is substantially more time-consuming or cost-expensive than in simulation models. Given the enhancement in convergence speed during the experimental setting optimisation, the requirement for multiple simulation model optimisations is justified in such contexts.

References

1. K. M. Jablonka, G. M. Jothiappan, S. Wang, B. Smit and B. Yoo, Bias free multiobjective active learning for materials design and discovery, *Nature Communications* **12**, p. 2312 (2021).
2. A. Hanuka, X. Huang, J. Shtalenkova, D. Kennedy, A. Edelen, Z. Zhang, V. R. Lalchand, D. Ratner and J. Duris, Physics model-informed Gaussian process for online optimization of particle accelerators, *Phys. Rev. Accel. Beams* **24**, p. 72802 (2021).
3. C. K. Williams and C. E. Rasmussen, Regression, in *Gaussian Processes for Machine Learning*, (The MIT Press, 2005)
4. G. Pang and G. E. Karniadakis, *Physics-Informed Learning Machines for Partial Differential Equations: Gaussian Processes Versus Neural Networks*, in *Emerging Frontiers in Nonlinear Science*, eds. P. G. Kevrekidis, J. Cuevas-Maraver and A. Saxena (Springer International Publishing, 2020), pp. 323–343.
5. J.-l. Wu, C. Michelén-ströfer and H. Xiao, Physics-informed covariance kernel for model-form uncertainty quantification with application to turbulent flows, *Computers and Fluids* **193**, p. 104292 (2019).
6. A. M. Tartakovsky, T. Ma and D. A. Barajas-solano, Physics-informed Gaussian process regression for states estimation and forecasting in power grids, *International Journal of Forecasting* **39**, 967 (2023).
7. D. Lee, M. Sugiyama, U. Luxburg, I. Guyon and R. Garnett (eds.), *Gaussian Process Bandit Optimisation with Multi-fidelity Evaluations* (Curran Associates, Inc., 2016).
8. D. R. Jones, M. Schonlau and W. J. Welch, Efficient Global Optimization of Expensive Black-Box Functions, *Journal of Global Optimization* **13**, 455 (1998).
9. B. Shahriari, K. Swersky, Z. Wang, R. P. Adams and N. de Freitas, Taking the human out of the loop: A review of bayesian optimization, *Proceedings of the IEEE* **104**, 148 (2016).
10. M. Styblinski and T.-S. Tang, Experiments in nonconvex optimization: Stochastic approximation with function smoothing and simulated annealing, *Neural Networks* **3**, 467 (1990).
11. J. Elder, Global r/sup d/ optimization when probes are expensive: the grope algorithm, in *Proc. 1992 IEEE Int. Conference on Systems, Man, and Cybernetics*, (Chicago, 1992).

Fault Detection-Oriented Gaussian Models of Wind Turbines

M. A. García-Vaca[*,¶], J. E. Sierra-García[†,||], M. Santos[‡,**] and R. Pandit[§,††]

[*]*Computer Sciences Faculty, University Complutense de Madrid*
Madrid 28040, Spain

[†]*Department of Digitalization, University of Burgos*
Burgos 09006, Spain

[‡]*Institute of Knowledge Technology, University Complutense of Madrid*
Madrid 28040, Spain

[§]*Centre for Life-cycle Engineering and Management, Cranfield University*
Cranfield, UK

[¶]*magvaca@ucm.es*
[||]*jesierra@ubu.es*
[**]*msantos@ucm.es*
[††]*Ravi.Pandit@cranfield.ac.uk*

The power curve is one of the most pertinent and helpful indicators in the analysis of wind turbines. It has been demonstrated that assessing turbine performance is critical to lowering operating and maintenance costs. This curve, which connects the electrical power produced by a turbine to wind speed, has been modeled and obtained using a variety of techniques. In this chapter, various probabilistic models of a wind turbine curve and the corresponding uncertainty are obtained using Gaussian process regression with different kernel functions. Specifically squared exponential, exponential and rational quadratic Gaussian models are compared. Results show how the models fit well the real data of a 2 MW wind turbine.

Keywords: Wind turbine; modeling; power curve; Gaussian process regression; kernel.

1. Introduction

The energy sector is considered one of the most important strategic sectors of any country and therefore has a significant future projection. In this sense, during the last decades, considerable efforts have been made to carry out a gradual transition

[¶]Corresponding author.

from the different traditional energy sources to more sustainable and environmentally cleaner options, as well as to provide them with technologies that improve their efficiency [1]. Among these energy sources, wind turbines (WT), both off-shore and on-shore, stand out [2]. However, in order to make electricity generation with these infrastructures profitable, it is necessary to provide them with tools to evaluate and, therefore, maximize their performance, thus reducing the costs associated with their operation and maintenance (O&M) [3–4].

It has been observed that one of the best indicators to evaluate the operating conditions of a turbine is its power curve [5], which relates the electrical power generated by the turbine to the wind available at each moment. This is a unique function for each specific turbine. However, modeling this curve is not a simple task as its shape is complex and not only nonlinear but also highly dependent on the environmental conditions at the site and the methods used to acquire the signals [6]. Therefore, a successful approach is the use of data-driven algorithms [7]. Modeling of the power curve, along with its uncertainty, enables us to detect failures which produce power reduction in turbine operation by monitoring the difference between expected and actual power.

In the scientific literature, there are numerous examples of data-driven modeling. Among other works, in Ref. [4], a review of latest algorithms is presented, including machine learning techniques. In Ref. [8], it is demonstrated how empirical copulas can model the probability density function of the power curve. In Ref. [9], the ε-SVR model is applied to predict the power curve and the uncertainty of this forecasting. In Ref. [10], a multilayer perceptron neural network is applied to model the power curve of a wind farm.

One of the techniques that also can be used to model this curve is the Gaussian Process Regression (GPR). One the advantages of GPR is the ability to select the most suitable kernel function for the specific data. This allows us to fit the model to the particular characteristics of the data and obtain more accurate predictions. In this chapter, different probabilistic models of a wind turbine curve are obtained with this technique, as well as the associated uncertainty. Specifically, we compare the models for three different kernel functions widely used in engineering problems [11]: squared exponential, exponential and rational quadratic. Results show how of the models better fit the data of a 2 MW wind turbine.

The structure of this chapter is as follows. In Section 2, the modeling based on GPR is presented. In Section 3, the dataset used in this chapter is described. Results are discussed in Section 4. This chapter ends with the conclusions and future works.

2. Models based on Gaussian Process Regression

Gaussian process regression (GPR) are kernel-based nonparametric probabilistic models [11]. A Gaussian process (GP) is a set of random variables such that any finite combination of them follows a joint Gaussian distribution. That is, if $f(x)$ is a GP, for all set of n observations of the process x_1, x_2,...x_n, the joint distribution of the random variables $f(x_1)$, $f(x_2)$,...$f(x_n)$ will be Gaussian.

A GP is defined by its mean $m(x)$ and covariance function $k(x,x')$. Therefore, if $f(x)$ is a GP then $E(f(x)) = m(x)$ and $Cov[f(x), f(x')] = k(x,x')$. The output of the GPR models is given by expression (1),

$$y(x) = h(x)^T\beta + f(x) \quad (1)$$

where $h(x)$ is a set of basis functions that transforms the input space, β is a vector of coefficients, $f(x)$ is a GP with zero mean, and $k(x,x')$ is the covariance function.

As the model is probabilistic, the output of the model y_i at instant i can be modeled by a gaussian probability with the expression (2),

$$P(y_i|f(x_i), x_i) \sim N(y_i|h(x_i)^T\beta + f(x_i), \sigma^2) \quad (2)$$

where N represents a gaussian distribution with media $h(x_i)^T\beta + f(x_i)$ and variance σ^2. Different basis functions $h(x)$ and kernel functions $k(x,x')$ can be considered. In this work linear basis functions have been selected and three different kernels have been evaluated: squared exponential, exponential, and rational quadratic.

2.1. *Squared exponential kernel*

This kernel is also called Radial Basis Function (RBF). It is an univeral approximator [12]. This kernel is infinitely differentiable, which implies that GPs with this kernel as covariance function have mean square derivatives of all orders, and are thus very smooth. It is given by the expression (3) and it is defined by two parameters: l and σ^2. The parameter l determines the width of the function. In general, the model will not be able to extrapolate more than l elements ahead. The output variance σ^2 determines the average distance of the function away from its mean. This parameter can be seen as a scale factor.

$$k(x,x') = \sigma^2 exp(-(x-x')^2/2l^2) \quad (3)$$

2.2. *Exponential kernel*

An exponential kernel function is used when the function to be modelled is assumed to be continuous, but non-differentiable in the meansquare sense and nor smooth [11]. It is given by the expression (4), very similar to (3), and it is affected by the same hyperparameters.

$$k(x,x') = \sigma^2 exp(-\sqrt{(x-x')^2}/l) \quad (4)$$

2.3. *Rational quadratic kernel*

When the function to be modeled is presumed to be continuous and differentiable in the mean-square sense, the rational quadratic covariance function can be used. That is, a function without any discontinuity or not being discontinuous in its first few derivatives. The rational quadratic covariance function can be seen as an infinite sum of squared exponential covariance functions with various characteristic length scales, or as a scale mixing [13]. It is given by (5).

$$k(x,x') = \sigma^2(1+(x-x')^2/2\alpha l^2)^{-\alpha} \quad (5)$$

3. Description of the Dataset

We use real data from an onshore wind farm in Penmanshiel, United Kingdom, [14]. The farm has pitch-regulated Senvion MM82 wind turbines. Table 1 lists its main features. A SCADA system was use for collecting data, with an average temporal duration of 10 minutes. The following variables, out of all those recorded by the SCADA system, have been selected: wind speed adjusted with air density (m/s), timestamp, and output power (kW). The research case has made use of 8352 data points collected from the turbine labeled as WT01 during January and February of 2017.

Table 1: Main features of the Senvion/ MM82 pitch regulated wind turbine.

Parameter	Value
Rated power (kw)	2050
Rotor diameter (m)	82
Hub height (m)	59
Rated wind speed (m/s)	14.5
Cut-out wind speed (m/s)	25
Cut-in wind speed (m/s)	2.5

Data are cleaned to remove outliers before power curve modeling is started. The data preprocessing is divided into two steps:

1. Removal of invalid data, i.e., data points with unknown values or with zero or negative output power.
2. Removal of outliers. The power curve is split into intervals of 0.5 m/s and the mean and standard deviation of these intervals are calculated according

to the international standard, IEC 61400-12-1:2017 [15]. An outlier is defined as a data point that deviates more than ±2.5σ from its appropriate interval; such data points are eliminated from the dataset.

Figure 1 shows the dataset after the elimination of any invalid data in the first step. The red continuous line indicates the interval mean values, and the error bars reflect the deviation of each interval mean, or ±2.5σ. As said, all points outside the error bars are considered outliers and deleted in the second step.

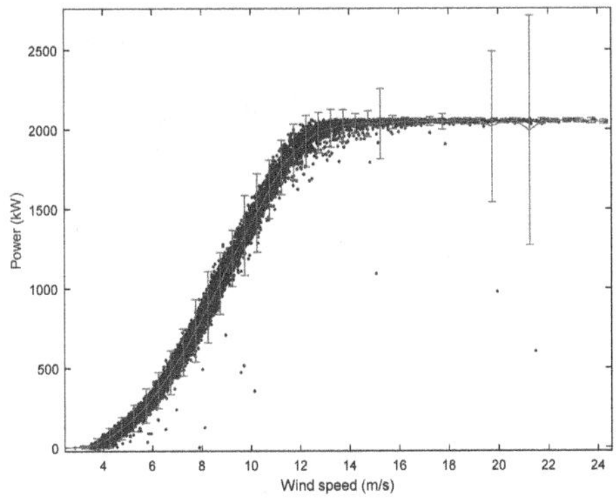

Fig. 1: Raw dataset after removal of data points with zero or negative generated power

4. Results Discussion

Figure 2 shows the results of the application of the GPR with different kernels considered. The ideal regression curve is shown as a red line overlaid on the dataset, and the black lines show the 95% confidence level range above and below the regression curve. It is possible to see how all models fit well the power curve. In terms of the confidence interval, it is evident that this interval overestimates the variance of the data in the first and last portions of the curve.

The main visual difference between the curves is that in the case of the exponential kernel, the ideal curve and the confidence levels are noisier. With the squared exponential or the rational quadratic kernel the lines are smoother.

Table 2 shows the parameters of the kernel functions identified during the training process. These parameters adjust the amplitude of the width of the kernel functions according to expressions (3-5). Figure 3 compares the shape of the kernels considering these parameters. The parameter σ affects the amplitude of the kernel; indeed, its amplitude is proportional to σ^2. The rational quadratic GP model has the largest σ (Table 2) and the largest amplitude (Figure 3). On the other hand, as the width of the kernel is determined by the parameter l, the exponential kernel has the largest l value (Table 2) and the largest width (Figure 3). This so large width may explain the noise behavior in Figure 2.

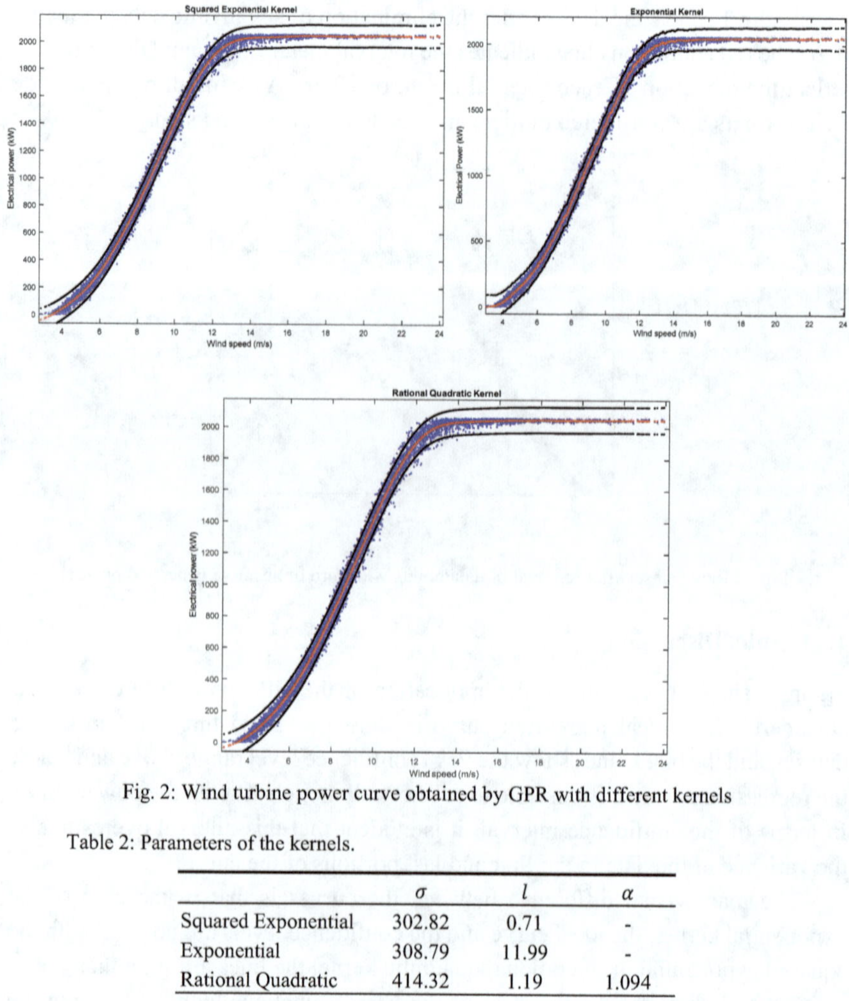

Fig. 2: Wind turbine power curve obtained by GPR with different kernels

Table 2: Parameters of the kernels.

	σ	l	α
Squared Exponential	302.82	0.71	-
Exponential	308.79	11.99	-
Rational Quadratic	414.32	1.19	1.094

In addition to the graphical comparison, the models have also been studied numerically. Table 3 shows the metrics used in the comparison. The best results have been boldfaced. The exponential kernel provides the best results in terms of S-MAPE, but the rational quadratic kernel provides the best RMSE and MAE. From this comparison together with the noisy behavior observed in Figure 2, we can conclude that the model with the rational quadratic kernel is slightly better than the others.

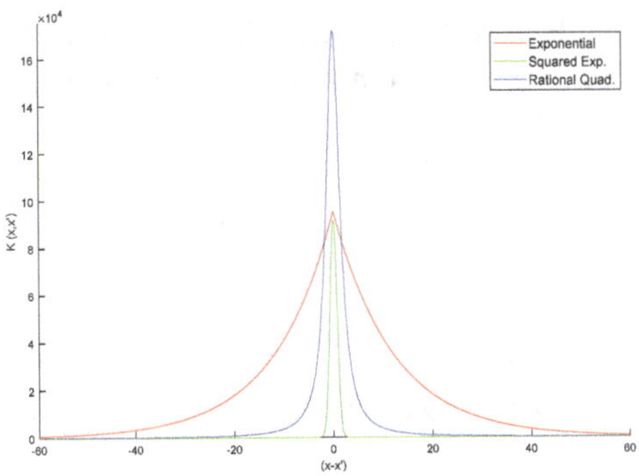

Fig. 3: Comparison of kernels functions used to obtain the models

Table 3: Performance evaluation of the models.

	S-MAPE [KW]	RMSE [KW]	MAE [KW]
Squared Exponential	8.27	41.416	29.768
Exponential	**7.8**	41.733	29.954
Rational Quadratic	8.18	**41.39**	**29.746**

5. Conclusions and Future Works

In this chapter, we have seen how Gaussian process regression can be successfully applied to model wind turbine power curves, and how they offer us the ability to faithfully capture the nonlinear and complex nature of this curve. Although all kernels studied can be used to model the curve, the rational quadratic one has given the best results.

Future work could include other variables to further improve the model, such as rotor speed, blade pitch or ambient temperature. In addition, this model could be applied to evaluate different types of turbine failures.

Acknowledgments

This work was partially supported by the Spanish Ministry of Science, Innovation and Universities under MCI/AEI/FEDER Project no. PID2021-123543OB-C21.

References

1. Strategic Energy Technology Information System (SETIS), 2023 https://setis.ec.europa.eu/index_, last accessed 2024/01/29.

2. B. Zhou, Z. Zhang, G. Li, D. Yang and M. Santos, Review of Key Technologies for Offshore Floating Wind Power Generation. *Energies* **16(2)**, 710 (2023).
3. Sierra-García, J. E., & Santos, M. (2021). Redes neuronales y aprendizaje por refuerzo en el control de turbinas eólicas. Revista Iberoamericana de Automática e Informática industrial, 18(4), 327-335.
4. R. Pandit, D. Astolfi, J. Hong, D. Infield and M. Santos, SCADA data for wind turbine data-driven condition/performance monitoring: A review on state-of-art, challenges and future trends. *Wind Engineering* **47(2),** 422-441 (2023).
5. O. Uluyol, G. Parthasarathy, W. Foslien and K. Kim, Power curve analytic for wind turbine performance monitoring and prognostics. In *Annual Conference of the PHM Society* (Vol. 3, No. 1) (2021)
6. R. Pandit, D. Infield and M. Santos, Accounting for environmental conditions in data-driven wind turbine power models. *IEEE Transactions on Sustainable Energy* **14(1),** 168-177 (2022).
7. R. Morrison, X. Liu and Z. Lin, Anomaly detection in wind turbine SCADA data for power curve cleaning. *Renewable Energy* **184**, 473-486 (2022).
8. S. Gill, B. Stephen and S. Galloway, Wind turbine condition assessment through power curve copula modeling. *IEEE Transactions on Sustainable Energy* (vol. 3, no. 1, pp. 94-101) (2012).
9. M.Á. García-Vaca, J.E. Sierra-García and M. Santos, Prediction and Uncertainty Estimation in Power Curves of Wind Turbines Using ε-SVR. *International Conference on Intelligent Data Engineering and Automated Learning* (pp. 509-517) (2023).
10. F. Pelletier, C. Masson, and A. Tahan, Wind turbine power curve modelling using artificial neural network, *Renewable Energy*, vol. 89, pp. 207-214(2016)
11. C. E. Rasmussen and C. K. Williams, *Gaussian processes for machine learning* (Vol. 1, p. 159). Cambridge, MA: MIT press (2006).
12. D. Duvenaud, Automatic model construction with Gaussian processes, PhD Thesis, University of Cambridge (2014)
13. J. Kocijan, *Modelling and control of dynamic systems using Gaussian process models* (pp. 33-38). Cham: Springer International Publishing (2016)
14. Penmanshiel Wind Farm Data, https://doi.org/10.5281/zenodo.5946808, last accessed 2024/01/29
15. International Electrotechnical Commission, Wind energy generation systems—Part 12-1: Power performance measurements of electricity producing wind turbines. *International Electrotechnical Commission (IEC), IEC Central Office*, vol. 3, pp. 2017-03 (2017).

Plane Path Following Control for an Underactuated Parafoil System Based on ILOS-LADRC Under Wind Disturbances

Yuemin Zheng*, Jin Tao*, Qinglin Sun*,¶, Hao Sun*, Zengqiang Chen*, Mingwei Sun*, Xiaolong Wu†,‡,‖, Hongyan Yang†,‡, Honggui Han† and Xianyi Zeng§

*College of Artificial Intelligence, Nankai University, Tianjin 300350, China

†Faculty of Information Technology, Beijing University of Technology, Beijing 100124, China

‡Beijing Key Laboratory of Computational Intelligence and Intelligent System, Beijing 100124, China

§Department of Textile Management and Product Design, Ecole Nationale Superieure des Arts et Industries Textiles, Roubaix 59100, France

¶sunql@nankai.edu.cn

‖lewis_wxl@sina.com

This chapter presents a plane path following control strategy for the parafoil system, utilizing the Integral Line-of-Sight (ILOS) guidance law-based linear active disturbance rejection control (LADRC) approach. The proposed method addresses the challenges associated with underactuation and external disturbances, aiming to achieve accurate and robust path-tracking performance. Simulation results demonstrate the effectiveness of the proposed approach in maintaining precise path following despite wind disturbances. This research contributes to enhancing the control capabilities of underactuated parafoil systems, especially in scenarios where wind-induced perturbations pose significant challenges.

Keywords: Parafoil system; linear active disturbance rejection control; integral line of sight; path following control.

1. Introduction

A parafoil system comprises a parafoil canopy, connecting rope, and payload. Its widespread use in aerospace, military, and civil fields is attributed to its superb gliding ability and payload characteristics. However, unpredictable wind disturbances would impact the dynamic performance. Hence,

¶Corresponding author.

overcoming these challenges and precisely controlling the parafoil to follow the desired path are critical for mission success.

The path following control of the parafoil system poses a challenge as it falls into the category of underactuated problems. This stems from the fact that achieving convergence in lateral error and yaw angle error is solely reliant on manipulating the parachute rope by pulling it to the left or right. Hence, it is common practice to employ a guidance law for the generation of the guidance yaw angle. Line-Of-Sight (LOS) is a commonly employed guidance law in ship navigation control.[1] It produces a guidance heading angle by leveraging the ratio of longitudinal tracking error to the forward sight distance. Building upon this concept, E. Borhaug *et al.* introduced the Integral Line-Of-Sight (ILOS),[2] which incorporates an integral term to counteract the impact of drift angle. Subsequently, a controller can be designed to track this guidance yaw angle, facilitating the accomplishment of path following control.

Various control strategies, including PID, sliding mode control *et al.*, have been proposed to track the guidance yaw angle. The LADRC, based on modern control theory and employing a Linear Extended State Observer (LESO) for disturbance estimation, is particularly noteworthy.[3] Utilizing Linear State Error Feedback (LSEF), this method efficiently compensates for observed disturbances. Therefore, this investigation utilizes ILOS for yaw angle guidance and employs LADRC for precise yaw angle control, thereby attaining effective path-following control for the parafoil system.

2. Studied 8-DOF Dynamic Model

Figure 1 depicts the parafoil system, which includes the parafoil canopy and payload. To facilitate analysis, three coordinate systems are defined: the ground coordinate system $O_d x_d y_d z_d$, the parafoil coordinate system $O_s x_s y_s z_s$, and the payload coordinate system $O_w x_w y_w z_w$.

In the parafoil coordinate system, we consider the velocity vector $V_s = [u_s, v_s, w_s]^T$ and the angular velocity vector $W_s = [p_s, q_s, r_s]^T$. Similarly, there are $V_w = [u_w, v_w, \dot{w_w}]^T$ and $W_w = [p_w, q_w, r_w]^T$ in the payload coordinate system. Take the state variable as $x_s = \left[V_w^T, W_w^T, V_s^T, W_s^T, \dot{\psi}_r, \dot{\theta}_r\right]^T$ with ψ_r and θ_r being the relative yaw angle and relative pitch angle between the parafoil canopy and payload, respectively, then the 8-DOF parafoil's dynamic equation can be expressed as

$$\dot{x_s} = f(x_s, u) \tag{1}$$

where u represent the pulling force.

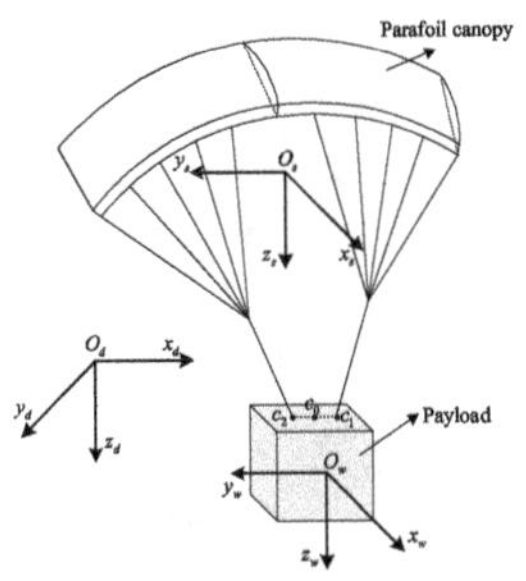

Fig. 1. Schematic diagram of powered parafoil system

Equation (2) represents the speed relationship between the two coordinate systems of $O_s x_s y_s z_s$ and $O_d x_d y_d z_d$,

$$\begin{bmatrix} \dot{x} \\ \dot{y} \\ \dot{z} \end{bmatrix} = \begin{bmatrix} \mathrm{c}\theta\mathrm{c}\psi & \mathrm{c}\theta\mathrm{s}\psi & -\mathrm{s}\theta \\ \mathrm{s}\phi\mathrm{s}\theta\mathrm{c}\psi - \mathrm{c}\phi\mathrm{s}\psi & \mathrm{s}\phi\mathrm{s}\theta\mathrm{s}\psi + \mathrm{c}\phi\mathrm{c}\psi & \mathrm{s}\phi\mathrm{c}\theta \\ \mathrm{c}\phi\mathrm{s}\theta\mathrm{c}\psi + \mathrm{s}\phi\mathrm{s}\psi & \mathrm{c}\phi\mathrm{s}\theta\mathrm{s}\psi - \mathrm{s}\phi\mathrm{c}\psi & \mathrm{c}\phi\mathrm{c}\theta \end{bmatrix}^T V_s \tag{2}$$

where sθ denotes the $sin\theta$, and cθ denotes the $cos\theta$. ϕ, θ, and ψ are three Euler angles consist of roll angle, pitch angle, and yaw angle, respectively. Besides, there is the following relationship between the Euler angles and the angular velocities,

$$\begin{bmatrix} \dot{\phi} \\ \dot{\theta} \\ \dot{\psi} \end{bmatrix} = \begin{bmatrix} 1 & \sin\phi tan\theta & \cos\phi tan\theta \\ 0 & \cos\phi & -\sin\phi \\ 0 & \sin\phi/\cos\theta & \cos\phi/\cos\theta \end{bmatrix} W_s \tag{3}$$

The detailed model construction process is not covered in this paper due to space constraints; readers can consult Ref. 4 for more information.

3. Design of ILOS-LADRC for Path Following Control

3.1. *ILOS guidance law*

Figure 2 illustrates the path guidance schematic diagram of the parafoil system. Here, OXY represents the geodetic coordinate system, $O_1X_1Y_1$ represents the coordinate system of the parafoil, and $O_2X_2Y_2$ corresponds to the target point coordinate system.

Then in the coordinate system of $O_2X_2Y_2$, there is

$$\begin{bmatrix} x_e \\ y_e \end{bmatrix} = \begin{bmatrix} \cos\psi_d & -\sin\psi_d \\ \sin\psi_d & \cos\psi_d \end{bmatrix}^T \begin{bmatrix} x - x_p \\ y - y_p \end{bmatrix} \tag{4}$$

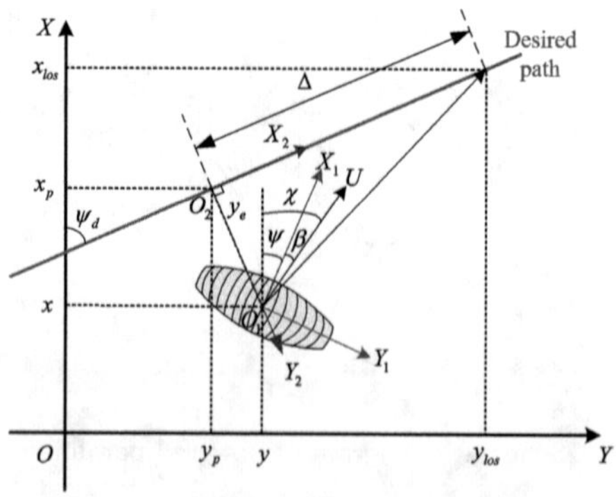

Fig. 2. LOS guidance geometry

with $x_e = 0$. $\psi_d = arc\tan(\dot{y}_p/\dot{x}_p)$ is the desired yaw angle, ψ is the actual yaw angle and β is the sideslip angle. Define $\chi = \psi + \beta$ as the new yaw angle, and the following ILOS guidance law[5] is adopted as

$$\chi_g = \psi_d - \arctan\left(\frac{y_e + ky_{\text{int}}}{\Delta}\right) \tag{5}$$

wiht

$$\dot{y}_{\text{int}} = \frac{Uy_e}{\sqrt{\Delta^2 + (y_e + ky_{\text{int}})^2}} \tag{6}$$

where Δ represents the look-ahead disturbance, U denotes the horizontal speed, and $k > 0$ refers to an adjustable parameter. The Equation (6) is designed in a way such that the influence of the integrator diminishes when the cross-track error increases.

Theorem 3.1. *For the system derived from Equation (4):*

$$\dot{y}_e = U\sin(\psi + \beta - \psi_d) \tag{7}$$

it is satisfied that when χ can track χ_g well, y_e can be asymptotically stable.

Proof. According to Equations (7) and (5), there is

$$\begin{aligned}\dot{y}_e &= U\sin(\psi + \beta - \psi_d) \\ &= U\sin(\chi - \psi_d) \\ &= U\sin\left(-\arctan\left(\frac{y_e + ky_{\text{int}}}{\Delta}\right)\right)\end{aligned} \tag{8}$$

Then construct the Lyapunov function as $V = \frac{1}{2}y_e^2 + \frac{1}{2}ky_{\text{int}}^2$, with its differential expression as

$$
\begin{aligned}
\dot{V} &= y_e\dot{y}_e + ky_{\text{int}}\dot{y}_{\text{int}} \\
&= y_e U \sin\left(-\arctan\left(\frac{y_e + ky_{\text{int}}}{\Delta}\right)\right) + \frac{ky_{\text{int}}Uy_e}{\sqrt{\Delta^2 + (y_e + ky_{\text{int}})^2}} \\
&= -\frac{Uy_e(y_e + ky_{\text{int}})}{\sqrt{\Delta^2 + (y_e + ky_{\text{int}})^2}} + \frac{ky_{\text{int}}Uy_e}{\sqrt{\Delta^2 + (y_e + ky_{\text{int}})^2}} \\
&= -\frac{Uy_e^2}{\sqrt{\Delta^2 + (y_e + ky_{\text{int}})^2}}
\end{aligned}
\tag{9}
$$

Since $U = \sqrt{u_s^2 + v_s^2} \geq 0$, there is $\dot{V} \leq 0$. As a result, the guidance law is is proven to be stable. □

Remark 3.1. The y_e in this paper is essentially the cross-track error, and the calculation process can be found in Ref. [6].

3.2. *LADRC yaw angle controller*

The design of the LADRC controller needs the order information of the controlled system, so, according to Equation (3), the following expression can be obtained,

$$
\begin{aligned}
\ddot{\psi} =& \frac{\sin\phi}{\cos\theta}\dot{q}_s + \frac{\cos\phi}{\cos\theta}\dot{r}_s + \frac{\sin\theta\sin 2\phi}{\cos^2\theta}q_s^2 - \frac{\sin\theta\sin 2\phi}{\cos^2\theta}r_s^2 \\
&+ \frac{\cos\phi}{\cos\theta}p_s q_s - \frac{\sin\phi}{\cos\theta}p_s r_s + \frac{2\sin\theta\cos 2\phi}{\cos^2\theta}q_s r_s
\end{aligned}
\tag{10}
$$

Combined with the parafoil system model, the following expression can be obtained

$$\ddot{\psi} = f_1(\cdot) + f_2(u), \tag{11}$$

where $f_1(\cdot)$ represents the disturbance, including the internal state and external disturbance information, and $f_2(u)$ is an expression of the control variable. u represents the deflection of the left and right trailing edges of the canopy.

In addition, since χ and ψ have the same order, we can rearrange Equation (11) as

$$\ddot{\chi} = f(\cdot) + f_2(u) - b_0 u + b_0 u = f + b_0 u, \tag{12}$$

where f can be regarded as the total disturbances of the parafoil system containing unmodeled dynamics inside the system and disturbances in the external environment, and b_0 is an adjustable parameter.

To estimate the unknown disturbance f, the third-order LESO is designed as

$$\begin{cases} \dot{\hat{x}} = A\hat{x} + Bu + L\left(\dot{\hat{y}} - \hat{y}\right) \\ y = C\hat{x} \end{cases}, \tag{13}$$

where $A = \begin{bmatrix} 0\,1\,0 \\ 0\,0\,0 \\ 0\,0\,0 \end{bmatrix}$, $B = \begin{bmatrix} 0 \\ b_0 \\ 0 \end{bmatrix}$, $L = \begin{bmatrix} \beta_{01} \\ \beta_{02} \\ \beta_{03} \end{bmatrix}$, $C = \begin{bmatrix} 1 \\ 0 \\ 0 \end{bmatrix}^T$. And $\hat{x} = \begin{bmatrix} \hat{\chi} \\ \hat{\dot{\chi}} \\ \hat{f} \end{bmatrix}$ is the estimated state vector for the true state vector $x = \begin{bmatrix} \chi \\ \dot{\chi} \\ f \end{bmatrix}$. In addition, L is the observer gain vector, which is the prerequisite for realizing the observed state $\hat{x}$ to estimate the true value of the state x. Usually, the pole configuration method is used to configure the observer gain at the pole $-\omega_o$, which is arranged as

$$|sI - (A - LC)| = s^3 + \beta_{01}s^2 + \beta_{02}s + \beta_{03} = (s + \omega_o)^3. \tag{14}$$

Thus, $\beta_{01} = 3\omega_o$, $\beta_{02} = 3\omega_o^2$, and $\beta_{03} = \omega_o^3$. In this way, by adjusting the parameter ω_o, the total system disturbance f can be estimated.

Further, the LSEF control law is designed to eliminate the estimated disturbance:

$$u = (k_p(\chi_g - \hat{x}(1)) - k_d\hat{x}(2) - \hat{x}(3))/b_0 \tag{15}$$

where k_p and k_d are the controller parameters.

4. Simulation Results

The parameters of the parafoil system used in this paper are shown in Table 1. And the initial velocity and angular velocity of the parafoil are $V_s = [14.9, 0, 2.1]^T$ m/s and $W_s = [0, 0, 0]^T$ rad/s, respectively.

To validate the efficacy of the aforementioned path-following approach, we investigated the straight-path following case. The designated path constitutes a straight line with a yaw angle of 45°, originating from the initial position of (0,0)m. The parafoil's initial position is situated at (0,80)m, and

Table 1. Parameters of the parafoil system

Parameter Description	Value	Parameter Description	Value
Wing span	4.5m	Mass of the payload	20 kg
Mean aerodynamic chord	1.3 m	Wing area	6.5 m^2
Mass of the parafoil	1.7 kg	Rope length	3 m

at 60 seconds, it encountered winds of varying magnitudes and directions. The outcomes of the simulation are illustrated in Figures 3, 4.

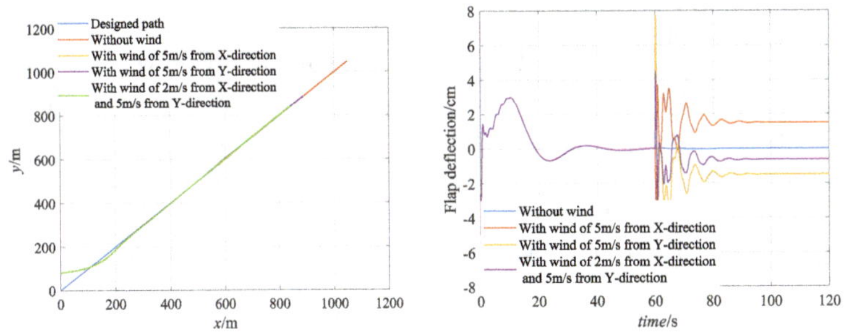

(a) Path tracing effect (b) Control variables

Fig. 3. Path following control results and control variables of parafoil system

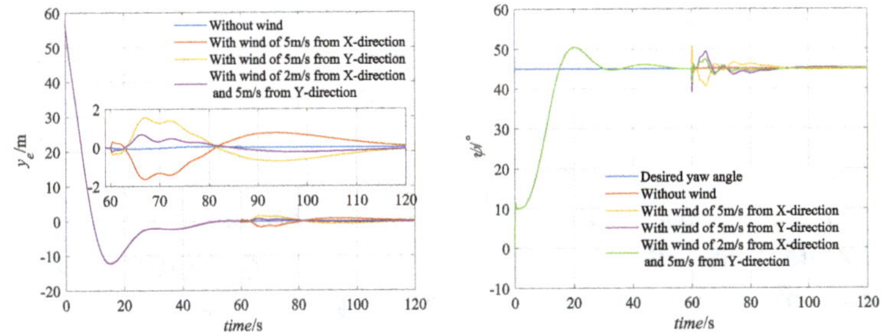

(a) Cross-track error (b) Yaw angle

Fig. 4. Cross-track error and yaw angle of parafoil system

The parameters involved in the control system are selected as: $\Delta = 50$, $k = 0.4$, $\omega_o = 1.2$, $b_0 = 0.2$, $k_p = 0.4$ and $k_d = 2.5$. And from the above results, it can be observed that the proposed method can overcome the influence of wind disturbance and achieve stable path tracking.

5. Conclusion

In this chapter, an ILOS-LADRC control approach is applied to the path following control of a parafoil system. First, the 8-DOF model of parafoil system is introduced. Then the ILOS is used to provide the guidance yaw angle, and the stability of ILOS is also analyzed. Further, the LADRC controller is designed to realize the yaw angle tracking, thereby completing the path following. Finally, simulation results verified the effectiveness.

Acknowledgments

This work was supported by the National Natural Science Foundation of China(Grant No.61973172, 61973175, 62003175, 62003177, 62125301, 62021003, 62103011, 62103012 and 61622301), the key Technologies Research and Development Program of Tianjin(Grant No.19JCZDJC32800), National Key Research and Development Project under Grants 2018YFC1900800-5 and Beijing Natural Science Foundation under Grant KZ202110005009.

References

1. TI. Fossen, M. Breivik, R. Skjetne, Line-of-sight path following of underactuated marine craft, *IFAC Proceedings Volumes* **36**, 211-216 (2003).
2. E. Borhaug, A. Pavlov, K. Pettersen, Integral LOS control for path following of underactuated marine surface vessels in the presence of constant ocean currents, in *2008 47th IEEE conference on decision and control*, (Cancun, Mexico, 2008).
3. J. Han, From PID to active disturbance rejection control, *IEEE transactions on Industrial Electronics* **56**, 900-906 (2009).
4. E. Zhu, Q. Sun, P. Tan, et al., Modeling of powered parafoil based on Kirchhoff motion equation, *Nonlinear Dynamics* **79**, 617-629 (2015).
5. A. Lekkas, TI. Fossen, Integral LOS Path Following for Curved Paths Based on a Monotone Cubic Hermite Spline Parametrization, *IEEE TRANSACTIONS ON CONTROL SYSTEMS TECHNOLOGY* **22**, 2287-2301 (2014).
6. Y. Zheng, J. Tao, Q. Sun, et al.DDPG-based active disturbance rejection 3D path-following control for powered parafoil under wind disturbances, *Nonlinear Dynamics* **111**, 11205-11221 (2023).

Flight Phase Fuzzy Identification for Airplanes High-lift Devices Management

E. Plaza[*,§], M. Santos[†,¶] and J. E. Sierra-García[‡,||]

[*]Computer Sciences Faculty, University Complutense de Madrid
Madrid 28040, Spain

[†]Institute of Knowledge Technology, University Complutense of Madrid
Madrid 28040, Spain

[‡]Department of Digitalization, University of Burgos
Burgos 09006, Spain

[¶]msantos@ucm.es

[§]elplaza@ucm.es

[||]jesierra@ubu.es

This chapter presents an intelligent system to automate the pilot's actions to control high-lift devices in commercial airplanes. The modular system designed uses fuzzy logic in the two main modules: flight phase identification and flap lever command, with the objective of safely adapting the aircraft speed envelope to assure flight safety. The fuzzy system decides when and how to change the position of the high-lift devices of the aircraft as a human pilot would do but automatically, based on the identified flight phase and the speed margins. The results in a simulated real scenario show the expected behavior and have been quantitatively evaluated using some metrics.

Keywords: Knowledge-based system; fuzzy logic; aircraft; flaps and slats; safety.

1. Introduction

This chapter intends to contribute to the Upset Prevention and Recovery strategy of the airplanes. It is related to how pilots must manage the aircraft to avoid abnormal situations and so to ensure a safe flight, which is of the utmost importance [1]. The development presented here is an intelligent decision system that automates some pilots' actions in order to increase the flight safety.

The proposal is focused on the management of high-lift devices fitted in commercial airplanes. They are moving surfaces placed on the wings, commonly

[§]Corresponding author.

named flaps (normally located on the trailing edge) and slats (located on the leading edge). During cruise phase, they are fully retracted to reduce the drag and avoid damage on the own surfaces (they are not designed to support the big loads caused by high speed). During take-off, first climb, approach and landing phases, they must be deployed to increase the amount of lift produce by the wing [2]. That is, with these surfaces extended, the airplane stalls at a higher angle of attack, therefore it can safety operate at lower speed (necessary for take-off and landing). In the event of pilots do not configure correctly the flaps/slats position, a plane crash may occur. Therefore, everything that can be done to automate the procedure and avoid human errors is relevant to safety [3].

Using a soft computing technique, specifically fuzzy logic, we here propose an intelligent decision system that is able to manage the flaps/slats configuration, based on the flight phase identification and the flap lever command decision. It has been developed for any aircraft. Simulation results using a real flight profile show how this modular fuzzy system can control the position of the high-lift devices. The results are as expected qualitatively and quantitatively.

The fundamental innovation of this chapter is the research into intelligent techniques that can be applied to problems in the aeronautical field, especially related to human (pilot) interpretations and decisions, which have not been widely explored yet or in which classical techniques are less effective.

This chapter is structured is as follows. In Section 2, a brief state of the art is commented. In Section 3, the problem is described. The proposed intelligent system of the flaps/slats positioning using fuzzy logics is presented in Section 4. Simulation results are discussed in Section 5. Conclusions end this chapter.

2. Brief State of the Art

There is not much specific literature on the problem addressed here. However, some related works can be found. For example, the problem of stall, which is one of the main sources of upset situations, is discussed in [4]. Patel et al. [5] study high-lift devices design focused on improving the aforementioned stall problem. Da Silva et al. [6] propose active flow control of high-lift devices to reduce noise. Yang et al. [7], investigate the aerodynamic characteristics of high-lift configurations without slats, focusing on stall and post-stall conditions at low speed. Mustafa et al. [8], present an architecture of the flaps/slats system. Unlike our approach, none of them propose automatic control of high-lift devices to improve safety.

Regarding the techniques used, Jing [9] applies fuzzy logic combined with Bayesian networks to detect failures in the extension and retraction of the landing gear of a commercial airplane. In [10], the flight phase is estimated using fuzzy logic, but our solution uses a rule base specifically focused on the flap/slat automation problem.

In [2], Airbus brings together pilot procedures and specific values for flap/slat operation, which have served to validate our results.

3. Description of the Problem Addressed

The high-lift system consists on aerodynamic surfaces (flaps and slats) attached on the wings, that are extended when the airplane flies at low speed and retracted at high speed. The pilots command the extension or retraction of these devices using a lever (called flap lever) located inside the cockpit, changing the flaps/slats surface configuration [11]. The possible configurations are discrete, from 0 (fully retracted), 1, 2, 3, to 4 (fully extended).

The architecture of the intelligent system that manages the high-lift devices is shown in Fig. 1. In this case, fuzzy logic is applied to the two main modules: flight phase estimation and flap lever command.

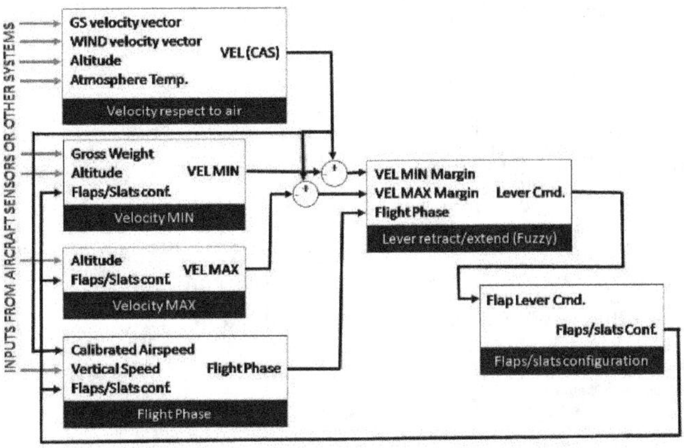

Fig. 1: General architecture of the intelligent decision system

The other modules and their implementation are as follows [12]: the velocity respect to the air is obtained using standard aeronautics formulas; the velocity MIN and MAX is calculated using tables built from known aircraft magnitudes; the flaps/slats configuration is estimated from the flap lever command taking into account the possible configurations and if the command is extended (increased configuration) or retract (decrease configuration).

4. Fuzzy Logic Modules Design and Implementation

4.1. *Flight phase estimation*

The flight phase estimation module uses as inputs the following variables: calibrated airspeed, vertical speed and flaps/slats configuration. The fuzzy sets assigned to them are as follows. Three triangular and trapezoidal memberships (low, medium, centered at 175, and high) for calibrated airspeed (range 0-300 kts); three trapezoidal memberships (neg, null, pos) for vertical speed input (-6000, 6000 ft/min); five fuzzy singleton for flap/slat configuration (0, 1, 2, 3 and 4). The output, flight phase estimation, is represented by five triangular membership functions (Ground, Take-Off, Climb, Other, Approach).

The fuzzy sets and rules have been designed by applying expert knowledge about the piloting procedures [2]. Table 1 shows an example.

Table 1: Fuzzy rules for flight phase examples.

If	Then
Calibrated Airspeed is MED and Vertical Speed is POS	Flight Phase is CLB
Calibrated Airspeed is MED and Vertical Speed is NULL and Flap Conf. is 1	Flight Phase is TO

4.2. *Flap lever command module*

Three different sets of membership functions (Tune-LOW, Tune-MED and Tune-HIGH) have been defined for the minimum and maximum margins (Fig. 2, left and right respectively), depending on how much speed margin is allowed to make the decision. The flap lever command output consists of three singleton: retract, maintain and extend, and is the input of the flap/slat configuration module (see Fig. 1). This configuration module is further explained in [12].

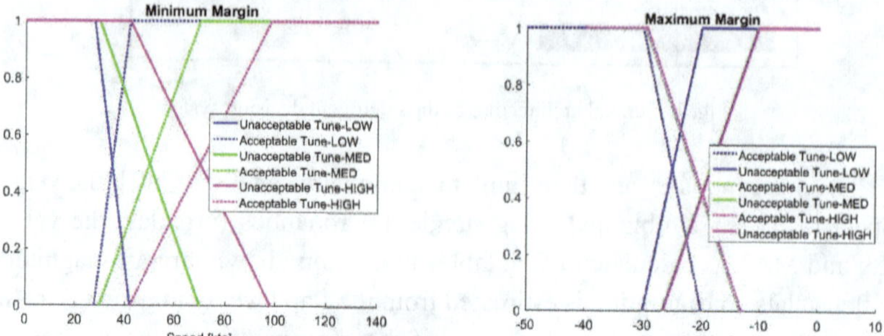

Fig. 2: Fuzzy sets for minimum margin (left) and for maximum margin (right).

5. Simulation Results and Discussion

Real flight profile taken from Eurocontrol database [13] is used to test the system. The recorded data are waypoints position (longitude, latitude, altitude) along the flight plan from the departure to the destination.

Assuming the wind is negligible, the calibrated airspeed profile obtained through the module "Velocity respect to air" (Fig. 1) is shown in Fig. 3 (left). The altitude profile is directly given by the recorded data (Fig. 3, middle). Deriving it with respect to the time the vertical speed profile is obtained (Fig. 3, right).

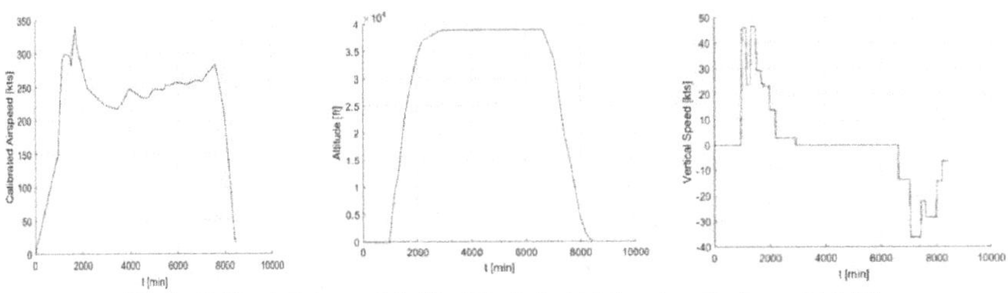

Fig. 3: Calibrated airspeed (left), altitude (middle) and vertical speed (right)

The flap/slat configuration evolution during the flight is presented in Fig. 4, left. Each simulation has been carried out with each set of membership functions (Tune-LOW, Tune-MED and Tune-HIGH). Fig. 4, right, shows a zoom, corresponding to the approach to landing phase, in order to better appreciate the flap/slat movements.

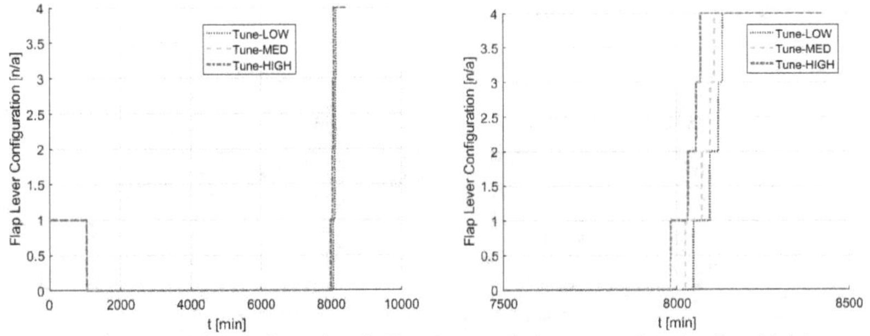

Fig. 4: Flap/slat configuration (left) and zoom during approach to landing (right).

The results are as expected for the three simulations: the flaps/slats are retracted to clean configuration after the take-off, maintained in clean configuration during most part of climbing, cruise and descent, and during the final approach to landing the flaps/slats are extended as the speed is reduced.

The fuzzy flight phase estimation is shown in Fig. 5, right, where: 1=Takeoff, 2=Climb, 3=Other (in this case, Cruise phase), 4=Approach/landing, 0=Ground. The continuous values of this signal are due to the flight fuzzy estimation, similarly as a pilot would intuitively give degrees of reliability to his/her observations. The output, flap lever command, is shown in Fig. 5, left.

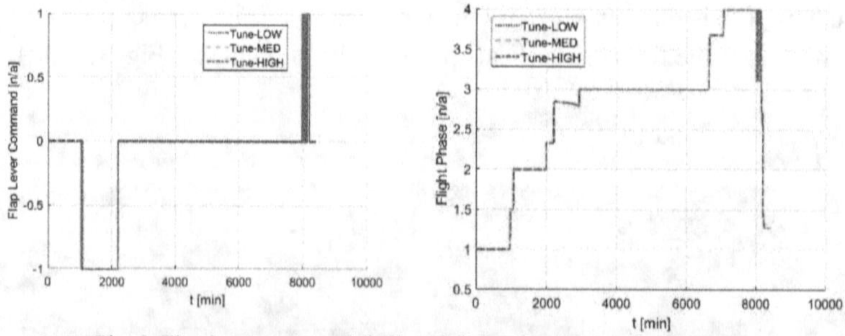

Fig. 5: Flap lever command (left) and flight phase estimation (right).

In order to quantitatively evaluate the performance of the system, two parameters have been defined, K and T. Following [2], the K parameter for each flaps/slats transition from configuration "n" to "m" is:

$$K_{(n\to m)} = V_{(n\to m)} / V_{MIN|Conf(j)} \tag{1}$$

Where $V_{(n\to m)}$ is the speed at which the retraction between configurations "n" and "m = n-1" or extension between "n" and "m = n+1" occurs. $V_{MIN|Conf(j)}$ is the minimum allowed speed in configuration "j" (see Table 2).

Table 2: Values of n, m, j and K range according to the procedure [2].

Flap transition	n	m	j	Min K	Max K
Retraction 1 to 0	1	0	0	1.21	1.25
Extension 0 to 1	0	1	0	1.21	1.25
Extension 1 to 2	1	2	0	1.21	1.23
Extension 2 to 3	2	3	2	1.38	1.47
Extension 3 to 4	3	4	3	1.32	1.36

The higher the K, the safer since it means that the flaps/slats are deployed at aircraft speed with a larger margin above the minimum allowed speed. However, if K is too large there may be a very small margin with respect to the maximum allowed speed. So, an optimal value of K would be the largest one within the indicated range for each transition.

Table 3 presents the optimal K and the resulted K at each simulation (best results bolded). Simulation time T corresponds to the instant when each flaps/slats transition occurs. The longer the flaps/slats are retracted the better because with flaps/slats extended the drag increases. However, if the flaps/slats transitions are delayed excessively, the aircraft could reach the maximum allowed speed in climbing phase or reach the minimum allowed speed in approach phase. So, the optimal T for this specific speed profile is obtained from Fig. 3, left, that is the time in which the speed is equal to $V_{(1 \to 0)}$ (for the acceleration part) or to $V_{(n \to n+1)}$ (for the deceleration part), resulting from applying equation (1) with the minimum value of the each K's range, according to the corresponding flap transition.

Respect to the flaps/slats extensions (transitions 0 to 1, 1 to 2, etc.), in the approaching phase, it is possible to see in Table 3 that the fuzzy flap lever command Tune-LOW gives a value of K too small compared to the optimal one while the Tune-HIGH gives a K much larger. Tune-MED gives good results in transitions 0 to 1, 2 to 3, and 3 to 4, but not good enough in transition 1 to 2.

In terms of T, Tune-LOW is the best for drag reduction and therefore fuel consumption because it lasts longer than the others without extending the flaps/slats. But compared to the optimal T, Tune-MED is generally better.

In the take-off and climbing phases, when flaps/slats are retracted (transition 1 to 0), the different fuzzy flap lever command tunings do not affect the result. The values of K are larger than the expected ones but on the safe side.

Table 3: K and T values obtained at each simulation versus the K, T optimal.

	Optimal		*Tune-LOW*		*Tune-MED*		*Tune-HIGH*	
Flap transition	K	T	K	T	K	T	K	T
1 to 0	1.25	1001	**1.50**	**1056**	1.51	1058	1.51	1059
0 to 1	1.25	8046	1.22	**8047**	**1.28**	8024	1.39	7982
1 to 2	1.23	8050	1.08	8095	1.15	8072	**1.26**	**8032**
2 to 3	1.47	8088	1.27	8119	1.35	**8096**	**1.49**	8056
3 to 4	1.36	8121	1.28	**8131**	**1.37**	8108	1.51	8069

6. Conclusions and Future Works

In this chapter, an intelligent decision system has been designed and implemented, based on fuzzy logic, which allows automating the deployment of the high-lift devices of a commercial airplane. The system behaves as expected according to the procedures a pilot would follow, as shown using a real flight profile.

In the approach phase for landing, a sufficiently good solution has been found, very close to the optimal one. However, it could be improved by expanding the definition of fuzzy sets for some input variables, such as the minimum and maximum speed range. Other intelligent techniques that enable metric-based optimization, such as reinforcement learning, could also be applied.

References

1. E. Plaza and M. Santos, Knowledge based approach to ground refueling optimization of commercial airplanes. *Expert Syst.* 38(2), e12631 (2021).
2. L. De Baudus and P. Castaigns, *Control your Speed Series, in: Safety First.* (Airbus, Special Edition, February 2018).
3. E. Plaza and M. Santos, Management and intelligent control of in-flight fuel distribution in a commercial aircraft. *Expert Syst.* e13075 (2022).
4. N. Fala, An analysis of fixed-wing stall-type accidents in the United States, *Aerospace* 9 (4), 178 (2022).
5. M. P. Patel, Z. H. Sowle, T .C. Corke, C. He, Autonomous sensing and control of wing stall using a smart plasma slat. *J. Aircr.* 44(2), 516–527, (2007).
6. G.P.G. Da Silva, J.P. Eguea, J.A.G. Croce, and F.M. Catalano, Slat aerodynamic noise reduction using dielectric barrier discharge plasma actuators. *Aerosp. Sci. Technol*, 97, 105642 (2020).
7. Z. Yang, J. Li, J. Jin, H. Zhang, and Y. Jiang, Investigation and improvement of stall characteristic of High-Lift configuration without slats. *Int. J. Aerosp. Eng*, 2019, 1–14 (2019).
8. N. Mustafa, Y. Labiche, and and D. Towey, Traceability in Systems Engineering: An Avionics Case Study, in *IEEE 42nd Annual Computer Software and Applications Conference* (COMPSAC), (Tokio, Japan, 2018).
9. Y. Jing, Fault Diagnosis of Landing Gear Extension and Retraction System Based on Fuzzy Fault Tree and Bayesian Network, in *IEEE 4th Int. Conf. Civil Aviation Safety and Information Technology* (ICCASIT), (Dali, China, 2022).
10. J. Sun, J. Ellerbroek, and J. Hoekstra, Large-scale flight phase identification from ADS-B data using machine learning methods, in 7th *Int. Conf. Research in Air Transportation* (ICRAT), (Philadelphia, USA, 2016)
11. M. Benarous, and I. Panella, Flap system power drive unit (PDU) architecture optimization, in *9th Int. Conf. Power Electronics, Machines and Drives* (PEMD), (Liverpool, UK, 2018).
12. E. Plaza, M. Santos, and J.E. Sierra-García, First Approach of an Intelligent Automatic System for Aircraft Flap/Slat Positioning, in *18th International Conference on Soft Computing Models in Industrial and Environmental Applications* (SOCO), (Salamanca, Spain, 2023).
13. Eurocontrol Aviation Data for Research (2021), *https://www.eurocontrol.int/dashboard/rnd-data-archive.*

Path Planning of Disinfection Robots Enhanced with Metaheuristics Techniques: A First Approach

M. Peñacoba[*,‡], J. Enrique Sierra-García[*,§] and M. Santos[†,¶]

[*]*Department of Digitalization, University of Burgos, 09006 Burgos, Spain*

[†]*Institute of Knowledge Technology,*
University Complutense of Madrid, 28040 Madrid, Spain

[‡]*mpenacoba@ubu.es*
[§]*jesierra@ubu.es*
[¶]*msantos@ucm.es*

The global spread of COVID-19 has prompted research on the use of mobile robotics and UV-C light to automate disinfection procedures. This attempt, however, is plagued with problems that demand nuanced solutions for successful implementation. This chapter addresses the issue by integrating a metaheuristic technique, specifically genetic algorithms, with a UV-C radiation estimator deployed along a predetermined trajectory and a mobile robot motion simulator. The primary goal is to achieve broad surface coverage with acceptable radiation levels for effective disinfection. Simultaneously, it is crucial to reduce radiation dose to conserve energy and resources while protecting the surrounding environment. The proposed solution leverages genetic algorithms for the strategic selection of waypoints, trajectory creation, radiation dose estimation, and iterative path refinement, aligning it with a predefined threshold to achieve optimal disinfection results. This chapter conclusively demonstrates that intelligent optimization significantly boosts the efficiency of the disinfection procedure across the treated area. Moreover, it ensures precision in the application of radiation within the designated space, effectively neutralizing the target microorganism and minimizing any adverse environmental impact linked to radiation exposure.

Keywords: Metaheuristic techniques; genetic algorithm; UV-C radiation; optimization; differential robot.

1. Introduction

The COVID-19 pandemic has reshaped the global landscape, emphasizing the crucial need for effective disinfection strategies in diverse environments. Ultraviolet-C (UV-C) light disinfection has emerged as a promising solution, capable of deactivating microorganisms [1]. While UV-C technology has been

[‡]Corresponding author.

used in controlled environments, the recent crisis has renewed interest in its widespread application [2].

This chapter introduces an innovative architecture for optimizing UV-C radiation from a disinfection luminaire along a defined trajectory in a given space. The approach involves a differential robot model as the automated transporter, a movement simulation environment, obstacle avoidance, and calculation of the radiation dose [3]. The research entails developing a software architecture that identifies key points, simulates the robot's trajectory, calculates UV radiation, and optimizes the process through the genetic algorithm. The excellence of this innovative approach lies in its ability to optimize the required radiation at each point in space, approaching the desired target radiation. This skill is made possible by the implementation of the intelligent solution provided by genetic algorithms (GA) [4]. Furthermore, the ultimate goal is practical implementation on a real mobile robot for disinfecting healthcare facilities [5].

2. Description of the Experimental Set Up: Robot and Radiation Scattering

The construction of the experimental environment requires the definition of the type of mobile robot, its movement and parameters to be controlled, as well as the model of the radiation and its dispersion throughout space. In this work, to obtain results close to reality, a simulation environment was created to replicate the real movement of the Hussar robot in a defined space. The kinematics of this robot is shown in Figure 1.

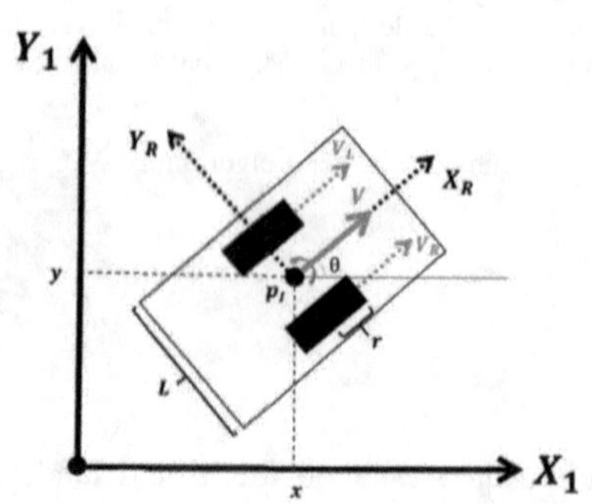

Fig. 1. Kinematic model of the Hussar robot.

In this figure, the robot's location is symbolized by its Cartesian coordinates and orientation, denoted as $p_I = \{x, y, \theta\}$ within the inertial reference frame $\{X_I, Y_I\}$. The absence of lateral slip is assumed, thus the transverse velocity, Y_R, is considered negligible, simplifying the model.

Within this framework, the longitudinal velocity (V) and angular velocity (W) come from the combination of linear velocities associated with each wheel, represented as V_R and V_L (where subscripts denote right and left, respectively).

$$V = \frac{V_R + V_L}{2}, W = \frac{V_R - V_L}{L} \tag{1}$$

If p_I is considered as an arbitrary pain (position, orientation) in the global inertia frame, the kinematic model corresponds to (2), where $\dot{x}$ and $\dot{y}$ are velocity in x and y respectively, and $\dot{\theta}$ is angular velocity.

$$\dot{p}_I = \begin{bmatrix} \dot{x} \\ \dot{y} \\ \dot{\theta} \end{bmatrix} = \begin{pmatrix} \cos(\theta) & -\sin(\theta) & 0 \\ \sin(\theta) & \cos(\theta) & 0 \\ 0 & 0 & 1 \end{pmatrix} * \begin{bmatrix} V \\ 0 \\ W \end{bmatrix} \tag{2}$$

The state vector is determined by $[x, y, \theta]$. All cartesian coordinates are in meters and angles are in radians. Likewise, linear velocities are expressed in meters per second and angular velocities in radians per second.

The specified disinfection lamp reference emits a radiation of 5.5 J/m^2 when situated at a one-meter distance for one second, assuming perpendicular exposure to the radiation. This radiation diminishes proportionally with the square of the distance (radius) [6].

The designated target radiation level aligns with the requirement to deactivate SARS-CoV-2 at a rate of 99%, which is approximately 500 J/m^2. Given these parameters and considering the radiation absorption coefficient, the radiation scattering from the lamp was computed using equation (3). It is important to note that this model simplifies the representation of ultraviolet radiation propagation.

$$R = \frac{P_L * k * t}{D^2} \tag{3}$$

Where: k is a scattering constant (m^2/J) dependent on the medium to be disinfected ($k = 0{,}71$), P_L is the lamp power (W), t is the disinfection time (s) and D is the distance between points (m).

Thus, the radiation matrix, portraying the virtual environment in which the robot operates is obtained using equation (4).

$$R_M(i, j, t) = P_L \cdot k \int \frac{1}{D_{i,j}^2(t)} dt \tag{4}$$

The disinfection system, which recreates a 207J High Energy Pulse lamp, has been simulated by successively creating straight lines with a defined pitch angle starting from a certain point (location of the lamp). The possible intersection of the radiation with any obstacle is calculated by segmenting each colliding line when it intersects with it. This methodology allows the radiation calculating program to dismiss the radiation overshadowed by the opaque obstacles of the environment.

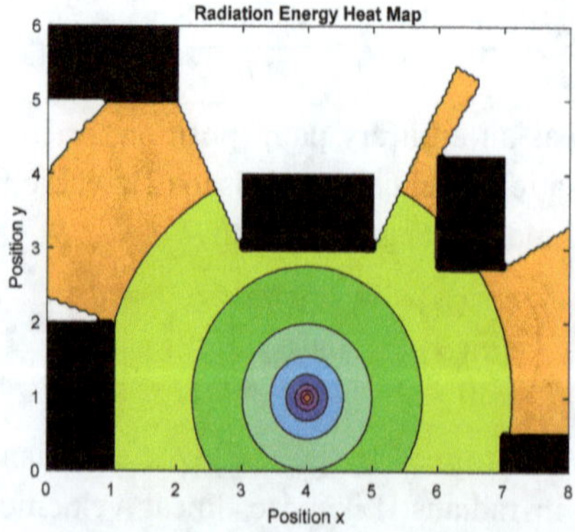

Figure 2. Radiation calculus.

Lines of radiation whose coordinates match the spatial matrix cells which are not overshadowed by opaque obstacles are classified as "seen" and radiation is, therefore, calculated. Figure 2 illustrates the modeling of the radiation with an extent of 5 meters. It has been considered that beyond 5 meters, the radiation is so close to 0 that it can be deemed negligible.

3. Path Optimization Methodology

The aim of this study is to identify optimal waypoints that define the disinfection trajectory, allowing a subsequently automated robot carrying a UV-C lamp to maximize the zones in the map that receive the required dose. To achieve this, a set of waypoints and potential stopping points must be defined. In other words, the robot recognizes a finite set of points as coordinates and sequentially passes through them. Additionally, a stop time has been defined at each point to ensure proper disinfection. Based on this procedure, the code simulating its movement has been developed. The state diagram is shown in Figure 3.

These stop points (State 2, Figure 3) are the ones we aim to optimize since, with the most efficient arrangement and durations. Each point consists of four reference parameters: $(x_r,\ y_r)$ coordinates, angle(θ_r), and time(t_r), that is $(x_r, y_r, \theta_r, t_r)$, where the subscript r indicates that these values are reference points for the robot. To avoid prolonging the program execution time excessively, we have fixed the angles and time. With these fixed values, the genetic algorithm provides the position coordinates (x_r, y_r) [7].

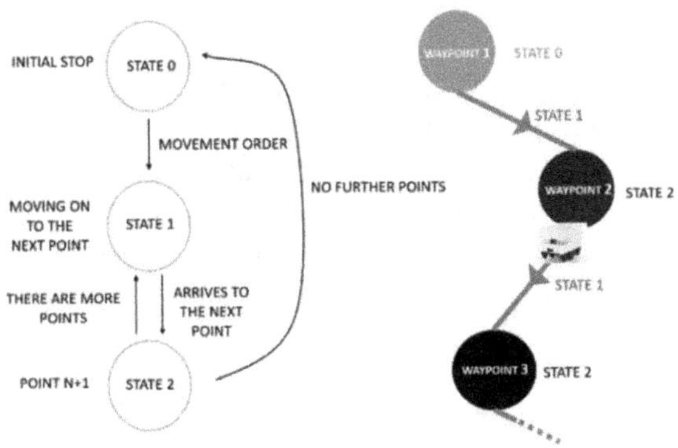

Figure 3. State diagram

The implementation of the software architecture (Figure 4) for integrating the genetic algorithm with the robot and radiation simulator has been as follows. The actual robot has a program that supports XML files with the parameters of the points mentioned in section 3, $\{x_r, y_r, \theta_r, t_r\}$. Therefore, a code has been programmed to convert these points from matrix format to this type of file.

Whenever the genetic algorithm provides a new combination of parameters, a program is called to convert it into XML. This file is read by the Python path simulation program (just like the actual robot would do). This simulator generates a log with the robot's position at each moment in time, and another program in Matlab uses this log to calculate the radiation matrix. Furthermore, this program also detects whether there has been a collision with an obstacle and calculates the distances between intersections. Then, these parameters are used in the cost function. In fact, the cost function of the optimization algorithm differentiates two cases: collision and non-collision. When there is not collision the radiation matrix is taken into account in the cost function.

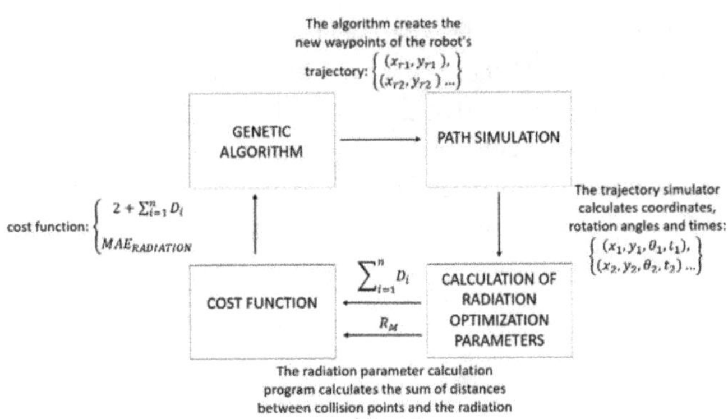

Figure 4. Software architecture.

3.1. *Optimization problem*

The optimization of the trajectory has been carried out by appropriately defining a piecewise cost function that the genetic algorithm seeks to minimize (5). The first row is used when there are collisions, and the second row when not. In this last case the Mean Absolute Error (MAE) of the matrix R_M is calculated. In other words, it considers the absolute difference between the actual $real\_rad$ and target radiation, Obj_rad. Loss function is shown in Equation 5.

$$f_c = \begin{cases} 2 + \sum_{i=1}^{n} D_i & if \sum_{i=1}^{n} D_i \neq 0 \\ \frac{1}{N} \cdot \sum_{i=1}^{N} | Obj\_rad(i) - real\_rad(i)| & if \sum_{i=1}^{n} D_i = 0 \end{cases} \quad (5)$$

In Equation 5, n is the number of obstacles, D_i is the distance that the trajectory of the vehicle crosses the obstacle i, and N is the number of cells in the radiation matrix.

4. Simulation Results

The method described in this article has been evaluated in two medical environments: the first one consists of an operating room of 48 m^2 with several obstacles [8]. More specifically, two medical closets, a stretcher, a table and a washbasin has been considered. The second and the complex one consists of a 192 m^2 hospital waiting room. It has nine obstacles: a reception desk, six seats and a column. Both environments has been defined with a matrix of 33 columns and 25 rows, i.e. a total of 825 cells of 0,0582 m^2 for the operating room and 0,234 m^2 for the waiting room. Therefore, as the target radiation to deactivate SARS-CoV-2 is 500 J/m^2, each cell should ideally receive: $R_C = 29{,}1\,J$ for the operating room and $R_C = 116{,}36\,J$ for the waiting room (R_C corresponds to the radiation per unit cell). Thus, the experiment was carried out in the environments shown in Figure 5 and 6. In those, an initial rectangular trajectory (blue) was set up and, after two hours of iterative execution, the genetic algorithm found better solutions. For the experiment execution, θ_r and t_r has been fixed at 0 and all radiation measurements have been conducted in decibels (dB). Figures 7 and 8 show the radiation heat map of the last, and therefore the best solution.

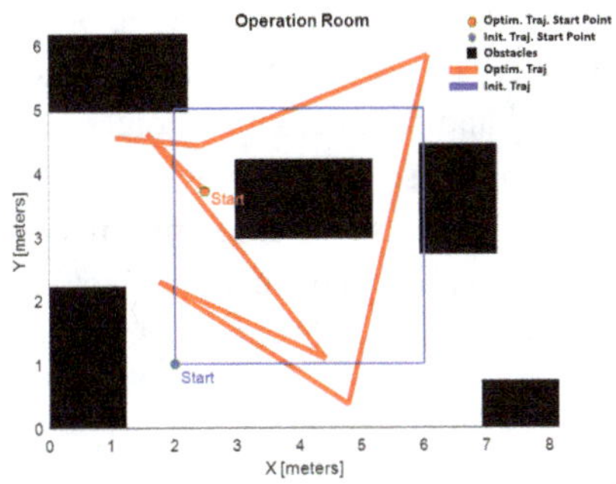

Figure 5. Operating Room.

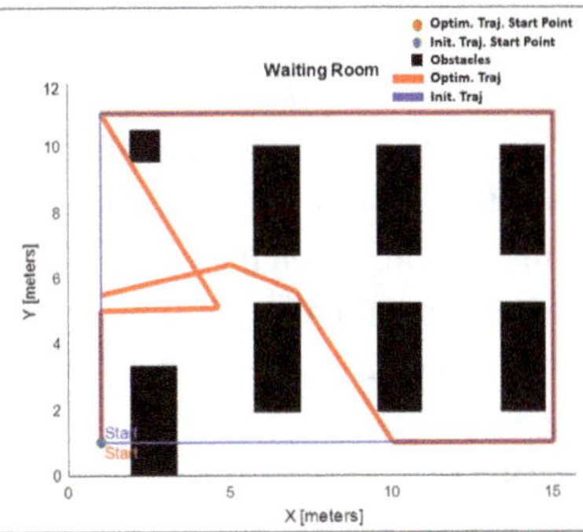

Figure 6. Waiting Room.

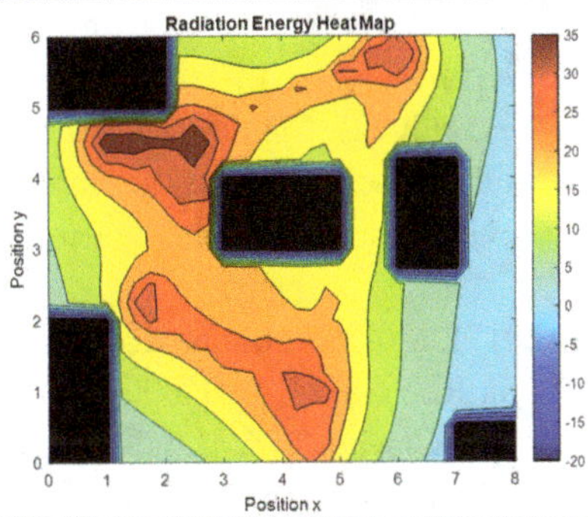

Figure 7. Solution Heat Map.

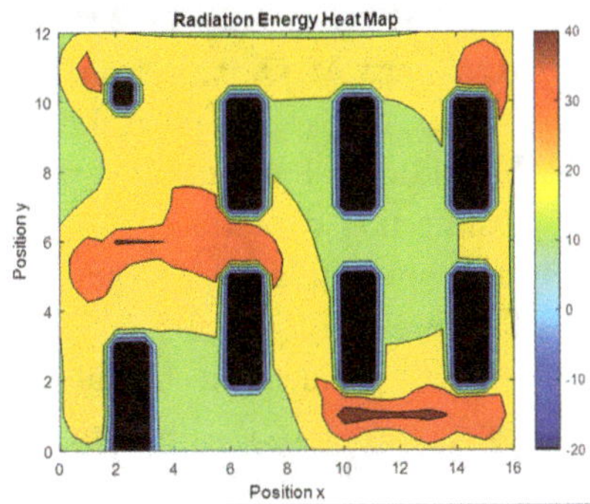

Figure 8. Solution Heat Map.

Table 1 shows how the genetic algorithm is able to improve the radiation spreading. That is, it makes it more homogeneous. Therefore, the MAE between the target and actual radiation decreases.

Table 1.MAE evolution.

Environment	MAE of initial trajectory $\frac{J}{cell}$	MAE of first trajectory without collisions $\frac{J}{cell}$	MAE of optimized trajectory without collisions $\frac{J}{cell}$
Operating Room	24,58	21,98	21.46
Waiting Room	102,9	99,77	97,87

5. Conclusions and Future Works

This chapter introduces an architecture for optimizing the radiation emitted by an automatic disinfection robot using genetic algorithms. The architecture integrates a differential robot simulator, an obstacle avoidance algorithm, a genetic algorithm, and a radiation calculator. When systematically executed in sequence, these components yield a trajectory that irradiates a sufficient amount to deactivate the target microorganisms effectively. To execute it, it one only needs to be aware of the target disinfection threshold and then define the total number of points and set the maximum space and stop times. This system will find the optimal solution with these parameters. For future works, this program is intended to be used on a real robot (Hussar robot) equipped with a disinfection lamp, and its application is anticipated in hospitals, clinics, and other healthcare facilities.

References

1. She, R. C., Chen, D., Pak, P., Armani, D. K., Schubert, A., & Armani, A. M. (2020). Lightweight UV-C disinfection system. Biomedical Optics Express, 11(8), 4326-4332.
2. Alvarenga, M. O. P., Veloso, S. R. M., Bezerra, A. L. C. A., Trindade, B. P., Gomes, A. S. L., & de Melo Monteiro, G. Q. (2022). COVID-19 outbreak: Should dental and medical practices consider uv-c technology to enhance disinfection on surfaces?–A systematic review. Journal of photochemistry and photobiology, 9, 100096.
3. Pandey, A., Pandey, S., & Parhi, D. R. (2017). Mobile robot navigation and obstacle avoidance techniques: A review. Int Rob Auto J, 2(3), 00022.
4. Blatchley III, E. R., Brenner, D. J., Claus, H., Cowan, T. E., Linden, K. G., Liu, Y., ... & Sliney, D. H. (2023). Far UV-C radiation: An emerging tool for pandemic control. Critical Reviews in Environmental Science and Technology, 53(6), 733-753.
5. Kramer, O., & Kramer, O. (2017). Genetic algorithms (pp. 11-19). Springer International Publishing.
6. Rodrigo, D. V., Sierra-García, J. E., & Santos, M. (2023). Glasius bio-inspired neural networks based UV-C disinfection path planning improved by preventive deadlock processing algorithm. Advances in Engineering Software, 175, 103330.
7. Peñacoba, M., Sierra-García, J. E., Santos, M., & Mariolis, I. (2023). Path Optimization Using Metaheuristic Techniques for a Surveillance Robot. Applied Sciences, 13(20), 11182.
8. Demeersseman, N., Saegeman, V., Cossey, V., Devriese, H., & Schuermans, A. (2023). Shedding a light on ultraviolet-C technologies in the hospital environment. Journal of Hospital Infection, 132, 85-92.

Generation of Pitch Signals for the Control of Wind Turbines with Neural Networks: A First Approach

C. Blanco[*,§], M. Santos[†,¶] and J. E. Sierra-García[‡,||]

[*]*ETSI Informática, UNED, 28040 Madrid, Spain*

[†]*Institute of Knowledge Technology, Complutense University of Madrid, 28040 Madrid, Spain*

[‡]*Department of Digitalization, University of Burgos, Burgos, Spain*

[§]*cblanco177@alumno.uned*

[¶]*msantos@ucm.es*

[||]*jesierra@ubu.es*

The importance and future prospects of offshore wind power generation invite great efforts and investments to make it an efficient technology. A crucial aspect is the development of efficient control strategies, which in many cases require models to identify time accurately the state of the turbine at a given moment. These models must be simple enough not to increase the computational complexity of the control algorithm while being able to capture the nonlinearity and coupling of the wind devices. In this chapter, we exploit the possibility of using neural networks to simulate the response of a model predictive controller. A sequential neural network model is trained, with satisfactory results in terms of accuracy.

Keywords: Neural networks; artificial neural networks; model predictive control; wind turbine; system identification; control-oriented model.

1. Introduction

Offshore wind energy is experiencing an important growth with an increase in this kind of installation, up to 21% in 2021, see Ref. [1]. Most of these offshore wind turbines are built in shallow water with fixed foundations, while floating turbines present significant challenges due to their greater complexity, according to Ref. [2]. This complexity requires the use of models and control algorithms able to capture all dynamic effects to design and operate optimally the turbines.

[§]Corresponding author.

Besides, these controllers must be computationally cost efficient to operate in real time, e.g. Ref. [3]. The main objectives of wind turbine control are to maintain safe operation, maximize power generation, mitigate fatigue loads on the structure, and avoid fault conditions, see Ref. [2].

Commercial wind turbines typically operate at variable speed and use different control strategies depending on the operating regime, e.g. Ref. [3]. The use of classical control technics like PID (Proportional-Integral-Derivative) regulators based on a single input single output (SISO) scheme requires multiple control loops to stabilize the structure. For more complex turbines, multiple-input multiple-output (MIMO) strategies that are able to capture the most critical dynamics are a must. It is in this context where the identification of the most important features of the turbine system is an important step, in order to design effective control strategies, such as MPC (Model Predictive Control) or State Space-based control.

Some researchers have worked on this approach. In Ref. [4], a MPC-based control strategy to stabilize the power generation and reduce the dynamic loads in the structure of the 5-MW floating turbine defined in Ref. [5], in the constant speed operation regime, was proposed. An internal linear model, identified from the results of a highly complex nonlineal model, is used to predict the system behavior and optimize the control signals. Another approach is to design a digital twin of a 1.5-MW turbine to monitor the state of the structure from the loads estimated with the twin, see Ref. [6]. The authors use a Kalman filter-based linear model designed with operational data complemented by nonlinear simulations. This model allows estimating the state of the system at an instant of time with limited information about it. In a similar way, in Ref. [7], a monitoring system of a gear box based on data obtained from different wind farms is presented. Different machine learning (ML) and deep learning (DL) methods are used to design predictive models. The model based on deep neural networks performs the best. The use of neural networks to estimate turbine power generation is an interesting topic. In Ref. [8], an artificial neural network (ANN) is used to estimate the power generated by a five-turbine wind farm and to propose optimized angles to reduce the wake effect on the turbines. Long short-term memory (LSTM) neural networks have been shown as an effective tool to predict the power generation of the turbines in a wind farm, see Ref. [9]. The training of the networks is done using historical wind data and turbine power generation. In Refs. [10] and [11], an effective wind speed estimator based on neural networks is successfully used to improve the performance of a classical controller. An example of an MIMO that uses a complex autoregressive model based on simple neural networks is presented in Ref. [12]. Moreover, the use of nonlinear autoregressive with exogenous input models (NARXs) based on artificial neural networks (ANNs) is successfully implemented, in Ref. [13], to perform the system identification of a 5-MW turbine.

In this chapter, a model predictive control strategy, based on a linearization of a complex nonlinear model of a wind turbine around an operating point, is implemented. Control data of this MPC strategy are used to train a nonlinear model based on artificial neural networks (ANNs). Next, the outputs of the MPC strategy and the neural network (NN) model are compared, using a set of test data, to study the performance of the neural model. It has been observed how the neural model replicates the output of the MPC. The idea behind this work is to train a single low-complexity neural network model using a bank of linear MPCs that cover the complete range of operation of the turbine and use it as the controller. The use of low-complexity neural networks will reduce the computational complexity of the controller and will enable its use in low-cost hardware platforms. In this chapter, the idea is tested with only one MPC as a first approach.

2. System Description

2.1. *Wind turbine*

The wind turbine used is described in [14]. It is an offshore wind turbine with a fixed-bottom monopile support structure. This reference wind turbine is a Class IB direct-drive machine, with a rotor diameter of 240 meters (m) and a hub height of 150 m. The main characteristics of this IEA 15-MW WT are the following: rated wind speed 10.59 m/s, cut out wind speed 25 m/s; rated rotor speed 7.55 rpm.

In order to design a suitable MPC strategy, a linear model is necessary. Openfast software is used to obtain a linearized model in an open loop configuration at equilibrium point. The linear model has two inputs, the wind speed (m/s) and the rotor speed error (rpm), and one output, the pitch angle in radians. In this study, the linearization was performed at around 15 m/s, generator speed of 7.55 rpm, and 11.39° pitch angle, which is the equilibrium point found by the Openfast linearizing algorithm.

The linearization was validated comparing the linear model vs. the non-linear Openfast model. The linear model is accurate up to a deviation of around 1 m/s from the nominal wind speed 15 m/s so the system will be limited to that range.

2.2. *Model predictive controller*

In this work, a model predictive controller is designed and implemented using Matlab, based on the linear model obtained previously. The main characteristics of the MPC are a sample time of 0.2 s; the manipulated variable (MV) is the pitch angle command (rad); the measured output (MO) is the rotor speed (rpm); a measured disturbance (MD) is introduced in the form of the wind speed (m/s). The prediction horizon consist of 20 samples, and the control horizon of 10 samples. The cost function is defined as follows:

$$J = 0.0367879 * MV + 2.71828 * MO \tag{1}$$

The complete system is simulated using Matlab. Fig. 1 shows a schematic view of the closed loop system used in the experiments.

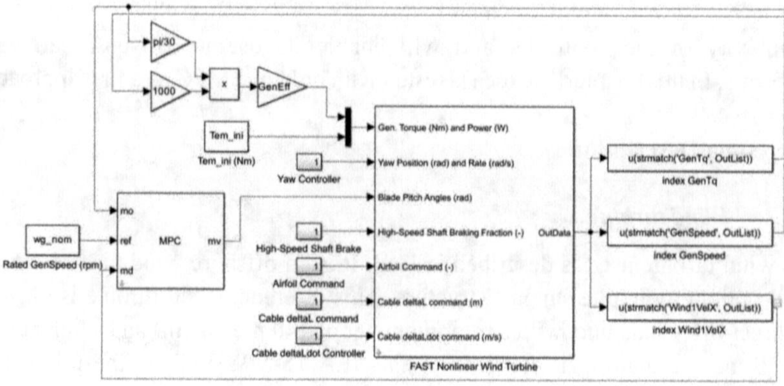

Fig. 1. MPC closed loop schematic.

Fig. 2 shows the response of the MPC against a PI controller used as benchmark (P = -1.0, I = -4.0, tuned with Ziegler–Nichols method) regarding the rotor speed. The first 100 seconds of the simulation are discarded due to irregular response of the Openfast turbine in that time.

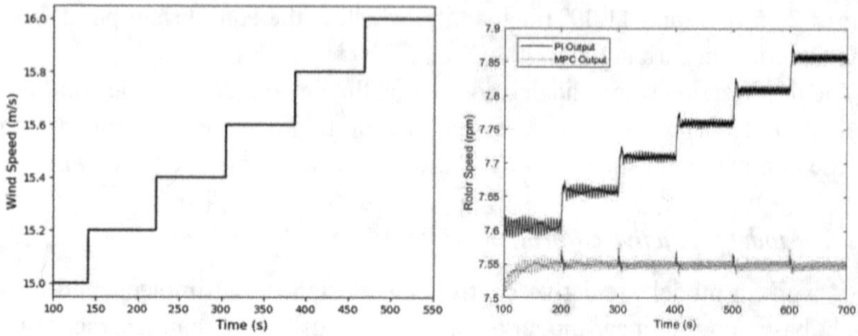

Fig. 2. MPC vs PI comparison output.

3. Neural Network Model

The neural networks used in the study are sequential feedback neural networks. The model is designed with Tensorflow. Since it is based in the data generated with the MPC, the macro structure needs to be adapted to capture the same information. In this case, two inputs are used, rotor speed error and wind speed. Both have the same structure, i.e. a vector with 20 samples, matching the prediction horizon of the MPC. Fig. 3 shows a view of the model. The network has two branches, one for each input feature, with the same structure. There is an input layer that serves to define the structure of the feature, and then two dense layers with ten units each for predictions. The output branch has a similar structure but instead of an input layer it has a concatenate layer to mix the predictions from the other branches, and a last layer with one unit, that is, the pitch angle.

The neural network training is performed using the MPC data, emulating the control strategy of the MPC, i.e. wind speed and rotor speed error as inputs, and pitch angle as output. A total of 600 s of data are used, with a sample time of 0.2 s. It is important to remark that the inputs are vectors which contain 20 data points. A 80/20 percent split between training and validation data is used. The testing data set is generated in a separate Openfast run.

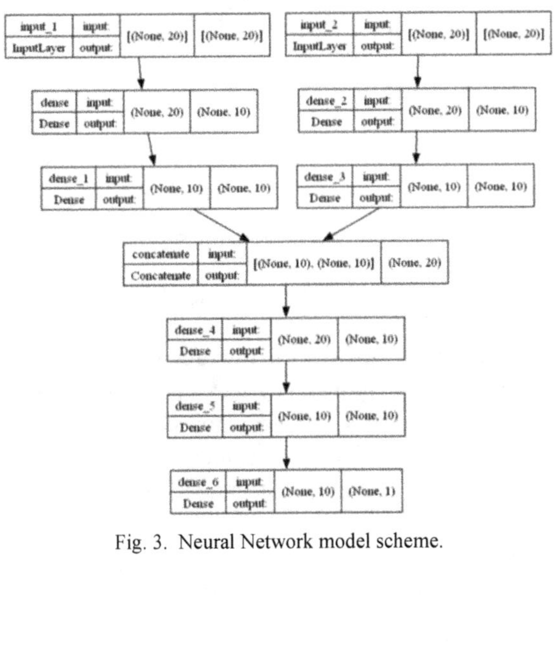

Fig. 3. Neural Network model scheme.

4. Results Discussion

The 15-MW WT is simulated with Openfast. A specific data set is used to test the neural network consisting of a total of 550 s. Fig. 4, left, shows the wind speed profile of the data. Fig. ,4, right shows a comparison between the output obtained with the MPC and the NN model.

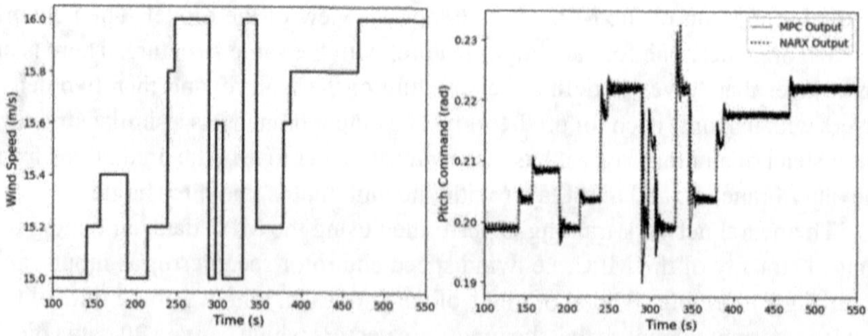

Fig. 4. Wind speed profile (left) and MPC and NN output comparison (right).

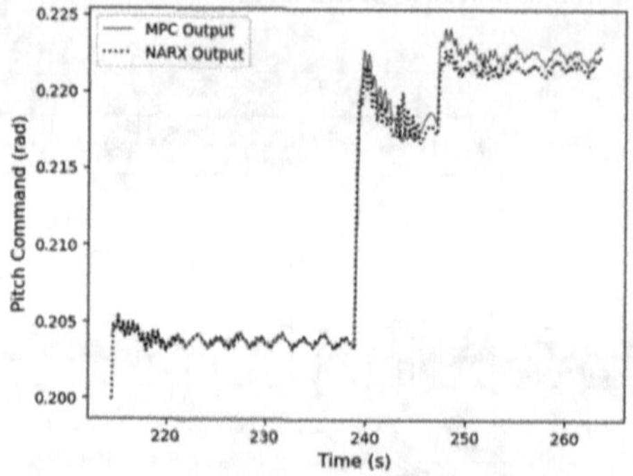

Fig. 5. MPC and NN output comparison detail

The first 100 seconds in Figure 4 are discarded due to irregular response of the Openfast turbine in that time. It can be seen that the neural network output is remarkably close to that of the MPC. Table 1 shows the metrics of the comparison, MAE (mean absolute error) and RMSE (root mean squared error), which allow to consider the NARX model as a good replacement to the MPC.

Table 1. Results Comparison Summary

Units	MAE	RMSE
[rad]	6.3E-4	2.2E-3
[deg]	0.03	0.10

5. Conclusions and Future Works

The suitability of nonlinear autoregressive with exogenous input models based on feedforward neural networks to substitute a model predictive control strategy is studied. A linear MPC is designed to control a 15 MW wind turbine, and data generated with this controller are used to train the neural network model. The controller based on the network model shows good accuracy compared to the MPC. It is important to note that the model hyper-parameters have not been thoroughly optimized, and an accurate enough network has been selected for the study.

The results are promising in cementing the idea of training neural networks to use as substitute of a model predictive control. Apart from the computational effort, that is transferred from the operation of the plant to the design of the controller, this design could be extended to train a neural network with datasets generated by several linear model predictive controllers to cover a more complete set of operating points. To implement this neural controller, we study its stability with Lyapunov's theory.

Acknowledgments

This work was partially supported by the Spanish Ministry of Science, Innovation and Universities under MCI/AEI/FEDER Project no. PID2021-123543OB-C21.

References

1. Global Wind Energy Council, "Global Wind Report 2022," GWEC, Bruselas, 2022.
2. R. Hu, L. Conghuan, H. Ding and P. Zhang, "Implementation and evaluation of control strategies based on an open controller for a 10 MW floating wind turbine," *Renewable Energy,* vol. 179, pp. 1751-1766, 2021.
3. A. D. Wright and L. J. Fingersh, "Advanced Control Design for Wind Turbines Part I: Control Design, Implementation, and Initial Tests," NREL, Golden, 2008.
4. T. Wakui, A. Nagamura and R. Yokoyama, "Stabilization of power output and platform motion of a floating offshore wind turbine-generator system using model predictive control based on previewed disturbances," *Renewable Energy,* vol. 173, pp. 105-127, 2021.
5. J. Jonkman, S. Butterfield, W. Musial and G. Scott, "Definition of a 5-MW Reference Wind Turbine for Offshore System Development," NREL, 2009.
6. E. Branlard, J. Jonkman, S. Dana and P. Doubrawa, "A digital twin based on OpenFAST linearizations for real-time load and fatigue estimation of

land-based turbines," *Journal of Physics: Conference Series,* no. 1618, 2020.

7. L. Wang, Z. Zhang, H. Long, J. Xu and R. Liu, "Wind Turbine Gearbox Failure Identification With Deep Neural Networks," *IEEE Transactions on Industrial Informatics,* vol. 13, no. 3, pp. 1360-1368, 2017.
8. H. Sun, C. Qiu, L. Lu, X. Gao, J. Chen and H. Yang, "Wind turbine power modelling and optimization using artificial neural network with wind field experimental data," *Applied Energy,* vol. 280, 2020.
9. J. Zhang, J. Yan, D. Infield, Y. Liu and F.-s. Lien, "Short-term forecasting and uncertainty analysis of wind turbine power based on long short-term memory network and Gaussian mixture model," *Applied Energy,* vol. 241, pp. 229-244, 2019.
10. J. E. Sierra-García and M. Santos, "Improving wind turbine pitch control by effective wind neuro-estimators," *IEEE Access,* vol. 9, pp. 10413-10425, 2021.
11. J. E. Sierra-García and M. Santos, "Redes neuronales y aprendizaje por refuerzo en el control de turbinas eólicas," *Revista Iberoamericana de Automática e Informática Industrial,* vol. 18, no. 4, pp. 327-335, 2021.
12. A. Alonso, A. Zabaljauregi, M. Larrea, E. Irigoyen and J. Sanchís, "Studying the Use of ANN to Estimate State-Space Variables for MIMO Systems in a NMPC Strategy," *17th International Conference on Soft Computing Models in Industrial and Environmental Applications (SOCO 2022). SOCO 2022. Lecture Notes in Networks and Systems,* vol. 531, 2022.
13. C. Blanco, M. Santos and S.-G. J. E., "Forecasting of wind turbine synthetic signals based on nonlinear autoregressive networks," in *19th International Conference on Artificial Intelligence Applications and Innovations*, León, 2023.
14. E. Gaertner, J. Rinker, L. Sethuraman, F. Zahle, B. Anderson, G. Barter, N. Abbas, F. Meng, P. Bortolotti, W. Skrzypinski, G. Scott, R. Feil, H. Bredmose, K. Dykes, M. Shields, C. Allen and A. Viselli, "IEA Wind TCP Task 37 - Definition of the IEA 15-Megawatt Offshore Reference Wind," NREL, 2020.

An Adaptive Stacking Method for Multiple Data Streams Learning under Concept Drift

Kun Wang, Jie Lu*, Anjin Liu and Guangquan Zhang

Australia Artificial Intelligence Institute (AAII), University of Technology Sydney, Sydney, NSW 2007, Australia

**jie.lu@uts.edu.au*

Multiple data streams learning has attracted high attention recently. However, different feature spaces and uncertain concept drift situations among each stream may lead to the learning decay of machine learning models. To address this issue, this chapter proposes an adaptive stacking method for multiple data stream learning. First, a stacking-based learning framework is built to handle multiple data streams with different feature spaces. Second, a selective retraining scheme is developed for concept drift adaptation. Finally, by testing the proposed method on five data scenarios and comparing it with three benchmarks, the experiment results show the efficiency.

Keywords: Concept drift; data stream; ensemble learning; machine learning.

1. Introduction

Data stream learning has been highly focused recently, especially dealing with concept drift under uncertain environments.[1–3] Streaming data widely exists in real-world life, and many of them are shown as multiple data streams, like daily weather records of multiple regions, and interest changes among different users. However, most previous studies focus on handling a single data stream without considering multiple data streams under concept drift, which is more complicated. Therefore, to further maintain the learning performance in such a complex scenario, it is needed to consider handling multiple streams under concept drift.

In addition, previous studies about multiple data stream learning include regression[4,5] and classification tasks.[6] For the regression task, the relationship among streams has been considered and extracted to support model learning. For the classification task, methods like transfer learning[7] and ensemble learning[8] have been conducted for learning adaptation. However, few of them consider the scenario of multiple data streams with different feature spaces and drift situations, as shown in Fig. 1. Motivated by this, this chapter aims to focus on dealing with multiple data streams with different feature spaces and help adapt to different concept drift situations.

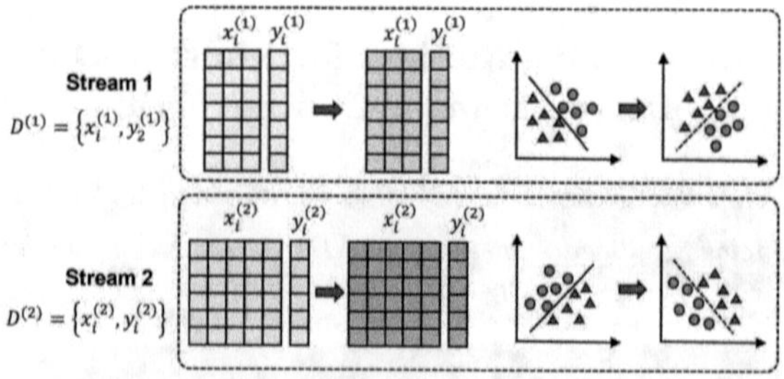

Fig. 1. An illustration of multiple data streams with concept drift.

The main contributions of this chapter are listed as follows:

- A stacking-based learning framework is given to deal with multiple data streams with different feature spaces.
- A selective retrain scheme has been developed to obtain the learning performance.
- Experiments on five data scenarios show the efficiency of the proposed method. The source code of the proposed method is available at `https://github.com/kunkun111/Adaptive-Stacking`.

The following contents of this chapter are organized as follows: Section 2 summarizes the recent studies. Our proposed method is introduced in Section 3, followed by Section 4, which outlines detailed experiment setting and discussion. Finally, Section 5 gives the conclusion and future study.

2. Related Work

2.1. *Data stream learning under concept drift*

Concept drift is known as the data distribution changes timely that may lead to a learning decay[1,2], denoted as $P_t(y|x) \neq P_{t+1}(y|x)$, where t and $t+1$ are two consecutive time steps, and $P(y|x)$ expresses the data distribution with data attributes x and data label y.

To handle real-time data stream under concept drift, sufficient methods have been developed. On the one hand, some previous studies aim to identify concept drift firstly to further support model learning. For example, competence models-based concept drift detection methods[9] and Detection

Delay Index (DD Index)[10] are recently proposed concept drift detection methods. On the other hand, sufficient drift adaptation methods have been proposed to obtain performance. As ensemble learning methods perform excellently, they have been chosen and redesigned for data stream learning under concept drift[11]. Methods like Adaptive Random Forests (ARF)[12], Streaming Random Patches (SRP)[13,14], pENsemble[15], and Graph ensemble boosting[16] are widely used ensemble learning methods.

2.2. *Mutiple data stream learning approaches*

To handle more complex learning scenarios, multiple data stream learning under concept drift attracts attention[17,18]. The supervised tasks for handling multiple data streams mainly include regression and classification. For multiple data stream regression, research[4] gives a fuzzy drift correlation matrix. In addition, a graph neural network-based method has been proposed to handle multiple data stream regression under concept drift[5].

For unsupervised tasks, research[19] obtains ideas of transfer learning for handling multiple data streams. In addition, focusing on the problem of unlabelled drifting streams, a framework called Learn-to-Adapt (L2A) has been developed to handle concept drift in multiple data streams[6]. Also, clustering methods have been used for multiple streams learning[20].

3. Methodology

3.1. *Stacking-based multiple data stream learning*

Given two data streams $S_1 = \{x_i^{S_1}, y_i^{S_1}\}_{i=1}^n$, $S_2 = \{x_i^{S_2}, y_i^{S_2}\}_{i=1}^n$, both of them consists of data chunks D with the same size at each time point. The data chunks of two streams are expressed as $\{D_t^{S_1}\}_{t=1}^T$ and $\{D_t^{S_2}\}_{t=1}^T$, where T is the total time steps. First, we initially train two base models f, g on chunks $D_t^{S_1}$ and $D_t^{S_2}$ sepatately, and then test them on $D_{t+1}^{S_1}$, $D_{t+1}^{S_2}$. Thus, we can get the initial prediction results of two chunks of data

$$\hat{y}_{t+1}^{S_1} = \{\hat{y}_{t+1}^{f,S_1}, \hat{y}_{t+1}^{g,S_1}\}, \quad \hat{y}_{t+1}^{S_2} = \{\hat{y}_{t+1}^{f,S_2}, \hat{y}_{t+1}^{g,S_2}\}. \tag{1}$$

Second, we aim to reconstruct a meta train set to combine the knowledge of two data streams together. So, we give a label α for each data stream, and reconstruct the meta train set as

$$D_{t+1}^{meta} = \{\alpha_{t+1}, \hat{y}_{t+1}, y\}, \tag{2}$$

where $\alpha = \{\alpha_{t+1}^{S_1} \cup \alpha_{t+1}^{S_2}\}$, $\hat{y} = \{\hat{y}_{t+1}^{S_1} \cup \hat{y}_{t+1}^{S_2}\}$. $y = \{y_{t+1}^{S_1} \cup y_{t+1}^{S_2}\}$ is the true label of two data chunks. Then, we train a new learning model on this meta training set, denoted as

$$\hat{y} = F(\alpha_{t+1}, \hat{y}_{t+1}) + \epsilon, \tag{3}$$

where ϵ is the learning error. This new learning model can be an ensemble model with several base learners.

3.2. *Selective retraining scheme for learning adaption*

To maintain learning performance under concept drift, a selective retraining scheme has been utilized. First, after testing the model at time $t+1$, the accuracy of model on each data stream at time t and $t+1$ can be obtained, denoted as $\{\varepsilon_t^{S_1}, \varepsilon_{t+1}^{S_1}\}, \{\varepsilon_t^{S_2}, \varepsilon_{t+1}^{S_2}\}$. Then, we retrain the base model of each stream once the accuracy decreases, denoted as

$$\begin{cases} \text{If } \varepsilon_t^{S_1} > \varepsilon_{t+1}^{S_1} \textbf{ or } \varepsilon_t^{S_2} > \varepsilon_{t+1}^{S_2}, \text{ retrain } f^{S_1}, g^{S_1} \textbf{ or } f^{S_2}, g^{S_2} \\ \text{Else, test } f^{S_1}, g^{S_1} \textbf{ and } f^{S_2}, g^{S_2} \end{cases}. \tag{4}$$

Then, to further reduce the impact of concept drift, we also consider updating the meta train set D_{t+1}^{meta} adaptively to help retrain the meta model. The meta train set updation depends on whether the accuracy is decreased, expressed as

$$\begin{cases} \text{If } \varepsilon_t^{S_1} > \varepsilon_{t+1}^{S_1}, \varepsilon_t^{S_2} \le \varepsilon_{t+1}^{S_2}, & D_{t+1}^{meta} = D_{t+1}^{meta,S_1} \cup D_{t+1}^{meta,S_2} \cup D_t^{meta,S_2} \\ \text{If } \varepsilon_t^{S_1} \le \varepsilon_{t+1}^{S_1}, \varepsilon_t^{S_2} > \varepsilon_{t+1}^{S_2}, & D_{t+1}^{meta} = D_{t+1}^{meta,S_1} \cup D_t^{meta,S_1} \cup D_{t+1}^{meta,S_2} \\ \text{If } \varepsilon_t^{S_1} \le \varepsilon_{t+1}^{S_1}, \varepsilon_t^{S_2} \le \varepsilon_{t+1}^{S_2}, & D_{t+1}^{meta} = D_{t+1}^{meta} \cup D_t^{meta} \\ \text{If } \varepsilon_t^{S_1} > \varepsilon_{t+1}^{S_1}, \varepsilon_t^{S_2} > \varepsilon_{t+1}^{S_2}, & D_{t+1}^{meta} = D_{t+1}^{meta} \end{cases}. \tag{5}$$

Finally, retrain the meta model on the updated meta train set, thus helping the learning model handle different types of concept drift.

4. Experiment

4.1. *Experiment setting*

Datasets To simulate the scenarios of multiple data streams with different drift situations, we choose six commonly used synthetic datasets (SEAa (sudden drift), RTG (no drift), RBF (incremental drift), RBFr (incremental drift), AGRa (sudden drift), HYP (incremental drift)) for the experiment. We simply deal with two data streams at the same time, five scenarios with

Table 1. Data scenarios

Scenarios	#Sample	#Feature	#Class	#Drift type
SEAa → RTG	10,000	3	2	No
SEAa → RBF	10,000	3	2	Incremental
SEAa → RBFr	10,000	3	2	Incremental
SEAa → AGRa	10,000	3	2	Sudden
SEAa → HYP	10,000	3	2	Incremental

different drift situations are given, as shown in Table 1. We repeat the experiment on each scenario 15 times and calculate the average results of accuracy and F1 score.

Benchmarks In the experiment, we compare our proposed self-adaptative Stacking method with three benchmark methods, as described below:

- **Baseline:** is the initial stacking-based framework, which is trained on the first chunk and tested on the rest of the chunks.
- **Retrain-B:** retrains base models in the stacking-based framework once the learning accuracy is reduced.
- **Retrain-M:** retrains the meta model in the stacking-based framework on the updated meta train set once the learning accuracy is degraded.
- **Adaptive Stacking:** combines the processes of Retrain-B and Retrain-M in the stacking-based framework.

The experiment is run under Python 3.10, the computing environment is 12th Gen Intel(R) Core(TM) i7-1255U 1.70 GHz with 16.0 GB RAM.

4.2. *Experiment results discussion*

We have evaluated the proposed method on five scenarios of multiple data streams, and collect the average accuracy and F1 score, as shown in Tables 2, 3. According to the experiment results, it can be shown that the Baseline method performs not ideal because of the lack of real-time adaptation. Then, by selectively retraining the base model or meta model in methods of Retrain-B and Retrain-M, the average performance has been improved respectively. Finally, the proposed Adaptive Stacking method got a relatively higher performance with a lower standard error. Such performance improvement is ideal in five scenarios, which can deal with multiple data streams with different feature spaces and drift situations.

Table 2. Average Accuracy of the Adaptive Stacking Method

Scenarios	Baseline	Retrain-B	Retrain-M	Adaptive Stacking
SEAa → RTG	57.86 ± 5.9	58.31 ± 6.4	59.25 ± 5.4	59.79 ± 5.3
SEAa → RBF	69.11 ± 5.2	77.83 ± 3.4	71.65 ± 4.6	78.16 ± 3.0
SEAa → RBFr	68.20 ± 5.0	72.60 ± 6.3	71.42 ± 3.5	75.21 ± 2.5
SEAa → AGRa	57.54 ± 1.3	77.04 ± 2.6	67.23 ± 2.4	78.75 ± 1.2
SEAa → HYP	72.02 ± 6.7	72.28 ± 6.3	74.56 ± 3.2	76.54 ± 0.5

Table 3. Average F1 score of the Adaptive Stacking Method

Scenarios	Baseline	Retrain-B	Retrain-M	Adaptive Stacking
SEAa → RTG	49.56 ± 7.5	50.74 ± 9.3	50.19 ± 6.8	53.19 ± 6.5
SEAa → RBF	66.98 ± 6.3	76.92 ± 3.3	70.27 ± 5.0	77.32 ± 2.9
SEAa → RBFr	65.58 ± 6.3	70.22 ± 8.7	69.84 ± 3.7	74.09 ± 2.8
SEAa → AGRa	57.14 ± 1.4	76.95 ± 2.8	67.22 ± 2.4	78.73 ± 1.2
SEAa → HYP	70.58 ± 10.5	70.89 ± 10.3	74.42 ± 3.3	76.53 ± 0.5

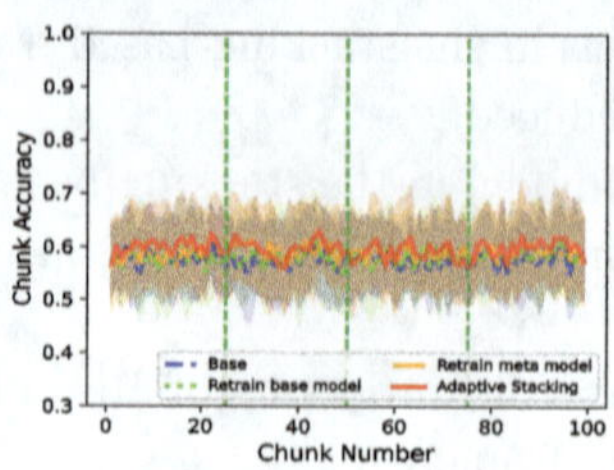

(a) SEAa → RTG

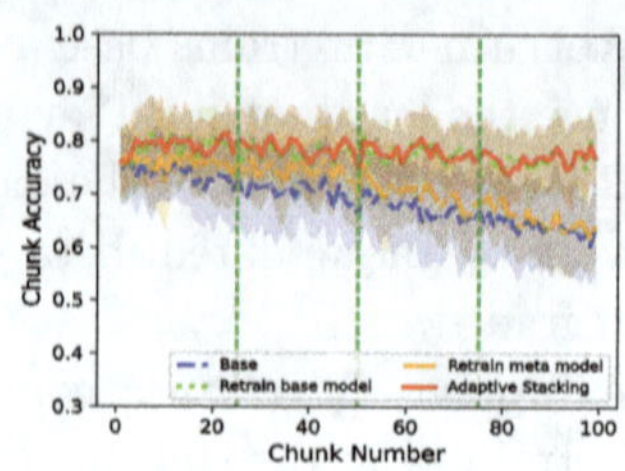

(b) SEAa → RBF

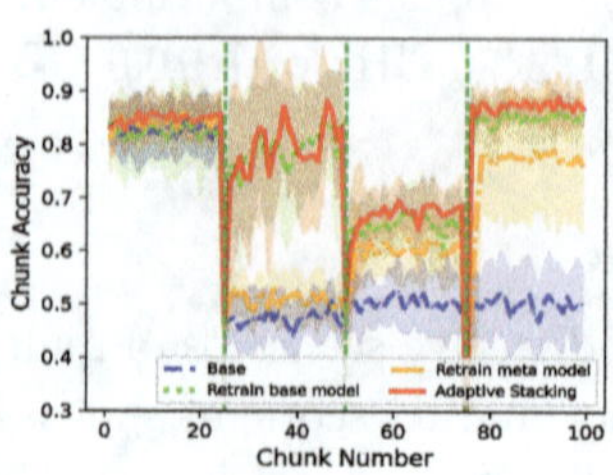

(c) SEAa → AGRa

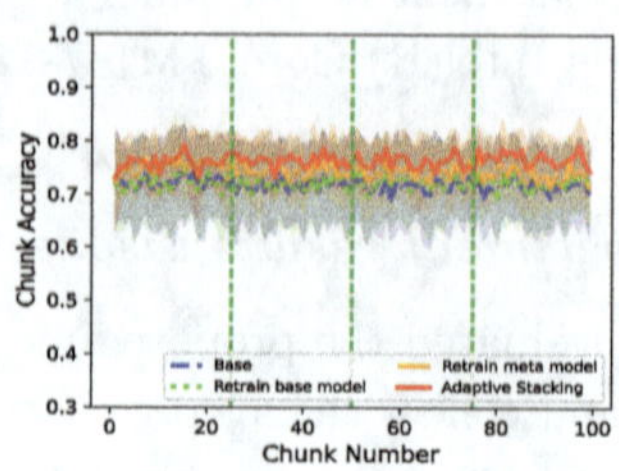

(d) SEAa → HYP

Fig. 2. A plot of the chunk accuracy of the proposed method with several datasets.

To further observe the experiment results, the average chunk accuracy of each benchmark method on different scenarios has been visualized, as shown in Figure 2. For handling sudden drift, the proposed method can help recover the learning performance better when drift occurs. For dealing with incremental drift, the Adaptive Stacking method got a higher performance.

5. Conclusion

This chapter develops an adaptive stacking method for dealing with multiple data streams under concept drift. A stacking-based learning framework has been built to deal with multiple data streams with different feature spaces and drift situations. Appropriately and selectively retrain the base model and meta model to help maintain the learning performance. The experiment results on five data scenarios show the efficiency of the proposed method. In our future study, we will continue to optimize the current method for better performance and robustness.

Acknowledgments

This work was supported by the Australian Research Council under Grant FL190100149.

References

1. J. Lu, A. Liu, F. Dong, F. Gu, J. Gama and G. Zhang, Learning under concept drift: A review, *IEEE Transactions on Knowledge and Data Engineering* **31**, 2346 (2019).
2. J. Gama, I. Zliobaite, A. Bifet, M. Pechenizkiy and A. Bouchachia, A survey on concept drift adaptation, *ACM Computing Surveys* **46**, 1 (2014).
3. J. Shao, Y. Tan, L. Gao, Q. Yang, C. Plant and I. Assent, Synchronization-based clustering on evolving data stream, *Information Sciences* **501**, 573 (2019).
4. Y. Song, G. Zhang, H. Lu and J. Lu, A fuzzy drift correlation matrix for multiple data stream regression, in *2020 IEEE International Conference on Fuzzy Systems*, (IEEE, 2020).
5. M. Zhou, J. Lu, Y. Song and G. Zhang, Multi-stream concept drift self-adaptation using graph neural network, *IEEE Transactions on Knowledge and Data Engineering* **35**, 12828 (2023).
6. E. Yu, Y. Song, G. Zhang and J. Lu, Learn-to-adapt: Concept drift adaptation for hybrid multiple streams, *Neurocomputing* **496**, 121 (2022).
7. A. Haque, Z. Wang, S. Chandra, B. Dong, L. Khan and K. W. Hamlen, Fusion: An online method for multistream classification, in *Proceedings of the 2017 ACM on Conference on Information and Knowledge Management*, (ACM, 2017).

8. K. Wang, J. Lu, A. Liu and G. Zhang, An augmented learning approach for multiple data streams under concept drift, in *36th Australasian Joint Conference on Artificial Intelligence, AI 2023*, (Springer, 2023).
9. N. Lu, G. Zhang and J. Lu, Concept drift detection via competence models, *Artificial Intelligence* **209**, 11 (2014).
10. A. Liu, J. Lu, Y. Song, J. Xuan and G. Zhang, Concept drift detection delay index, *IEEE Transactions on Knowledge and Data Engineering* **35**, 4585 (2022).
11. B. Krawczyk, L. L. Minku, J. Gama, J. Stefanowski and M. Woźniak, Ensemble learning for data stream analysis: A survey, *Information Fusion* **37**, 132 (2017).
12. H. M. Gomes, A. Bifet, J. Read, J. P. Barddal, F. Enembreck, B. Pfharinger, G. Holmes and T. Abdessalem, Adaptive random forests for evolving data stream classification, *Machine Learning* **106**, 1469 (2017).
13. H. M. Gomes, J. Read and A. Bifet, Streaming random patches for evolving data stream classification, in *2019 IEEE International Conference on Data Mining*, (IEEE, 2019).
14. H. M. Gomes, J. Read, A. Bifet and R. J. Durrant, Learning from evolving data streams through ensembles of random patches, *Knowledge and Information Systems* **63**, 1597 (2021).
15. M. Pratama, W. Pedrycz and E. Lughofer, Evolving ensemble fuzzy classifier, *IEEE Transactions on Fuzzy Systems* **26**, 2552 (2018).
16. S. Pan, J. Wu, X. Zhu and C. Zhang, Graph ensemble boosting for imbalanced noisy graph stream classification, *IEEE Transactions on Cybernetics* **45**, 954 (2014).
17. J. W. Sangma, V. Pal, N. Kumar, R. Kushwaha *et al.*, Fhc-nds: Fuzzy hierarchical clustering of multiple nominal data streams, *IEEE Transactions on Fuzzy Systems* **31**, 786 (2022).
18. H. Yu, W. Liu, J. Lu, Y. Wen, X. Luo and G. Zhang, Detecting group concept drift from multiple data streams, *Pattern Recognition* **134**, p. 109113 (2023).
19. S. Chandra, A. Haque, L. Khan and C. Aggarwal, An adaptive framework for multistream classification, in *Proceedings of the 25th ACM International Conference on Information and Knowledge Management*, (ACM, 2016).
20. A. Otero, P. Félix, D. G. Márquez, C. A. García and G. Caffarena, A fault-tolerant clustering algorithm for processing data from multiple streams, *Information Sciences* **584**, 649 (2022).

A Convolutional Recurrent Neural Network with Spatial Feature Fusion for Environmental Sound Classification

Meehir Mhatre*, Li Zhang* and Arjun Panesar†

*Department of Computer Science, Royal Holloway, University of London
Surrey, TW20 0EX, UK
†DDM Health Ltd., University of Warwick, CV4 7EZ, UK

This chapter proposes a new Convolutional Recurrent Neural Network (CRNN) model with spatial feature fusion for environmental sound classification. Besides data preprocessing such as spectrogram transformation and data augmentation, customized deep networks, i.e. VGG19, ResNet152, and EfficientNetB0, with additional layers, are also proposed for audio classification. Specifically, the proposed CRNN model embeds ResNet152 and EfficientNetB0 in the encoder where spatial features extracted by both networks are concatenated. A Long Short-Term Memory (LSTM) component is used as the decoder in the proposed CRNN for temporal feature extraction. Evaluated using the ESC-50 dataset, the proposed CRNN model with a multi-channel spatial feature fusion outperforms the customized VGG19, ResNet152, and EfficientNetB0 networks as well as existing studies, significantly. The spatial feature fusion in conjunction with LSTM-based sequential feature extraction accounts for the superiority of the proposed CRNN model for environmental sound classification.

Keywords: Sound classification; spatial feature fusion; long short-term memory.

1. Introduction

Deep Convolutional Neural Networks (CNNs), such as VGG19, ResNet152, and EfficientNetB0, have revolutionized image classification tasks. As deep learning models continue to evolve, there is a compelling motivation to explore their applications in other application domains. Due to the exponential daily growth in the volume of audio data, recent studies also exploited CNNs for audio classification tasks, such as Environmental Sound Classification (ESC). CNN models have demonstrated significant advancement owing to their superior capabilities in audio spectrogram processing.

This research proposes a novel Convolutional Recurrent Neural Network (CRNN) with multi-channel spatial feature fusion in conjunction with Long Short-Term Memory (LSTM) for temporal feature extraction with respect to environmental sound classification. Specifically, it concatenates spatial features extracted by ResNet152 and EfficientNetB0, with temporal sequential features examined by an LSTM component. Besides the CRNN model with spatial feature fusion, customized VGG19, ResNet152, and EfficientNetB0, each with additional layers, are also proposed. Specifically, additional dense layers, batch normalization layers, and dropout layers are added to each of these networks to enhance feature extraction and generalization.

The key processes of this chapter include spectrogram transformation from audio files, data augmentation, and audio classification using the proposed CRNN and three customized networks. First, fast Fourier transform is used to convert the original audio files into 224×224×3 spectrogram images, followed by data augmentation and normalization. The next phase encompasses optimizing hyperparameters, fine-tuning architectures, and applying effective training techniques to maximize model performance. Performance evaluation using metrics such as accuracy, log loss, precision, recall, and F1-score provides a comprehensive assessment of the proposed models. A comparative analysis with existing state-of-the-art identifies the strengths and weaknesses of our proposed models, contributing valuable insights into their suitability for sound classification tasks.

2. Related Work

We introduce recent state-of-the-arts for audio classification in this section. Piczak [1] proposed a CNN with 14 layers using log-mel spectrogram as input. Such spectrogram inputs were computed for each audio frame, representing the squared magnitude of each frequency area. The work used a two-channel function by applying the log-mel spectrogram and its delta information as inputs to the CNN model. Evaluated using the ESC-50 dataset, it outperformed Random Forest (RF) by 20.5%. Takahashi et al. [2] used the log-mel spectrogram and respective delta and delta-delta information as a three-channel input in a similar manner to RGB. Agrawal et al. [3] used gamma-tone spectrograms integrated with a CNN architecture similar to those used by Piczak [1]. Their work achieved accuracy rates of 79.1% for ESC-50 and 83.34% for UrbanSound8K. Moreover, existing studies also used pre-trained models for performance enhancement for audio classification tasks [4-6]. Choi et al. [4] proposed a transfer learning network for music-related audio classification tasks. Using multilayer concatenated feature maps of pre-trained CNNs as efficient transferable knowledge, the work showed that transfer learning with concatenated features improved performance on a variety of music-related audio classification tasks. Chen et al. [5] proposed a deep network with dilated convolution for evaluating UrbanSound8K and achieved 78% accuracy. Boddapati et al. [6] studied transfer learning models based on GoogleNet and AlexNet for audio classification. GoogleNet with spectral images as inputs achieved remarkable results with an accuracy rate of 86% on ESC-10, 73% on ESC-50 and 93% for UrbanSound8k. These related works paved the way for the development of proposed models with enhanced performance in this research.

3. The Proposed Methodology

3.1. *Pre-processing*

A spectrogram is a visual way of representing the signal strength over time at various frequencies present in a particular waveform [7]. The audio to spectrogram conversion process involves several stages to prepare audio data for subsequent classification. Initially, the ESC-50 dataset is transformed into spectrogram images. Feature extraction, using the librosa library, samples audio data at a rate of 44.1 kHz, capturing 44,100 samples per second. The mel spectrogram is then computed using the Fast Fourier Transform, representing the audio signal's frequency content evolution over time. After that, a power-to-decibel (dB) conversion is applied, enhancing the representation by converting power values to a logarithmic scale, crucial for capturing audio signal dynamic ranges. The mel spectrograms are then resized to a consistent 224×224 pixel shape to fulfil CNN input size requirements. To adapt the data for CNNs, the single-channel spectrogram is replicated into three identical channels, providing effective learning of spatial features. This transformation process offers a standardized and structured input for deep networks for subsequent classification.

3.1.1. *Data augmentation*

Data Augmentation is a crucial step to increase the diversity and quantity of training data. This aids in improving the model's generalization and robustness. Several techniques are employed for the process of data augmentation, e.g. Width and Height Shift, Shear Intensity, Zooming, Flipping, Brightness Adjustment, Channel Shifting, and Fill Mode. For each class, 50 additional images are generated through the augmentation process and then these augmented images are merged with the original spectrogram images for model training and validation.

3.2. *Proposed models*

3.2.1. *The proposed CRNN model with spatial feature fusion*

A novel CRNN model is proposed which employs a spatial feature encoder integrating ResNet152 and EfficientNetB0, and an LSTM decoder. Figure 1 depicts the proposed model architecture. To be specific, the discriminative spatial features extracted by ResNet152 and EfficientNetB0 are concatenated in the encoding process. These merged features are then channeled through an LSTM component in the decoding process, allowing the model to capture intricate temporal patterns in the audio data. Strategic dropout layers and densely connected units further enhance its robustness and ability to interpret complex sound patterns. The final softmax layer is used for sound classification.

Parameter tuning and loss functions are pivotal elements in the optimization of sound classification models. For the proposed CRNN model, we employ a batch size of 64, 100 training epochs, Adam optimizer with learning rate of 0.00001 with a cross-entropy loss. The above settings are also used for the subsequently proposed customized VGG19, ResNet152, and EfficientNetB0 networks, except for a batch size of 32, to strike a balance between computational efficiency and performance.

3.2.2. *Customized VGG19, efficientNetB0 and ResNet152 with additional layer sequences*

We subsequently propose customized VGG-19, EfficientNetB0 and ResNet152 models with additional layer sequences for sound classification. The ImageNet pre-trained VGG19 consists of 16 convolution layers and 3 fully connected layers. EfficientNetB0 employs a scaling of resolution, width, and depth to enhance performance with light-weight computational cost. ResNet152 is equipped with residual connections to better tackle vanishing gradient.

To adapt these three networks for sound classification tasks, we add the following additional layer sequence on top of each pre-trained network, i.e. Dense Layer (512 units, ReLU Activation), Dropout Layer (Dropout Rate: 0.5), Dense Layer (256 units, ReLU Activation), Flatten Layer, Dense Layer (224 units, ReLU Activation), Dropout Layer (Dropout Rate: 0.5), and Output Layer (Softmax Activation). The insertion of these additional layers is identified by trials-and-errors. These customized networks are subsequently used for sound classification.

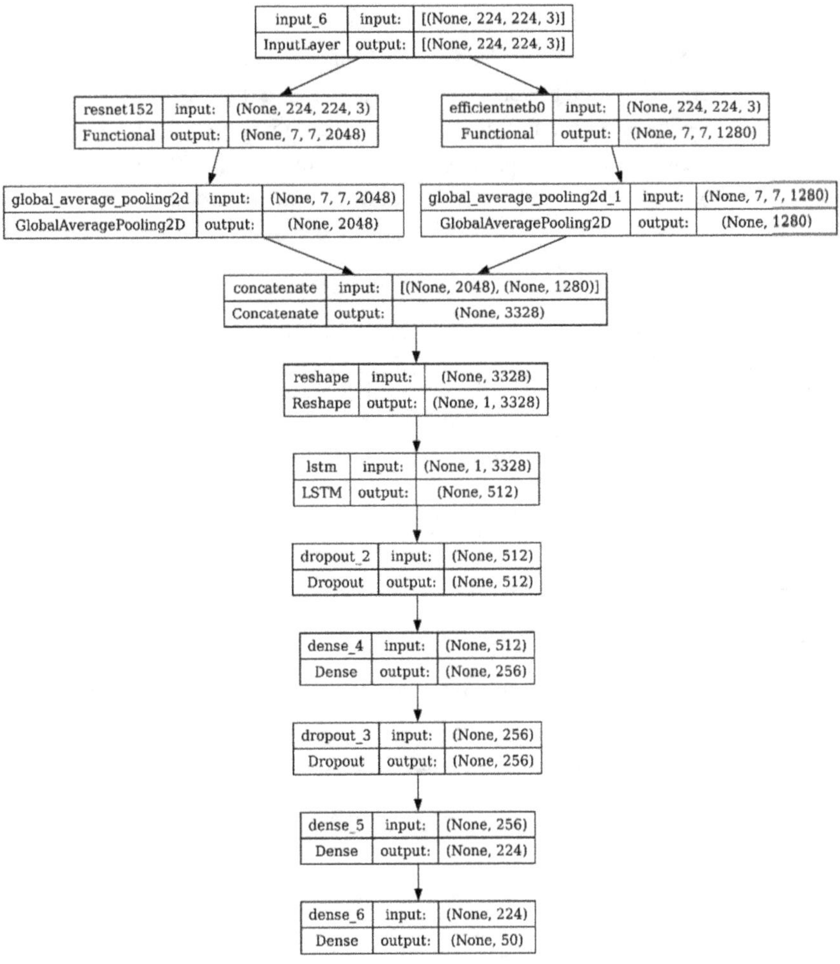

Fig. 1. The proposed CRNN model with an encoder integrating ResNet152 and EfficientNetB0, and an LSTM decoder

4. Evaluation

We employ the proposed four models for experimental studies using ESC-50. A 5-fold cross validation is employed as recommended by existing studies. The detailed evaluation results are provided in Table 1. As indicated in Table 1, the proposed CRNN model with spatial feature fusion achieves the highest performance, in terms of accuracy, precision, recall and F1 scores, in comparison with those of the customized VGG19, EfficientNetB0 and

ResNet152 networks. In addition, the customized ResNet152 outperforms the customized EfficientNetB0 and VGG19 by a significant margin. The proposed encoder integrating EfficientNetB0 and ResNet152 and the LSTM decoder show significantly enhanced capabilities in spatial-temporal feature learning, and account for the superior capabilities of the proposed CRNN model for distinguishing different sound classes. The attachment of additional layers in customized networks also enhances the performance of existing architectures.

Table 1. Evaluation results of the proposed models using ESC-50

Models	Accuracy	Precision	Recall	F1-Score
VGG19 with customized layers	0.9467	0.9510	0.9467	0.9470
EfficientNetB0 with customized layers	0.9567	0.9602	0.9567	0.9567
ResNet152 with customized layers	0.9600	0.9657	0.9600	0.9604
The proposed CRNN with spatial feature fusion	0.9678	0.9710	0.9678	0.9676

5. Comparison with Existing Studies

The comparison with existing approaches highlights the significance of the propose models. A variety of existing models are presented in Table 2, including ACRNN [8], RACNN [9], Pro-CNN [10], AemNet [11], Audio Spectrogram Transformer (AST) [12], Pyramid-Combined CNN [13], and CLAP [14].

Table 2. Comparison with existing studies using ESC-50

Models	Accuracy	Trainable Parameters	Features
ACRNN [8]	86.10%	3.81M	Log Gammatone
RACNN [9]	85.65%	531.8K	Log-Mel
Pro-CNN [10]	82.80	-	Log-Mel + Raw Data
AemNet [11]	92.32%	14.4M	E2E Log-Mel
AST [12]	95.70%	87M	Log-Mel
Pyramid-Combined CNN [13]	81.40%	-	Spectrogram
CLAP [14]	96.70%	80.8M + 110M	Audio + Text
Customized VGG19 (ours)	**94.67%**	**3.2M**	**Log-Mel**
Customized EfficientNetB0 (ours)	**95.67%**	**3.61M**	**Log-Mel**
Customized ResNet152 (ours)	**96.00%**	**4.00M**	**Log-Mel**
Prop. CRNN with Spatial Feature Fusion (ours)	**96.78%**	**8.05M**	**Log-Mel**

In comparison with the above existing studies, the proposed CRNN model and the customized EfficientNetB0 and ResNet152 networks obtain better performance. In particular, the proposed CRNN model, with a spatial feature encoder integrating ResNet152 and EfficientNetB0, and an LSTM decoder, achieves the best accuracy rate of 96.78% for 5-fold cross validation, followed by the customized ResNet152 with an accuracy rate of 96%. The proposed CRNN model, integrating ResNet152 and EfficientNetB0 for spatial feature fusion and an LSTM for sequential feature extraction, harnesses the collective wisdom of multiple architectures. The embedded distinctive individual encoder and decoder models complement one another, reducing biases and errors, while enhancing overall accuracy for distinguishing different sound classes.

6. Conclusion

This chapter has proposed a CRNN model integrating ResNet152 and EfficientNetB0 for spatial feature fusion and an LSTM for sequential feature extraction. Three customized VGG19, EfficientNetB0, and ResNet152 networks with additional layer sequences are also exploited to further improve feature learning capabilities. Evaluated using ESC-50, the empirical results indicate that the proposed CRNN model and the customized ResNet152 excelled in identifying 50 sound categories and showcased exceptional test accuracy. In particular, the CRNN model with spatial feature fusion harnesses diverse model strengths, minimizing biases, and maximizing overall performance. In future directions, reinforcement learning algorithms will be exploited for hyperparameter optimization [15, 16] to further enhance the performance of the proposed models. We also deploy the proposed models to other sound classification tasks, e.g. lung/heart disease diagnosis, speech emotion recognition, and speaker identification, to further test their efficiency.

References

1. K.J. Piczak. 2015. ESC: Dataset for environmental sound classification. In *Proceedings of the 23rd ACM international conference on Multimedia*, pp.1015-1018.
2. N. Takahashi, M. Gygli, B. Pfister, and L.V. Gool. 2016. Deep convolutional neural networks and data augmentation for acoustic event detection. *arXiv preprint arXiv:1604.07160.*
3. D.M. Agrawal, H.B. Sailor, M.H. Soni, and H.A. Patil, 2017. Novel TEO-based Gammatone features for environmental sound classification. In *Proceedings of European Signal Processing Conference,* pp.1809-1813.

4. K. Choi, G. Fazekas, M. Sandler, and K. Cho. 2017. Transfer learning for music classification and regression tasks. In *Proceedings of the 18th International Society for Music Information Retrieval Conference.*
5. Y. Chen, Q. Guo, X. Liang, J. Wang, and Y. Qian. 2019. Environmental sound classification with dilated convolutions. *Applied Acoustics*, 148, pp.123-132.
6. V. Boddapati, A. Petef, J. Rasmusson, and L. Lundberg. 2017. Classifying environmental sounds using image recognition networks. *Procedia computer science*, 112, pp.2048-2056.
7. L. Zhang, C.P. Lim, Y. Yu, and M. Jiang. Sound classification using evolving ensemble models and particle swarm optimization. *Applied soft computing*, vol. 116, p. 108322, 2022.
8. Z. Zhang, S. Xu, S. Zhang, T. Qiao, and S. Cao. 2021. Attention based convolutional recurrent neural network for environmental sound classification. *Neurocomputing, 453*, pp.896-903.
9. Z. Fang, B. Yin, Z. Du, and X. Huang. 2022. Fast environmental sound classification based on resource adaptive convolutional neural network. *Scientific Reports*, 12(1), 6599.
10. S. Li, Y. Yao, J. Hu, G. Liu, X. Yao, and J. Hu. 2018. An Ensemble Stacked Convolutional Neural Network Model for Environmental Event Sound Recognition. *Applied Sciences*, 8(7), 1152.
11. P. Lopez-Meyer, J.A. del Hoyo Ontiveros, H. Lu, and G. Stemmer. 2021, June. Efficient end-to-end audio embeddings generation for audio classification on target applications. In *Proceedings of IEEE International Conference on Acoustics, Speech and Signal Processing*, pp.601-605.
12. Y. Gong, Y.A. Chung, and J. Glass. 2021. AST: Audio spectrogram transformer. *arXiv preprint arXiv:2104.01778.*
13. F. Demir, M. Turkoglu, M. Aslan, and A. Sengur. 2020. A new pyramidal concatenated CNN approach for environmental sound classification. *Applied Acoustics*, 170, 107520.
14. B. Elizalde, S. Deshmukh, M. Al Ismail, and H. Wang. 2023. Clap learning audio concepts from natural language supervision. In *Proceedings of IEEE International Conference on Acoustics, Speech and Signal Processing.*
15. L. Zhang, C.P. Lim, and C. Liu. 2023. Enhanced Bare-Bones Particle Swarm Optimization based Evolving Deep Neural Networks. *Expert Systems with Applications*, p.120642.
16. S. Slade, L. Zhang, H. Huang, et al. 2023. Neural inference search for multiloss segmentation models. *IEEE Transactions on Neural Networks and Learning Systems*.

Speech Emotion Recognition Using Convolutional Recurrent Neural Networks

Abhishek Gangani[*], Li Zhang[*,‡] and Ming Jiang[†]
[*]*Department of Computer Science, Royal Holloway, University of London*
Surrey, TW20 0EX, UK
[†]*School of Computer Science, University of Sunderland, UK*
[‡]*li.zhang@rhul.ac.uk*

Research suggests that various machine learning and deep learning models can be used for implementation of speech emotion recognition (SER) using different acoustic properties, such as voice, pitch, loudness, intensity, Mel-frequency cepstral coefficients, and spectral characteristics. This chapter conducts speech emotion recognition using deep neural networks, such as long short-term memory, gated recurrent units, and convolutional recurrent neural network. These different acoustic features are investigated in our studies owing to their great efficiency in representing key events in audio representations. A cross-validation evaluation has been conducted with the data from different actors for model evaluation to check the robustness of each proposed network. The proposed models show impressive performances in comparison with those of existing state-of-the-art methods for evaluating several speech emotion datasets.

Keywords: Speech emotion recognition; long short-term memory; gated recurrent units; convolutional recurrent neural network.

1. Introduction

Emotion plays a significant role in human communication, influencing perception, and decision-making in many ways. The practice of interpretation of human emotions is the fundamental of effective communication, fostering empathy and understanding. Speech emotion recognition (SER) is a complex problem where identifying the human emotions from speech pattern remains a significant challenge. It has many day-to-day applications, such as mental health, independent living, and robotic applications. It can use various acoustic features to leverage computational models to decode the affective states embedded in human vocal attributes. Using SER, the limitations of text-based sentiment analysis can be surpassed. Though implementation of SER can have critical challenges, such as the complexity of the problem domains, data privacy, and security issues. An SER system is made up of various components that include feature selection and extraction, feature classification, acoustic modeling, recognition per unit, and most importantly language-based modeling. The traditional SER systems typically incorporate various classification models, such as Gaussian mixture models (GMMs) and hidden Markov models (HMMs).

Deep learning methods are composed of various nonlinear components that perform computation on a parallel basis. In this research, several deep learning mdoels, such as Recurrent Neural Network (RNN), Long Short-Term Memory (LSTM), Gated Recurrent Unit (GRU), Convolutional Neural Network (CNN), and Convolutional Recurrent Neural Network (CRNN), are proposed for classifying audio emotion inputs. A variety of acoustic features are also extracted to test network efficiency. Several well-known speech emotion datasets are investigated in our experimental studies.

2. Related Work

SER has witnessed substantial advancements with the integration of deep learning techniques. As an example, Khalil et al. [1] investigated the superiority of deep learning models over traditional machine learning models in handling complex patterns in speech emotion detection. The study of Sudhakar and Anil [2] explored various speech features used in emotion detection including pitch, intensity, format frequencies, and speech rate. Their work discussed different feature extraction techniques and their effectiveness in capturing emotion-related information. Schuller et al. [3] employed an HMM-based approach for speech emotion recognition. The study demonstrated the effectiveness of HMMs in capturing temporal information in speech signals, making them suitable for emotion recognition tasks. The combination of various acoustic features, such as pitch, energy, and formants, were used to improve the performance of the HMM-based SER system. Keren et al. [4] conducted a comparative analysis of various deep learning architectures, including CNN, RNN, and LSTM networks, for emotion recognition from speech. The advantages and limitations of each architecture was exploited and multiple benchmark datasets were used for model evaluation. Trigeorgis et al. [5] studied an end-to-end SER system using a deep convolutional recurrent network (DCRN) that combines the strengths of both CNN and RNN architectures. They proposed a novel approach that eliminated the need for handcrafted feature extraction, directly learning representations from raw speech signals. There are also other studies focusing on state-of-the-art deep networks for speech emotion, sound classification, video/audio caption generation, and other video/image/audio processing tasks [6-15].

3. The Proposed Models for SER

Owing to the effectiveness of spatial-temporal feature learning capabilities of GRU, CRNN, and LSTM, these networks are exploited in this research for SER.

3.1. *GRU*

GRUs are a type of recurrent neural network (RNN) known for their efficiency in sequence modelling. They consist of gating mechanisms which are the update gates and the reset gates which regulate the flow of information. Update gate decides how much of the past information needs to be passed along the future. It is crucial for the model to retain long-term dependencies. Reset gate determines how much information to forget, which is essential in learning representations from the speech data where only certain parts of the sequence may be relevant for emotion detection. We employ GRUs for SER, as it can easily adapt to capturing temporal dynamics in speech, crucial for recognizing emotions that unfold over time.

3.2. *LSTM*

LSTMs are a type of recurrent neural network designed to learn long-term dependencies. They are composed of a cell state and three gates: input, output, and forget. Input gate updates the cell state with new information from the current input, which is vital for capturing the nuances in the speech for emotion detection. Output gate determines the next hidden state based on the current input and the updated cell state, contributing to the prediction of emotions. The cell state acts as a 'memory' of the network, carrying relevant information throughout the sequence of data, making it suitable for modeling the emotional context in speech. Forget Gate decides what information should be discarded from the cell state, allowing the model to forget irrelevant parts of the previous state for better emotion prediction. LSTMs excel in scenarios where the emotional context spans over a longer duration in the speech signal, enabling the model to recognize emotions from more extended speech utterances.

3.3. *CRNN*

CRNNs integrate the spatial feature extraction capabilities of CNNs with the sequential learning abilities of RNNs. The initial layers are convolutional, often followed by pooling layers. The CNN layers extract and down sample the spectral features from the speech signal. The output from the CNN layers is fed into RNN layers. These layers model the temporal dependencies in the extracted features, capturing the dynamics of emotional expressions over time. The network typically ends with fully connected layers, which integrate the learned spatial and temporal features to make final emotion predictions. CRNNs are particularly effective in scenarios where both the spectral content and the temporal evolution of speech are critical for emotion recognition.

4. Model Implementation

4.1. *Preparation of dataset*

A combination of RAVDESS and TESS has been selected for this research. Mel-Frequency Cepstral Coefficients (MFCCs) are extracted from the audio data. The labels are encoded using One hot encoder. The dataset is then split into training/validation and testing datasets using a splitting factor of 67-33.

4.2. *Proposed architecture for GRU model*

The GRU model consists of **Layer 1 (GRU)**, a GRU layer with 123 hidden units; **Layers 2,4 & 6 (Dense),** fully connected neural network layers with 64 and 32 units, respectively, and **Layers 3 & 5 (Dropout)**, dropout layers, randomly setting a fraction of the input units to 0 at each update during training time to help prevent overfitting.

Table 1. The proposed GRU architecture

Layer No.	Layer Type	Details	No. of Parameters
1	GRU	Number of hidden units: 123; Return sequences: False Input shape: (40, 1)	46494
2	Dense	Number of units: 64; Activation: 'reLu'	7936
3	Dropout	Dropout rate: 0.2	0
4	Dense	Number of units: 32; Activation: 'reLu'	2080
5	Dropout	Dropout rate: 0.2	0
6	Dense	Number of units: 9; Activation: 'SoftMax'	264

4.3. *Proposed architecture for LSTM model*

The LSTM model has a similar architecture that of GRU. It consists of, **Layer 1 (LSTM)**, an LSTM layer with 123 hidden units; **Layers 2,4 & 6 (Dense)**, fully connected neural network layers with 64 and 32 units, respectively; and **Layers 3 & 5 (Dropout)**, dropout layers, randomly setting a fraction of the input units to 0 at each update during training time.

Table 2. The proposed LSTM architecture

Layer No.	Layer Type	Details	No. of Parameters
1	LSTM	Number of hidden units: 123; Return sequences: False Input shape: (40, 1)	46494
2	Dense	Number of units: 64; Activation: 'reLu'	7936
3	Dropout	Dropout rate: 0.2	0
4	Dense	Number of units: 32; Activation: 'reLu'	2080
5	Dropout	Dropout rate: 0.2	0
6	Dense	Number of units: 9; Activation: 'SoftMax'	264

4.4. *Proposed architecture for CRNN model*

Table 3. The proposed CRNN architecture

Layer No.	Layer Type	Details	No. of Parameters
1	Conv1D	Filters: 64, Kernel size: 3, Activation: 'reLu', Input shape: (40, 1)	256
2	MaxPooling1D	Pool size: 2	0
3	Dropout	Dropout rate: 0.2	0
4	Conv1D	Filters: 128, Kernel size: 3, Activation: 'reLu'	24704
5	MaxPooling1D	Pool size: 2	0
6	Dropout	Dropout rate: 0.2	0
7	LSTM	Units: 123, Return sequences: False	123984
8	Dense	Units: 64, Activation: 'reLu'	7936
9	Dropout	Dropout rate: 0.2	0
10	Dense	Units: 32, Activation: 'reLu'	2080
11	Dropout	Dropout rate: 0.2	0
12	Dense	Units: 9, Activation: 'SoftMax'	297

The CRNN model is composed of **Layers 1 & 4 (Conv1D)**, convolutional layers with 64 and 128 filters, a kernel size of 3, and reLu activation for sequential data processing; **Layers 2 & 5 (MaxPooling1D),** Max Pooling layers with a pool size of 2, to reduce the dimensionality of the data by taking the maximum value over a window of 2 time steps; **Layers 3,6,9 & 12 (Dropout)**: dropout layers; **Layer**

7 (LSTM), an LSTM layer with 123 hidden units; and **Layers 8,10 & 12 (Dense)**, fully connected neural network layers with 64 and 32 units, respectively.

5. Evaluation

The detailed classification results for the unseen test set for GRU, CRNN, and LSTM are provided below. The weighted average is calculated by taking the number of audio clips for each emotion category into account, while the macro average is calculated by considering all emotion categories equally.

5.1. *GRU*

Using GRU, as indicated in Table 4, we obtain a weighted mean F1-score of 84% for all emotion categories.

Table 4 . Classification results of GRU model

	Angry	Calm	Disgust	Fear	Happy	Neutral	Sad	Sleepy	Surprise	Macro Avg	Weighted Avg
Precision	0.88	0.47	0.87	0.92	0.81	0.88	0.86	0.96	0.43	0.79	0.84
Recall	0.85	0.69	0.86	0.86	0.81	0.88	0.83	0.84	0.54	0.80	0.83
F1-score	0.86	0.56	0.87	0.89	0.81	0.88	0.84	0.90	0.48	0.79	0.84

5.2. *LSTM*

As shown in Table 5, using LSTM, we obtain a weighted mean F1-score of 83% for all emotion categories.

Table 5. Classification results of LSTM model

	Angry	Calm	Disgust	Fear	Happy	Neutral	Sad	Sleepy	Surprise	Macro Avg	Weighted Avg
Precision	0.82	0.39	0.83	0.87	0.82	0.94	0.91	0.93	0.44	0.77	0.84
Recall	0.85	0.73	0.83	0.85	0.77	0.87	0.81	0.98	0.48	0.80	0.82
F1-score	0.84	0.51	0.83	0.86	0.79	0.91	0.85	0.95	0.46	0.78	0.83

5.3. *CRNN*

Using CRNN, as indicated in Table 6, we obtain a weighted mean F1-score of 84% for all emotion categories.

Table 6. Classification results of CRNN model

	Angry	Calm	Disgust	Fear	Happy	Neutral	Sad	Sleepy	Surprise	Macro Avg	Weighted Avg
Precision	0.84	0.44	0.84	0.88	0.83	0.94	0.89	0.94	0.51	0.79	0.85
Recall	0.84	0.80	0.85	0.87	0.78	0.88	0.84	0.96	0.52	0.80	0.83
F1-score	0.84	0.57	0.85	0.88	0.80	0.91	0.87	0.95	0.52	0.80	0.84

6. Conclusion

The selection of GRU, LSTM, and CRNN for an SER system depends on the specific requirements of the task. GRUs offer a balance between model complexity and temporal modeling capability, making them suitable for less computationally intensive applications. CRNNs are ideal for scenarios requiring detailed spectral and temporal analysis. LSTMs are best suited for contexts where long-term temporal dependencies are crucial for emotion recognition. The choice among these models can also be influenced by factors such as the available dataset sizes, computational resources, and specific nuances of the speech data in the SER tasks. Future work will explore these and other networks' adaptability to different speech accents, languages, and cultural contexts. Hyper-parameter optimization will also be exploited to increase model robustness [16–20].

References

1. R.A. Khalil, E. Jones, M.I. Babar, T. Jan, M.H. Zafar, and T. Alhussain. (2019). Speech Emotion Recognition Using Deep Learning Techniques: A Review. *IEEE Access*, 7, 117327-117345.
2. R.S. Sudhakar and M.C. Anil. (2015). Analysis of Speech Features for Emotion Detection: A Review. In *Proceedings of International Conference on Computing Communication Control and Automation*, India, pp.661-664.
3. B. Schuller, G. Rigoll, and M. Lang. (2003). Hidden Markov model-based speech emotion recognition. In *Proceedings of IEEE International Conference on Acoustics, Speech, and Signal Processing*.
4. G. Keren, B. Schuller, and P. De Chazal. (2018). Emotion Recognition from Speech Using Deep Learning: A Comparative Study. In *Proceedings of IEEE ICASSP*, Calgary, AB, pp. 5139-5143.
5. G. Trigeorgis, F. Ringeval, R. Brueckner, E. Marchi, M.A. Nicolaou, B. Schuller, and S. Zafeiriou. (2016). Adieu features? End-to-end speech emotion recognition using a deep convolutional recurrent network. In *Proceedings of IEEE ICASSP*, Shanghai, pp. 5200-5204.

6. L. Zhang, C.P. Lim, Y. Yu, and M. Jiang, 2022. Sound classification using evolving ensemble models and Particle Swarm Optimization. *Applied Soft Computing*, *116*, p.108322.
7. T.Y. Tan, L. Zhang, C.P. Lim, B. Fielding, Y. Yu, and E. Anderson. 2019. Evolving ensemble models for image segmentation using enhanced particle swarm optimization. *IEEE Access*, *7*, pp.34004-34019.
8. L. Zhang and C.P. Lim, 2020. Intelligent optic disc segmentation using improved particle swarm optimization and evolving ensemble models. *Applied Soft Computing*, *92*, p.106328.
9. Xie, H., Zhang, L. and Lim, C.P., 2020. Evolving CNN-LSTM models for time series prediction using enhanced grey wolf optimizer. *IEEE Access*.
10. P. Kinghorn, L. Zhang and L. Shao. A Hierarchical and Regional Deep Learning Architecture for Image Description Generation, *Pattern Recognition Letters*. 2019.
11. P. Kinghorn, L. Zhang and L. Shao. A region-based image caption generator with refined descriptions, *Neurocomputing*. 272 (2018) 416-424.
12. T. Lawrence and L. Zhang. 2019. IoTNet: An Efficient and Accurate Convolutional Neural Network for IoT Devices. *Sensors*, 19, 5541.
13. D. Pandit, L. Zhang, S. Chattopadhyay, C.P. Lim, and C. Liu. 2018. *A Scattering and Repulsive Swarm Intelligence Algorithm for Solving Global Optimization Problems*. Knowledge-Based Systems.
14. L. Zhang, K. Mistry, C.P. Lim and S.C. Neoh. 2018. Feature selection using firefly optimization for classification and regression models. *Decision Support Systems*. 106 (2018) 64–85.
15. L. Zhang, W. Srisukkham, S.C. Neoh, C.P. Lim, and D. Pandit. 2018. Classifier ensemble reduction using a modified firefly algorithm: An empirical evaluation. *Expert Systems with Applications*, 93. pp. 395-422.
16. Y. Zhang, L. Zhang, S.C. Neoh, K. Mistry, and A. Hossain. 2015. Intelligent affect regression for bodily expressions using hybrid particle swarm optimization and adaptive ensembles. *Expert Systems with Applications*, 42 (22). pp. 8678-8697.
17. L. Zhang, C.P. Lim, and Y. Yu. 2021. Intelligent human action recognition using an ensemble model of evolving deep networks with swarm-based optimization. *Knowledge-Based Systems*, *220*, p.106918.
18. L. Zhang, C.P. Lim, and C. Liu, 2023. Enhanced Bare-Bones Particle Swarm Optimization based Evolving Deep Neural Networks. *Expert Systems with Applications*, p.120642.
19. Zhang, L., Zhao, D., Lim, C.P., Asadi, H., Huang, H., Yu, Y. and Gao, R., 2024. Video Deepfake Classification Using Particle Swarm Optimization-based Evolving Ensemble Models. *Knowledge-Based Systems*, p.111461
20. T.Y. Tan, L. Zhang, and C.P. Lim, 2020. Adaptive melanoma diagnosis using evolving clustering, ensemble and deep neural networks. *Knowledge-Based Systems*, *187*, p.104807.

A 1D Convolutional Neural Network for Spam Classification

Jingtong Chen[*,‡], Li Zhang[*,§] and Ming Jiang[†]

[*]*Department of Computer Science, Royal Holloway, University of London Surrey, TW20 0EX, UK*

[†]*School of Computer Science, University of Sunderland, UK*

[‡]*Jingtong.Chen.2023@live.rhul.ac.uk*

[§]*li.zhang@rhul.ac.uk*

Emails and short messages are favored for their communication efficiency. However, a substantial volume of spam is communicated even with recipient consent. To tackle such problems, models like TextCNN are employed for spam detection due to their simple network structures, swift training speed, and commendable performance. While TextCNN excels in spam classification, it has several drawbacks. For example, it utilized multi-channels with limited improvement and reduced explainability. It also lacks the capability to extract long-range features. Therefore, we introduce a novel 1D convolutional neural network, overcoming TextCNN limitations by removing channels and employing an extended kernel. We assess both the proposed and the existing TextCNN models on a large-scale mixed dataset. The empirical results indicate a superior performance of the proposed model over TextCNN with more efficient computational cost, marking a significant advancement in spam classification.

Keywords: 1D convolutional neural network; nature language processing; spam classification.

1. Introduction

Emails and short messages (like SMS) are incredibly popular and widely used forms of communication to ensure efficiency. However, a significant amount of phishing emails are also sent to trick recipients into providing sensitive information [4]. These kinds of spams will reduce productivity and cause security risks. As a result, a method for distinguishing legitimate emails and junk emails should be advanced.

Spam classification remains an important and evolving research subject. Previously, there are many studies focusing on developing machine learning techniques for spam filtering [5] [7–9]. However, traditional text classification methods based on machine learning have many disadvantages, such as

[‡]Corresponding author.

dimension explosion, data sparsity, and limited generalization ability [1]. As a result, researchers shift their attention to deep learning methods to obtain better performance.

Applying Convolutional Neural Network (CNN) to NLP including text classification has drawn many interests in recent years. It has been shown that CNNs can be directly applied to embeddings of words [3]. Among all of the methods, a kind of multi-channel CNN architecture, which is known as TextCNN [2], shows great strengths in text classification tasks and also proved to be a competitive method in tackling spam detection tasks [6]. However, the existing TextCNN model still has disadvantages. First, compared with models with one channel, the multi-channel does not improve the outcome obviously [2] but decreases the explainability. Second, TextCNN uses small convolution kernels, which has limited feature extraction capabilities.

The aim of this chapter is to explore the possible improvements of CNN architecture for spam classification with SMSSpamCollection dataset and self-collected data and compare its performance metrics with the existing methods. The main contributions of this research include the following:

- the elimination of the channels in TextCNN, transitioning to 1D convolution,
- the addition of Dropout and Batch Normalization in each block for reducing overfitting and better generalization,
- the employment of a large convolution kernel in the second pipeline to extract long-range features.

2. Related Work

2.1. *Machine learning*

Throughout recent years, researchers have persistently dedicated efforts to refine and elevate the methodologies employed within the improvements of Machine Learning. For instance, Ning et al. [7] investigate a naïve Bayes algorithm based on multi-two-classification to deal with Spam Message Classification, which outperforms other traditional machine learning algorithms. In addition, Singh et al. [8] evaluate the performance of Non Linear SVM based classifiers with two different kernel functions over SpamAssasin Public Corpus Dataset. In addition, Trivedi et al. [9] present a comparative study between various decision tree classifiers (such as AD tree, decision stump and REP tree) with different boosting algorithms and compare each model's performance.

2.2. *Deep learning*

Deep learning has gained significant popularity within Natural Language Processing (NLP) due to its ability to process myriad domains. For NLP researchers, the primary deep learning models are Recurrent Neural Network (RNN), Long Short-Term Memory Network (LSTM) and Convolutional Neural Network (CNNs). In 2004, Yoon Kim's [2] diverse CNN model variants proposed for text classification set a benchmark in utilizing CNNs for processing natural language data. In another study [10], Raj et al. introduce a novel method Long Short-Term Memory (LSTMs) with gating mechanism including memory cells to provide 97.5% using the SMS dataset. In [11], Taheri et al. introduce a novel RNN-based approach for distinguishing variable-length sequences of ham and spam, achieving an accuracy of 98.11%.

3. The Proposed Methodology

3.1. *Dataset and pre-processing*

The dataset we used in this world is a mix of SMS Spam Collections [12] and self-collected short emails dataset. The final dataset contains 16237 messages with 6078 spams and 10159 non-spam messages. Next, We implement pre-processing including cleaning, standardizing text, removing stop words. All texts are transformed into numerical sequences using Tokenizer, limited to 32-length.

3.2. *Model 1 – The existing TextCNN model*

The existing TextCNN model [2] overview is depicted in Fig. 1, while its structural details are presented in Table 1. The Tokenizer processes a total of 20,000 distinct words, generating sequences with a fixed length of 32 tokens. Each word undergoes embedding, transforming it into a 50-token vector.

The model comprises three parallel blocks following the embedding layer. In the first block, a 2D convolution with 2 channels and a kernel shape of (2, 50) is employed. After the convolution, we implement the squeezing on the last dimension, which is due to the width of the kernel is 50, the same as the length of one embedded sequence. Subsequently, a max pooling operation with a kernel size of 31 is performed, compressing the last dimension and yielding a 2×1 feature.

Similar operations are conducted in block 2 and block 3, using kernel sizes of (3, 50) and (4, 50), respectively. Once all three blocks conclude, the resulting three 2×1 features are concatenated to produce a 6×1 feature. Finally, a dense layer functions as the classifier for this feature representation.

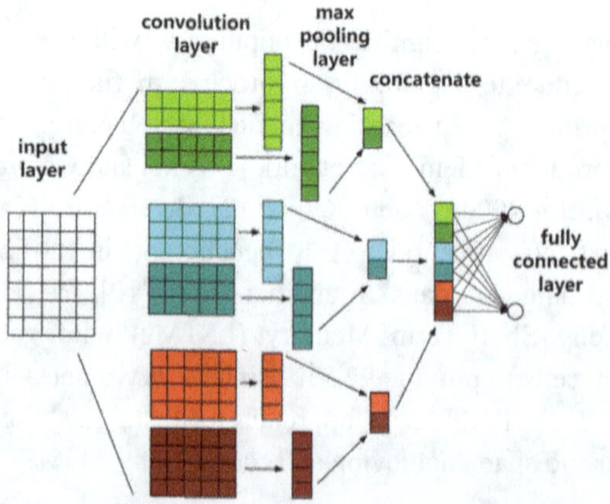

Fig. 1. Overview of the TextCNN model [2]

Table 1 Detailed layer description for TextCNN model [2]

Layer Description		
Embedding(20000, 50)		
block 1	**block 2**	**block 3**
Conv2d(1, 2, kernel_size=(2, 50), stride=(1, 1))	Conv2d(1, 2, kernel_size=(3, 50), stride=(1, 1))	Conv2d(1, 2, kernel_size=(4, 50), stride=(1, 1))
ReLU()	ReLU()	ReLU()
MaxPool1d(kernel_size=31, stride=31, padding=0, dilation=1, ceil_mode=False)	MaxPool1d(kernel_size=30, stride=30, padding=0, dilation=1, ceil_mode=False)	MaxPool1d(kernel_size=29, stride=29, padding=0, dilation=1, ceil_mode=False)
Concatenate		
Linear(in_features=6, out_features=2, bias=True)		

3.3. *Model 2 – The proposed 1D CNN model*

The proposed CNN model overview is illustrated in Fig. 2. and the layer description is shown in Table 2. As discussed earlier, TextCNN uses 2 channels when implementing convolution but does not gain obvious improvements [2]. As a result, we change the 2D convolution into 1D convolution to estimate the channel dimension. In the beginning, each words is also embedded into a 50-token vector. Next, we add a Dropout layer before performing convolution to yield robustness to variations of sentences and reduce overfitting [13].

There are two pipelines in the proposed model. The first pipeline starts with a 1D convolution. In this convolution, we treat the word dimension as the channels (different from the channels in Section 3.2) and convolute across the sentence dimension. With an output channel set to 50 and a kernel size of 5, we maintain the 50-token representation for each word. Subsequently, ReLU, Maxpooling, and Dropout layers are applied sequentially. Next, we add a Batch Normalization to end this block for faster convergence and better generalization [14]. The block 2 is almost a repetition of block 1 while we change the output channel (the embedded word dimension) to 1. Finally, we obtain a 5×1 output.

The second pipeline contains only 1 block. Above all, the convolution kernel size is set to 23, aiming to extract longer features. Next, ReLu, Maxpooling, Dropout and Batch Normalization layers are implemented one by one, leading to a 5×1 output. After concatenation and flatten, we put the features into a fully connected layer, ending with a Softmax function for a binary classification task.

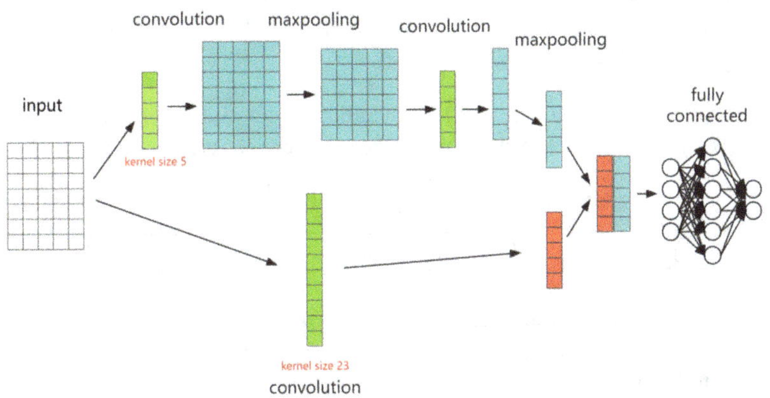

Fig. 2. Overview of the proposed model

Table 2 Detailed layer description for the proposed model

<table>
<tr><th colspan="2">Layer Description</th></tr>
<tr><td colspan="2">Embedding(20000, 50)</td></tr>
<tr><td colspan="2">Dropout(p=0.5, inplace=False)</td></tr>
<tr><td>block 1</td><td>block 2</td></tr>
<tr><td>Conv1d(50, 50, kernel_size=(5,), stride=(1,))</td><td rowspan="3">Conv1d(50, 1, kernel_size=(23,), stride=(1,))</td></tr>
<tr><td>ReLU()</td></tr>
<tr><td>MaxPool1d(kernel_size=2, stride=2, padding=0, dilation=1, ceil_mode=False)</td></tr>
<tr><td>Dropout(p=0.5, inplace=False)</td><td rowspan="2">ReLU()</td></tr>
<tr><td>BatchNorm1d(50, eps=1e-05, momentum=0.1, affine=True, track_running_stats=True)</td></tr>
<tr><td>block 3</td><td rowspan="3">MaxPool1d(kernel_size=2, stride=2, padding=0, dilation=1, ceil_mode=False)</td></tr>
<tr><td>Conv1d(50, 1, kernel_size=(5,), stride=(1,))</td></tr>
<tr><td>ReLU()</td></tr>
</table>

<table>
<tr><td>MaxPool1d(kernel_size=2, stride=2,
padding=0, dilation=1,
ceil_mode=False)</td><td>Dropout(p=0.5, inplace=False)</td></tr>
<tr><td>Dropout(p=0.5, inplace=False)
BatchNorm1d(1, eps=1e-05, momentum=0.1,
affine=True,
track_running_stats=True)</td><td>BatchNorm1d(1, eps=1e-05, momentum=0.1,
affine=True,
track_running_stats=True)</td></tr>
<tr><td colspan="2">Flatten(start_dim=1, end_dim=-1)</td></tr>
<tr><td colspan="2">Linear(in_features=10, out_features=128, bias=True)</td></tr>
<tr><td colspan="2">ReLU()</td></tr>
<tr><td colspan="2">Linear(in_features=128, out_features=2, bias=True)</td></tr>
<tr><td colspan="2">Softmax(dim=-1)</td></tr>
</table>

4. Evaluation

4.1. *Parameters and results*

The dataset is split into training set, validation set and testing set as shown in Table 3. All the datasets are pre-processed as we discussed in Section 3.1.

Table 3 Dataset splitting

Training set	Validation set	Testing set
13,151	1,462	1,624

The configuration of each model's hyper-parameters is shown in Table 4. We use a different learning rate in TextCNN model because it converts very slowly if the rate is also 0.005.

Table 4 Hyper-parameters

	TextCNN	Proposed model
Loss	Cross-Entropy Loss	Cross-Entropy Loss
Learning rate	0.01	0.005
Momentum	0.9	0.9

We adopt the four metrics to evaluate the models: Accuracy, Precision, Recall, and F1 Score. We choose the models' weights in least loss for evaluating the testing set. Table 5 illustrates the outcomes of both models.

Table 5 Performance comparison between the proposed model and TextCNN

	TextCNN	Proposed model
Accuracy	0.7688	**0.7954**
Precision	0.6765	**0.7073**
Recall	0.6682	**0.7537**
F1 Score	0.6644	**0.7159**

In the meantime, we calculate the parameters numbers and used GPU resources.

Table 6 Computational cost

	TextCNN	Proposed model
Parameters	1112,220	1015,722
Occupied GPU	1.1 GB	1.1 GB

4.2. *Analysis*

As shown in Table 5, the proposed model has obtained a higher accuracy, precision, recall and F1 scores, than those of the existing TextCNN model, which suggests the efficiency of the new model in enhancing spam classification accuracy. Moreover, from Table 6, our proposed model has fewer parameters and contribute to more efficient computational cost. The superiority is attributed to our better model architectures, such as, (1) the replacement of the channels in TextCNN with 1D convolution, (2) introduction of Dropout and Batch Normalization within each block to mitigate overfitting and enhance generalization, and (3) the implementation of a larger convolution kernel to extract extended features in the second pipeline.

5. Conclusion

In this chapter, we introduce a CNN architecture and conduct a comparative analysis against the existing TextCNN model. The findings showcase promising results, affirming our model's superior performance in the spam classification tasks. Looking ahead, we aim to integrate potent word embedding models, such as word2vec [15–20], to further enhance the capabilities of our approach.

References

1. Gibson, S., Issac, B., Zhang, L. and Jacob, S.M., 2020. Detecting spam email with machine learning optimized with bio-inspired metaheuristic algorithms. *IEEE Access*, *8*, pp.187914-187932.
2. Kim, Y.: Convolutional neural networks for sentence classification. *arXiv preprint arXiv:1408.5882*
3. Soni, S., Chouhan, S.S. and Rathore, S.S. TextConvoNet: a convolutional neural network based architecture for text classification. *Applied Intelligence,* 53, 14249–14268 (2023).
4. Yu, B., and Xu, Z. B. (2008). A comparative study for content-based dynamic spam classification using four machine learning algorithms. *Knowledge-Based Systems*, 21(4), 355-362.
5. Kinghorn, P., Zhang, L. and Shao, L., 2019. A hierarchical and regional deep learning architecture for image description generation. *Pattern Recognition Letters*, 119, pp.77-85.
6. Waja, G., Patil, G., Mehta, C., and Patil, S. (2023). How AI Can be Used for Governance of Messaging Services: A Study on Spam Classification Leveraging Multi-Channel Convolutional Neural Network. *International Journal of Information Management Data Insights*, 3(1), 100147.

7. Ning, B., Junwei, W., and Feng, H. (2019). Spam message classification based on the Naïve Bayes classification algorithm. *IAENG International Journal of Computer Science*, 46(1), 46-53.
8. Singh, M., and Pamula, R. (2018). Email spam classification by support vector machine. In *International Conference on Computing, Power and Communication Technologies (GUCON)* (pp. 878-882). IEEE.
9. Trivedi, S. K., and Panigrahi, P. K. (2018). Spam classification: a comparative analysis of different boosted decision tree approaches. *Journal of Systems and Information Technology*, 20(3), 298-105.
10. Raj, H., Weihong, Y., Banbhrani, S. K., and Dino, S. P. (2018). LSTM based short message service (SMS) modeling for spam classification. In *Proceedings of International Conference on Machine Learning Technologies.*
11. Taheri, R., and Javidan, R. (2017). Spam filtering in SMS using recurrent neural networks. In *Artificial Intelligence and Signal Processing Conference.*
12. Almeida, T. A., Hidalgo, J. M. G., and Yamakami, A. (2012). Sms spam collection data set. UCI Machine Learning Repository.
13. Zhang, L., Lim, C.P., Yu, Y. and Jiang, M., (2022). Sound classification using evolving ensemble models and Particle Swarm Optimization. *Applied Soft Computing*, 116, p.108322.
14. Xie, H., Zhang, L. and Lim, C.P. 2020. Evolving CNN-LSTM Models for Time Series Prediction Using Enhanced Grey Wolf Optimizer, *IEEE Access*.
15. Mikolov, T., Chen, K., Corrado, G., and Dean, J. (2013). Efficient estimation of word representations in vector space. *arXiv preprint arXiv:1301.3781*.
16. Slade, S., Zhang, L., Yu, Y. and Lim, C.P., (2022). An evolving ensemble model of multi-stream convolutional neural networks for human action recognition in still images. *Neural computing and applications*.
17. Zhang, L., Zhao, D., Lim, C.P., Asadi, H., Huang, H., Yu, Y. and Gao, R., (2024). Video Deepfake Classification Using Particle Swarm Optimization-based Evolving Ensemble Models. *Knowledge-Based Systems*, p.111461.
18. Zhang, L., Slade, S., Lim, C.P., et al., (2023). Semantic segmentation using Firefly Algorithm-based evolving ensemble deep neural networks. *Knowledge-Based Systems*, 277, p.110828.
19. Zhang, L., Lim, C.P. and Liu, C., 2023. Enhanced Bare-Bones Particle Swarm Optimization based Evolving Deep Neural Networks. *Expert Systems with Applications*, p.120642.
20. Zhang, L., Lim, C.P. and Yu, Y., 2021. Intelligent human action recognition using an ensemble model of evolving deep networks with swarm-based optimization. Knowledge-based systems, 220, p.106918.

Supervised and Unsupervised Machine Learning Algorithms: An Empirical Evaluation

Rajarajan Rajkumar*, Li Zhang*,‡, Vivian Sedov* and Kamlesh Mistry†

*Department of Computer Science, Royal Holloway, University of London
Surrey, TW20 0EX, UK

†Department of Computer and Information Sciences, Northumbria University
Newcastle, NE1 8ST, UK

‡li.zhang@rhul.ac.uk

Machine Learning (ML) algorithms are a subset of artificial intelligence that are applied to data with a primary focus of improving its accuracy over time by replicating and imitating the learning styles of human beings. Within this framework, several supervised and unsupervised learning algorithms are studied through different scenarios. The advantages and disadvantages of these algorithms are analysed through these case studies.

Keywords: Supervised and unsupervised learning algorithms; classification; clustering.

1. Introduction

The aim of this chapter is to understand which machine learning (ML) algorithm is more effective under different testing scenarios. ML is the use and development of computer systems that are able to learn and adapt without following explicit instructions, by using algorithms and statistical models to analyse and draw inferences from patterns in data. They have become popular due to their capability in executing complex tasks that are difficult to program and the ability to adapt to changes in circumstances through self-learning. Hence, it is crucial to differentiate and understand between algorithms that can be utilised to enhance data processing efficiency in our daily routines. This chapter aims to exploit distinctive learning behaviours of several supervised and unsupervised algorithms when tackling different classification/clustering tasks. The algorithms in the scope of this chapter are as follows: supervised learning — K-nearest neighbour (KNN), decision tree learning and Support Vector Machines (SVM), and unsupervised learning — K-means and Agglomerative Hierarchical

Clustering (AHC). These algorithms were applied and tested on four different datasets from various sectors: Medical, Finance, Ornithology and Outdoor Image datasets.

2. Related Work

The study of [1] compared various supervised machine learning classification methods using a Diabetes data set with 786 instances and eight attributes as independent variables. Seven different algorithms were tested and SVM was found to be the most effective, followed by Naive Bayes and Random Forest. Sathya and Abraham [2] presented a comparative study of unsupervised and supervised learning models and their classification performance on higher education scenarios. Kohonen map of unsupervised learning model offers the most efficient solutions. Khanam [3] discussed the use of machine learning algorithms such as neural networks for early detection of diabetes. The Pima Indian Diabetes dataset was used, which contains information on 768 patients and nine attributes. Seven machine learning algorithms were studied and the combination of logistic regression and SVM was found to be the most effective for diabetes prediction. A neural network model with two hidden layers provided an accuracy rate of 88.6%. Perols [4] compared the performance of six different machine learning models in detecting financial statement fraud with different misclassification costs and fraud ratios. Logistic regression and SVMs were found to be the most effective methods, and only six predictors were consistently used across the algorithms. Abas [5] examined different data clustering algorithms, including K-Means, hierarchical clustering, self-organising maps, and expectation maximisation. The algorithms were evaluated based on factors such as dataset size, number of clusters, and dataset type.

3. The Proposed Methodologies

For the model to give an effective output within reasonable performance, a pipeline must be created. The machine learning pipeline is an end-to-end framework which manages the data flow. This study was conducted by following the pipeline model recommended by [6], i.e. missing value checking, along with various data pre-processing steps.

3.1. *Dataset description*

The datasets used in this research are extracted from Kaggle and UCI data repositories including (1) Medical Dataset [7], a labelled binary classification dataset, with 303 instances and 13 columns for heart attack prediction, (2) Finance Dataset [8] from the UCI data repository with 1372 samples with 5 attributes; (3) Ornithology Dataset [9] for bird species classification using bone structures and shapes as inputs, with 413 samples and 12 features. (4) Outdoor Image Dataset [10] with 2310 samples with 15 attributes.

3.2. *Data pre-processing*

Data pre-processing includes four main techniques: data cleaning, data integration, data transformation, and data reduction. Data cleaning involves filling in missing values, smoothing noisy data, resolving inconsistencies, and removing outliers. Data integration merges data from multiple sources into a single data store, while data transformation involves generalisation, normalisation, attribute selection, and aggregation.

4. Evaluation of Supervised Learning Algorithms

4.1. *K-nearest neighbour*

The KNN algorithm classifies items based on the distance between an item and its neighbouring points. The K value can affect accuracy and needs to be chosen carefully depending on the specific characteristics of the data with the attempt to secure a better accuracy [11]. Hyperparameters, such as the K value and distance metrics, are also investigated. Table 1 shows the best accuracy result obtained for each sector and corresponding K value which was tested between 1 to 25.

Table 1 KNN Results

Sectors	Self-built Model *K-Value,Accuracy*	Sklearn Model *K-Value,Accuracy*
Medical	k-15, 0.868	k-6, 0.868
Finance	k-1, 1.0	k-1, 1.0
Ornithology	k-1, 0.808	k-1, 0.808
Outdoor Image	k-6, 0.917	k-1, 0.912

4.2. *Decision tree learning*

Decision Tree is a supervised machine learning algorithm that uses a tree structure to predict and classify data based on a set of identified most effective attributes and their respective values. The algorithm selects the best attribute at each node using either Entropy and Information Gain or Gini Impurity. In this research, pre-pruning was used for Decision Tree where the hyper-parameters were tuned before the training [12]. Advantages of the model include the following. (1) It can handle both nominal and numeric input attributes, providing flexibility in data handling. (2) It can handle datasets that may have errors or missing values, providing robustness in data handling. (3) When classification cost is high, decision trees may be particularly attractive, as they only require values of the features along a single path from the root to a leaf, simplifying the classification process. Table 2 shows the results obtained from prepruning of decision tree across different sectors. The main pruning was on the depth and criteria of decision tree.

Table 2 Decision Tree Learning Results

Sectors	Pre Pruning *Depth,Criteria*	Self-built Model *Accuracy*	Sklearn Model *Accuracy*
Medical	D-3, Gini	0.684	0.737
Medical	D-5, Entropy	0.829	0.803
Finance	D-3, Gini	0.883	0.918
Finance	D-5, Entropy	0.924	0.956
Ornithology	D-3, Gini	0.490	0.510
Ornithology	D-5, Entropy	0.548	0.529
Outdoor Image	D-3, Gini	0.407	0.559
Outdoor Image	D-5, Entropy	0.818	0.817

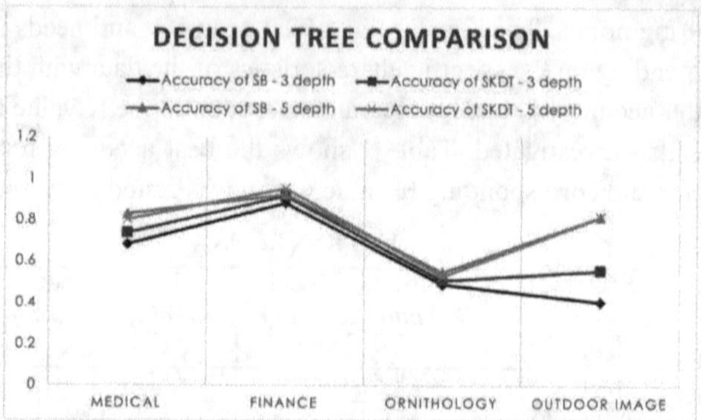

Fig. 1. Decision Tree result comparison

Reducing the depth and changing the criteria of the decision tree model resulted in a decrease in accuracy as shown in Figure 1. The above process helps to determine the most effective model topologies and configurations.

4.3. *SVM*

SVM is a supervised learning algorithm used for classification and regression tasks. The main goal is to find the optimal hyperplane that can best separate different classes of data points. Kernel functions are used in SVM to simplify complex calculations and enable smoother computations, even in higher dimensional space. The linear kernel and Radial basis function (RBF) kernel are commonly used kernel functions. Both kernels were used in this research to test performance along with the multi-class strategy of one vs one (ovo) or one vs rest (ovr) [13]. Advantages are, (1) SVM provides regularisation techniques and hence has good generalisation performance and is less prone to overfitting. (2) It can handle both linear and nonlinear data effectively, using the kernel trick to map the data into a higher dimensional space where it can be linearly separable. (3) It is less sensitive to outliers compared to other classification algorithms due to the use of support vectors located near the decision boundary.

Table 3 SVM results

Sectors	Kernel	Self-built Model *Strategy, Accuracy*	Sklearn Model *Strategy, Accuracy*
Medical	linear	0.882	0.882
Finance	rbf	0.945	0.991
Ornithology	rbf	ovo, 0.490	ovr, 0.654
Outdoor Image	rbf	ovo, 0.801	ovr, 0.888

Table 3 shows the best results obtained from SVM across different sectors. The use of different strategy and kernel was also tested. The lowest performance was on the bird species as expected. RBF kernel performed better on all the datasets except for the Heart Attack prediction where the linear kernel outperformed the RBF kernel. The results of sklearn show similar trends to those of the model built across the datasets. Table 4 provides a comprehensive overview of the accuracy results across all sectors, as well as how each model performed within those sectors.

Table 4 Supervised learning results

Model Name	**Best Accuracy and AUC across Sectors**			
	Medical	***Finance***	***Ornithology***	***Image***
Self-built KNN	0.868, 0.85	1.0, 0.98	0.808	0.917
Sklearn KNN	0.868, 0.86	1.0, 0.98	0.808	0.912
Self-built DT	0.823, 0.84	0.924, 0.93	0.548	0.818
Sklearn DT	0.75, 0.76	0.956, 0.96	0.538	0.816
Self-built SVM	0.882, 0.89	0.845, 0.85	0.548	0.810
Sklearn SVM	0.829, 0.84	0.962, 0.97	0.654	0.888

Figure 2 shows the summary of supervised learning algorithm performance. As seen from the bar graph, KNN algorithm outperformed Decision Tree and SVM in both the self-built model as well as the sklearn model in terms of predictability. In the self-built model decision tree edged over SVM by a slight margin whereas, in the sklearn model, SVM came clear second best behind

KNN with Decision Tree being last. Ornithology sector performed the least compared to other sectors due to data imbalance.

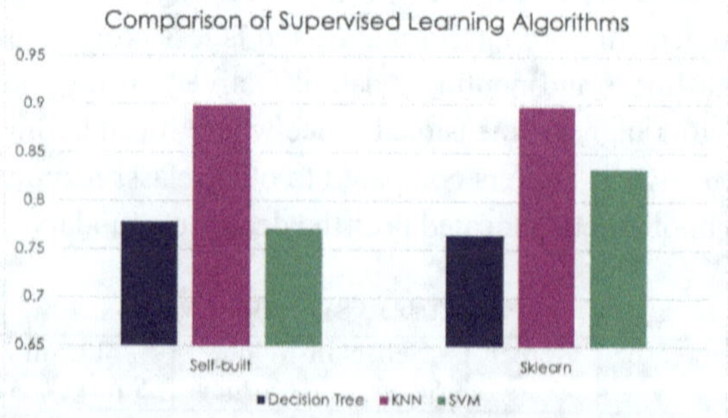

Fig. 2. Comparison of different supervised learning algorithms

5. Evaluation of Unsupervised Learning Algorithms

Unsupervised learning is used in situations where dataset labelling is time-consuming or unavailable. This research studies unsupervised clustering algorithms and utilizes four previous datasets without labels to evaluate the accuracy of the respective algorithms. K-means clustering is an unsupervised learning algorithm used to find a fixed number of clusters in a given dataset. It is a distance-based algorithm that assigns points to clusters based on their similarity. Agglomerative Clustering (AHC) is a bottom-up technique that merges the closest clusters based on distance until a stopping criterion is met. It results in a smaller number of clusters and can be visualised using a dendrogram [14, 15]. Tables 5 and 6 shows the detailed results for both K-Means and AHC clustering models.

Table 5. K-Means Results

Sectors	Self-built Model *Accuracy*	Sklearn Model *Accuracy*
Medical	0.894	0.815
Finance	0.921	0.920
Ornithology	0.666	0.719
Outdoor Image	0.529	0.540

Table 6. AHC Results

Sectors	Self-built Model *Accuracy*	Sklearn Model *Accuracy*
Medical	0.974	0.868
Finance	–NA–	0.735
Ornithology	0.845	0.845
Outdoor Image	–NA–	0.478

6. Conclusions

Both supervised and unsupervised learning algorithms are studied in this chapter, since they aid in effective prediction and reveal hidden data patterns. Our research revealed that KNN outperforms decision trees and SVM in classification problems. On the other hand, K-means and AHC showed almost equal proficiency in solving unsupervised clustering problems. This chapter shows the comparison between different algorithms for a wide spectrum of datasets. The results open up pathways for better selection of algorithms for time efficiency in critical real-world testing scenarios. The deployment of hyperparameter optimization and unsupervised learning methods for image segmentation will also be exploited in future studies [16–25].

References

1. Osisanwo, F.Y., Akinsola, J.E.T., Awodele, O., Hinmikaiye, J.O., Olakanmi, O. and Akinjobi, J., 2017. Supervised machine learning algorithms: classification and comparison. *International Journal of Computer Trends and Technology*, 48(3), pp.128-138.
2. Sathya, R. and Abraham, A., 2013. Comparison of supervised and unsupervised learning algorithms for pattern classification. *International Journal of Advanced Research in Artificial Intelligence*, 2(2), pp.34-38.
3. Khanam, J.J. and Foo, S.Y., 2021. A comparison of machine learning algorithms for diabetes prediction. *ICT Express*, 7(4), pp.432-439.
4. Perols, J., 2011. Financial statement fraud detection: An analysis of statistical and machine learning algorithms. *Auditing: A Journal of Practice and Theory*, 30(2), pp.19-50.
5. Abbas, O.A., 2008. Comparisons between data clustering algorithms. *International Arab Journal of Information Technology*, 5(3).
6. Hapke, H. and Nelson, C., 2020. Building machine learning pipelines. O'Reilly Media.
7. Rahman, Rashik. Heart Attack Analysis and Prediction Dataset. Kaggle. [Online] 2019. https://www.kaggle.com/datasets/rashikrahmanpritom/heart-attackanalysis-prediction-dataset.
8. Dua, D. and Graff, C. banknote authentication Data Set. UCI Machine Learning Repository. [Online] 2019. [accessed on 23 Jan 2024.] https://archive.ics.uci.edu/ml/datasets/banknote+authentication.
9. ZLF. Birds' Bones and Living Habits. Kaggle. [Online] 2017. https://www.kaggle.com/datasets/zhangjuefei/birdsbones-and-living-habits.
10. Image Segmentation Data Set. UCI Machine Learning Repository. [Online] 2019. https://archive.ics.uci.edu/ml/datasets/Image+Segmentation.
11. Shalev-Shwartz, S. and Ben-David, S., 2014. Understanding machine learning: From theory to algorithms. Cambridge university press.
12. Maimon, O.Z. and Rokach, L., 2014. Data mining with decision trees: theory and applications (Vol. 81). World scientific.
13. Moreira, J., Carvalho, A. and Horvath, T., 2018. A general introduction to data analytics. John Wiley and Sons.

14. Sisodia, D., Singh, L., Sisodia, S. and Saxena, K., 2012. Clustering techniques: a brief survey of different clustering algorithms. *International Journal of Latest Trends in Engineering and Technology*, 1(3), pp.82-87.
15. S.C. Neoh, W. Srisukkham, L. Zhang, S. Todryk, B. Greystoke, C.P. Lim, A. Hossain and N. Aslam. An Intelligent Decision Support System for Leukaemia Diagnosis using Microscopic Blood Images, *Scientific Reports*, 5 (14938). 2015.
16. Lawrence, T., Zhang, L., Rogage, K. and Lim, C.P., 2021. Evolving Deep Architecture Generation with Residual Connections for Image Classification Using Particle Swarm Optimization. *Sensors*, 21(23), p.7936.
17. Lawrence, T., Zhang, L., Lim, C.P. and Phillips, E.J., 2021. Particle swarm optimization for automatically evolving convolutional neural networks for image classification. *IEEE Access*, 9, pp.14369-14386.
18. P. Kinghorn, L. Zhang and L. Shao. A Hierarchical and Regional Deep Learning Architecture for Image Description Generation, *Pattern Recognition Letters*. 2019.
19. T.Y. Tan, L. Zhang, and C.P. Lim. Adaptive melanoma diagnosis using evolving clustering, ensemble and deep neural networks. *Knowledge-based Systems*, 187:104807, 2020.
20. Slade, S., Zhang, L., Yu, Y. and Lim, C.P., 2022. An evolving ensemble model of multi-stream convolutional neural networks for human action recognition in still images. *Neural computing and applications*, 34(11), pp.9205-9231.
21. Zhang, L., Zhao, D., Lim, C.P., Asadi, H., Huang, H., Yu, Y. and Gao, R., 2024. Video Deepfake Classification Using Particle Swarm Optimization-based Evolving Ensemble Models. *Knowledge-Based Systems*, p.111461.
22. Zhang, L., Lim, C.P., Yu, Y. and Jiang, M., 2022. Sound classification using evolving ensemble models and Particle Swarm Optimization. *Applied Soft Computing*, 116, p.108322.
23. Zhang, L., Lim, C.P. and Yu, Y., 2021. Intelligent human action recognition using an ensemble model of evolving deep networks with swarm-based optimization. *Knowledge-based Systems*, 220, p.106918.
24. Xie, H., Zhang, L. and Lim, C.P. 2020. Evolving CNN-LSTM Models for Time Series Prediction Using Enhanced Grey Wolf Optimizer, *IEEE Access*. 8, p. 161519-161541.
25. Zhang, L., Lim, C.P. and Liu, C., 2023. Enhanced Bare-Bones Particle Swarm Optimization based Evolving Deep Neural Networks. *Expert Systems with Applications*, p.120642.

Protecting the Dataset Ownership against Unauthorized Utilization by Generative Models via Spatial Watermark

ChangQin*, RuipingYin† and ZhenYang‡

Faculty of Information Technology BJUT, Beijing University of Technology, Beijing 100124, China

* *qinchang@emails.bjut.edu.cn*

† *yinruiping@bjut.edu.cn*

‡ *yangzhen@bjut.edu.cn*

In recent years, generative models have made remarkable progress in image synthesis, prompting concerns about potential intellectual property infringement while training. This concern focuses on the unauthorized utilization of a dataset by generative models. In order to protect the legitimate rights of dataset owners, this chapter proposes a novel approach to safeguard dataset ownership by introducing a watermark signal that generative models can memorize. To implement this, we select a subset of samples from the dataset and apply a spatial watermark to create a protected dataset comprising both watermarked and non-selected samples. Dataset owners can then detect the watermark information in suspicious generative model outputs to determine whether the model used the protected dataset during training. The efficacy of the proposed method is validated using a VAE model on the CelebA dataset.

Keywords: Dataset protection; dataset watermarking; dataset ownership verification.

1. Introduction

In recent years, generative models have made significant progress in the field of image generation,[1] including facial synthesis,[2] generation of images with arbitrary styles,[3–6] and generating images based on textual prompts.[7] The rapid development of generative models relies not only on powerful algorithms and excellent computing resources but also on high-quality datasets.

The pivotal role of datasets in the evolution of generative models has prompted significant apprehensions regarding the unauthorized utilization of these datasets by models. For instance, a model could directly employ a collection of an artist's works for training to generate images with the artist's distinctive style, all without requiring the artist's authorization.[8–10]

†Corresponding author.

Therefore, detecting unauthorized usage of datasets by generative models becomes crucial for dataset owners.

We focus on the protection of dataset ownership for generative models. Specifically, it explores how dataset owners can, under the constraint of only querying the generative model, determine if a suspicious generative model used their protected dataset during training.

Currently, there are two categories of approaches for protecting the ownership of datasets: one is based on backdoor attacks,[11–14] and the other is based on unlearnable example.[15–20] The backdoor attack approach establishes a connection between watermark samples and labels in the dataset by adding a backdoor watermark. Models trained on this dataset will learn this connection, and during validation, ownership of the dataset is verified by detecting this connection. The unlearnable approach protects dataset ownership by preventing the model from generating samples with styles similar to those in the dataset. However, the backdoor attack approach is primarily used for classification models, while the unlearnable manner hinders the authorized use of the dataset.

Based on the idea of providing a signal that can be memorized by generative models for datasets, we propose a method that involves adding a spatial watermark to protect dataset ownership. By adding a watermark signal that generative models can memorize to the protected dataset, dataset owners can detect the watermark in images generated by suspicious models. Using this method, we confirm the effectiveness of our approach on the CelebA dataset with a VAE model.

Our main contributions are as follows:

(1) We proposed a method for protecting dataset ownership against unauthorized utilization by generative models. It consists of dataset watermarking by spatial watermark and detecting the dataset unauthorized utilization.
(2) Experiments on VAE demonstrate the effectiveness of our method.

2. Related Work

2.1. *Backdoor watermark*

Liy et al. first proposed a method for protecting open-source datasets using a backdoor watermarking approach[11]. This method used poison-only backdoor attacks with poisoned label[21–23]. Liy et al. designed a untargeted backdoor watermark and utilized it for harmless and covert dataset ownership verification[12]. Tang et al. introduced a framework for backdoor watermarking technology that incorporates clean labels, using imperceptible perturbations to replace incorrectly labeled samples[14].

2.2. *Unlearnable example*

Huang et al. advocated the use of Sample-wise error-minimizing noise for crafting unlearnable samples, thereby deceiving the model into perceiving "nothing" to learn from the examples[16]. Ren et al. introduced the Class-wise Separability Discriminant to generate optimizable linear-separable perturbations[17]. Shan et al. employed subtle perturbations applied to the image, causing its representation in the feature space of the generative model to shift to another style[15]. Salman et al. proposed the addition of carefully designed imperceptible perturbations to specific images, rendering them resistant to AI manipulation[18].

3. Method

In this section, we present our proposed method for detecting unauthorized usage of protected datasets by image generation models.

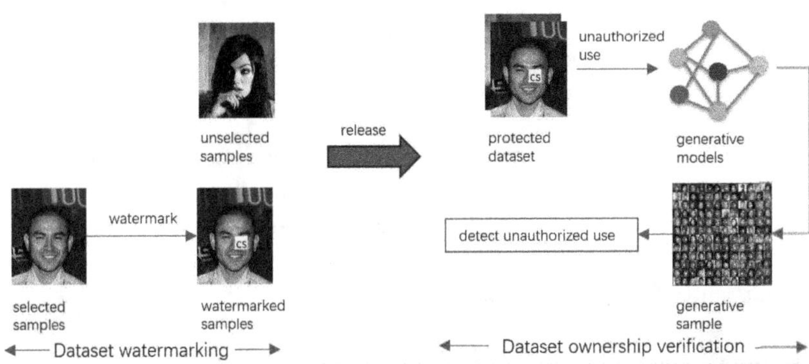

Fig. 1. The watermarked samples when $\gamma = 0$

3.1. *Problem definition*

We focuses on empowering dataset owners to make strong judgments when violators use their datasets without authorization, enabling them to take legal actions.

As shown in Fig.1, the method has two stage: dataset watermarking and dataset ownership verification. In the dataset watermarking stage, given the initial dataset D, the dataset size N, the watermark template t and the sample selection radio p, the target is to select $p \times N$ samples f_{select}, merge the watermark template with selected sample $f_{mark} = f_{select} + t$ and combine it with the unselected samples $f_{unselect}$ to form a protected dataset D_p. In the dataset ownership verification stage, given the suspicious generative model G, the target is to detect whether G unauthorized used the dataset D_p.

3.2. *Dataset watermarking*

Watermarking is a classic method for protecting dataset ownership. We choose a template as the watermark template. According to Formula 1, we merge the watermark template with the original sample images. A subset of benign samples from the dataset is selected to add the watermark. Then the samples with added watermarks and the benign samples not chosen compose the protected dataset for distribution.

Formula 1:

$$f_{mark}(x,y) = \begin{cases} f_{\text{select}}(x,y) \cdot (1-\gamma) + t(i,j)\ cdot\gamma, \\ \quad p \leq x \leq (w+p),\ q \leq y \leq (h+q),\ 0 \leq i \leq w,\ 0 \leq j \leq h \\ f_{\text{select}}(x,y), \\ \quad \text{else} \end{cases}$$

where $f_{\text{mark}}(x,y)$ represents the sample after adding the watermark template, $f_{\text{select}}(x,y)$ represents the selected sample, $t(i,j)$ is the watermark template, w and h are the width and height of the watermark template, and p and q are the insertion positions of the top-left corner of the watermark template in the selected sample.

3.3. *Dataset ownership verification*

The ownership verification of dataset can be very easy. Data owners can just let suspicious generative model generate enough samples.By detecting the watermark in the samples to determine the suspicious generative model use the protected dataset in the training process.

4. Experiment

In this section, we first introduce our experimental setup and then validate the effectiveness of our proposed method in protecting the ownership of datasets through experiments.

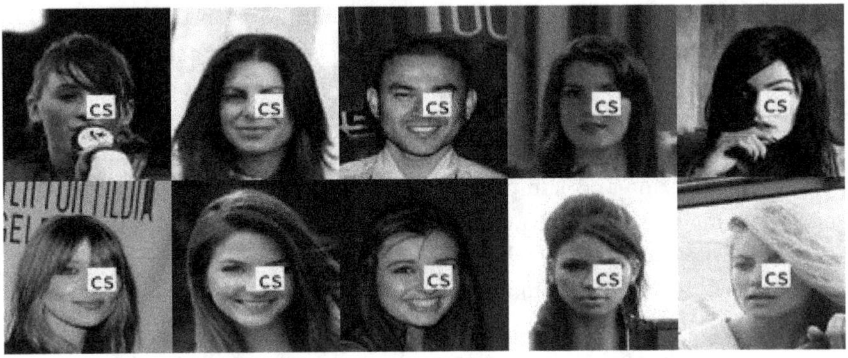

Fig. 2. The watermarked samples when $\gamma = 0$

4.1. *Experimental setup*

Our method is implemented in Python 3.9 and PyTorch 2.1.2. All experiments are conducted on a server running Ubuntu 20.04. This server is equipped with two GPUs.

Models and Datasets: We selected the CelebA dataset as the protected dataset, consisting of 200,000 annotated face images of 10,000 celebrities. Each image is annotated with 40 attributes, including gender, hair color, age, and smile status. We chose the VAE model to simulate violators who use the protected dataset without authorizationon the.

Fig. 3. Generative samples from VAE trained from proctected dataset, which watermark rate is 100%.

Fig. 4. Generative samples from VAE trained from proctected dataset, which watermark the "smiling" attribute.

Evaluation Metrics: We used detection accuracy as the evaluation metric. Accuracy is defined as the probability that dataset owners can correctly distinguish a given set of generative models, including models trained on the protected dataset and models trained without the protected dataset.

Implementation Details: We explored the impact of selecting the proportion of watermarked samples in relation to the entire dataset on the generative model's ability to memorize the watermark signal. The subset of watermarked samples are shown in Fig.2.

Given the diverse attributes of facial images in the CelebA dataset, particularly those associated with facial expressions such as smiling, we introduced watermarks to facial samples with the smiling attribute. In Fig.4, we show the samples generates from suspicious generative model.

4.2. *Effectiveness*

In this section, we demonstrate the effectiveness of our method. As shown in Fig.3 and in Fig.4, when watermark rate is set to 100% and watermark attribute is set to "smiling" respectively, we can see watermark clearly. We also find that as the frequency of selecting additional samples increases, the detection accuracy gradually improves.

5. Conclusion

To safeguard the rightful ownership of datasets and detect unauthorized utilization of protected datasets by generative models, we present a dataset protection method centered around spatial watermarking. This method entails embedding spatial watermarks into a subset of benign samples, thereby generating watermark samples. These watermark samples are then integrated with the remaining unselected samples to construct the protected dataset. Dataset owners have the capability to ascertain whether a suspicious model has used protected dataset without authorization by scrutinizing the model's behavior. Experiments on VAE confirms the efficacy of our proposed approach.

Acknowledgments

This work is supported by the National Key R&D Program of China under Grant number 2022YFB3103100, and the R&D Program of Beijing Municipal Education Commission (KM202310005028).

References

1. R. Rombach, A. Blattmann, D. Lorenz, P. Esser and B. Ommer, High-resolution image synthesis with latent diffusion models, in *Proceedings of the IEEE/CVF conference on computer vision and pattern recognition*, (New Orleans, USA, 2022).
2. J. Sun, Q. Deng, Q. Li, M. Sun, M. Ren and Z. Sun, Anyface: Free-style text-to-face synthesis and manipulation, in *Proceedings of the IEEE/CVF Conference on Computer Vision and Pattern Recognition*, (New Orleans, USA, 2022).
3. Z. Pan, X. Zhou and H. Tian, Arbitrary style guidance for enhanced diffusion-based text-to-image generation, in *Proceedings of the IEEE/CVF Winter Conference on Applications of Computer Vision*, (Vancouver, Canada, 2023).
4. R. Gal, Y. Alaluf, Y. Atzmon, O. Patashnik, A. H. Bermano, G. Chechik and D. Cohen-Or, An image is worth one word: Personalizing text-to-image generation using textual inversion, *arXiv preprint arXiv:2208.01618* (2022).
5. N. Ruiz, Y. Li, V. Jampani, Y. Pritch, M. Rubinstein and K. Aberman, Dreambooth: Fine tuning text-to-image diffusion models for subject-driven generation, in *Proceedings of the IEEE/CVF Conference on Computer Vision and Pattern Recognition*, (Vancouver, Canada, 2023).
6. C. Meng, Y. He, Y. Song, J. Song, J. Wu, J.-Y. Zhu and S. Ermon, Sdedit: Guided image synthesis and editing with stochastic differential equations, *arXiv preprint arXiv:2108.01073* (2021).
7. C. Saharia, W. Chan, S. Saxena, L. Li, J. Whang, E. L. Denton, K. Ghasemipour, R. Gontijo Lopes, B. Karagol Ayan, T. Salimans *et al.*, Photorealistic text-to-image diffusion models with deep language understanding, *Advances in Neural Information Processing Systems* **35**, 36479 (2022).
8. K. Jones, Will art generated by artificial intelligence replace the role of a graphic designer?, PhD thesis, The University of Memphis, (Tennessee, USA, 2023).
9. C. Chen, J. Fu and L. Lyu, A pathway towards responsible ai generated content, *arXiv preprint arXiv:2303.01325* (2023).
10. W. Zhuang, C. Chen and L. Lyu, When foundation model meets federated learning: Motivations, challenges, and future directions, *arXiv preprint arXiv:2306.15546* (2023).
11. Y. Li, Z. Zhang, J. Bai, B. Wu, Y. Jiang and S.-T. Xia, Open-

sourced dataset protection via backdoor watermarking, *arXiv preprint arXiv:2010.05821* (2020).

12. Y. Li, Y. Bai, Y. Jiang, Y. Yang, S.-T. Xia and B. Li, Untargeted backdoor watermark: Towards harmless and stealthy dataset copyright protection, *Advances in Neural Information Processing Systems* **35**, 13238 (2022).
13. Y. Li, M. Zhu, X. Yang, Y. Jiang, T. Wei and S.-T. Xia, Black-box dataset ownership verification via backdoor watermarking, *IEEE Transactions on Information Forensics and Security* (2023).
14. R. Tang, Q. Feng, N. Liu, F. Yang and X. Hu, Did you train on my dataset? towards public dataset protection with clean-label backdoor watermarking, *arXiv preprint arXiv:2303.11470* (2023).
15. S. Shan, J. Cryan, E. Wenger, H. Zheng, R. Hanocka and B. Y. Zhao, Glaze: Protecting artists from style mimicry by text-to-image models, *arXiv preprint arXiv:2302.04222* (2023).
16. H. Huang, X. Ma, S. M. Erfani, J. Bailey and Y. Wang, Unlearnable examples: Making personal data unexploitable, *arXiv preprint arXiv:2101.04898* (2021).
17. J. Ren, H. Xu, Y. Wan, X. Ma, L. Sun and J. Tang, Transferable unlearnable examples, *arXiv preprint arXiv:2210.10114* (2022).
18. H. Salman, A. Khaddaj, G. Leclerc, A. Ilyas and A. Madry, Raising the cost of malicious ai-powered image editing, *arXiv preprint arXiv:2302.06588* (2023).
19. C. Liang, X. Wu, Y. Hua, J. Zhang, Y. Xue, T. Song, Z. Xue, R. Ma and H. Guan, Adversarial example does good: Preventing painting imitation from diffusion models via adversarial examples, in *International Conference on Machine Learning*, (Honolulu, Hawaii, USA, 2023).
20. T. Van Le, H. Phung, T. H. Nguyen, Q. Dao, N. N. Tran and A. Tran, Anti-dreambooth: Protecting users from personalized text-to-image synthesis, in *Proceedings of the IEEE/CVF International Conference on Computer Vision*, (Vancouver, Canada, 2023).
21. T. Gu, K. Liu, B. Dolan-Gavitt and S. Garg, Badnets: Evaluating backdooring attacks on deep neural networks, *IEEE Access* **7**, 47230 (2019).
22. A. Turner, D. Tsipras and A. Madry, Label-consistent backdoor attacks, *arXiv preprint arXiv:1912.02771* (2019).
23. X. Chen, C. Liu, B. Li, K. Lu and D. Song, Targeted backdoor attacks on deep learning systems using data poisoning, *arXiv preprint arXiv:1712.05526* (2017).

Detection of Crops from Satellite Images for Biomass Fuzzy Estimation

Camilo Franco[*,‡], Rocío Sierra[†,§] and Joan Macías[†,¶]

[*] *Joint Research Centre, European Commission, Ispra (VA), 21027, Italy*
[†] *Faculty of Engineering, University of the Andes 111711, Bogotá*
[‡] *camilo.franco-de-los-rios@ec.europa.eu*
[§] *rsierra@uniandes.edu.co*
[¶] *jd.macias@uniandes.edu.co*

Ensuring access to safe, affordable, and non-polluting energy sources is an important goal for sustainable development. One such source is residual biomass, which can be used for the production of biogas through anaerobic digestion methods. To aid the adoption of green energy production, this chapter developed and implemented an open-data and remote sensing methodology for the automatic detection of crops, which should enable the corresponding estimation of their residual biomass potential through a fuzzy formulation. This chapter explores the use of neural networks jointly with RGB images taken by the Sentinel 2 satellite over different regions in Colombia. As a result, convolutional neural networks achieved a validation accuracy in the detection of crops of 97.7%, generating reliable knowledge for estimating their fuzzy potential for residual biomass production.

Keywords: Remote sensing; convolutional neural nets; fuzzy estimation; residual biomass.

1. Introduction

The world's current dependence on fossil fuels as the main source of energy is a significant issue to overcome for sustainable development, climate change, and the overall resilience of the world economy. Therefore, it is necessary to search for new renewable sources that are friendly to the environment. One promising solution to this problem is the use of biomass in residual form to obtain biogas, which can be used as fuel to obtain energy with major benefits.[1] An appropriate bioenergy policy can help decarbonize the economy, enhance the reliability of the energy supply, and revitalize rural areas.[2]

[†]Corresponding author.

This proposal aims to develop a fast and reliable methodology for detecting crops, which are the main source of residual biomass, using open-data and true-color (RGB) Sentinel satellite images. The Sentinel satellite constellation is one of the platforms that provide free satellite images, characterized by having six different satellites that update their images every 3 days, approximately. The signals and images provided by this constellation can be used to detect areas with greater potential to generate residual biomass. The proposed methodology is demonstrated for a case study in Colombia, taking as *residual biomass* any organic residual generated from agricultural activities.[3]

Current uses of machine and deep learning on satellite imagery for agricultural research mainly deal with crop segmentation, yield prediction, and residual biomass potential estimation, using previously identified geographical areas.[4–7] In this same line of research, but exploring the potential of satellite remote sensing technology for automatically detecting suitable areas for biomass recollection and new green energy projects,[8] the methodology presented here addresses a previous step with respect to the image-based estimation of yield and crop characteristics, namely, the automatic detection of the residual biomass sources from agricultural fields, performed entirely by machine learning and remote sensing techniques.

It is stressed that those studies mainly use field measurements or available inventory data as response variables, and high-resolution satellite images or vegetation indices as explanatory variables, to develop color-based or statistical estimation methods. However, the present study aims to automatically detect crops with important residual biomass potential without requiring pre-existing crop-data inventories, relying only on expert annotations on the available satellite images. The proposed methodology combines the human ability to detect anthropogenic agricultural interventions with a broad knowledge of the geographical location of relevant crops, enabling the development of a robust automatic detection methodology to support the development of biomass estimation systems.

This chapter also highlights the possibility of applying machine learning methods for the automatic detection of crops that cope with the inherent uncertainty of mapping biomass cover and/or cover change related to temporal and spatial resolution or the frequency of available images when dealing with satellite imagery of a fixed location (a common issue for satellite-based remote sensing approaches to estimating above-ground biomass[9]). In this sense, the identification of crops is developed directly on the available satellite images, enabling the dynamic construction of statistical detection models while reducing the above-mentioned uncertainty.

The main objective of this chapter consists in developing a methodology that automatically detects crops, as primary agricultural biomass sources, from Sentinel satellite true-color (RGB) images, which in turn can be used to estimate their mean potential for energy production. The computation of the biomass energy potential is dependent on the availability of other relevant data. In this way, as it is explained in Section 2, images from the Sentinel constellation are extracted, processed, and used for building the models, identifying the best available model according to the prediction error under a double cross-validation scheme. In Section 3, the results are presented for the case study in Colombia. This chapter concludes with Section 4, offering some final comments and lines for future research.

2. Materials and Methods

2.1. *Satellite open-data*

Satellite images were gathered from the band 8 of the Sentinel 2 satellite (*product S2MSI2A*). For this, the official page of the *Copernicus*[a] program was used, which gives free access to high resolution (10 meters per pixel) satellite images. Specifically, RGB images of 1 km^2 were taken throughout 2020, between the months of January and March, belonging mainly to the geopolitical areas (*departments*) of Meta, Caquetá, Guaviare, the Cauca Valley and Guajira in Colombia.

2.2. *Extraction of examples*

In order to properly detect crops in satellite images, a set of three classes was considered, namely {*Forest, Crop, Other*}. Hence, the images obtained from the Sentinel SciHub were extracted taking into account that a representative sample was required for each one for the classes. Here, it is stressed the importance of taking care that the composition of the sample is fair and representative of the crops that should be detected. This is a key aspect to properly train the models, as all the knowledge from where the models can draw any conclusions is implicit in such a sample.

In this way, there were 290 images of forests, 250 images of crops, and 310 images containing other sort of entities such as cities, clouds, rivers, desserts, swamps or sea. The raw images, as they were extracted from the Sentinel satellite 2A, were then segmented (without overlapping) into sub-images or cells of 20 × 20 pixels. This was the size of the objects that the

[a]https://scihub.copernicus.eu/.

models were expected to handle, and at the same time, defined the size of the crops that the models should be able to detect. Therefore, if the cell was 200 × 200 meters, the minimum size of a crop that can be detected by the model would be (greater than) 100 m^2, as long as the cell was classified as a crop if more than 50% of its content actually belonged to a crop.

The different examples of 20 × 20 pixels were then randomly assigned to some training and validation sets in order to build the predictive classification models. The complete validation scheme corresponded to a double-cross-validation procedure[10], where the construction of the training (containing 70% of the cells), first and second validation (each one holding 15% of the cells) sets was randomly repeated multiple times, in order to have an unbiased estimation for the prediction error and the performance of the models (as reported for the second validation). Here, performance was measured by the predictive error of the models.

2.3. *Crop recognition by logistic and neural network models*

The different models used in this study for detecting crops took from simple linear classifiers, such as regularized logistic regression, to simple (single-layer) feed-forward nets (universal approximators[11]) and convolutional neural networks (CNN)[12]. In this way, a reference-benchmark linear model was established, to properly assess the performance of the more complex models. In the experiments, the number of neurons in the hidden layer of simple neural nets varied. Parameters were randomly initialized for each repetition of the double cross-validation procedure (from a standard-Normal distribution with 0.05 deviation). The *early-stopping* technique was used together with the stochastic gradient descent algorithm with momentum, with a constant 0.01 error rate and a 0.01 momentum, identifying the best epoch accomplishing the minimum error in validation. Lastly, the best model was chosen after optimizing the number of epochs and neurons accomplishing the maximum prediction accuracy, and satisfying the prediction recall. This same procedure was implemented for the respective experiments performed with convolutional neural nets (CNN).

On the other hand, CNN were composed by different layers with convolutional filters, each one of them focusing on low-level attributes[12]. For the network's output containing the last convolutional layer, a *flattened*, vector was obtained that was then fed to a fully connected layer. For such a layer, the experiments took from 1 to 15 neurons that were activated by the sigmoid function, passing the resulting output into the final soft-max layer. In order to find the optimal set of parameters, the cross entropy loss function was minimized by using the *Adam* algorithm[13,14].

After carrying the methodology previously described, concerning the architecture of the network and the optimization of its parameters with respect to the training data, a qualitative test of the best model was carried out on an image of size 500 × 1000 pixels. Thus, visualizing the expected performance of the model in a situation when it would be facing completely new data. Simulating the production phase, the image was scanned taking 20 × 20 size cells, obtaining a binary response from the model, in case of it having a crop or not.

In consequence, based on the model's output, an estimation of the residual biomass energy potential can be formulated, focusing on the maximum energy contents that can be obtained from burning all of the available dry biomass[15]. A fuzzy formulation is considered here aiming at capturing the possible interaction between the weight and lower calorific value of the dry-matter biomass residue. This can done by means of a symmetric aggregation function $f : [0.1]^2 \to [0,1]$, such that $f(0,0) = 0$ and $f(1,1) = 1$. Some examples for estimating the fuzzy biomass potential can be taken as $f = product$, $f = min$ or $f = max$.

In this way, for every cell classified as crop, where $c_i = 1$ holds, otherwise $c_i = 0$, the potential energy production is estimated as in

$$b(X) = \sum_{i}^{|X|} c_i * f(W_i, P_i), \qquad (1)$$

taking into account the normalized values for the weight and lower calorific value of the dry-matter residue (W and P, respectively) of the specific crop being detected in cell i, for every sub-image (x_i) in the whole body of available images (X). Therefore, it is stressed that the residual biomass potential estimation function $b(X)$, is dependent on the specific type of crop being detected, due to the necessary input parameters of the weight of the dry-matter and the lower calorific value of the residue.

3. Results

The results obtained for the logistic regression showed a low sensitivity for the crop class, and the models were prone to mistakenly classifying images that were forests as crops. Concerning simple neural nets, the best configuration used 50 neurons in its hidden layer, improving the logistic regression benchmark. The sensitivity against each class had a value close to the total accuracy (90 %). Nonetheless, the crop class continued to have the lowest performance regarding sensitivity.

Overall, the model with highest accuracy in validation was obtained with a CNN model of 2 convolutional layers and 12 neurons in the hidden, densely connected layer. The validation performance can be examined with the resulting confusion matrix. In this way, it obtained an accuracy for validation of 97.7% (as shown in Table 1), where the sensitivity for the forest class was 98.19%, for crops it obtained 96.88% and for the others, 97.51%.

Table 1. Confusion matrix in validation for the three predicted classes.

	Prediction		
Real	Forest	Crop	Other
Forest	1088	11	9
Crop	15	903	14
Other	17	11	1101

Once the best CNN model was identified, it was evaluated on the image presented in Fig. 1, obtaining the results shown to its right. The violet color marks the crops, while blue is for the others and yellow is for forests. In the image, it can be seen that the model properly distinguished crops form urban sectors, and that it reliable detected crops, although many crops were omitted or mis-classified. Hence, some evident difficulty existed in detecting crops with high precision, with the subsequent consequence of sub-estimating the cultivated area and the potential energy production.

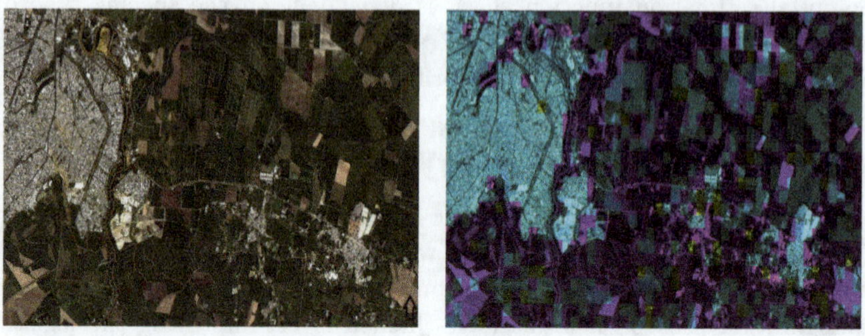

Fig. 1. On the left, the original image used for external validation. On the right, the results after applying the CNN model on the image.

4. Conclusions

The evidence recollected in this chapter exemplifies the benefits of building models that take advantage of the information provided by satellite true-color images, in particular from the Sentinel constellation, for the detection of cultivation areas with biomass production potential. In this case, and the application to different regions of Colombia, the best model corresponded to a convolutional neural network with 2 convolutional layers followed each with a respective "Max pooling" layer, a densely connected layer of 12 sigmoid activation neurons, and a Softmax activation output layer. Such a model obtained an accuracy in validation of 97.7%.

For future research, the limitations of the models should be addressed, particularly for detecting specific types of crops under different weather and atmospheric conditions. It is also of interest to include more regions and information from other satellites of the Sentinel constellation, in addition to the band 8 of the Sentinel 2 satellite, such as the Synthetic Aperture Radar (SAR) data from Sentinel 1.[16] Finally, the energy potential of the cultivars detected by the model should be directly computed, after obtaining reliable estimates of the necessary input parameters included in Eq. (1), regarding the specific crop being detected and its dry-matter residual weight and calorific value.

Acknowledments

The views expressed are those of the authors and cannot be regarded as stating an official position of the European Commission.

References

1. S. Shapiro-Bengtsen, L. Hamelin, L. Bregnbaek, L. Zou and M. Munster, Should residual biomass be used for fuels, power and heat, or materials? assessing costs and environmental impacts for china in 2035, *Energy Environ. Sci.* **15**, 1950 (2022).
2. A. Paiano and G. Lagioia, Energy potential from residual biomass towards meeting the eu renewable energy and climate targets. the italian case., *Energy Policy* **91**, 161 (2016).
3. D. Jiménez and G. Siabato, *Atlas georreferenciado del potencial energético de la biomasa residual en zonas no interconectadas de Colombia* (Universidad Distrital Francisco José de Caldas, Bogota, Colombia, 2021).

4. B. Victor, Z. He and A. Nibali, A systematic review of the use of deep learning in satellite imagery for agriculture, *arXiv:2210.01272* (2022).
5. A. Chakraborty, A. Biswal, V. Pandey, S. Shadab, K. Kalyandeep, C. Murthy, ... and S. Chowdhury, Developing a spatial information system of biomass potential from crop residues over india: A decision support for planning and establishment of biofuel/biomass power plant., *Renewable and Sustainable Energy Reviews* **165**, p. 112575 (2022).
6. L. Geng, T. Che, M. Ma, J. Tan and H. Wang, Corn biomass estimation by integrating remote sensing and long-term observation data based on machine learning techniques., *Remote Sensing* **13**, p. 2352 (2021).
7. A. Bharti, M. V. R. Paritosh, K., A. Chawade and V. Vivekanand, Gis application for the estimation of bioenergy potential from agriculture residues: An overview., *Energies* **14**, p. 898 (2021).
8. M. Nezhad, B. Nastasi, D. Groppi, M. Lamagna, G. Piras and D. Astiaso Garcia, Green energy sources assessment using sentinel-1 satellite remote sensing, *Frontiers in Energy Research* **9**, p. 649305 (2021).
9. S. Abbas, M. Wong, J. Wu, N. Shahzad and S. Irteza, Approaches of satellite remote sensing for the assessment of above-ground biomass across tropical forests: Pan-tropical to national scales., *Remote Sensing* **12**, p. 3351 (2020).
10. P. Filzmoser, B. Liebmann and K. Varmuza, Repeated double cross validation., *Journal of Chemometrics* **23**, 60 (2009).
11. K. Hornik, M. Stinchcombe and H. White, Multilayer feedforward networks are universal approximators, *Neural Networks* **2**, 359 (1989).
12. I. Goodfellow, Y. Bengio and A. Courville, *Deep learning* 2016.
13. D. Kingma and J. Ba, Adam: A method for stochastic optimization, *3rd International Conference for Learning Representations, arXiv:1412.6980* , 359 (2015).
14. S. Reddi, S. Kale and S. Kumar, On the convergence of adam and beyond, *arXiv:1904.09237* (2019).
15. L. Cáceres-Martínez, D. Guío-Pérez and S. Rincón, Potencial energético teórico y técnico de biomasa residual disponible en colombia para el aprovechamiento en procesos de trasformación termoquímica, in *3er Congreso de Energía sostenible*, 2016.
16. M. Meroni, R. d'Andrimont, A. Vrieling, D. Fasbender, G. Lemoine, F. Rembold, ... and A. Verhegghen, Comparing land surface phenology of major european crops as derived from sar and multispectral data of sentinel-1 and -2, *Remote Sensing of Environment* **253**, p. 112232 (2021).

ML4MDS: A Machine Leaning Platform for Multiple Data Streams

Bin Zhang, Jie Lu*, Kun Wang and Guangquan Zhang

Australia Artificial Intelligence Institute (AAII), University of Technology Sydney, Sydney, NSW 2007, Australia

**jie.lu@uts.edu.au*

Concept drift refers to the change of distribution of data streams. The correlations between data streams can also change over time. To deal with multiple data streams, a machine learning platform called `ML4MDS` is proposed for multiple data streams learning. It provides ways to evaluate algorithms designed for multi-stream problems and visualize the results and correlations between data streams dynamically. It is flexible and extensible to add new datasets and new algorithms. The API's design is inspired by `scikit-multiflow`, and the browser-based graphical user interface is based on `streamlit`. It is released under BSD 3-Clause License. The source code is available at `https://www.github.com/ml4mds/ml4mds`.

Keywords: Multiple data stream; concept drift; machine learning; online learning.

1. Introduction

One of the key challenges in data stream mining research is concept drift, a phenomenon where data distribution changes over time. Once concept drift is detected, the learning model will be updated. Many concept drift adaptation methods have been developed in the previous literature.[1,2] To perform algorithms designed for data streams, several open-source software are developed. Massive Online Analysis (MOA)[3] is one of the most popular open-source frameworks for machine learning in data streams. After the growing of deep learning, Python becomes the most popular programming language in machine learning area. The `scikit-multiflow`[4] was proposed for algorithms implemented in Python. It adopts a `scikit-learn`-like API and is designed to leverage the efficient Python numerical and scientific packages `NumPy`[a] and `SciPy`,[b] which quickly attracts a large number of

[a]https://numpy.org.

[b]https://scipy.org/.

users. Then `river`[5] merged the scikit-multiflow as well as another open-source Python library, `creme`.[6]

So far, the design and testing of most concept drift methods are conducted in a *single stream* scenario. However, practical applications often handle *massive* and *correlated* data streams. When concept drift is detected in one data stream, similar concept drift might have occurred in another correlated data stream a while ago, which means that data from these correlated data streams can reflect the new concept of the data after concept drift. But few research works leverage the information of correlations between data streams to address the concept drift problem. The `MOA` and `creme` are designed for single data stream. The `scikit-multiflow` and `rivier` include multi-output learning problems in data streams setting, ignoring correlations between multiple data streams.

Thus, we developed `ML4MDS`, a machine learning platform for multiple data streams. It permits to evaluate algorithms designed for multi-stream problems and visualize the correlations between data streams dynamically.

2. Related Work

To handle more complex learning scenarios, multiple data stream learning under concept drift attracts attention[7]. The supervised tasks for handling multiple data streams mainly include regression[8,9] and classification[10]. For unsupervised tasks, research[11] obtains ideas of transfer learning for handling multiple data streams. Further studies on this setting include Learn-to-Adapt[12] and Online Boosting Adaptive Learning[13]. The multi-output learning problem is also worth mentioning. Multi-output learning addresses the problem that the outputs of machine learning models are multiple[14]. The key point of the multi-output learning problem is to track the correlations between multiple outputs, sharing a similar idea to that of multiple data stream learning.

In recent years, various libraries have been released for data stream learning. As one of the most popular open source framework, Massive online analysis (MOA) was early designed by JAVA for data stream learning, many related machine learning algorithms are included[3]. To meet the requirements of different users, `scikit-multiflow` extends the learning packages in the Python ecosystem[15]. Next, `creme`[6] is released as a Python library for online machine learning. Interestingly, to be the most user-friendly library for doing machine learning on streaming data, `river`[16] merges the libraries of creme and scikit-multiflow.

3. Architecture of the Machine Learning Platform for Multiple Data Streams

This section illustrates the framework of the proposed multiple data stream learning platform. The architecture of `ML4MDS` can be divided into two components: back-end component and graphical user interface component. The architecture could be shown in Fig. 1.

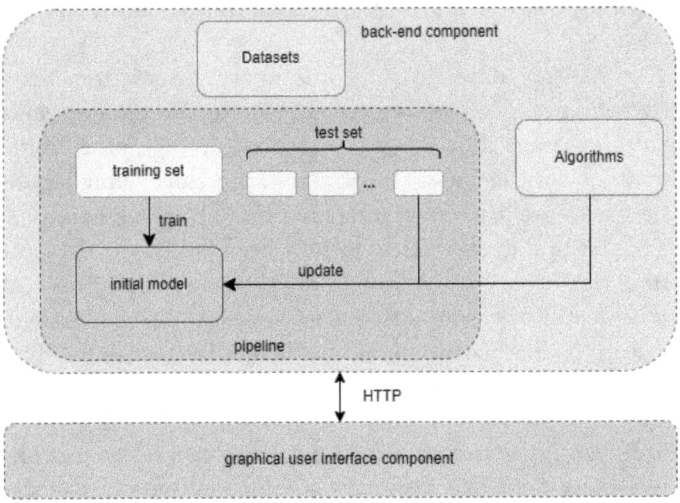

Fig. 1: The illustration of the architecture of ML4MDS.

3.1. *Back-end component*

The back-end component includes the implementation of the evaluation of algorithms. There are two functions that the back-end component would deal with:

- To load the dataset and feed the data to algorithms chunk by chunk.
- To wrap the algorithms designed for multiple data streams and provide unify APIs.

Datasets All data are stored in the form of 3-dimension ($n \times m \times d$) array after loading to memory, where n, m, d are the number of samples, the number of streams and the dimension of samples respectively. A quickly index function for a specific chunk is implemented to generate a 3-dimension

$(l \times m \times d)$ array as a chunk of data, where l is the size of the data chunk and m and d are the same with above. The `ML4MDS` has included three multiple data streams datasets:

- The Train data set[17] records train dwelling information of eight stations in New South Wales. There are eight data streams in the data set. Each data stream is corresponding to a train station. The task is to predict the load change of carriages during the train dwelling time, so that staff can organize the crowd and avoid congestion.
- The Weather data set[18] records weather data from ten weather stations near Beijing China. There are ten data streams in the data set. Each data stream corresponds to a weather station. The weather data includes temperature, pressure, humidity, etc.. The task is to predict the accumulated rainfall in one hour.
- The sensor dataset[19] records data from 54 sensors deployed in the Intel Berkeley Research Lab from February 28th to April 5th, 2004. We selected 6 sensors with few missing data as 6 data streams. The feature include the temperature, humidity and light. The task is to predict the voltage of the sensor.

Algorithms All implemented algorithms are wrapped to provide unified APIs. The design of APIs is inspired by scikit-multiflow[4]. The algorithms have four main functions:

- `fit` - train a model from scratch
- `partial_fit` - train a model incrementally
- `predict` - predict target's value
- `score` - predict target's value and calculate a specific metric

Pipeline All algorithms are evaluated in a fixed pipeline, which is shown in Fig. 1. The datasets are divided into two parts, training set and test set. The training set is used to train the initial model. The test set is sliced into multiple chunks of a fixed size. Then data are feed to model chunk by chunk, and the algorithms would update the model according to the newly arrived data chunk.

3.2. *Graphical user interface component*

As the development of web technology, more and more applications use browser as their interface as its cross-platform nature and separation of back-end and interface. The latter feature allow to deploy computation intensive applications on servers and let users access remotely by browser. These examples in machine learning areas includes Jupyter Notebook[c] and TensorBoard.[d] Thus, we follow this trend and develop the graphical user interface based on Streamlit.[e]

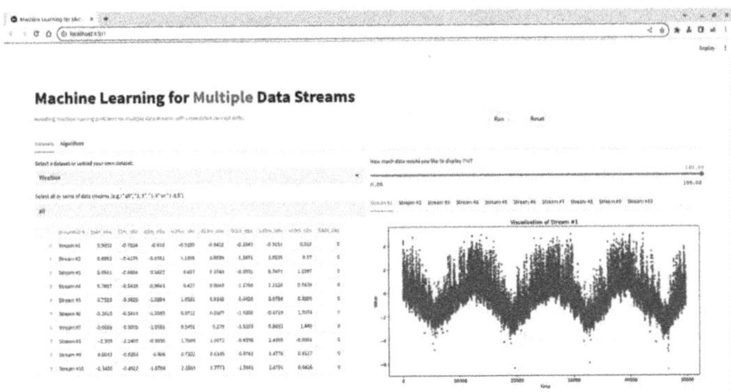

(a) Dataset tab.

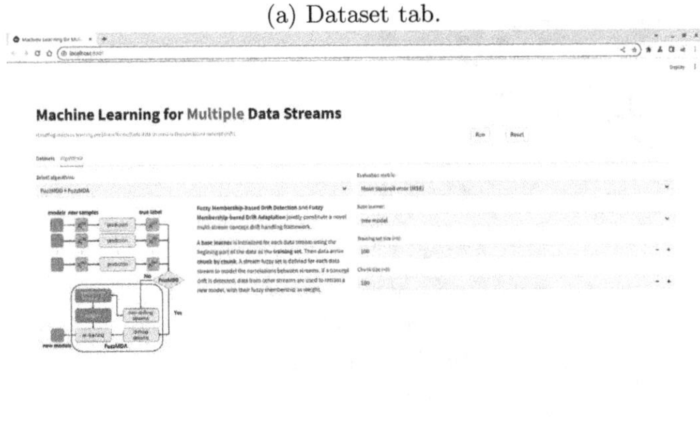

(b) Algorithm tab.

Fig. 2: A screenshot of the front page. In the first tab, a quick view of the dataset and its visualization would be given. In the second tab, a short introduction to the algorithm and its parameters are shown.

---

[c]https://jupyter.org.
[d]https://www.tensorflow.org/tensorboard.
[e]https://streamlit.io.

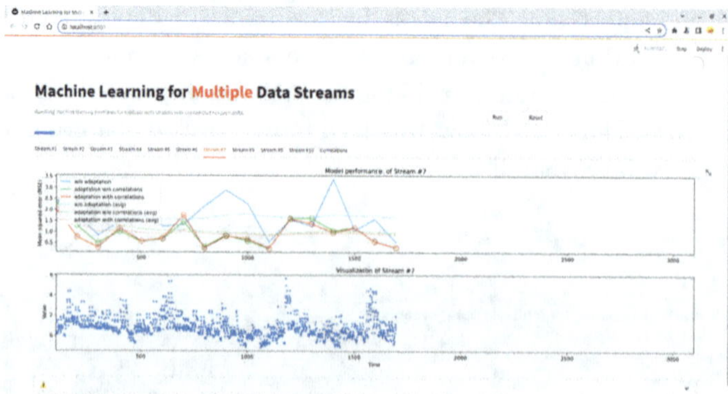

(a) Evaluation of a data stream.

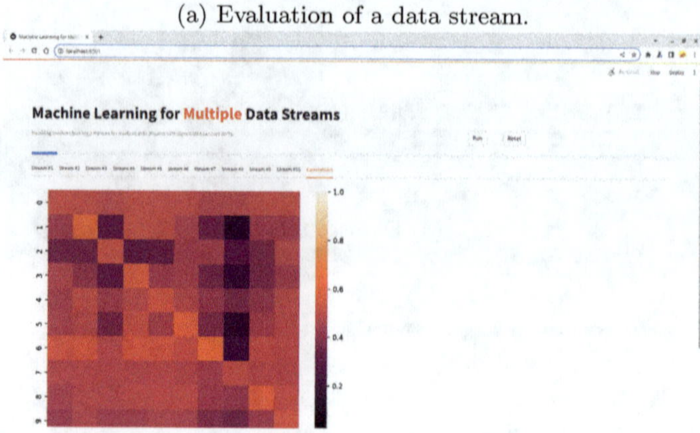

(b) Visualization of correlations.

Fig. 3: A screenshot of the evaluation page. For each data stream, the evaluation results of all algorithms are plotted in one chart. A visualization of data is also shown below to illustrate the concept drift. The correlations between data streams are visualized in a heat map.

The front page is designed for setting up. You can select which algorithms to be evaluated on which datasets. And some hyper parameters of specific algorithms can also be set up. Fig. 2 shows the interface for setting up. After clicking the `run` button, the evaluation results would be shown dynamically. In addition, we design an API to visualize the correlations between data streams. The API accepts a matrix $M \in \mathbb{R}^{m \times m}$ as input, where the element $M[i, j]$ means the correlation between the ith data stream and the jth data stream. The definition of correlations between data streams varies depending on the algorithms. The interface of evaluation results and correlations are shown in Fig. 3.

4. Conclusion

This chapter develops ML4MDS, a machine learning platform for multiple data stream learning. The algorithm designed for multiple data streams is wrapped and can be real-time tested on embedded datasets. And the real-time experiment results as well as the visualization of correlations can be shown dynamically. In our future study, we will focus on further optimization on running efficiency, visualization, and system maintenance for better usage.

Acknowledgments

This work was supported by the Australian Research Council under Grant FL190100149.

References

1. J. Gama, I. Zliobaite, A. Bifet, M. Pechenizkiy and A. Bouchachia, A survey on concept drift adaptation, *ACM Computing Surveys* **46**, 1 (2014).
2. J. Lu, A. Liu, F. Dong, F. Gu, J. Gama and G. Zhang, Learning under concept drift: A review, *IEEE Transactions on Knowledge and Data Engineering* **31**, 2346 (2019).
3. A. Bifet, G. Holmes, R. Kirkby and B. Pfahringer, MOA: Massive online analysis, *Journal of Machine Learning Research* **11**, 1601 (2010).
4. J. Montiel, J. Read, A. Bifet and T. Abdessalem, Scikit-multiflow: A multi-output streaming framework, *Journal of Machine Learning Research* **19**, 1 (2018).
5. J. Montiel, M. Halford, S. M. Mastelini, G. Bolmier, R. Sourty, R. Vaysse, A. Zouitine, H. M. Gomes, J. Read, T. Abdessalem and A. Bifet, River: machine learning for streaming data in python, *Journal of Machine Learning Research* **22**, 1 (2021).
6. M. Halford, G. Bolmier, R. Sourty, R. Vaysse and A. Zouitine, creme, a Python library for online machine learning (2019).
7. H. Yu, W. Liu, J. Lu, Y. Wen, X. Luo and G. Zhang, Detecting group concept drift from multiple data streams, *Pattern Recognition* **134**, p. 109113 (2023).

8. Y. Song, G. Zhang, H. Lu and J. Lu, A fuzzy drift correlation matrix for multiple data stream regression, in *2020 IEEE International Conference on Fuzzy Systems*, (IEEE, 2020).
9. M. Zhou, J. Lu, Y. Song and G. Zhang, Multi-stream concept drift self-adaptation using graph neural network, *IEEE Transactions on Knowledge and Data Engineering* **35**, 12828 (2023).
10. K. Wang, J. Lu, A. Liu and G. Zhang, An augmented learning approach for multiple data streams under concept drift, in *36th Australasian Joint Conference on Artificial Intelligence, AI 2023*, (Springer, 2023).
11. S. Chandra, A. Haque, L. Khan and C. Aggarwal, An adaptive framework for multistream classification, in *Proceedings of the 25th ACM International Conference on Information and Knowledge Management*, (ACM, 2016).
12. E. Yu, Y. Song, G. Zhang and J. Lu, Learn-to-adapt: Concept drift adaptation for hybrid multiple streams, *Neurocomputing* **496**, 121 (2022).
13. E. Yu, J. Lu, B. Zhang and G. Zhang, Online boosting adaptive learning under concept drift for multistream classification, *arXiv preprint arXiv:2312.10841* (2023).
14. H. Yu, J. Lu and G. Zhang, Morstreaming: A multioutput regression system for streaming data, *IEEE Transactions on Systems, Man, and Cybernetics: Systems* **52**, 4862 (2022).
15. J. Montiel, J. Read, A. Bifet and T. Abdessalem, Scikit-multiflow: A multi-output streaming framework, *The Journal of Machine Learning Research* **19**, 2915 (2018).
16. J. Montiel, M. Halford, S. M. Mastelini, G. Bolmier, R. Sourty, R. Vaysse, A. Zouitine, H. M. Gomes, J. Read, T. Abdessalem *et al.*, River: machine learning for streaming data in python, *The Journal of Machine Learning Research* **22**, 4945 (2021).
17. H. Yu, A. Liu, B. Wang, R. Li, G. Zhang and J. Lu, Real-time decision making for train carriage load prediction via multi-stream learning, in *AI 2020: Advances in Artificial Intelligence: 33rd Australasian Joint Conference*, (Springer, 2020).
18. B. Wang, J. Lu, Z. Yan, H. Luo, T. Li, Y. Zheng and G. Zhang, Deep uncertainty quantification: A machine learning approach for weather forecasting, in *Proceedings of the 25th ACM SIGKDD International Conference on Knowledge Discovery & Data Mining*, (ACM, 2019).
19. Intel Lab Data.

Paper Break Fault Recognition in Long Process Papermaking Process Based on Autoencoder

Dingding Hu[*], Guojian Chen[*], Zhenglei He[*,‡], Kim-Phuc Tran[†] and Xianyi Zeng[†]

[*]*State Key Lab of Pulp and Paper Engineering, South China University of Technology, Guangzhou, 510640, China*

[†]*Univ. Lille, ENSAIT, GEMTEX–Laboratoire de Génie et Matériaux Textiles, F-59000 Lille, France*

[‡]*hezhenglei@scut. edu.cn*

The equipment operational state detection is important in modern industry for reducing production costs and improving production efficiency. However, due to the complex and continuous nature of the papermaking process, analyzing paper break faults remains challenging. This chapter proposes a paper break fault classification and identification method based on a Stacked Autoencoder (SAE) and a Softmax classifier. A stacked autoencoder model is established and trained to extract deep features from the data, and a Softmax classifier is employed to identify paper breakage faults based on the extracted features. A case study with data collected from a real paper mill demonstrates that the classification model based on SAE feature extraction and Softmax classifier can effectively achieve paper break fault recognition.

Keywords: Softmax; fault recognition; stacked autoencoder; papermaking; process.

1. Introduction

Papermaking consists of a continuous production process composed of a variety of mechanical equipment working together to achieve the production of paper. With the growing development of the economy, the competition in the papermaking industry is becoming increasingly fierce.[1] Despite technological advances, various factors such as stock characteristics, chemical usage, and equipment conditions can lead to surface defects or failures that can impact productivity and profitability.[2] Early signs of failure are difficult to detect with traditional walk-through inspections, and as more and more equipment is integrated into the production line, there is a need to proactively predict failures.[3]

Analyzing operational data holds great promise for predicting potential failures but lacks depth in analyzing failure types and causes. Machine learning-based fault diagnosis can make a difference in reducing maintenance frequency and improving accuracy.[4] Previous studies have successfully applied unsupervised feature learning[5] and self-coding methods[6] to fault diagnosis, utilizing techniques such as multiscale feature extraction and convolutional neural networks with Softmax classifiers.[7]

In this chapter, we employ a self-encoder for unsupervised feature learning and combine it with a Softmax classifier to identify paper break faults in a complex and interconnected system of papermaking machines.

2. Self-Encoder Based Fault Classification Model

2.1. *Self-encoder network structure*

The structure of the self-encoder is shown in Fig. 1:

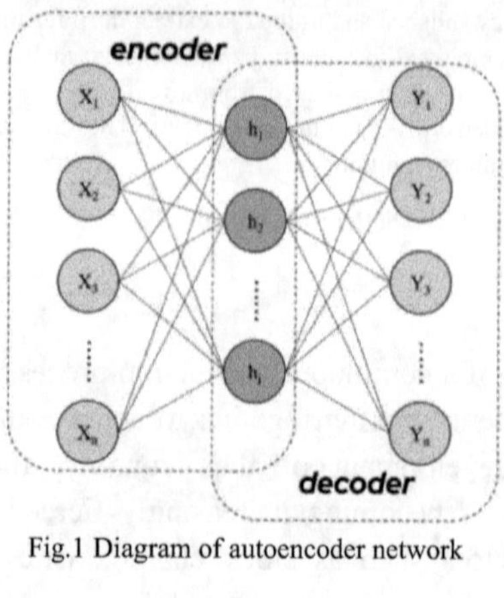

Fig.1 Diagram of autoencoder network

Self-encoder has two parts: the encoding phase and the decoding phase, in the encoding phase, the input samples are encoded, thereby obtaining the features of the hidden layer. In the decoding phase, the features of the hidden layer are decoded, and the input sequence is reconstructed. The operations in the encoding and decoding phases are as follows:

$$h = f(W_1X + b_1) \tag{1}$$

$$Y = f(W_2h + b_2) \tag{2}$$

In Eq. (1) and (2), X is the input sequence, h is the hidden layer feature, and Y is the sequence obtained by reconstruction; W_1 and b_1 are the weights and bias values for the encoding process, and W_2 and b_2 are the weights and bias values for the decoding process, respectively. The self-encoder obtains the reconstructed sequence through the encoding and decoding processes, and the model's training is achieved by minimizing the error between the input sequence and the revamped sequence.

Self-encoders are shallow neural network models with weak feature extraction capability for high-dimensional data. The stack self-encoder improves the nonlinear mapping learning ability by increasing the number of hidden layers. The number of layers of encoding and decoding layers is the same, and reasonable parameter settings can accelerate the training speed and reduce the risk of overfitting. After the training is completed, the data feature parameters are extracted through the encoding layer. Its schematic diagram with three hidden layers is shown in Fig. 2:

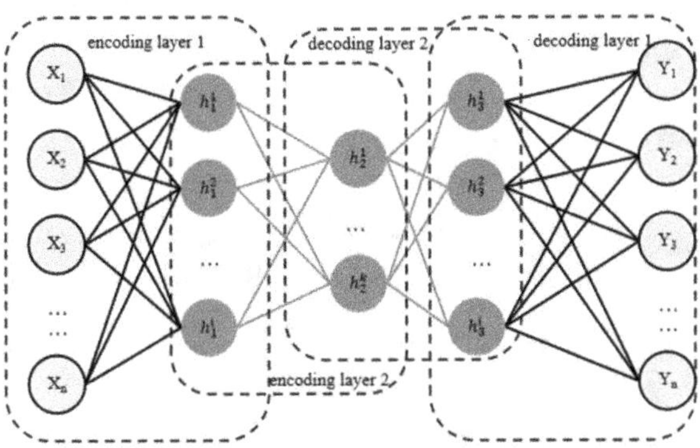

Fig.2 Diagram of stacked autoencoder network

2.2. *Softmax classifier*

Paper break fault is diverse in terms of performance and causes, and is difficult to accurately locate, increases the toughness of process diagnosis. In the multi-classification task, unlike general classifiers, the Softmax classifier outputs the probability of classification, and by exponentiating and normalizing the raw scores of each category, the scores of the categories can be mapped to a range between 0 and 1 to obtain the probability of different faults. This is denoted as[8] :

$$f(x_i) = \frac{e^{x_i}}{\sum_{n=1}^{N} e^{x_k}} \quad i = 1, 2, \& \ldots, N. \tag{3}$$

where x_i denotes the raw score of the ith category in the input, and $f(x_i)$ denotes the probability that the sample belongs to the ith category.

2.3. *Algorithm flow*

In this paper, model modeling and effect verification are carried out using actual production data of a paper mill, and different fault types are summarized and classified through historical data and expert experience. The causes of paper breaks are complex, and the same parameter change may lead to different faults, while different parameter changes can also cause the same fault. Therefore, on this basis, the self-encoder model is combined with the Softmax classifier, which retains the coding part of the self-encoder as a feature extraction module. The Softmax classifier is used as a categorization module to implement the multiclassification process based on its characteristics, which allows mapping the input data to the categories and outputting the categorization probability values with a probability sum of one. The most likely type of fault can be obtained by sorting the obtained fault category probabilities. The fault identification scheme of this study is shown in Fig.3:

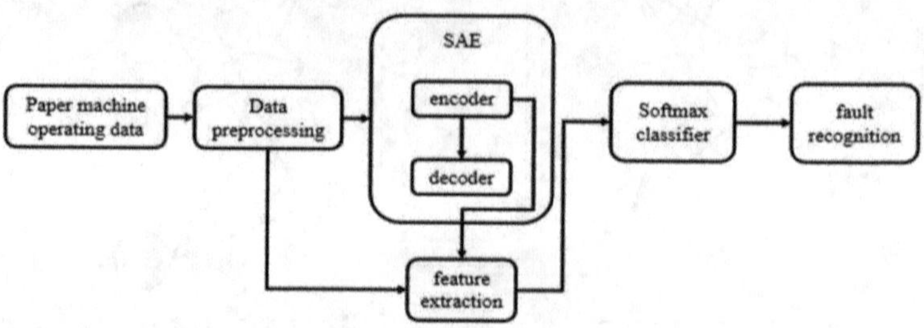

Fig.3 Flowchart of fault recognition

3. Experimental

This paper combines mechanism and data analysis to establish a fault classification model. It mainly includes mechanism analysis, data processing, modeling, and model validation. Firstly, it summarizes the paper break failure mechanism and empirical knowledge to find the relevant process parameters; then processed the feature data through correlation analysis and machine learning tools[9];the key parameters are used to establish a failure prediction model afterward; and finally, a case study is employed to verify the feasibility of the model through data.

3.1. *Data processing*

The historical data of a paper machine in a paper mill is used to verify the fault classification method in this paper. In the papermaking process, multi-dimensional condition monitoring needs to be carried out at various process locations, and directly obtained the monitoring data has a high dimensionality and variable redundancy, there may be outliers, so it is necessary to remove the outliers and normalize the data to eliminate the influence of singular samples[10]. Combining mechanism analysis and feature screening methods[11] to analyze the features, which is considered to remove redundant variables, and reduce the dimensionality of the features.

3.2. *Failure analysis*

Paper breaks in production lines concern a multitude of factors, involving the nature of the pulp, process flow, equipment condition, operation, external environment, and so on. The pulp-long fiber ratio affects the wet paper web strength, rotting pulp or pulp also affects the nature of the pulp. The grinding process with over-high a kowtow degree would reduce dehydration efficiency, and affect paper strength. Transmission system fluctuations affect transmission synchronization. Uneven distribution of moisture and dosing leads to inconsistent paper strength. Vacuum fluctuations affect paper strength. Chemicals, and filler buildup cause paper breakage. The temperature and humidity steady state of the drying section is also important, as the dew dripping may occur, causing paper breakage faults.

The occurrence of paper break faults goes through the process of condition generation to paper web break. By summarizing the causes of paper breaks to get the characteristic parameters, and carrying out the characteristic degradation and clustering analysis, the type of paper breaks and related variables can be obtained. Analyzing the fluctuation of these variables can identify the causes of paper breaks, quickly locate the fault location, and give identification results.

3.3. *Failure analysis*

In this experiment, the model evaluation metrics chosen were Accuracy, Precision, and Recall[12]:

$$\text{Accuracy} = \frac{TP + TN}{TP + TN + FP + TN} \times 100\%, \tag{4}$$

$$\text{Precision} = \frac{TP}{TP + FP} \times 100\%, \tag{5}$$

$$\text{Recall} = \frac{TP}{TP + FN} \times 100\%, \tag{6}$$

where *TP* is the number of positive samples correctly identified, *FP* is the number of false negative samples, *TN* is the number of negative samples correctly identified, and *FN* is the number of positive samples missed.

3.4. *Analysis of results*

In the operation of the paper machine, different quantities and types of paper represent different working conditions, requiring the use of different pulp raw materials and different process parameters. A tissue paper machine is taken as an example analysis, the distribution of the actual setting value and the cylinder drying ratio are shown in Fig. 4:

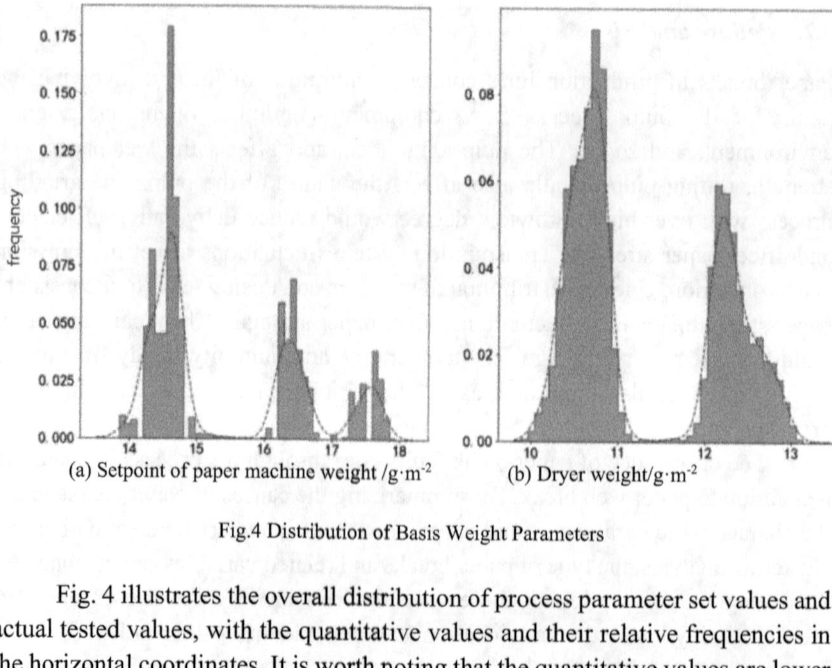

(a) Setpoint of paper machine weight /g·m^{-2} (b) Dryer weight/g·m^{-2}

Fig.4 Distribution of Basis Weight Parameters

Fig. 4 illustrates the overall distribution of process parameter set values and actual tested values, with the quantitative values and their relative frequencies in the horizontal coordinates. It is worth noting that the quantitative values are lower for household paper, with set values in the range of 13.8 g·m^2 to 17.9 g·m^2, and in the range of 9.8 g·m^2 to 13.6 g·m^2 for the upper cylinder. Both distributions tend to be multi-peaked, with three sharp peaks in the left panel and only two in the right panel, as the second peak in the right panel shows a left bias. The quantitative set value is higher than the upper cylinder quantitative because chemicals are added after the upper cylinder, making the lower volume quantitative closer to the set value. By adjusting the process parameters to keep the actual down-roll quota stable, the up-cylinder quota can be used as a reference.

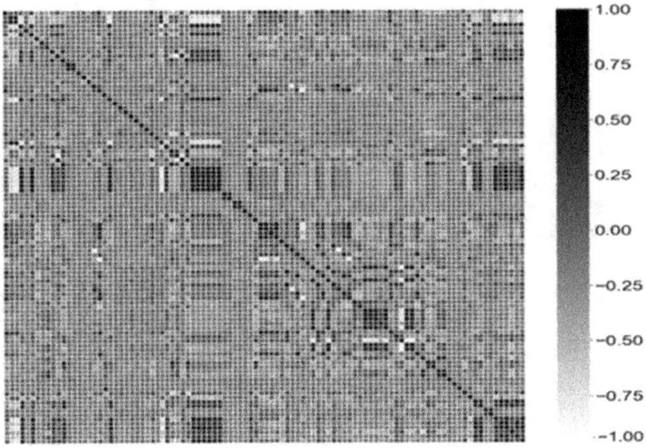

Fig.5 Result of correlation analysis

Table 1 Correlation analysis result for partial variables

correlation coefficient	Paper machine speed	Forming roll speed	Press roll speed	Fabric roll speed	Web speed
Paper machine speed	1	0.998	0.994	0.995	0.999
Forming roll speed	0.998	1	0.997	0.990	0.989
Press roll speed	0.994	0.997	1	0.999	0.999
Fabric roll speed	0.995	0.990	0.999	1	0.999
Web speed	0.999	0.989	0.999	0.999	1

The data were initially screened through the combination of mechanism analysis and data analysis, and a total of 100 dimensions were selected from the entire production process of the paper machine for correlation analysis, the results of which are shown in Fig. 5. The color in the figure reflects the strength and direction of correlation: black for strong positive correlation, grey for weak correlation, and white for strong negative correlation. From the results of the correlation analysis, it can be seen that the overall correlation between the parameter variables part of the correlation between the process variables has a strong correlation. Taking the correlation analysis of the paper machine transmission system as an example, the results are shown in Table 1. It can be seen that the correlation between the speed of the paper machine and the operating speed of the various parts of the paper machine is high.

Combined with the existing data to analyze the faults of paper breaks, the faults of paper breaks are clustered, and the results are shown in Fig. 6:

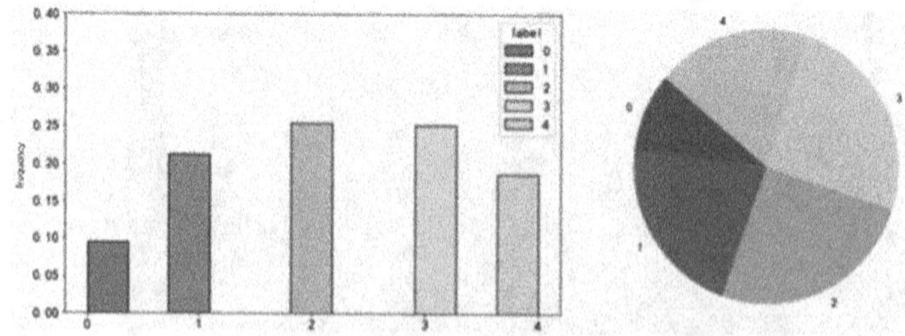

Fig.6 Types and distribution of paper break faults

Through the analysis of paper break failures, we have identified five different types of causes. The left and right graphs in Fig. 6 respectively illustrate the percentage of these fault types. The results indicate that the proportion of each type of fault is relatively balanced, with no obvious dominant type. This suggests that paper break failures are influenced by a variety of factors, highlighting the critical nature of all aspects of the production process for stability. This categorical labeling provides a foundation for further analysis of parametric variables

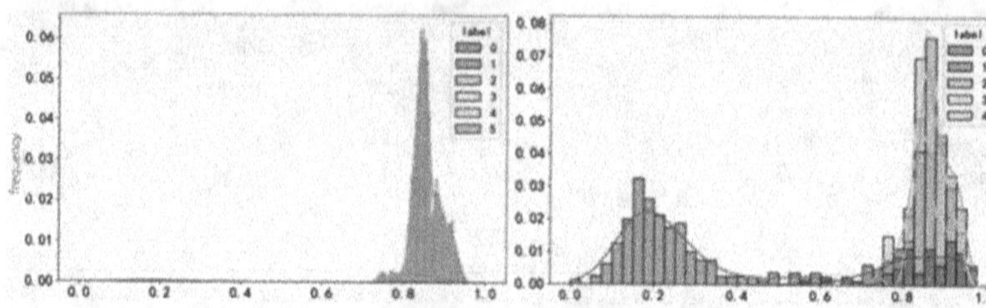

Fig.7 Distribution of wet end air and steam pressure parameters under different states

The distribution of the wet section water vapor wind pressure parameters normalized to normal and different fault types is shown in Fig. 7. The left figure shows the overall distribution, showing that the wet section water vapor wind pressure values are relatively high under normal operation. The right figure demonstrates the distribution of the parameters under different fault types. It is observed that a fault labeled as 1 may be associated with the wet section water vapor wind pressure parameter. These analyses help to understand the correlation between the parameters and the fault types, and thus infer the possible fault locations and causes.

In this study, by comparing the effects of different self-encoder network layers and feature dimensions, we finally chose the SAE model containing three hidden layers, where the hidden layer dimensions of the first, second, and third layers are 16, 5, and 16, respectively. After completing the training of the SAE model, we retained the parameter weights of the coding layer for feature extraction and input these features into the Softmax classifier in the Softmax classifier. In this study, the proposed method is compared with logistic regression, support vector machine, random forest, and Softmax model based on neural network specialization[13]. As shown in Fig. 8.

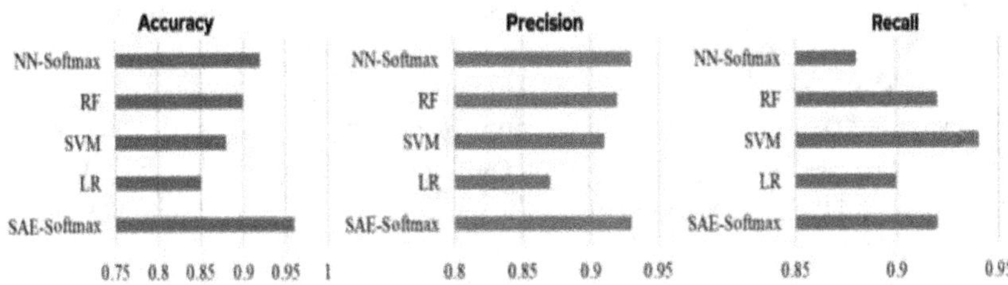

Fig.8 Comparison of different models on the test set

Combining the data from multiple evaluation metrics, the comprehensive performance of the Softmax classification method based on SAE feature extraction proposed in this paper outperforms the Softmax classification method based on neural network feature extraction with commonly used machine learning algorithms.

4. Conclusion

Paper machine break fault analysis and identification are complex and require integrated mechanisms and data analysis. In this chapter, a stack self-encoder combined with a Softmax classifier is proposed for extracting process parameter features and identifying faults. The faults are comprehensively analyzed, classified, and identified through integrated mechanism analysis and a data-driven approach. The results show that the proposed model works better than other commonly used models. Subsequent studies will analyze the causes of each failure in depth and enhance the consideration of different production conditions to improve the accuracy and generalizability of the fault diagnosis methods.

Acknowledgments

This work is supported by the Science and Technology Program of Guangzhou (202201010356, 2023A04J1367), the State Key Laboratory of Pulp and Paper Engineering (2022ZD02), and the Fundamental Research Funds for the Central Universities (2023ZYGXZR100).

References

1. Zhang S J. Research on Energy Saving and Low Carbon Development Technology Route of Paper Industry in Guangdong Province [D]. South China University of Technology, 2016.
2. Wang Y J. The Main Reasons and Protective Measures of Papermaking Machinery Equipment Fault [J]. Papermaking Equipment & Materials, 2020,49(05):8-10.1.
3. Zhang K S, Guan K K. Paper Defect Detection Method Based on Improved YOLOv5 [J]. China Pulp & Paper, 2022,41(10):79-86.
4. Li S Y. Research on Reliability and Fault Diagnosis Methods in Mechanical Equipment[J]. Mechanical Management and Development, 2022,37(02):135-137.
5. Lei Y G, Yang B, Jiang X W, et al. Applications of machine learning to machine fault diagnosis: A review and roadmap[J]. Mechanical Systems and Signal Processing, 2020,138: 106587.
6. Lei Y G, Jia F, Lin J, et al. An intelligent fault diagnosis method using unsupervised feature learning towards mechanical big data[J].IEEE Transactions on Industrial Electronics, 2016,63(5):3137-3147.
7. Sun M, Wang H, Liu P, et.al. A sparse stacked denoising autoencoder with optimized transfer learning applied to the fault diagnosis of rolling bearings[J]. Measurement: Journal of the International Measurement Confederation, 2019,146:305-314.
8. Gao F, Li B, Chen L, et al. A softmax classifier for high-precision classification of ultrasonic similar signals[J]. Ultrasonics, 2021,112: 106344.
9. Meng X L, Jiang W, Li X F, Medical diagnosis based on enhanced sparse autocoder and Softmax regression[J]. Electronic Measurement Technology, 2021,44(11):114-119.
10. Li Z, Gao X, Lu D. Correlation analysis and statistical assessment of early hydration characteristics and compressive strength for multi-composite cement paste[J]. Construction and Building Materials, 2021,310: 125260.
11. Pei X, Zhao Y H, Chen L, et al. Robustness of machine learning to color, size change, normalization, and image enhancement on micrograph datasets with large sample differences[J]. Materials & Design, 2023,232: 112086.
12. Zhou P, Wang X, Du L. Bi-level ensemble method for unsupervised feature selection[J]. Information Fusion, 2023,100: 101910.
13. Duan Z, Xu H, Wang Y, et al. Multivariate time-series classification with hierarchical variational graph pooling[J]. Neural Networks, 2022,154: 481-490.
14. Wimhurst J J, Greene J S, Koch J. Predicting commercial wind farm site suitability in the conterminous United States using a logistic regression model[J]. Applied Energy, 2023,352: 121880.

Abductive Learning for Short-term Traffic Flow Prediction of Highways

Yuanfu Guo[*,‡], Yan Yang[*,†,§] and Zhuyang Xie[*,¶]

[*] *School of Computing and Artificial Intelligence, Southwest Jiaotong University, Chengdu 610031, China*

[†] *Engineering Research Center of Sustainable Urban Intelligent Transportation, Ministry of Education, China*

[‡] *guoyuanfu@my.swjtu.edu.cn*

[§] *yyang@swjtu.edu.cn*

[¶] *zyxie@my.swjtu.edu.cn*

Accurate prediction of highway traffic flow during holidays is crucial for traffic management and control. Traditional traffic flow prediction methods mainly rely on historical data and statistical models, which are difficult to cope with the complex and changeable traffic conditions during holidays. This chapter proposes a short-term highway traffic flow prediction method based on abductive learning for holidays. The method combines machine learning and logical reasoning, learns traffic flow features from historical data, and utilizes expert knowledge to construct logical rules to constrain and optimize the prediction results. Meanwhile, semi-supervised learning is introduced to improve the generalization ability of the model by using unlabeled data. Experimental results show that the method can effectively improve the accuracy of short-term highway traffic flow prediction during holidays and provide reliable decision support for traffic management.

Keywords: Traffic flow prediction; abductive learning; logical reasoning; domain knowledge.

1. Introduction

Accurate prediction of short-term traffic flow during holidays is significantly important for alleviating traffic congestion and ensuring safety. However, the inherent unpredictability of holiday traffic poses challenges for traditional prediction methods like time series and machine learning models. Recently, abductive learning, which combines data-driven modeling and knowledge-based logical reasoning, has shown promise. This chapter proposes an abductive learning methodology to enhance holiday traffic flow

[‡]Corresponding author.

forecasts. It first leverages LSTM to learn spatial-temporal traffic patterns from historical data. Subsequently, expert traffic domain rules refine the forecasts by constraining illogical or unlikely predictions. Further accuracy gains are achieved by semi-supervised incorporation of unlabeled data. Through abductive reasoning's iterative optimization using deductive and inductive processes, the integrated model meaningfully improves prediction capabilities versus alternatives. This demonstrates abductive learning's robustness for addressing complex real-world prediction tasks like volatile holiday traffic flows.

2. Related Work

2.1. *Traffic prediction techniques*

Traffic flow prediction has been a long-standing research topic, and various prediction methods have been proposed over the past decades. Early methods were primarily based on statistical models, such as historical average models, time series models, and others [1]. These methods analyze historical data to capture the periodicity and trend features of traffic flow and utilize these features for prediction. However, these traditional methods lack the ability to model complex nonlinear patterns, making it difficult to meet the demands of modern transportation systems. In recent years, machine learning-based methods have received widespread attention in the field of traffic flow prediction. Common ma chine learning models include artificial neural networks [2], support vector machines [3], decision trees [4], and others. These models can automatically extract features from data and model complex traffic patterns, achieving better prediction performance. In particular, deep learning models have demonstrated outstanding performanc in traffic flow prediction tasks due to their powerful feature extraction and pattern recognition capabilities [5]. Commonly used deep learning models include convolutional neural networks (CNN), recurrent neural networks (RNN), and their variants, such as long short-term memory networks (LSTM). Despite the progress made by machine learning-based methods, they still face several challenges, especially in the task of short-term traffic flow prediction during holidays. The primary challenge is the lack of annotated data. Due to the dynamic changes in traffic conditions and the occurrence of anomalies during holidays, obtaining accurate annotated data is costly and difficult. Moreover, existing methods often overlook the professional knowledge and experience from the traffic management domain, which is essential for improving prediction models.

2.2. *Abductive learning*

Abductive learning, a groundbreaking paradigm in machine learning, synergizes logical reasoning with conventional data-driven methodologies to mitigate the paucity of annotated data. By harnessing background knowledge, abductive learning empowers models to derive preliminary rules from limited labeled data, forecast outcomes on unlabeled data, and iteratively refine the rules based on discrepancies between predictions and actual observations [6-8]. This self-correcting mechanism enables models to converge towards optimal performance. While abductive learning has been triumphantly applied in diverse domains such as natural language processing [6], computer vision [7], and knowledge graph construction [8], its potential in the transportation domain remains untapped. This paper pioneers the application of abductive learning to short-term traffic flow prediction during holidays, surmounting the challenges posed by limited annotated data and incorporating domain-specific knowledge to augment prediction accuracy.

3. Methodology

This section details the architecture of our proposed abductive learning framework for short-term traffic flow prediction.

3.1. *Problem formulation*

Consider a traffic network represented as a directed graph $G = (V, E)$, where V denotes the set of N sensor nodes and E denotes the set of M road segments. $X = \{x_1, x_2, \ldots, x_t\}$ denote the multivariate time-series traffic flow measurements collected from the sensor network over T discrete time steps, where $x_t \in \mathbb{R}^{N \times D}$ contains measurements from N nodes, each as a D-dimensional feature vector. The traffic forecasting problem involves using observed historical sensor data $X_{1:t} = \{x_1, x_2, \ldots, x_t\}$ to predict future traffic state $X_{(t+1):(t+h)}$ over a forecast horizon h, formulated as:

$$X_{(t+1):(t+h)} = f\left(X_{1:t}\right) \tag{1}$$

where $f(X_{1:t})$ is a predictive function to be learned.

Recent deep learning models using CNN, RNN, LSTM, Graph Networks etc. have shown promise by automatically learning feature representations from traffic data. For example, a LSTM network processes sequential data as:

$$h_t = \text{LSTM}(h_{t-1}, x_t) \tag{2}$$

where h_t is the hidden state at time t, encapsulating past context. Graph networks operate on road topology and are able to maintain spatial structure. However, a limitation of data driven approaches is they do not incorporate domain knowledge and theory which can improve generalization.

3.2. *Framework*

Fig.1 shows the overall framework of the proposed prediction method. The framework consists of the main modules:

Factor Identification Model: Identifies key factors influencing traffic flow, categorized into shallow normalized factors (e.g., holidays) and deep semantic factors (e.g., adverse weather). Uses regular methods for shallow factors and multi-label classification for deep semantic factors.

Domain Knowledge Base: Contains traffic rules, matching rules, common sense, and learnable parameters to support traffic flow predictions. Used to narrow down the prediction search space and improve efficiency.

Abductive Reasoning: Employs machine learning classification methods to predict unlabeled traffic data and generate pseudo-labels. Predicts future traffic flow using knowledge base rules, pseudo-labels, and historical flow data.

Abductive Correction: Compares predicted and actual flow, triggering correction if significant differences exist. Generates hypotheses about pseudo-labels, filters valid ones using knowledge base rules, and selects the most accurate hypothesis. Corrected labels are returned to the classifier for retraining.

3.3. *Workflow*

Construct a comprehensive model that includes influencing factor identification, flow prediction, and logical abduction. The specific steps are as follows:

(1) Feed the traffic flow data back to the influencing factor identification model to identify the key factors that affect traffic flow.
(2) Calculate the basic flow prediction value based on historical traffic flow data, and adjust the basic prediction value according to the identified key influencing factors to obtain the final predicted flow.
(3) Since the labeled data in the transportation field may also be insufficient or erroneous, the supervised prediction results may not be accurate enough. Therefore, we can introduce unlabeled data, compare

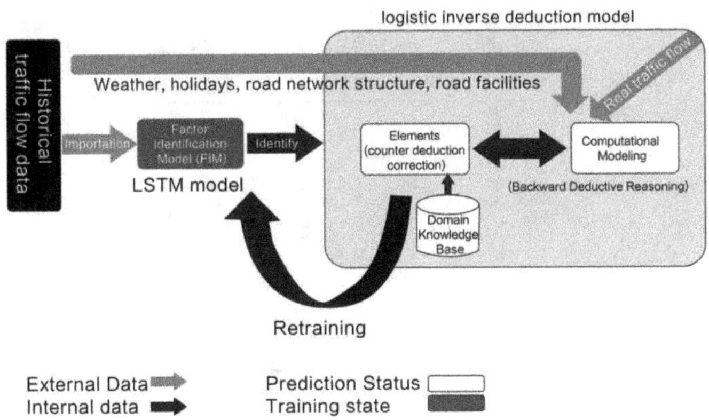

Fig. 1. Abductive Traffic Prediction Framework

the model's predicted flow with the actual flow, and if a significant difference is found, it indicates that the identified key influencing factors may be incorrect.

4. Experiments

The study utilized three traffic datasets, namely PeMSD7, PeMS04, and METR-LA, which contain multivariate sensor measurements structured as 3D tensors. To evaluate the performance of the proposed spatio-temporal model, several state-of-the-art baseline models were implemented for comparison, including HA, ARIMA, FC-LSTM, GCN, DCRNN, and STGCN.

PeMSD7: A traffic flow dataset collected by the California Department of Transportation (Caltrans) Performance Measurement System (PeMS). The data is sampled at 5-minute intervals and records information such as vehicle flow, speed, and occupancy for each detector station. This dataset is commonly used for research on traffic flow prediction and anomaly detection.

PeMS04: The PeMS04 dataset provides rich traffic flow information that can be used for research in traffic flow prediction, congestion analysis, and transportation planning.

METR-LA: A traffic speed dataset collected by the Los Angeles County transportation department. The METR-LA dataset is commonly used for research on traffic speed prediction and transportation network analysis.

4.1. *Performance results*

Table 1. Performance on PeMSD4 dataset.

Model	RMSE	MAPE
HA	21.12	12.45
ARIMA	18.92	10.23
FC-LSTM	16.23	8.92
GCN	15.12	7.86
DCRNN	14.56	7.45
STGCN	13.92	6.95
Ours	**12.38**	**6.12**

Table 2. Performance on PeMSD4 dataset.

Model	RMSE	MAPE
HA	29.36	14.56
ARIMA	26.74	12.38
FC-LSTM	23.45	10.26
GCN	21.53	9.35
DCRNN	20.36	8.95
STGCN	19.82	8.56
Ours	**17.49**	**7.69**

Table 3. Performance on METR-LA dataset.

Model	RMSE	MAPE
HA	38.64	16.48
ARIMA	34.92	14.86
FC-LSTM	31.54	13.59
GCN	29.46	12.67
DCRNN	27.81	11.95
STGCN	26.35	11.26
Ours	**23.84**	**10.15**

Tables 1–3 present the RMSE and MAPE values of the Abductive model compared to baselines on the PeMSD7, PeMS04, and METR-LA datasets, respectively, across different prediction horizons. The Abductive model

achieves the best performance on all datasets and prediction horizons, demonstrating its superiority in capturing the spatial-temporal dynamics of traffic flow. Compared to other models, the RMSE and MAPE metrics of the Abductive model are reduced by an average of 10%-20%, which can be attributed to the introduction of the knowledge base and abductive reasoning module in the model, enabling it to fully utilize domain knowledge and data features for continuous self-correction and optimization.

4.2. *Component contributions to 1-hour traffic prediction accuracy*

Ablation experiments on the PeMSD7 dataset, presented in Table 4, showed that disabling any component of our architecture degraded accuracy metrics for 1-hour prediction horizon, highlighting their individual contributions. The knowledge base provided contextual domain intelligence, the spatio-temporal model learned data-specific correlations, the correction module allowed evolving the knowledge, and their combination provided robustness.

Table 4. Data for the ablation analysis.

Excluded Component	RMSE	MAPE
Full Model	**23.56**	**11.23**
Spatio-Temporal Model	26.15	12.62
Knowledge Base	25.36	11.96
Correction Modu	24.92	11.56

Table 5. Data for the deductive revision stage.

Learned Rule	Prediction	Actual
May13	9% lower	31% higher
MayDay	14% lower	28% higher
The Duanwu Festival	5% lower	18% higher

We further analyzed the contributions of each component of the abductive learning model. As shown in Table 4, experiments were conducted on the PeMSD7 dataset for a 1-hour prediction time span. Removing any component led to a performance degradation, indicating that each component played an important role in the prediction. Specifically, the knowledge base provided contextual domain information for the model, the spatio-temporal model learned data-specific correlations, the correction module allowed knowledge evolution, and their combination provided robustness.

4.3. *Qualitative examples*

Table 5 shows sample logical rules automatically learned and added to the knowledge base for holiday weekends based on predictions on the PeMSD7 dataset and their comparisons with ground truth data. These rules, learned by the correction module from inconsistencies between predictions and actual observations, demonstrate how the knowledge base is augmented with data-specific knowledge.

We demonstrated the logical rules for holidays that the abductive learning model automatically learned and added to the knowledge base on the PeMSD7 dataset through qualitative examples. As shown in Table 5, these rules were learned by the correction module from inconsistencies between predictions and actual observations, showing how the knowledge base is expanded with data-specific knowledge. For example, the model predicted a 14% decrease in traffic flow during the "MayDay" holiday, while the actual observation was a 28% increase. Through the correction module, the model learned the rule that traffic flow usually increases during the "May Day" holiday and added it to the knowledge base for optimizing subsequent predictions.

5. Conclusion

This chapter proposes an abductive learning framework integrating machine learning and logical reasoning for traffic forecasting. The architecture combines a spatio-temporal neural network. Experiments demonstrated significant accuracy gains over state-of-the-art methods. Key contributions include formulating traffic forecasting as spatio-temporal sequence prediction and enabling inductive–deductive reasoning synergy.

References

1. J. Zhang, Y. Zheng, D. Qi, DL-Transformer: Attention-based deep learning model for traffic flow prediction, IEEE Trans. Intell. Transp. Syst., vol. 23, no. 4, pp. 3399-3410, Apr. 2021.
2. S. Yin, S. Li, and M. Karim, Traffic Flow Prediction with Graph Neural Networks: A Survey, IEEE Trans. Intell. Transp. Syst., 2022, doi: 10.1109/TITS.2022.3214954.

3. P. Duan, J. Kang, X. He, Convolutional sequence-to-sequence model for highway traffic flow prediction, IEEE Trans. Intell. Transp. Syst., vol. 23, no. 2, pp. 1146-1157, Feb. 2021.
4. Z. Cui, R. Ke, Z. Pu, Stacked convolutional LSTM network for urban short-term traffic flow prediction, IET Intell. Transp. Syst., vol. 15, no. 9, pp. 1265-1274, Sep. 2021.
5. W. Luo, C. Xu, H. Ren, Machine learning for traffic flow prediction: A review and recent advances, Transp. Res. Part C Emerg. Technol., vol. 135, p. 103521, Feb. 2022.
6. L. Xie, W. Xing, M. Hu, Incorporating domain knowledge for text-to-SQL with abductive reasoning, arXiv preprint arXiv:2211.09607, 2022.
7. S. Li, N. Corso, G. Yang, Abductive reasoning for explanations in medical computer vision, Pattern Recognit., vol. 119, p. 108082, Nov. 2021.
8. Y. Zhao, H. Mao, C. Zhang, Abductive reasoning for neural program repair, in Proc. 45th Int. Conf. Softw. Eng., 2023, pp. 1-13.

Neural Network-based Fault Detection in Floating Wind Turbines

G. Andrade[*,‡], S. Esteban[*,§] and M. Santos[†,¶]

[*] *Department of Computer Architecture and Automatic Control, Faculty of Physics, Universidad Complutense de Madrid,*

[†] *Institute of Knowledge Engineering, Universidad Complutense de Madrid*

[‡] *giordyan@ucm.es*

[§] *sesteban@ucm.es*

[¶] *msantos@ucm.es*

It is imperative to tackle the issue of intermittent power generation in wind energy, particularly for floating devices that have more stringent environmental requirements. Floating turbine failures or malfunctions can be caused by the aggressiveness of the waves, so knowledge of the cause of the failure is particularly important to minimise both its frequency and severity. This chapter deals with a 5 MW offshore floating wind turbine and simulates breakdowns of several structural sections using NREL OpenFAST software. Neural networks are used to identify these anomalies, as well as their distinctive response, which would enable a quick and accurate failure diagnosis.

Keywords: Fault detection; neural networks; floating offshore wind turbines.

1. Introduction

Wind energy is firmly on the list of renewables in growing demand, with floating technology finding a clear avenue for both technological and economic expansion.[1] Access to stronger and more stable winds,[2] coupled with flexibility of location and reduced environmental restrictions,[3] which together allow for the construction of large and powerful wind turbines,[4] make floating wind an ambitious but attractive option for the future of renewable energy.[5] On the other hand, operating and maintaining the machines, both due to their size and complex environmental operating conditions, becomes one of the main challenges of this technology.

In fact, wind turbines (WT) can experience failures that affect production doubly, on the one hand, because of the cost of repair and, on the other hand, because of the economic loss due to machine downtime. In addition,

[‡]Corresponding author.

the larger size of floating offshore wind turbines (FOWT) compared to their onshore counterparts increases the frequency and severity of structural failures.[6]

At the same time, the literature on the study of structural failures in floating wind turbines, such as rotor blades and mooring lines, is scarce. The published works in the area focus mainly on failures in the pitch control, the gearbox, the bearings, and the electrical system.[7–9] The use of neural networks for fault detection in wind turbines has been explored in[10] and[11], while its applications in offshore scenarios are few, outstanding the work of[12], which focuses on detection and monitoring of gearbox faults.

This chapter presents a study of three potential serious failures in an offshore wind turbine: a rupture of one of the blades, a breakage of the anchoring system to the seabed, and a failure of the pitch actuator of one of the blades, simulated with NREL OpenFAST software, which is a good option due to the lack of available real data.[13]

The ways of detecting these anomalies are analysed, as well as the perturbations with respect to the normal behaviour that would be observed in the degrees of freedom, namely in the surge and yaw of the platform, in the oscillations in the direction fore-aft of the nacelle, in the rotor rotation speed and in the angle of pitch of each of the blades determined by the control.[14] We apply a multilayer perceptron (MLP) neural network (NN) to identify these faults.

This paper is organised as follows. Section 2 explains the simulation of the faulty scenarios. Section 3 describes the results of the application of the NN to fault detection. The paper ends with the conclusions an future works.

2. Simulation of FOWT Faults

There is currently no benchmark that includes rotor blade or mooring lines failure scenarios. To overcome this, in this paper a novel methodology is used to generate them synthetically, altering the structural properties of the wind turbine to simulate each of the failures. The simulation starts in nominal mode and then the failure is generated continuing the simulation with the faulty machine, splicing the final conditions of the first simulation with the initial conditions of the second.

In the present work, the NREL 5-MW floating Wind Turbine[15] with a ITI barge-type foundation is used[16]. Implemented in OpenFAST, the environmental conditions chosen were calm seas and uniform longitudinal wind of 18 m/s, around the nominal power production values of the wind turbine.

Initially, the system produces power at its nominal value, with a collective pitch control of variable speed, and no torque control. In order to obtain the correct identification of wind turbine faults, it has been decided to work in the third region of operation of a wind turbine (pitch region, nominal wind). For this reason, torque control is not present in the simulations carried out. The complete simulation consists of 1000 seconds for each signal, and the faults are implemented at 500 second.

The three failures have been simulated:

- Blade breakage is recreated by drastically decreasing the blade mass from the middle of the blade to the tip. To prevent convergence problems, defining the blade mass as zero is avoided.
- The disengagement of the anchor line is recreated by elongation of the mooring line of which the malfunctioning is to be analysed.
- The fault in the pitch actuator is recreated by inserting a white noise signal to the pitch angle reference signal. Therefore, the controller erroneously makes corrections to the blade angle, producing instability in the system.

2.1. *Faulty scenarios simulation*

As previously mentioned, the neural network uses training data obtained from simulations performed with the OpenFast software. This is why the neural network does not work with real-time data, as OpenFast does not allow access to the wind turbine information during the simulation. Once the simulations have been carried out, the signals are filtered with a moving average filter. On the other hand, the signals have been normalised according to the specific range of each wind turbine output variable to prevent over-fitting problems during the training of the neural network.

The seven signals mentioned have been analysed, but for the sake of brevity we will analyse three that are of interest: the rotor revolutions per minute, the pitch angle of one of the blades, and oscillations in the fore-aft direction of the nacelle (Figure 1). Those faults have been analysed and the following conclusions can be drawn:

- The blade breakage leads to a precession movement in the rotor, which is observed in a periodic envelope in the pitch signal that alters the rotational speed at a high frequency. As a result of this

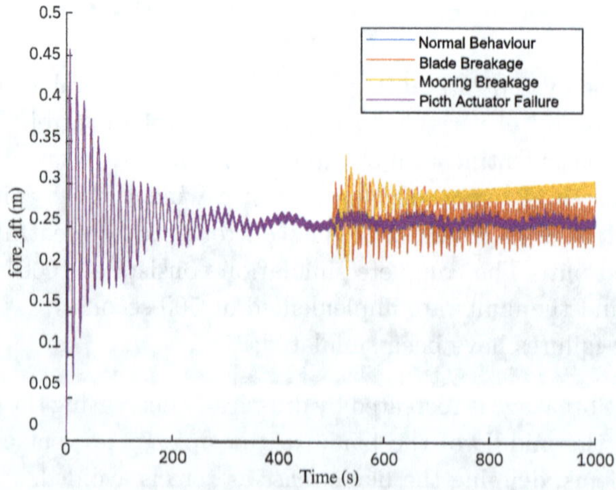

Fig. 1. Comparison of the nominal behaviour of the nacelle oscillations in the fore-aft direction against different faults.

decompensation, the nacelle oscillations increase symmetrically to the nominal ones.

- Breaking the mooring lines means moving away from the wind source, so the pitch actuator compensates appropriately by decreasing the pitch to maintain rotor speed. On the other hand, the nacelle oscillations increase progressively due to the lack of restoring force to dampen them from the base.
- The failure in the pitch actuator leads, as expected, to the non-stabilisation of the blade pitch actuators, even if that blade is undamaged, generating increased fore-aft oscillations. As for the rotor speed, it generates an extra fast frequency component on top of the slow oscillation of nominal behaviour.

3. MLP Neural Network Fault Detection and Results

A neural network can be defined as a set of interconnected nodes whose purpose is to process the information that arrives at its input and send it through its output connections[17]. There are rules governing the dynamics of sending information between the units, also known as weights, which allow learning or adaptation to be provided to the system[18].

A MLP neural network with fifteen neurons in the hidden layer is designed with Matlab software . As inputs to the network, the aforementioned seven outputs of the floating wind turbine have been considered.

For each fault, three examples have been incorporated by changing the time of occurrence of the fault. A total of 160004 samples have been used to train the neural network. The samples are split into training, validation and test sets in a ratio of 70/15/15. So, 112003 samples were used for training, 24001 for validation and 24000 for testing.

The activation function used in the hidden layer of the neural network is the hyperbolic tangent function `tansig` which output is in the range -1 to 1. The activation function used in the output layer is the linear function `purelin` so the output can be any real value.

The performance of the neural network has been tested for each of the examples provided by the simulation. It should be remembered that the main motivation for the development and design of the neural network is the real-time identification of the state of a floating wind turbine.

Figure 2 depicts the outputs of the neural network for each known failure case and for normal behaviour over the simulation time. The following state identification is established for each of the turbine conditions: For a fault-free behaviour, a wind turbine state is equal to state 1, while for blade breakage, mooring breakage and pitch actuator failure the state is 2, 3 and 4, respectively.

For better visualisation, 500 seconds of simulation time has been adjusted to the time at which the faults occur (dashed line), in order to be able to compare the NN output for each of the faults produced.

A correct identification of each of the faults is observed in the figure, starting from a normal operating state that is maintained until reaching the time instant where the faults have been caused in the wind turbine. It can be clearly seen how the designed neural network is able to identify the type of failure that has occurred almost instantly. It is noticeable, however, that the detection of the mooring failure is slower than the other failures. This delay in detection is evident, as the consequences of mooring failure (such as surge and yaw of the platform) do not manifest themselves as quickly and with the same intensity as other types of failures. Therefore, additional time is needed for these effects to become apparent and be identified.

However, it can be observed how the signals differ at the initial instant even if no fault occurs. This is due to the fact that although the training data have been generated from simulations with a uniform wind, simulations of the wind turbine with a random turbulent wind have been used for the verification of the neural network. At the same time, it can be seen that the network maintains the normal operating value considerably well in the initial state of each of the tests.

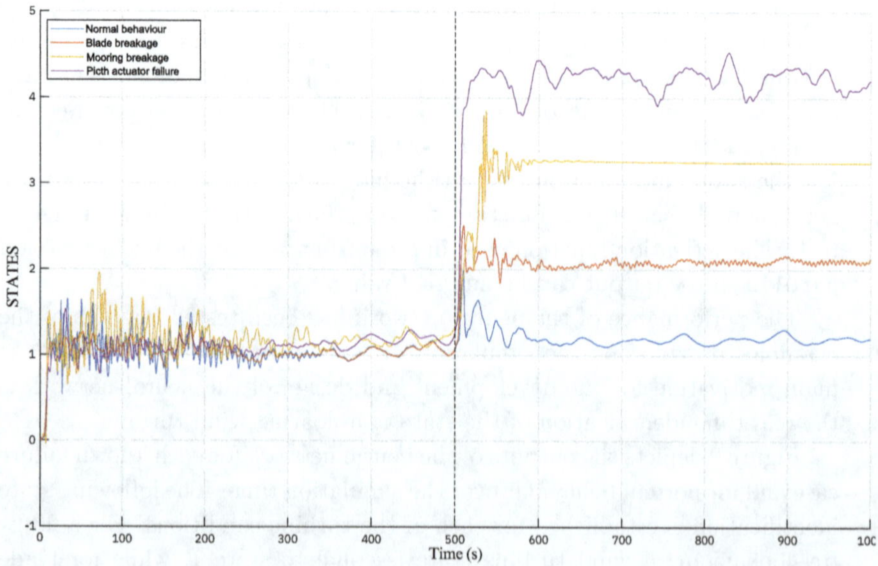

Fig. 2. Output response of the neural network for each of the cases. The dotted line indicates the time at which failures occur.

The network is able to return a signal very close to the desired one, achieving a very effective real-time identification of the wind turbine status.

Similarly, a change in the dynamics of the normal behaviour of the floating wind turbine can be seen at the intersection of the two simulations. This is due to the fact that the OpenFast software does not include by default the possibility of modifying the wind turbine parameters in the middle of the simulation.

4. Conclusions and Future work

A novel technique developed for the intelligent and efficient detection of faults in simulated floating wind turbine models has been demonstrated throughout this chapter. This research shows the benefits of the use of neural networks for the detection of faults in offshore wind turbines. Results are promising.

The difficulty yields in the lack of data on FOWT failures, especially structural. This may be due to the youth of the technology and the lack of available public data.[19] Besides, wind turbine simulation software does not possess the ability to faithfully recreate these failure scenarios, so a compromise needs to be made with the available tools.

As future works, it would be interesting to incorporate cases in which the other operating regions of the wind turbine are present, as well as torque control, into the fault detection. It is also possible to test the effectiveness of other types of neural networks.

Acknowledgments

This work has been partially financed by the Spanish Ministry of Science and Innovation in the framework of the project MCI/AEI/FEDER number PID2021-123543OB-C21.

References

1. M. Tomás-Rodríguez and M. Santos, Modelling and control of floating offshore wind turbines, *Revista Iberoamericana de Automática e Informática Industrial* **16** (2019).
2. J. McMorland, M. Collu, D. McMillan and J. Carroll, Operation and maintenance for floating wind turbines: A review, *Renewable and Sustainable Energy Reviews* **163**, p. 112499 (2022).
3. B. Snyder and M. J. Kaiser, Ecological and economic cost-benefit analysis of offshore wind energy, *Renewable Energy* **34**, 1567 (2009).
4. L. Li, Y. Liu, Z. Yuan and Y. Gao, Wind field effect on the power generation and aerodynamic performance of offshore floating wind turbines, *Energy* **157**, 379 (2018).
5. I. G. García, S. E. San Román and M. S. Peñas, Identificación preliminar de la dinámica de la barcaza de una turbina flotante, 25 (2023).
6. S. Faulstich, B. Hahn and P. J. Tavner, Wind turbine downtime and its importance for offshore deployment, *Wind Energy* **14**, 327 (2011).
7. H. Li, H. Díaz and C. G. Soares, A failure analysis of floating offshore wind turbines using ahp-fmea methodology, *Ocean Engineering* **234**, p. 109261 (2021).
8. S. Cho, Z. Gao and T. Moan, Model-based fault detection, fault isolation and fault-tolerant control of a blade pitch system in floating wind turbines, *Renewable energy* **120**, 306 (2018).
9. A. R. Nejad, P. F. Odgaard and T. Moan, Conceptual study of a gearbox fault detection method applied on a 5-mw spar-type floating wind turbine, *Wind Energy* **21**, 1064 (2018).
10. S. Cho, M. Choi, Z. Gao and T. Moan, Fault detection and diagnosis of a blade pitch system in a floating wind turbine based on kalman filters and artificial neural networks, *Renewable Energy* **169**, 1 (2021).
11. B. Yang, A. Cai and W. Lin, Analysis of early fault vibration detection and analysis of offshore wind power transmission based on deep neural network, *Connection Science* **34**, 1005 (2022).

12. B. W. Shaheen and I. Németh, Performance monitoring of wind turbines gearbox utilising artificial neural networks—steps toward successful implementation of predictive maintenance strategy, *Processes* **11**, p. 269 (2023).
13. N. R. E. Lab, Openfast: A multi-physics, multi-fidelity tool for simulating the coupled dynamic response of wind turbines (2021).
14. J. Kang, L. Sun and C. G. Soares, Fault tree analysis of floating offshore wind turbines, *Renewable energy* **133**, 1455 (2019).
15. J. Jonkman, S. Butterfield, W. Musial and G. Scott, *Definition of a 5-MW reference wind turbine for offshore system development*, tech. rep., National Renewable Energy Lab.(NREL), Golden, CO (United States) (2009).
16. J. M. Jonkman, *Dynamics modeling and loads analysis of an offshore floating wind turbine* (University of Colorado at Boulder, 2007).
17. O. I. Abiodun, A. Jantan, A. E. Omolara, K. V. Dada, A. M. Umar, O. U. Linus, H. Arshad, A. A. Kazaure, U. Gana and M. U. Kiru, Comprehensive review of artificial neural network applications to pattern recognition, *IEEE access* **7**, 158820 (2019).
18. M. Schlechtingen and I. F. Santos, Comparative analysis of neural network and regression based condition monitoring approaches for wind turbine fault detection, *Mechanical systems and signal processing* **25**, 1849 (2011).
19. R. Pandit, D. Astolfi, J. Hong, D. Infield and M. Santos, Scada data for wind turbine data-driven condition/performance monitoring: A review on state-of-art, challenges and future trends, *Wind Engineering* **47** (2023).

Solving the Shortest Path Problem with Intervals as Costs Through Aggregation Functions

Pelayo S. Dosantos*, Agustina Bouchet, Irene Mariñas-Collado and Susana Montes

Department of Statistics and Operation Research and Mathematics Didactics, 33007 Oviedo, Spain

**suarezdpelayo@uniovi.es*

The interval shortest path problem emerges as a variant of the shortest path problem when cost uncertainties are present, introducing a range of possible values instead of fixed costs. This variation of the problem allows an appropriate adjustment to real-world situations. However, it is important to recognize that operating with intervals is not trivial. To solve the interval shortest path problem, this chapter proposes a variation of Dijkstra's algorithm that includes the use of either averaging functions or admissible orders. These techniques, along with the usual interval arithmetic, provide good solutions to the problem and can be adapted according to the context to best deal with the corresponding imprecision.

Keywords: Interval shortest path problem; averaging functions; aggregation functions; admissible orders.

1. Introduction

The Shortest Path Problem (SPP) is one of the most studied optimization problems due to its versatility in several situations. It is commonly applied in logistics, route planning, electronic circuit design and social network analysis, among others (see, for example, Refs. 1 and 2). The mathematical models that solve it in each context take into account several factors related to their corresponding situations. However, the inherent imprecision in real-world data is often overlooked. This uncertainty can be integrated directly into the edge costs.

In addressing imprecise edge costs within the SPP, various methods have emerged. While fuzzy numbers are one approach, intervals represent another common strategy for handling this uncertainty. This chapter concentrates on intervals, i.e., the SPP with interval costs or Interval SPP (ISPP), due to their clear interpretability. Following are some of the approaches found in the literature to solve this problem.

In Ref. 3, the robust shortest path is chosen as the optimal solution. This is the path that minimizes the difference in terms of cost with the shortest path in the worst case scenario, where all the edges have their minimum value except for the ones in the selected path, which take their maximum. In Ref. 4, a process based on fuzzy logic is used to compare intervals. The sorting method is used at every step of a dynamic algorithm to solve the problem. In Ref. 5, an acceptability index is built to quantify the sentence "an interval is inferior to another". The main foundation of the index is based on the midpoints of the intervals, and in case of coincidence, the choice is left to a decision-maker (DM) who selects preferences based on their degree of pessimism. A variation of Dijkstra's algorithm is presented in Ref. 6 to solve the ISPP using the acceptability index and compare the performance with random instances of the networks obtaining similar results.

As it can be seen, all the approaches have something in common. It is required to define a way to compare intervals. This can be done either by summarizing the information into one point and making the comparison based on the typical ordering in the real numbers, or by making a multiple step process to decide which is the lowest. This chapter proposes a variation of Dijkstra's algorithm to include the comparison between intervals through averaging functions or admissible orders.

The rest of this chapter is organized as follows. Section 2 presents the preliminary interval notions used in this chapter. Section 3 details the proposed approach to solve the ISPP. In Section 4, the methodology is applied to a case study and, in Section 5, some conclusions are drawn.

2. Preliminaries

This section covers the definitions, concepts and notations necessary in the approach to solve the ISPP proposed in this work. Without loss of generality, the costs of the edges of the considered networks will be restricted to the interval $[0, 1]$ since by an easy standardization, all the cases can be covered with this subcase.

The set of the closed intervals contained in $[0, 1]$ is denoted as $L([0, 1])$. That is, $[a_1, a_2] \in L([0, 1]) \iff 0 \leq a_1 \leq a_2 \leq 1$.

The sum of two intervals $[a_1, a_2], [b_1, b_2]$ is defined as

$$[a_1, a_2] \oplus [b_1, b_2] = [a_1 + b_1, a_2 + b_2].$$

Although only intervals in $L([0, 1])$ are considered, it is not guaranteed that the sum of any two intervals is also in $L([0, 1])$. To ensure this, all the intervals in the considered ISPP are divided by the sum of the upper bounds of every cost in the graph of the problem. That is, given the set of intervals in $L([0,1])$, $P = \{[a_{11}, a_{12}], \dots, [a_{n1}, a_{n2}]\}$, that define the costs of the edges in a graph and calling $S = a_{12} + \dots + a_{n2}$, the sum of any two intervals in the set is computed as follows:

$$[a_{i1}, a_{i2}] \oplus_P [a_{j1}, a_{j2}] = \left[\frac{a_{i1} + a_{j1}}{S}, \frac{a_{i2} + a_{j2}}{S}\right] \quad \forall i, j \in \{1, \dots, n\}$$

Notice that this is equivalent to divide each interval by S (the bounds get divided by S) and compute the sum using the usual $\oplus$ operator.

Observe that if we call $K([0,1]) = \{(x,y) \in [0,1]^2 : x \leq y\}$, there is a natural bijection between that set and $L([0,1])$. This allows the application of the techniques designed to work on $[0,1]^2$, such as the usual partial order in $\mathbb{R}^2$, for the task of sorting intervals. Recall that given a non-empty set S, a partial order $\preceq$ is a binary relation satisfying reflexivity, antisymmetry and transitivity. For example, in $L([0,1])$, the relation $\preceq_{LO}$ given by

$$[a_1, a_2] \preceq_{LO} [b_1, b_2] \iff a_1 \leq b_1 \ \wedge \ a_2 \leq b_2$$

is a partial order. This relation is known as the *lattice order.*

The issue that arises with partial orders is that, given any two elements from a set, they may not be comparable. This desirable property is satisfied by total orders (Ref 7).

Definition 2.1. A partial order $\preceq$ on $L([0,1])$ is called an admissible order if it is a total order on $L([0,1])$ and it refines the lattice order on $L([0,1])$

In Ref. 7, it is established how to create admissible orders from aggregation functions.

Definition 2.2. Let $n>1$, an aggregation function of n arguments on $[0,1]^n$ is a mapping $f : [0,1]^n \to [0,1]$, satisfying the following properties:

(1) (bound preservation) $f(0,\dots,0) = 0 \ \wedge \ f(1,\dots,1) = 1$.
(2) (monotonicity) if $\mathbf{x} \leq \mathbf{y} \Rightarrow f(\mathbf{x}) \leq f(\mathbf{y}) \ \forall \mathbf{x}, \mathbf{y} \in [0,1]^n$.

where $\mathbf{x} \leq \mathbf{y}$ means $x_i \leq y_i \ \forall i = 1, \dots, n$.

A type of aggregation functions that have special interest in the ISPP, are the averaging functions.

Definition 2.3. An aggregation function $g : [0,1]^n \to [0,1]$ is said to be averaging if it is bounded by the minimum and maximum of its arguments:

$$min(\mathbf{x}) \leq g(\mathbf{x}) \leq max(\mathbf{x}), \ \forall \mathbf{x} \in [0,1]^n.$$

Taking into account that intervals in the ISPP mean that the real cost of a path is between their bounds, the use of averaging functions of $n = 2$ arguments has been considered a sensible approach for dealing with this kind of imprecision.

Example 2.1. Consider the functions $g_1, g_2, g_3 : [0, 1]^2 \to [0, 1]$ given by:

$$g_1(a_1, a_2) = \frac{a_1 + a_2}{2}, \quad g_2(a_1, a_2) = 0.7a_1 + 0.3a_2, \quad g_3(a_1, a_2) = \sqrt{\frac{a_1^2 + a_2^2}{2}}.$$

It is easy to prove that they are averaging functions. Figure 1 shows their behaviour on several intervals.

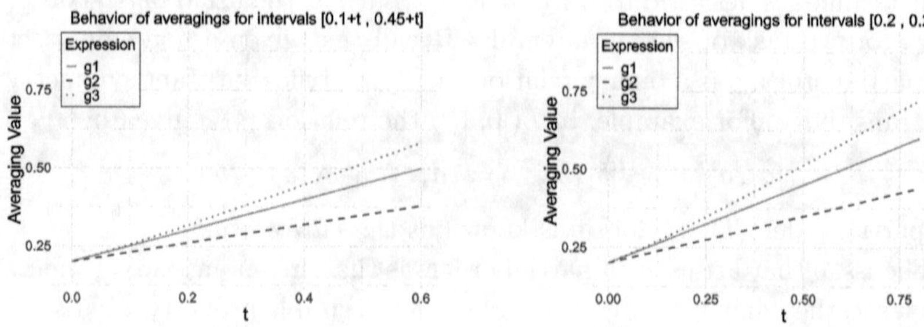

(a) Intervals with equal width. (b) Intervals with increasing width.

Fig. 1: Averaging functions from Example 2.1 evaluated across the extremes of different intervals.

As it can be seen in Ref. 7, a particular way to generate admissible orders using aggregations functions is the following:

Let $\alpha \in [0, 1]$ and $K_\alpha : [0, 1]^2 \to [0, 1]$ be the aggregation function defined by $K_\alpha(x_1, x_2) = (1 - \alpha)\, x_1 + \alpha\, x_2$. Analogously, $K_\beta(x_1, x_2) = (1 - \beta)\, x_1 + \beta\, x_2$ for $\alpha \neq \beta \in [0, 1]$. The order $\leq_{\alpha,\beta}$ defined by:

$$[a_1, a_2] \leq_{\alpha,\beta} [b_1, b_2] \iff \begin{cases} K_\alpha(a_1, a_2) \leq K_\alpha(b_1, b_2) \\ \text{or} \\ K_\alpha(a_1, a_2) = K_\alpha(b_1, b_2) \wedge K_\beta(a_1, a_2) \leq K_\beta(b_1, b_2) \end{cases}$$

is an admissible order on $L([0, 1])$.

3. An Approach to Solve the ISPP

In order to model the ISPP, a graph $G = (V, E)$ is considered, where $V = \{v_1, \ldots, v_m\}$ is the set of vertices and $E \subseteq V \times V$ is the set of edges. Each edge $(v_i, v_j) \in E$ in the graph has an associated cost expressed as an interval $C_{ij} = [\underline{C_{ij}}, \overline{C_{ij}}]$. The objective is to find the path from a source node $v_s \in V$ to another target node $v_t \in V$ in such a way that the sum of the costs of the edges along that path is minimized. To solve the classic SPP, Dijkstra's algorithm is commonly used. However, it involves the computation of the minimums of the costs of the paths to each vertex in the graph, so it cannot be directly applied to the ISPP. To adapt the algorithm, it is necessary to establish a manner of comparing intervals. Then, the minimum referred to that comparison can be calculated, making it possible to proceed with Dijkstra's algorithm.

To compare intervals and decide which one is smaller, the use of averaging functions is proposed, as they are a sensible way of reducing an interval to one point. Once this is done, the standard order in $[0, 1]$ is applied and it is extended to the intervals involved. Another natural procedure that has also been considered to compare intervals is the application of admissible orders. Notice that, although some admissible orders are generated through aggregation functions, this is not always the case. Thus, proposals based on aggregation functions and admissible orders are not equivalent in general. It is true that if the admissible order is generated by aggregation functions, both approaches are closely related. Nevertheless, even in this case, they are considered a distinct method from the averaging functions approach, since the behavior is not the same for intervals whose aggregation is the same. Algorithm 3.1 shows the pseudocode of the variation of Dijkstra's one that has been used in this work. The function receives as inputs a graph $G = (V, E)$, a source node $v_s \in V$ and either an averaging function $* \equiv g$ or and admissible order $* \equiv \leq_{\alpha,\beta}$. The main differences with the classical approach are that the distance to each of the vertices is now an interval, and the minimum is computed in terms of the chosen averaging function or admissible order. Notice that the interval costs are standardised at the beginning to compute the additions in terms of the $\oplus$ operator defined in Section 2. The algorithm returns the shortest paths and their cost from v_s to every node in the network, although the aim of the problem is to find just the shortest path from v_s to v_t.

In order to choose the most appropriate averaging function or admissible order to be included in the process, the decision is left to an expert DM who is supposed to know the context of the problem. For example, if the DM is categorized as optimistic, α and β values will be closer to 0, as that reflects a low point in the intervals.

Algorithm 3.1 Interval Dijkstra

```
function IDIJKSTRA(G, v_s, *)
    S ← Σ_{i,j∈{1,...,m}} overline{C_ij}
    for all (v_i, v_j) ∈ E do
        C_ij ← C_ij / S
    end for
    for all v_i ∈ V do
        distance[v_i] ← [∞, ∞]
        predecessors[v_i] ← NULL
    end for
    distance[v_s] ← [0, 0]
    Q ← {v_s}
    while Q is not empty do
        v_k ← ExtractMinUsing * from Q
        Q ← Q \ v_k
        for all unvisited neighbor v_j of v_k do
            tempDistance ← distance[v_k] ⊕ C_kj
            if tempDistance < distance[v_j] then
                distance[v_j] ← tempDistance
                previous[v_j] ← v_k
                Q ← Q ∪ v_j
            end if
        end for
    end while
    return distance[], previous[]
end function
```

4. Case Study

In this section, a brief example is presented to illustrate the application of the approach in solving an ISPP. The considered graph is shown in Figure 2. The objective is to find the shortest path that allows to go from v_s to v_t. The intervals in the edges represent the cost of travelling along that arc of the graph.

First of all, each interval needs to be divided by the sum of all the upper bounds. In this case, that sum is 16.1. To solve this ISPP, the averaging functions g_1, g_2 and g_3, shown in Example 2.1, are used in Algorithm 3.1. Additionally, admissible orders $\leq_{\alpha,\beta}$ for different values of α have also been considered. The results obtained for each of these options can be seen in Table 1.

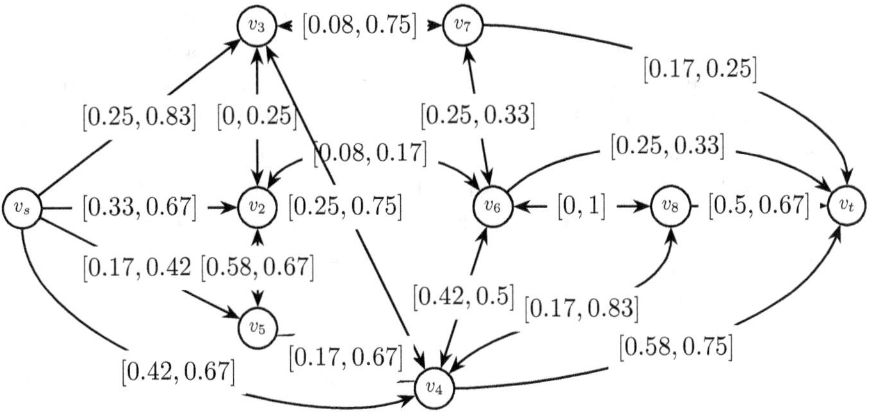

Fig. 2: Graph representation of the ISPP considered to go from v_s to v_t.

Criterion	Shortest path	Total cost
$g_1(a_1, a_2) = (a_1 + a_2)/2$	$v_s \to v_2 \to v_6 \to v_7 \to v_t$	$[0.83, 1.42]$
$g_2(a_1, a_2) = 0.7a_1 + 0.3a_2$	$v_s \to v_3 \to v_7 \to v_t$	$[0.5, 1.83]$
$g_3(a_1, a_2) = \sqrt{(a_1^2 + a_2^2)/2}$	$v_s \to v_2 \to v_6 \to v_7 \to v_t$	$[0.83, 1.42]$
$\leq_{\alpha,\beta}$: $\alpha < 0.446$	$v_s \to v_3 \to v_7 \to v_t$	$[0.5, 1.83]$
$\leq_{\alpha,\beta}$: $\alpha \geq 0.446$	$v_s \to v_2 \to v_6 \to v_7 \to v_t$	$[0.83, 1.42]$

Table 1: Shortest paths of the ISPP displayed on Figure 2 for different criteria.

It can be seen from Table 1 that depending on the chosen criterion, the shortest path changes. However, a form of consistency in the identified solutions is also discernible, as only two instances have emerged among all conceivable paths. Furthermore, the considered averaging functions have produced minimal paths according to the $\leq_{\alpha,\beta}$ admissible orders.

5. Conclusions

The ISPP serves to model a multitude of situations consistently and, therefore, it is essential to have a sensible way to solve it. A variation of Dijkstra's algorithm has been developed and it includes two different ways to compare intervals: averaging functions and admissible orders. Moreover, as a sort of validation, it has been observed that both techniques offer similar results. With this procedure, not only can the ISPP be solved in real-life contexts,

but it also allows easy interpretation for individuals involved in the problem. In this way, not only an optimized process is presented, but more information about the solution is transmitted, enabling feedback from an expert in the problem, to make changes to the model if needed, or choose a better averaging function or admissible order.
The future research goals include analyzing the impact of the choice of averaging functions and admissible orders on ISPP solutions to characterize and focus the search on the most desirable ones.

Acknowledgments

Authors would like to express their gratitude for the support from the Spanish Ministry of Science and Innovation (fund reference: PREP2022-000355. Project MCINN-23-PID2022-139886NB-I00).

References

1. Z. Wei, High performance computing simulation of intelligent logistics management based on shortest path algorithm, *Computational Intelligence and Neuroscience* **2022** (2022).
2. M. Gong, G. Li, Z. Wang, L. Ma and D. Tian, An efficient shortest path approach for social networks based on community structure, *CAAI Transactions on Intelligence Technology* **1**, 114 (2016).
3. R. Montemanni and L. M. Gambardella, The robust shortest path problem with interval data via benders decomposition, *4or* **3**, 315 (2005).
4. A. Sengupta and T. K. Pal, Solving the shortest path problem with interval arcs, *Fuzzy Optimization and Decision Making* **5**, 71 (2006).
5. A. Ebrahimnejad, An acceptability index based approach for solving shortest path problem on a network with interval weights, *RAIRO-Operations Research* **55**, S1767 (2021).
6. S. M. A. Nayeem and M. Pal, Shortest path problem on a network with imprecise edge weight, *Fuzzy Optimization and Decision Making* **4**, 293 (2005).
7. H. Bustince, J. Fernández, A. Kolesárová and R. Mesiar, Generation of linear orders for intervals by means of aggregation functions, *Fuzzy Sets and Systems* **220**, 69 (2013).

Decoding Employee Attrition: A Unified Approach with XAI and AHP

Gabriel Marín Díaz* and José Javier Galán Hernández
Faculty of Statistics, Complutense University Madrid 28040
gmarin03@ucm.es

In the face of escalating employee attrition challenges, organizations are increasingly relying on artificial intelligence (AI) to predict and address turnover. This chapter explores the application of explainable AI (XAI) to identify potential employee turnover, analyzing its impact on organizational productivity and stability. The second section focuses on AI techniques that leverage historical data to forecast attrition, enabling proactive interventions. The third part introduces XAI to enhance model transparency, providing HR professionals with deeper insights to develop targeted retention strategies aligned with individual employee needs. Integrating the Analytic Hierarchy Process (AHP) model becomes imperative to assign weights to criteria identified by AI as significant. This incorporation aims to introduce the human factor into decision-making.

Keywords: XAI; artificial intelligence; AHP; human resources; talent attraction.

1. Introduction

The contemporary labor market, particularly within Human Resources (HR), is undergoing rapid evolution, marked by the pervasive challenge of 'involuntary turnover'—employee departures contrary to employer intentions [1]. This phenomenon, accelerated by the pandemic and the 'Great Resignation' since 2021, poses multidimensional challenges for enterprises globally, resulting in tangible and intangible costs.

Recent findings from Randstad Research's "Labour Turnover Report in Spain" [2] reveal a significant 38.5% increase in turnover rates for 38.5% of Spanish enterprises in the past twelve months. While 77% attribute turnover to employees pursuing opportunities elsewhere, secondary factors include workforce concerns arising from recent crises, aspirational salary expectations, and a demand for increased flexibility.

Addressing this, AI emerges as a promising solution, particularly through Explainable AI (XAI) methodologies [3]. XAI not only sheds light on predictive models but also identifies critical factors contributing to predictions [4]. This knowledge empowers proactive implementation of targeted retention strategies, aligning with employee needs [5].

Incorporating AHP into the methodology allows for the assignment of weights to the characteristics of the learning model, considering the context of interpretability [6]. This strategic weighting enhances the overall effectiveness of the model in addressing the nuances of employee turnover.

The proposed approach facilitates actionable insights, allowing organizations to understand, intervene, and reshape strategies to mitigate turnover risks.

By emphasizing the interpretability of the XAI model, decision-makers can pinpoint pivotal factors driving attrition within their context, leading to targeted interventions, such as fostering a supportive culture and addressing compensation concerns. The methodology also enhances recruitment by identifying attributes associated with employee longevity, guiding efforts toward successful and enduring employment relationships.

Advocating for a paradigm shift, this chapter envisions organizations proactively implementing strategies for enhanced employee retention and optimized recruitment processes. The subsequent sections detail the development and application of the XAI model, reviewing current XAI methodologies, discussing the methodological framework, applying the XAI model in a business context, integrating AHP to assign weights to learning model characteristics for enhanced interpretability, and presenting discussions, conclusions, and future work.

2. Related Work

Studies related to employee attrition using machine learning techniques are scarce, with a total of 26 studies identified based on the following search variables in Web of Science: TS = (ATTRITION) AND TS = (EMPLOYEE) AND TS = (MACHINE LEARNING), Figure 1. These studies do not address algorithm interpretability; rather, they focus on incorporating predictive models to facilitate early detection of employee attrition. However, it is noteworthy that interpretability is not discussed or employed in these studies.

An important research frontier is unfolding within corporate settings, encompassing the integration of predictive models and interpretability into decision-making processes. Moreover, the inclusion of a pivotal determinant in decision-making, the human factor (AHP), provides a distinctive opportunity for exploration and advancement in this domain.

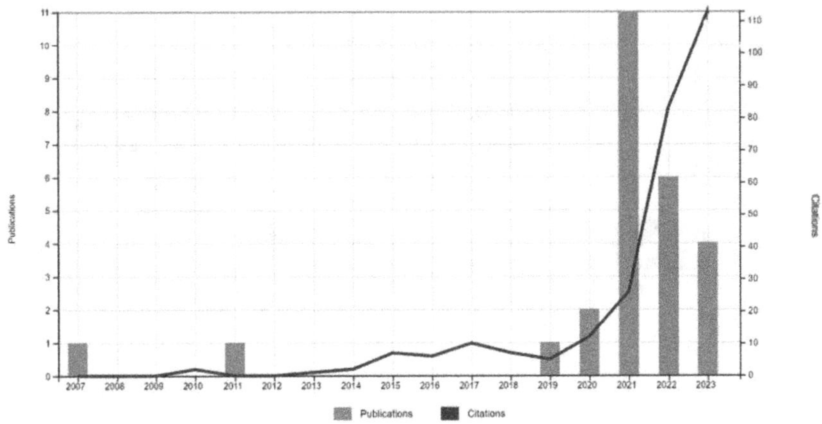

Fig. 1. Publications and citations (26). TS = (ATTRITION) AND TS = (EMPLOYEE) AND TS = (MACHINE LEARNING).

The predominant approach at present emphasizes the adoption of model-agnostic interpretability tools. The strategy involves separating the interpretation method from the employed model, facilitating the automation of interpretability. The use of agnostic methods allows for the replacement of both the learning model and the interpretation method, resulting in significantly scalable capabilities [7].

3. Methodology

The methodology we are going to present is based on the Knowledge Discovery in Databases (KDD) framework [8], which employs principles to extract actionable insights from data, with a particular focus on the variables determining employee departures. Subsequently, aligning hiring and promotion strategies with these identified variables allows for a targeted and informed approach to human resource management. The incorporation of XAI and AHP emphasizes the commitment to transparency and understanding in decision-making processes.

Explainable AI contributes to the interpretability of models, ensuring that decisions driven by artificial intelligence are accessible and understandable to human decision-makers. Simultaneously, AHP, as a decision-making model, provides a systematic and mathematical foundation for evaluating and prioritizing variables, incorporating both psychological and mathematical principles into the decision-making process. The fusion of these methodologies aims to create a comprehensive and intelligible framework for informed and strategic decision-making in the realm of human resource management, Figure 2.

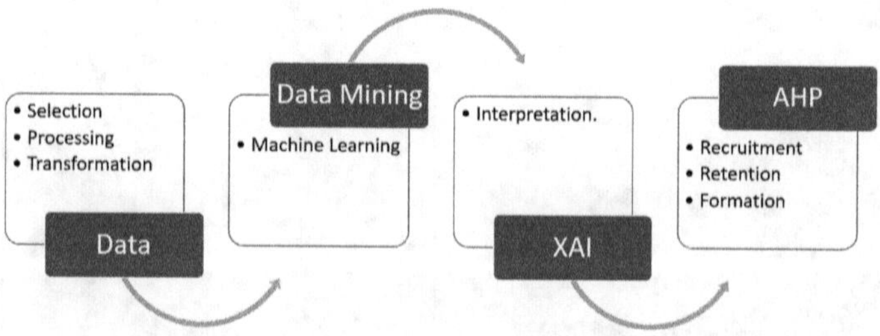

Fig. 2. KDD - Employee attrition analysis and HR recruitment, retention and training plans.

3.1. *Data object*

The initial phase in the Knowledge Discovery in Databases (KDD) process is the "Data" stage [9]. This stage is primarily focused on comprehending the domain and establishing the fundamental groundwork for the entire data analysis undertaking.

Data processing encompasses the tasks of cleaning, transforming, and organizing raw data into a format suitable for subsequent analysis [10]. This critical step addresses challenges such as handling missing values, identifying outliers, and mitigating noise within the dataset.

This process entails transforming raw data into an appropriate format that enhances its quality and facilitates effective analysis [11]. Techniques employed in this stage include normalization, aggregation, discretization, and attribute construction.

3.2. *Data mining*

In this employee attrition case study, we'll follow a structured workflow, applying various classification algorithms like Logistic Regression, Decision Trees, Random Forest, Support Vector Machines, or Gradient Boosting [3]. Each algorithm will predict an employee's likelihood of leaving based on relevant features. After predictions are generated, we'll evaluate each employee's attrition tendency, measuring model accuracy and constructing AUC/ROC curves for assessing predictive ability. Additionally, a detailed confusion matrix will be created for each algorithm, offering insights through metrics like precision, recall, specificity, and F1 score.

3.3. *XAI*

Throughout time, algorithms have primarily concentrated on exploratory data analysis to support decision-making processes. Although acquiring and analyzing data is often adequate for well-informed decisions, there are situations where validating decisions based on patterns becomes essential. In this regard, AI holds particular importance. Historically, machine learning algorithms have placed a premium on attaining high precision in results, progressively gravitating towards black-box algorithms. As a result, the interpretability of these decision-making algorithms has been restricted, leading to the emergence of interpretable algorithms [7], Figure 3. These interpretable models are explored through various methods outlined below.

1. Inherent Interpretable Models [12]: Models such as Linear Regression, Decision Trees, and Logistic Regression feature transparent structures and explicit relationships between features and predictions, making them inherently interpretable.
2. Rule-based Interpretable Models [13]: This category aims to simplify intricate patterns into clear-cut rules, promoting transparency in decision-making. Models like decision trees and rule-based systems utilize simple, understandable rules, establishing direct connections between input features and predictions.
3. Local Interpretable Models [13]: Focusing on explaining predictions at the instance level, these models offer a detailed understanding of how the model arrived at a specific prediction for a particular data point. Techniques like LIME fall into this category, providing insights into the model's behavior around a specific data point.
4. Model-Agnostic Approaches [7]: Model-agnostic methods, including SHAP, LIME, and Permutation Feature Importance, offer interpretability for a wide range of models without relying on their internal architecture, enhancing transparency across diverse machine learning models.
5. Feature Importance Techniques [14]: Techniques like ELI5 and Partial Dependence Plot (PDP) highlight the significance of each feature in the model's predictions, offering a straightforward understanding of feature contributions.

6. Surrogate Models [15]: Simpler and more interpretable models trained to approximate the predictions of complex models, surrogate models serve as a bridge between highly complex, less interpretable models and the need for comprehensible insights into their functioning.
7. Visualizations and Plots [16]: Techniques like partial dependence plots, SHAP summary plots, and feature contribution plots offer intuitive graphical representations of the model's behavior and feature influences.

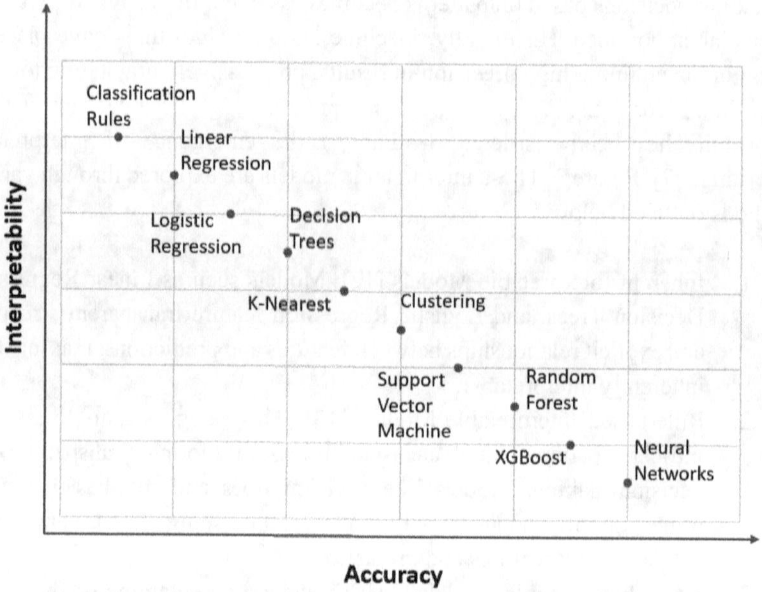

Fig. 3. Accuracy vs Interpretability.

3.4. *AHP*

The Analytic Hierarchy Process (AHP) is a decision-making methodology that can significantly contribute to the personnel selection process by enhancing interpretability. AHP involves structuring complex decision problems into a hierarchical framework, allowing decision-makers to systematically evaluate and prioritize various factors. This structured approach aids in understanding the relationships and importance of different criteria in the decision-making process. In the context of personnel selection, AHP can be employed by first defining the overarching goal, which, in this case, is selecting the most suitable candidate for a position. The decision criteria, such as skills, experience, cultural fit, and potential for growth, are then identified and organized hierarchically, to summarize, the criteria that the Interpretability model has deemed predominant in the employee attrition process will be crucial in the company's recruitment, retention, and personnel training processes. Subsequently, pairwise comparisons are conducted to determine the relative importance of each criterion in relation to the goal. These comparisons are facilitated by decision-makers, providing a qualitative assessment of the significance of one criterion over another, Figure 4. The pairwise comparisons generate a set of numerical values that represent the relative weights of each criterion. A mathematical process is then applied to derive a consistent set of weights that reflect the decision-makers' preferences. The final outcome is a prioritized list of criteria based on their importance in achieving the overall goal.

The AHP offers a valuable contribution to the personnel selection process, leveraging the significant criteria identified through AI interpretability models. AHP provides a structured and coherent framework, introducing a human-centric approach to decision-making.

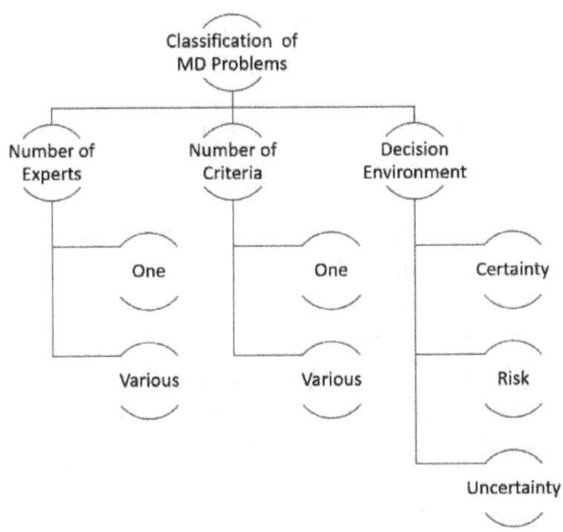

Fig. 4. Classification of MD Problems.

4. Conclusions

The methodology employed, leveraging eXplainable Artificial Intelligence (XAI) to identify critical employee attrition criteria and subsequently applying Analytic Hierarchy Process (AHP) decision theory, holds promising implications for diverse organizational processes.

The utilization of XAI in discerning key attrition criteria not only enhances transparency in decision-making but also provides valuable insights applicable to various facets of human resource management. The identified criteria, proven to be significant in predicting employee turnover, can be effectively translated and incorporated into distinct operational domains, such as personnel selection, retention strategies, and employee training programs.

Furthermore, the detailed process elucidated in this chapter is not confined to a specific industry or sector; rather, it serves as a versatile framework that can be extended to any business context. By systematically integrating XAI and AHP, along with fuzzy linguistic models, organizations gain a comprehensive understanding of the factors influencing workforce dynamics. This holistic approach not only aids in the formulation of targeted and effective strategies for personnel-related processes but also promotes adaptability, ensuring its applicability across diverse business environments.

References

[1] S. Rughoobur-Seetah, "The Unprecedented Lockdown: The consequences of job loss," *Zagreb Int. Rev. Econ. Bus.*, vol. 24, no. 2, pp. 1–23, 2021, doi: 10.2478/zireb-2021-0008.

[2] Randstad Research, "Informe de rotación laboral en España," *Randstad Res.*, 2022.

[3] G. Marín Díaz, J. J. Galán, and R. A. Carrasco, "XAI for Churn Prediction in B2B Models: A Use Case in an Enterprise Software Company," *Mathematics*, vol. 10, no. 20, 2022, doi: 10.3390/math10203896.

[4] P. Linardatos, V. Papastefanopoulos, and S. Kotsiantis, "Explainable ai: A review of machine learning interpretability methods," *Entropy*, vol. 23, no. 1, pp. 1–45, 2021, doi: 10.3390/e23010018.

[5] D. Mishra, "Review of literature on factors influencing attrition and retention," *Int. J. Organ. Behav. Manag. Perspect.*, vol. 2, no. 3, pp. 435–445, 2013.

[6] O. S. Vaidya and S. Kumar, "Analytic hierarchy process: An overview of applications," *Eur. J. Oper. Res.*, vol. 169, no. 1, pp. 1–29, 2006, doi: 10.1016/j.ejor.2004.04.028.

[7] C. Molnar, "Interpretable Machine Learning. A Guide for Making Black Box Models Explainable.," *Book*, p. 247, 2019, [Online]. Available: https://christophm.github.io/interpretable-ml-book.

[8] U. Shafique and H. Qaiser, "A Comparative Study of Data Mining Process Models (KDD , CRISP-DM and SEMMA)," *Int. J. Innov. Sci. Res.*, vol. 12, no. 1, pp. 217–222, 2014, [Online]. Available: http://www.ijisr.issr-journals.org/.

[9] S. Kaufman, S. Rosset, C. Perlich, and O. Stitelman, "Leakage in data mining: Formulation, detection, and avoidance," *ACM Trans. Knowl. Discov. Data*, vol. 6, no. 4, pp. 1–21, 2012, doi: 10.1145/2382577.2382579.

[10] F. Kamiran and T. Calders, *Data preprocessing techniques for classification without discrimination*, vol. 33, no. 1. 2012.

[11] S. A. Alasadi and W. S. Bhaya, "Review of Data Preprocessing Techniques," *Journal of Engineering and Applie Sciences*, vol. 12, no. 16. pp. 4102–4107, 2017.

[12] Z. C. Lipton, "The mythos of model interpretability," *Commun. ACM*, vol. 61, no. 10, pp. 35–43, 2018, doi: 10.1145/3233231.

[13] M. T. Ribeiro, S. Singh, and C. Guestrin, "'Why Should I Trust You?' Explaining the Predictions of Any Classifier," *NAACL-HLT 2016 - 2016 Conf. North Am. Chapter Assoc. Comput. Linguist. Hum. Lang. Technol. Proc. Demonstr. Sess.*, pp. 97–101, 2016, doi: 10.18653/v1/n16-3020.

[14] Y. Lou, R. Caruana, J. Gehrke, and G. Hooker, "Accurate intelligible models with pairwise interactions," *Proc. ACM SIGKDD Int. Conf. Knowl. Discov. Data Min.*, vol. Part F1288, pp. 623–631, 2013, doi: 10.1145/2487575.2487579.

[15] L. Breiman, "Stacked regressions," *Mach. Learn.*, vol. 24, no. 1, pp. 49–64, 1996, doi: 10.1023/A:1018046112532.

[16] S. M. Lundberg and S. I. Lee, "A unified approach to interpreting model predictions," *Adv. Neural Inf. Process. Syst.*, vol. 2017-Decem, no. Section 2, pp. 4766–4775, 2017.

Segregation Index, Analysis and Proposal for Non-binary Variables

P. Olaso*,§, K. Rojas*,¶, D. Gómez†,‖ and J. Montero‡,**

*Faculty of Mathematical Sciences, Complutense University of Madrid, Madrid, Spain
†Department of Computer Science and Statistics, Rey Juan Carlos University, Móstoles, Spain
‡Faculty of Statistics, Complutense University of Madrid, Madrid, Spain
§polaso@ucm.es
¶karinah.rpatuelli@urjc.es
‖dagomez@estad.ucm.es
**monty@ucm.es

Segregation is a phenomenon that has been widely studied in the social sciences, where two dimensions are intertwined: one social and the other class. However, there are no indexes that allow the use of nominal variables with more than two categories in the social dimension, leading researchers to force the use of indexes in order to adapt them to the objectives of their studies. This chapter proposes a segregation index that can be applied when the social dimension has more than two categories. The proposed indicator, chi dissimilarity index, has a behavior similar to the frequently used Duncan dissimilarity index but can be applied to problems with nominal variables in general, not only dichotomous ones. Finally, an application with real data from a school in Madrid is presented, in order to visualize its usefulness in school segregation studies.

Keywords: Segregation index; dissimilarity index; chi-squared statistic.

1. Introduction

Segregation is a phenomenon that has been widely studied in the social sciences. Initially called residential segregation, considering the geographic areas of a territory as natural groupings of people, social categories would be those that entail a condition of social inequality. One of the first definitions of residential segregation is "the relationship between the physical distance and the social distance of certain groups or individuals" (Park, 1926).

§Corresponding author.

Focusing on the phenomenon of segregation in the educational field, several authors define school segregation as "the unequal distribution of students in educational centers according to their personal or social characteristics" (Bellei, 2013; Dupriez, 2010; Ireson and Hallam, 2001), thus causing a situation of social inequality in the school population. The differences in the distribution of two or more demographic groups (gender and racial/ethnic groups) in a set of organizational units (Ricardo Mora and Javier Ruiz-Castillo). Over the years, various segregation studies have used different segregation indices, the most widely used of which are the Duncan, Gorard, square root and isolation indices. In all these studies, it is possible to identify two intertwined elements: a social dimension, which refers to a social, political or cultural condition of subjects by which they are segregated, and another element, named class dimension, referring to a grouping of subjects in which there is an unbalanced distribution across groups with respect to the overall distribution according to that social condition. Thus, the social dimension can be represented, for example, by variables, such as gender, race, socioeconomic level, and migrant status, and, on the other hand, the class dimension, which is where segregation occurs, such as educational centers, socioeconomic level, and neighborhood of residence, among others. It is evident the need to use a combination of binary, nominal and/or ordinal variables when studying this phenomenon. However, there are no indexes that reflect the level of flexibility with respect to the type of variable required to study segregation, leading researchers to force the use of indexes in order to adapt them to the objectives of their studies, as is the case of Duncan's dissimilarity index. This chapter advances in the mathematical analysis of Duncan's dissimilarity index, its association with another index frequently used in statistical inference and in the generation of a new index that facilitates its use in segregation studies, especially when there are more than two categories in the social dimension of the phenomenon. Finally, an application with real data from a school in Madrid is presented in order to visualize its usefulness in school segregation studies.

This paper advances in the mathematical analysis of Duncan's dissimilarity index, its association with another index frequently used in statistical inference and the generation of a new index that facilitates its use in segregation studies, especially when there are more than two categories in the social dimension of the phenomenon.

2. A New Segregation Index based on Chi-squared Measure

In this section we introduce the two segregation indexes that we are going to analyze, the well-known Duncan Dissimilarity Index and our proposal based on the Chi-Square index. Let P be a population with $|P| = T$ the total number of subjects under study. And consider that the social dimension is characterized by a two-label variable, A and B, and the class dimension by a series of groups K_i with $i \in \{1, \ldots, k\}$. Let x_i, t_i also be the number of subjects labeled as A and the total number of subjects, respectively, in the group $i, \forall i \in \{1, \ldots, k\}$.

Then, the Duncan's dissimilarity index, D, is obtained as follows.

$$D = \frac{1}{2} \sum_{i=1}^{k} |\frac{x_i}{X} - \frac{t_i - x_i}{T - X}| \tag{1}$$

This index takes values between 0 and 1, where the null value represents the non-existence of segregation and the unit value, the maximum level of segregation. When the index takes the value zero, it can be interpreted that the distribution of the variable representing the class dimension is distributed in the same way in all the groups of the class dimension and, therefore, also at the global level. This indicates, therefore, that there is no association between the two dimensions. While when it takes the unit value, the entire population that has a certain social status is grouped in only one group of the class dimension. Therefore, the value of Duncan's dissimilarity index represents the percentage of the population that must be "relocated" so that there is no segregation, that is, so that independence between both dimensions occurs.

Using the usual notation for contingency tables and considering the variables X and Y to represent the social and class dimensions respectively, we will note as n_{ij} the joint absolute frequency corresponding to the number of people in social group X_i and class Y_j. Note that the indices now vary as follows: $i \in 1, 2$, $j \in \{1, \ldots, k\}$. In addition, the marginal absolute frequencies of these variables are represented by $n_{i\cdot}$ and $n_{\cdot j}$ for the social and class dimension respectively. Finally, the total number of people in the population, will be $n_{\cdot\cdot}$.

Thus, the Duncan Dissimilarity Index, D, is as follows:

$$D = \frac{1}{2} \sum_{j=1}^{k} |\frac{n_{2j}}{n_{2\cdot}} - \frac{n_{1j}}{n_{1\cdot}}| \tag{2}$$

First, let us note that the above index is only defined when the variable X (social dimension) takes values in two categories. However, there are many interesting real situations where the social dimension is modeled by a nominal variable with more than two categories (e.g., nationalities, more than two sexual or gender orientations, different religions, etc.). Moreover, the problem becomes even more complicated if this social variable is ordinal, where different labels allow some sort of ordering (e.g., considering different socioeconomic levels, income or educational levels...).

For those situations, we can extended previous notation, assuming now that social variable X takes values in $\{X_1, \ldots, X_r\}$ and taking into account that variable Y cluster the population into groups Y_j, for j in $\{1, \ldots, k\}$. Following previous notation, n_{ij} denotes the number of subjects with X_i value in the social dimension that belong to the group Y_j.

The association between the concepts of segregation and statistical independence between nominal variables X and Y is undeniable, insofar as the level of segregation indicates that belonging to a given social label is more or less probable according to class membership. It is null when belonging to a class has no effect on belonging to a social label, that is, when both dimensions, social and class, are independent. Taking this into consideration, in this work we propose the use of the classical Chi-squared index as a segregation index that we will name as *Chi Dissimilarity index*, D_{Ch}.

$$D_{Ch} = \sum_{j=1}^{k}[(n_{2j} - \frac{n_{2.} \cdot n_{.j}}{n_{..}})^2(\frac{n_{2.} \cdot n_{.j}}{n_{..}})^{-1} + (n_{1j} - \frac{n_{1.} \cdot n_{.j}}{n_{..}})^2(\frac{n_{1.} \cdot n_{.j}}{n_{..}})^{-1}] \tag{3}$$

It is possible to see that $D'_{Ch} = \frac{D_{Ch}}{n_{..}}$ takes values between 0 and 1, this being the segregation index that is finally proposed.

3. Application: School Segregation in a Neighborhood of Madrid

This section presents an application with real data extracted from an educational center in the Community of Madrid, in order to visualize the usefulness of the Chi dissimilarity index in school segregation studies.

A mixed study was carried out in different public secondary schools located in two neighborhoods in the southern area of Madrid with a high level of inequality and a high percentage of migrant population. The information was collected during the months of May and June 2023, through self-administered online surveys conducted in the classroom of students aged between 12 and 14 years old in the first year of secondary education. For these purposes, information is used from two of the centers in this study, in which information was collected from 110 and 81 students distributed in four classrooms in the case of the first center and three classrooms in the second, as shown in Table 1:

Classrooms	School 1	School 2
A	36	26
B	26	28
C	30	27
D	24	—
Total	110	81

It is observed that there is a 70% of students from migrant families in School 1 and 56% in School 2. Two types of segregation analysis have been taken into account: intraschool segregation and interschool segregation. The former measures the level of segregation within the school according to the classrooms attended by the students. The second measures the level of segregation between schools in the same neighborhood. In both cases, the social dimension is considered to be the migrant status of the students, while the class dimension corresponds to the center's classroom for intra-school segregation and to the center where the student is enrolled for interschool segregation.

The Duncan dissimilarity index for intra-school educational segregation is 0.13 and 0.25 for schools 1 and 2, respectively. Meanwhile, the Chi dissimilarity index is 0.02 and 0.07, indicating low levels of intra-school segregation regarding migrant status in both centers, though slightly higher in school 2. Regarding the level of school segregation between centers, the Duncan dissimilarity index provides a level of 0.15%, and the Chi dissimilarity index is 0.03%.

However, it is interesting to visualize how the values of both indexes change as the level of segregation increases. In order to visualize it in this concrete example, we consider school 1, in which the number of migrant students in classroom A has been decreasing one by one, starting with 24 migrant students and ending with only 5. Figure 1 shows this behavior, showing the similar behavior of both indexes; however, there is a smoother trend in the Dissimilarity-Chi index and permanently lower values.

This confirms what has already been seen theoretically, that the curves grow and decrease in the same regions of the domain. It is possible to

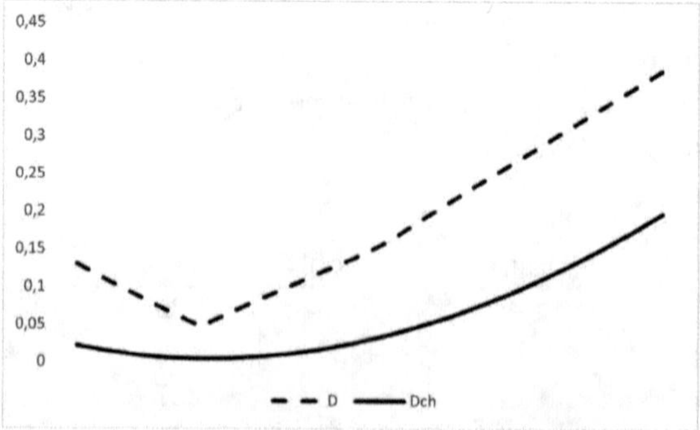

Fig. 1. Performance of both indexes .

apply the index proposed here when the social dimension is represented by a variable containing more than two labels, as long as the statistic allows it.

Continuing with the same example, but now considering as a class dimension the perception of academic achievement in their first year of high school. In this case the Chi dissimilarity index gives a segregation level of 0.36 under the following distribution of students in school 1, shown in Table 2:

Primary vs. secondary	A	B	C	D
Now I'm doing better than before	16%	20%	7%	21%
Now I'm doing much better than before	0%	0%	20%	4%
Things were much better for me before	20%	32%	40%	21%
Things were a bit better for me before	40%	28%	20%	25%
It's the same for me as before	24%	20%	13%	29%
Total	100%	100%	100%	100%

4. Conclusions and Final Remarks

This chapter demonstrates the similarity between the Duncan dissimilarity index and the proposed chi dissimilarity index, both in theoretical and applied terms. It is observed that the curves grow and decay in the same regions of the domain, however, the proposed index has a smoother behavior and takes lower values. It is therefore a more flexible index than the frequently used Duncan's dissimilarity index, in that it reflects the level of flexibility with respect to the type of variable required by studies of social segregation.

References

1. Bellei, C. LATEX,*El estudio de la segregación socioeconómica y académica de la educación chilena.* Estudios Pedagógicos, 39(1), 325-345 (2013). doi:10.4067/s0718-07052013000100019
2. Nancy A. Denton and Douglas S. Massey. LATEX,*Residential segregation of blacks, hispanics and asians by socioeconomic status and generation.* Social Science Quarterly, 69, 4, 797-817 (1988). https://www.jstor.org/stable/42864168
3. Duncan, O. B. y Duncan, B. LATEX, *A methodological analysis of segregation indexes.* American Sociological Review, 20(2), 210-217 (1955). doi:10.2307/2088328
4. Dupriez, V. LATEX, *Methods of grouping learners at school.* París: UNESCO (2010).
5. Gorard, S. y Taylor, C. LATEX, *What is segregation? A comparison of measures in terms of 'strong' and 'weak' compositional invariance.* Sociology, 36(4), 875-895 (2002). doi:10.1177/003803850203600405
6. Guinea-Martin, Daniel; Mora, Ricardo; Ruiz-Castillo, Javier LATEX, *Beyond occupation : the evolution of gender segregation over the life course.* DE - Working Papers. Economics. WE 840, 16-08
7. Ireson, J. y Hallam, S. LATEX, *Ability grouping in education.* Londres: Paul Chapman Publishing (2001).
8. Park, R. E. LATEX,*The urban community as a spatial pattern and a moral order.* The Urban Community, 2, 3-18 (1926).

Evaluating a Custom CNN Model for Traffic Sign Classification Against the Pre-trained VGG19

Salah Eddine Laouar*, J. Enrique Sierra-Garcia†, Salim Ouchtati*, Jean Sequeira‡ and Segundo Esteban San Romn§

*Electronics Research Laboratory of Skikda (LRES), University of Skikda, Algeria
†Department of Digitalization, University of Burgos, Spain
‡Image, Information & Knowledge (2IK), Roquevaire, France
§Computers Architecture and Automation, Complutense University of Madrid, Spain

This chapter addresses the critical need for accurate identification and categorization of road traffic signs within intelligent transportation systems, autonomous vehicles, and advanced driver assistance systems. We introduce a novel Convolutional Neural Network (CNN) architecture specifically designed for road traffic sign recognition, a significant advancement in the application of deep learning to transportation safety and efficiency. Our first contribution is the development of a custom CNN model, which is tailored to meet the unique challenges of traffic sign classification in diverse real-world scenarios. Additionally, we present an extensive comparative study, where we benchmark our CNN model against the well-established VGG19 pre-trained model. This comparison not only underscores the robustness of our model but also illustrates the impact of dataset augmentation on improving its performance. Employing the German Traffic Sign Recognition Benchmark (GTSRB) dataset, we conduct a comprehensive evaluation to assess and contrast the performance of both models. This comparative study provides valuable insights into the efficacy and practical applications of these models, contributing significantly to the field of traffic sign recognition and its integration into future intelligent transportation systems.

Keywords: Convolutional neural network; traffic signal classification; VGG19; GTSRB dataset.

1. Introduction

In recent years, the identification and categorization of road traffic signs [1,2] have become a crucial component of intelligent transportation systems,[3,4] autonomous cars,[5,6] and improved driver assistance systems. In order to navigate safely and comply with traffic regulations, it is crucial to possess the skill of promptly and precisely comprehending traffic signs. Various approaches exist for categorizing traffic signs, including traditional computer vision techniques and state-of-the-art deep learning algorithms.[7],[8] For the recognition of traffic signs, typical computer vision approaches including template matching, feature extraction with Support Vector Machines (SVM),[9] and decision tree-based classifiers[10] have been applied. These approaches, however, usually fall short of offering high classification accuracy, especially when confronted with tough circumstances involving shifts in lighting conditions, occlusion, and scale. As a result, deep learning-based techniques have become more popular, with Convolutional Neural Networks (CNNs) as the best option for classifying traffic signs to deliver improved accuracy by employing a large dataset, certain sophisticated pre-trained CNN models are applied. In this view, various studies based on deep learning approaches have been proposed. For example,[11] introduced LeNet-5 and VGGNet for medical picture classification, which detects diabetic retinopathy vision loss and distinguishes emphysema. In this chapter, we propose a customized CNN architecture for traffic sign recognition. The suggested model was evaluated against the pre-trained VGG19 model to highlight its effectiveness. To enhance our model's performance in recognizing traffic signs, we augmented our dataset. This step exposed the model to a broader range of scenarios, improving its adaptability and accuracy.

In the remainder of this chapter, we explore several key aspects of our research. Section 2 compares our CNN model with the VGG19 model, focusing on their design and capabilities. Section 3 details our approach to data processing, model implementation, and testing. Section 4 presents our findings and interprets their significance in the context of traffic sign recognition. Finally, Section 5 summarizes our key conclusions and outlines potential avenues for future research in this domain.

2. Architectures Comparison

2.1. *Our proposed CNN model*

In this work, we use a lightweight CNN-based model for road traffic sign classification in autonomous vehicle applications, and we used Keras library[12] which providing a powerful and flexible framework for creating models layer-by-layer. Specifically, our model for traffic sign recognition consists three convolutional layers, two pooling layers, one flatten layer, and two fully connected layers. The input image is reshaped into 30*30*3 and it is given as the input to our model. The kernel size of the first convolutional layers is 5*5 and the quantity is 32 and for the second convolutional

```
Model: "sequential"
_________________________________________________________________
 Layer (type)                Output Shape              Param #
=================================================================
 conv2d (Conv2D)             (None, 26, 26, 32)        2432

 conv2d_1 (Conv2D)           (None, 24, 24, 64)        18496

 max_pooling2d (MaxPooling2D  (None, 12, 12, 64)       0
 )

 dropout (Dropout)           (None, 12, 12, 64)        0

 conv2d_2 (Conv2D)           (None, 10, 10, 64)        36928

 max_pooling2d_1 (MaxPooling  (None, 5, 5, 64)         0
 2D)

 dropout_1 (Dropout)         (None, 5, 5, 64)          0

 flatten (Flatten)           (None, 1600)              0

 dense (Dense)               (None, 256)               409856

 dropout_2 (Dropout)         (None, 256)               0

 dense_1 (Dense)             (None, 43)                11051

=================================================================
Total params: 478,763
Trainable params: 478,763
Non-trainable params: 0
```

Fig. 1. Our Model Summary

layers we apply a 64 3*3 filters convolutional layers convolute the input image and a ReLU activation function is applied after each convolutional layer. The Kernel size of the first max-pooling layer is 2*2. The convolutional layer followed by the max-pooling layer focuses on only the most relevant patterns. It is evident in Fig 1 that the Max pooling layers have 0 parameters to learn. Max pooling layer can lead to faster convergence, select superior invariant features, and also enhance generalization[13]. The convolutional layer with the ReLU activation function and pooling layer can discover the patterns.

A dropout of 25% is added after the last max-pooling layer. The neurons which are dropped out do not participate in forward pass and backward propagation. Without dropout our network exhibits substantial overfitting. The kernel size of the third convolutional layer is 3*3 and the quantity is 64. The kernel size of the last maxpooling layer is 2*2. Fully Connected layers are intended to perform classification. A dropout of 25% is added after the last max-pooling layer. The hidden nodes of the full connection layer are 256, and the quantity of hidden nodes of the last output layer is 43. A dropout of 50% is added after the first dense layer. A softmax function is added to the last dense layer, which consists of 43 neurons which is equal to the number of classes in the dataset. When an input image is given, the softmax layer gives the probability of that image in 43 classes. Class with the highest probability will become the output of traffic sign classification using a deep learning approach.

2.2. *VGG19*

The VGG19 architecture, known for its depth and effectiveness in image processing, has been a significant point of reference in recent studies focusing on traffic sign recognition and classification. As shown in Figure 2, VGG19 consists of 19 layers, including 16 convolutional layers with small receptive fields and 3 fully connected layers.

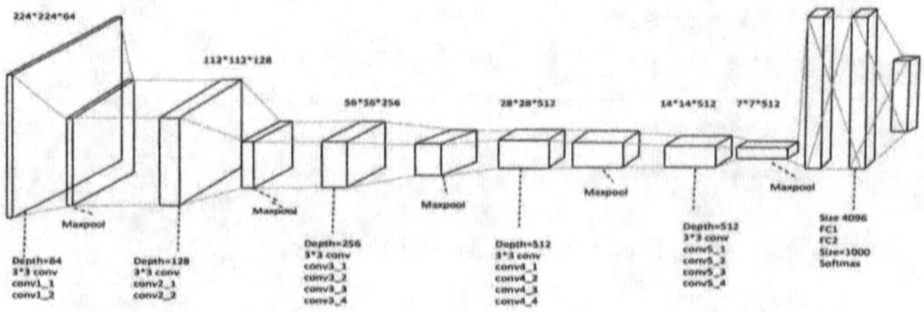

Fig. 2. VGG19 Architecture

3. Methodology

3.1. *GTSRB dataset*

In this comparison we used the German Traffic Sign Recognition Benchmark (GTSRB) dataset because it is considered the key resource in computer vision and machine learning for developing and evaluating traffic sign recognition algorithms. With 51,840 images representing over 1,700 instances across 43 classes, it showcases diverse traffic sign characteristics, including resolutions, lighting conditions, motion blur, and perspectives.

3.2. *Data augmentation*

To enhance the diversity of the training dataset, a combination of preprocessing and data augmentation techniques was applied. This involved subjecting the dataset to various transformations, illustrating the decrease in loss values as the model progressively learns during the training process. By simulating diverse scenarios and mitigating overfitting, this improvement heightened the model's capacity to generalize, resulting in an elevated categorization accuracy for traffic signs. The model's adaptability to changes in real-world data was reinforced by generating new training samples through the application of multiple transformations to existing photos, including rotation, scaling, and translation. Key methods employed in this process encompassed normalization, scaling, rotation, translation, and randomization.

3.3. *Data preprocessing*

Normalization of the images was carried out by dividing each pixel value by the maximum possible value of 255. This approach, known as min-max normalization, scales the pixel values to a range between 0 and 1. By doing so, the network operates on a standardized scale, which aids in balancing the calibration of its shared weights across different features. This type of normalization is crucial for consistent and effective convergence of the cost function, as it mitigates the risk of having certain features dominate the learning process due to their larger numeric range. Without this step, the model's convergence would likely be slower and less efficient as the un-normalized values could introduce discrepancies in the way the network interprets different features.

3.4. *Data augmentation with ImageDataGenerator*

The size of the training dataset was increased by producing augmented images using the ImageDataGenerator function From Keras Sequential model that was built on TensorFlow Library. To create new training samples, we used a variety of augmentation techniques in our approach, including rotation, zooming, shifting, and shearing. The function can rotate images up to 10 degrees at random because the rotation range parameter was set to 10. Images could be randomly zoomed up to 15% by using a zoom range of 0.15. The function was able to shift images horizontally and vertically by up to 10% because the width shift range and height shift range parameters were both set to 0.1. The function was configured to randomly shear images up to 15 degrees with a shear range of 0.15. We opted not to use horizontal or vertical flipping and employed the nearest fill mode to fill any gaps that might appear in the augmented images. By using the ImageDataGenerator function with these settings, we successfully generated a larger and more diverse dataset that helped mitigate overfitting and enhance the models accuracy.

4. Results and Discussion

We have conducted a comparative analysis of two distinct deep learning models, the pre-trained CNN model and the non-pretrained model of CNN. The implementation and evaluation took place on the free version of the Google Colab platform, utilizing its GPU resources. The training parameters used are shown in Table 1.

Table 1. Models Training parameters

parameters	Non-pretrained	Pre-trained
Loss	categorical crossentropy	categorical crossentropy
Optimizer	Adam	Adam
Epochs	150	100
Batch size	128	16

Due to their training on a broader and more diverse dataset, pre-trained models are expected to excel in feature extraction and exhibit superior initial performance compared to newly initialized models. Surprisingly, experimental results did not strongly support this expectation. Upon monitoring batch-wise model performance during the training, it was observed that the model without pre-training demonstrated significantly better performance than its newly trained counterpart. The unexpected performance superiority of the non-pretrained model over the pre-trained model in our experiments challenges conventional beliefs about the efficacy of pre-training. Generally, pre-trained models are believed to benefit from their exposure to diverse and extensive datasets, which is thought to enhance their ability to extract features and provide a strong starting point for further training. However, our observations suggest that in certain scenarios, starting with a fresh, non-pretrained model can be more advantageous. This could be attributed to the non-pretrained model's ability to learn features more specific and relevant to the task at hand, without any potential biases or limitations imposed by the pre-existing weights of a pre-trained model. Such a scenario indicates the need for a more nuanced understanding of when and why pre-training aids or hinders performance, particularly in specialized applications like traffic sign recognition.

Table 2 shows the metrics comparison of our experiments. Among all models tested, the non-pretrained model achieved the best average accuracy recognition, a 99.07% in 43 classes of traffic sign. Even in terms of training time and time to recognition for one image, the result was also in favor of the non-pretrained model. Therefore, the traffic sign recognition based on the proposed deep learning scheme shows excellent performance. As a result, the non-pretrained CNN model provides superior performance comparatively to the pre-trained model in terms of accuracy, learning time, and image recognition time. This outcome, suggests that a less complex, non-pretrained model can be more effective in specific tasks like traffic sign recognition and classification.

Table 2. Performance metrics of compared CNN Models

Models	Accuracy	Training Time	Recognition Time
CNN + data augment.	99.07%	49.83 Minute	0.06093 s
Transf learning + data augment.	98.28%	120 Minute	0.06725 s

5. Conclusions and Future Work

Our study, which involved a comparative analysis of a pre-trained CNN model and a non-pretrained CNN model for road traffic sign recognition using the GTSRB dataset, has yielded important insights. The research revealed a surprising superiority of the non-pretrained model over the pre-trained model in several key performance metrics, including accuracy, learning time, and image recognition time. This outcome challenges the traditional belief in the inherent advantage of pre-trained models, which are usually expected to perform better due to their training on diverse datasets. Our findings suggest that for specific applications like traffic sign recognition, starting from scratch with a non-pretrained model can sometimes be more beneficial, allowing the model to focus more directly on task-specific features without the influence of prior training.

The findings present new opportunities for future research in deep learning applications, particularly in developing customized models tailored to tasks requiring data specificity and task relevance. This can enhance efficiency and accuracy. Diversifying training datasets to encompass various geographical and environmental conditions may further strengthen the models' robustness.

References

1. A. de la Escalera, J. Armingol and M. Mata, Traffic sign recognition and analysis for intelligent vehicles, *Image and Vision Computing* **21**, 247 (2003).
2. A. de la Escalera, J. Armingol, J. Pastor and F. Rodrguez, Visual sign information extraction and identification by deformable models for

intelligent vehicles, *IEEE Transactions on Intelligent Transportation Systems* (2004).
3. A. Kumar, T. Singh and D. Vishwakarma, Intelligent transport system: Classification of traffic signs using deep neural networks in real time, in *Advances in Manufacturing and Industrial Engineering*, Lecture Notes in Mechanical Engineering (Springer, Singapore, 2021).
4. B. Le *et al.*, A deep learning based traffic sign detection for intelligent transportation systems, in *Computational Data and Social Networks*, , Lecture Notes in Computer Science Vol. 13116 (Springer, Cham, 2021).
5. R. Kukreja, S. Rinchen, B. Vaidya and H. Mouftah, Evaluating traffic signs detection using faster r-cnn for autonomous driving, in *2020 IEEE 25th International Workshop on Computer Aided Modeling and Design of Communication Links and Networks (CAMAD)*, (IEEE, 2020).
6. Y. Fan and W. Zhang, Traffic sign detection and classification for advanced driver assistant systems, in *2015 12th International Conference on Fuzzy Systems and Knowledge Discovery (FSKD)*, (IEEE, 2015).
7. Y. Saadna and A. Behloul, An overview of traffic sign detection and classification methods, *International Journal of Multimedia Information Retrieval* **6**, 193 (2017).
8. H. H. Aghdam and E. J. Heravi, Classification of traffic signs, in *Guide to Convolutional Neural Networks*, (Springer, Cham, 2017)
9. S. Maldonado-Bascon, S. Lafuente-Arroyo, P. Gil-Jimenez, H. Gomez-Moreno and F. Lopez-Ferreras, Road-sign detection and recognition based on support vector machines, *IEEE Transactions on Intelligent Transportation Systems* **8**, 264 (2007).
10. B. Jijo and A. M. Abdulazeez, Classification based on decision tree algorithm for machine learning, *Journal of Applied Science and Technology Trends* **2**, 20 (2021).
11. S. Dutta, B. C. Manideep, S. Rai and V. Vijayarajan, A comparative study of deep learning models for medical image classification, *IOP Conference Series: Materials Science and Engineering* **263**, p. 042097 (2017).
12. A. Krizhevsky, I. Sutskever and G. E. Hinton, Imagenet classification with deep convolutional neural networks, in *Advances in Neural Information Processing Systems*, 2012.
13. A. Voulodimos, N. Doulamis, A. Doulamis and E. Protopapadakis, Deep learning for computer vision: A brief review, *Computational Intelligence and Neuroscience* **2018** (2018).

A Novel 2-Tuple Linguistic Integrated WINGS Approach to Uncover Circular Economy Barriers for Smart Agriculture

Gülçin Büyüközkan*,‡ and Deniz Uztürk†,§

*Department of Industrial Engineering, Galatasaray University, Ortaköy, Istanbul 34349, Türkiye

†Department of Business Administration, Business Research Center, Galatasaray University, Ortaköy, Istanbul 34349, Türkiye

‡gbuyukozkan@gsu.edu.tr

§duzturk@gsu.edu.tr

This chapter tackles the critical barriers hindering the integration of Circular Economy (CE) and Smart Agriculture (SA). Employing a novel 2-Tuple Linguistic (2TL) model integrated WINGS (Weighted Influence Non-linear Gauge System) methodology within a Group Decision-Making (GDM) framework for the first time, we unveil the complex cause-and-effect relationships between these barriers. By capturing expert assessments with nuanced linguistic terms and information differences, our approach surpasses traditional methods in its accuracy and interpretability. The identified barriers pave the way for targeted interventions and advancements toward a more sustainable and resilient agricultural future. A case study with Turkish experts and a comparative analysis are provided to validate the novel suggested methodology.

Keywords: 2-Tuple linguistic model; circular economy; smart agriculture; GDM; WINGS.

1. Introduction

Efficient decision-making plays a vital role in Smart Agriculture (SA) and the advancement of circular economy (CE), aiming to maximize resource efficiency and bolster sustainability. This chapter explores decision-making in SA, highlighting the role of linguistic variables in facilitating informed processes. While data-driven technologies empower stakeholders with vast information, nuanced decisions require human expertise beyond algorithmic calculations.

The 2-tuple linguistic (2TL) model, introduced by Herrera and Martinez in 2000, presents a robust solution to the complexities inherent in decision-making processes [1]. The 2TL model enhances decision-making accuracy by capturing finer nuances and reducing information loss through its extension of traditional

‡Corresponding author.

linguistic variables [2]. The model's increased accuracy, interpretability, and broader applicability validate its usefulness as a valuable tool for addressing barriers in SA and CE integration.

This chapter aims to address two research questions (RQs) — RQ1: What are the primary barriers concerning the integration of CE principles in SA? RQ2: How linguistic Multi-Criteria Decision-Making (MCDM) techniques can be utilized to assess and prioritize these barriers effectively? Through an extensive review of literature and validation by field experts, the main barriers are derived, forming a foundational basis to enhance the mutual integration of CE and SA. Plus, by integrating the 2TL model with group decision-making (GDM)-WINGS (Weighted Influence Non-linear Gauge System) technique [23] for the first time, this chapter presents a novel approach.

In summary, this chapter contributes to the MCDM and linguistic decision-making literature in several ways. First, it introduces the integration of the 2TL model with GDM-WINGS technique, providing a novel methodological framework for addressing complex decision-making problems. Second, the identification and validation of primary barriers for CE in SA contribute to a deeper understanding of the challenges. Lastly, the comparative analyses conducted provide empirical evidence to validate the proposed method's effectiveness.

This chapter is organized as follows. Section 2 presents the literature review, Section 3 outlines the suggested 2TL-GDM-WINGS methodology, and Section 4 details the case study conducted with Turkish experts and its results. The final section include discussions and concluding remarks.

2. Literature Review

In this section, the literature review will evaluate existing research on MCDM, the 2TL model, as well as literature related to SA and CE.

2.1. *MCDM and 2TL*

In this section, we will review recent studies from 2020 onwards that have incorporated the 2TL model into MCDM methodologies. The selected studies will be evaluated based on their objectives, application areas, inclusion of case studies and GDM methodology, integrated MCDM techniques, and extensions of the 2TL model. Analyzing the presented studies reveals a compelling trend: the 2TL emerges as a powerful tool for incorporating expert insights and nuanced linguistic assessments within MCDM frameworks. This consistent application across a range of case studies strongly suggests that 2TL offers a potent solution for enhancing the accuracy, interpretability, and overall effectiveness of MCDM processes.

Table 1. The 2TL-MCDM studies from 2020.

Year and Ref.	Objective of the study	Integrated MCDM technique	Case Study	GDM	2TL Extension
2024 [3]	Creating an enhanced framework for hotel categorization by incorporating expert insights	CRITIC, AHP	✓	✓	X
2023 [4]	Developing a comprehensive framework for linguistic information processing in MCDM.	X	✓	✓	2-TCLE
2023 [5]	Analysis of influential factors impacting cloud computing technology service providers	DEMATEL	✓	✓	X
2023 [6]	Clustering for hotel segmentation	X	✓	✓	X
2023 [7]	Measuring how satisfied customers are with luxury hotels in London	AHP	✓	✓	X
2022 [8]	Suggesting a novel method for choosing transportation initiatives in mid-sized urban areas.	AHP-TOPSIS	✓	✓	X
2022 [9]	Enhancing the linguistic interpretability of the Quality of Life Index (QOLI) using linguistic quantifiers	OWA	✓	X	X
2022 [10]	Assessing the sustainability of renewable energy	AHP, Grey relation analysis	✓	✓	X
2020 [11]	Choosing Sustainable Urban Logistics Fleet Vehicles	SAW-VIKOR	✓	✓	X

Despite the demonstrated effectiveness of 2TL-MCDM in diverse fields, a notable gap exists in its application within SA. Therefore, this paper proposes the utilization of the 2TL-GDM-WINGS methodology to explore barriers and their interrelations of CE in the realm of SA.

2.2. *CE and SA*

The CE challenges the traditional "take-make-waste" model by advocating for closed-loop systems that minimize waste, regenerate resources, and optimize resource utilization [12,13]. This paradigm shift is crucial for environmental sustainability discourse [15]. SA, defined by Zhang [16], applies advanced technologies and data-driven solutions to enhance agricultural sustainability, aligning with CE's goals [17], [18]. The 2TL model, incorporating expert judgment and linguistic assessments, offers a comprehensive evaluation of barriers, as demonstrated by [6], [10]. WINGS methodology, with its 2TL extension, enhances accuracy and interpretability, revealing deeper cause-and-effect relationships. This combined approach provides a nuanced understanding of the complexities hindering CE and SA integration. This study explores the multifaceted barriers hindering the integration of CE principles within SA. We examine five key dimensions: Product and Operations (waste generation, resource recovery) [18,19], Supply Chain and Collaboration (transparency, logistics) [18,19], Governance and Business Models (policy support, economic viability)

[20], Technology and Skills (infrastructure, knowledge gaps) [20,21], and Stakeholder and Cooperation (producer-consumer engagement, information sharing) [18,21]. Analyzing these dimensions is crucial for understanding the challenges faced in transitioning towards a closed-loop agricultural system that leverages smart technologies for a more sustainable future.

Table 2. The CE barriers for SA. ([18], [19], [20], [21])

Main Dimensions	Barriers	Main Dimensions	Barriers
Product and Operations (PO)	Design (PO1)	Technology and Skills (TS)	Lack of access to technology (TS1)
	Production (PO2)		Data management challenges (TS2)
	Logistics (PO3)		Lack of workforce skills (TS3)
Supply Chain and Collaboration (SCC)	Fragmentation (SCC1)	Stakeholders and Cooperation (SC)	Weak engagement (SC1)
	Lack of trust (SCC2)		Conflicting interests (SC2)
	Limited Market access (SCC3)		Lack of collaboration (SC3)
Governance and Business Models (GBM)	Policies and regulations (GBM1)		
	Financing (GBM2)		
	Pricing (GBM3)		

3. Proposed Methodology

The 2TL model is a fuzzy logic-based model [22], minimizes information loss when converting linguistic data into numerical form. It employs a 2-Tuple form of *(s, α)*, where *'s'* is a linguistic label and *'α'* is a numerical value [1]. For the necessary and basic definitions, readers can refer to [2]. The leading translation equation of 2TL is given as follows:

$$\Delta_s : [0, g] \rightarrow \overline{S}$$
$$\Delta_s(\beta) = (S_i, \alpha), with \begin{cases} i = round(\beta) \\ \alpha = \beta - i \end{cases} \tag{1}$$

A *Linguistic Hierarchy (LH)* is the union of all levels *t*, where each level *t* corresponds to a linguistic term set symmetrically distributed with an odd granularity [2]. The transformation function to translate a linguistic term set with granularity *n(t)* to a linguistic term set having granularity *n(t')* is as follows:

$$TF_{t'}^{t} = (S_i^{n(t)}, \alpha^{n(t)}) = \Delta\left(\frac{\Delta^{-1}((S_i^{n(t)}, \alpha^{n(t)}) \times (n(t') - 1)}{n(t) - 1}\right) \tag{2}$$

WINGS, an advancement from DEMATEL, considers both "influence intensity" and "standard strength" to analyze factor relationships. This approach offers a refined method for solving management issues by comprehensively assessing factor impacts [23]. For details on the WINGS procedure, refer to [24]. The steps of the proposed novel 2TL-GDM-WINGS methodology are given as follows:

Step 1: In the WINGS model, each component's strength represents its importance and interconnections, provided as input by the decision-maker (DM) [24]. Components vary in strength, with some starting as negligible (zero) but may gain value through interactions, similar to DEMATEL [24]. The 2TL-GDM-WINGS methodology employs 2TL linguistic sets for DMs to evaluate strengths and influences, offering two levels based on their field experience. The LH aspect of 2TL assists in consolidating diverse granulated sets into the highest one. The proposed linguistic sets are detailed as follows:

S^5: No (N)-Low(L)- Medium (M)- High(H)-Extremely High (EH);

S^9: No (N)-Very Low (VL)-Low(L)-Almost Medium (AM)- Medium (M)-Almost High (AH)-High(H)- Very High (VH)-Extremely High (EH).

To maintain consistency in evaluations, DMs are requested to assess strengths and influences using the same linguistic sets assigned to them, with the additional labels "influence" and "strength." Barriers lacking evaluations for both aspects will be categorized as "no strength" and "no influence".

Step 2: Aggregation of DMs assessment.

After applying Step 1, the evaluations of strengths and importance for each component are obtained, the visual causal map is one of the effective ways to show them. After obtaining each assessment from DMs, first step is to unify the different granulated assessments. By using Eq. (2) the variated linguistic sets are uniformed and afterwards, by using the following aggregation operator (*Weighted Average Operator (WAO)*) the DMs' opinions are aggregated.

$$\vec{x} = \left(\frac{\sum_{i=1}^{n} \Delta^{-1}(r_i^k, \alpha_i) \times \Delta^{-1}(w_k, \alpha_k)}{\sum_{k=1}^{n} \Delta^{-1}(w_k, \alpha_k)}\right) = \Delta\left(\frac{\sum_{i=1}^{n} \beta_i^k \times w_k}{\sum_{k=0}^{n} w_k}\right) \tag{3}$$

Here r_i^k refers to the assessment of *kth* DM on *ith* component. and w_k refers to the weight assigned to the *kth* DM.

Step 3: The aggregated assessments turned into beta values as in Eq. (1) and inserted into the direct strength-influence *matrix S*. This is $n\, x\, n$ matrix with elements s_{ij}. 2TL beta strength values for components are placed along the principal diagonal (s_{ii}= strength of component i), while influence 2TL beta values are inserted as s_{ij} for the influence of component i on component j.

Step 4: *Matrix S* is adjusted according to the next relation:

$$C = {}^{1}/_{L} . S \tag{4}$$

where *L* is defined as sum of all elements of *matrix S*.

Step 5: Calculation of total strength-influence *matrix T* with the following relation:

$$\mathrm{T} = \mathrm{C} + \mathrm{C}^2 + \mathrm{C}^3 + \cdots = {}^{\mathrm{C}}\!/_{\mathrm{I} - \mathrm{C}} \tag{5}$$

Here is the *I* refers to the identity matrix, *C* is the calibrated (normalized) direct strength-influence *matrix.*

Step 6: For each barrier the row sum r_i and column sum c_j of *matrix T* are calculated. For each barrier $r_i + c_j$ and $r_i - c_j$ values are calculated.

Step 7: The r_i and c_j represent the overall impact and receptivity of a component. The sum $r_i + c_j$ indicates the total involvement of the component in the system, with the sign indicating its role or position within the system: a positive value signifies membership in the influencing group, while a negative value indicates membership in the influenced group.

4. Case Study

In this case study, five experts were engaged, including three academics specializing in agriculture, supply chain, and sustainability, alongside two consultants focused on CE and sustainability. To accommodate varying expertise levels, nine scaled linguistic sets were provided to academics and five to consultants. After presenting a list of derived barriers to each expert, they were tasked with refining them based on their insights. All experts unanimously accepted the model (Table 2) and assigned strength values to identified barriers. These values were aggregated to construct a comprehensive causal map, illustrating barrier interrelations. Despite page restrictions limiting detailed sub-dimensions, the methodology employed remains adaptable for further application.

The causal graph representation of the main barriers are given in the Figure 1. The barriers identified as influential for CE in SA are Technology and Skills (TS) and Stakeholders and Cooperation (SC). Therefore, future research efforts should focus on understanding how these barriers impact other aspects and devising strategies to overcome them.

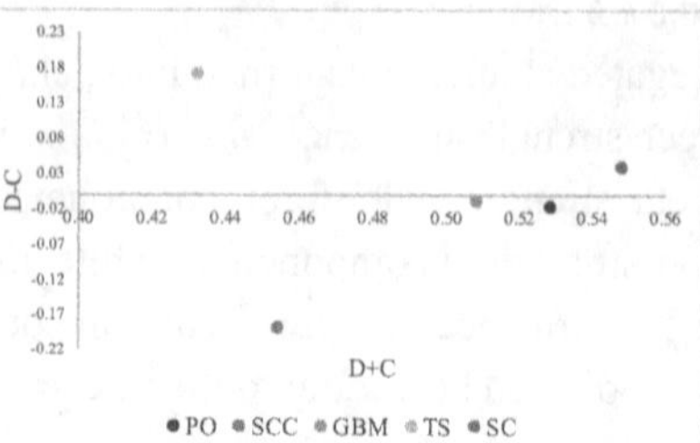

Figure 1. Auxiliary chart for the results of 2TL-GDM-WINGS application.

5. Discussions and Concluding Remarks

Table 3 provides a comparative analysis with the application of 2TL-DEMATEL. As the results indicate, 2TL-GDM-WINGS yields comparable outcomes to

2TL-DEMATEL, though with slight variations in the assessment of barrier influence. While the cause/effect clustering remains consistent, the influence levels differ between the two methodologies. Additional insights are provided during the presentation. Our findings highlight two critical barriers to achieving a CE within SA: TS and SC. The limitations in TS can hinder the adoption and effective use of precision farming tools and data analysis for optimizing resource use and minimizing waste [18, 20]. Additionally, inadequate knowledge and training for farmers regarding CE principles and practices within SA can impede progress [18]. Furthermore, the lack of SC, particularly in areas like producer–consumer collaboration and information sharing across the value chain, can limit the development of robust circular business models and effective waste management strategies.

Table 3. Comparative analysis.

	2TL-GDM-WINGS		2TL-GDM-DEMATEL	
Dimensions	D+C	D-C	D+R	D-R
PO	0.53	-0.02	3.49	-0.11
SCC	0.45	-0.19	3.62	-1.50
GBM	0.51	-0.01	3.16	-0.07
TS	0.43	0.17	3.69	1.39
SC	0.55	0.04	3.74	0.29

Through a case study, the methodology identifies primary barriers for CE in SA and highlights the WINGS methodology's capability to provide influential levels, enhancing the understanding of barrier relationships. While similar outcomes to DEMATEL are observed, WINGS extends the analysis by identifying not only direct influences between barriers but also unidirectional and indirect influences, enabling tailored solutions that address root causes rather than immediate symptoms. However, reliance on expert-based case studies involving Turkish experts may limit generalizability to other cultures, emphasizing the need for future research to explore leveraging these insights for developing targeted mitigation strategies.

Acknowledgments

The authors extend their sincere gratitude to the experts for their invaluable guidance and assistance. Also the authors thanks to Galatasaray University for their support by the Scientific Research Projects Commission.

References

1. F. Herrera and L. Martínez, "A 2-tuple fuzzy linguistic representation model for computing with words," *IEEE Transactions on fuzzy systems*, vol. 8, no. 6, pp. 746–752, 2000.
2. L. Martínez, R. M. Rodriguez, and F. Herrera, *The 2-tuple Linguistic Model.* Cham: Springer International Publishing, 2015.
3. Z. Shu, R. A. Carrasco González, J. P. García-Miguel, and M. Sánchez-Montañés, "A Model Integrating the 2-Tuple Linguistic Model and the CRITIC-AHP Method for Hotel Classification," *SN COMPUT. SCI.*, vol. 5, no. 1, p. 9, Nov. 2023.
4. Y. Liu, H. Zhou, R. M. Rodríguez, and L. Martínez, "Fuzzy encoding and decoding approaches for 2-TCLE and their applications in multi-criteria decision making," *Information Sciences*, vol. 640, p. 119033, Sep. 2023.
5. G. Büyüközkan, D. Uztürk, and A. Maden, "Influential factor analysis for cloud computing technology service provider," *Technological Forecasting and Social Change*, vol. 192, p. 122531, Jul. 2023.
6. Z. Shu, R. A. Carrasco González, J. P. García-Miguel, and M. Sánchez-Montañés, "Clustering using ordered weighted averaging operator and 2-tuple linguistic model for hotel segmentation: The case of TripAdvisor," *Expert Systems with Applications*, vol. 213, p. 118922, Mar. 2023.
7. Z. Shu, M. H. Torralba, R. A. Carrasco, and M. F. B. López, "Assessing customer satisfaction of London luxury hotels with the AHP method and the SERVPERF scale: a case study of customer reviews on TripAdvisor," *Procedia Computer Science*, vol. 221, pp. 73–80, Jan. 2023.
8. R. R. da Silva, G. D. Santos, and D. Setti, "A multi-criteria approach for urban mobility project selection in medium-sized cities," *Sustainable Cities and Society*, vol. 86, p. 104096, Nov. 2022.
9. Z. Shu, R. A. Carrasco, J. P. García-Miguel, and M. Sánchez-Montañés, "Multiple Scenarios of Quality of Life Index Using Fuzzy Linguistic Quantifiers: The Case of 85 Countries in Numbeo," *Mathematics*, vol. 10, no. 12, Art. no. 12, Jan. 2022.
10. S. Li and Y. Hu, "A Multi-Criteria Framework to Evaluate the Sustainability of Renewable Energy: A 2-Tuple Linguistic Grey Relation Model from the Perspective of the Prospect Theory," *Sustainability*, vol. 14, no. 8, Art. no. 8, Jan. 2022.
11. G. Büyüközkan and D. Uztürk, "Fleet Vehicle Selection for Sustainable Urban Logistics," in *Proceedings of the 2020 The 9th International Conference on Informatics, Environment, Energy and Applications*, Amsterdam Netherlands: ACM, Mar. 2020, pp. 116–120.
12. S. Faizi, T. Rashid, and S. Zafar, "TODIM approach based on score function under hesitant 2-tuple linguistic environment," *Journal of Intelligent & Fuzzy Systems*, vol. 38, no. 1, pp. 663–673, Jan. 2020.
13. R. Agrawal, V. A. Wankhede, A. Kumar, S. Luthra, and D. Huisingh, "Progress and trends in integrating Industry 4.0 within Circular Economy: A comprehensive

literature review and future research propositions," *Bus Strat Env*, vol. 31, no. 1, pp. 559–579, Jan. 2022.

14. S. L. Ngan *et al.*, "Prioritization of sustainability indicators for promoting the circular economy: The case of developing countries," *Renewable and Sustainable Energy Reviews*, vol. 111, pp. 314–331, Sep. 2019.
15. S. Kumar, R. D. Raut, K. Nayal, S. Kraus, V. S. Yadav, and B. E. Narkhede, "To identify industry 4.0 and circular economy adoption barriers in the agriculture supply chain by using ISM-ANP," *Journal of Cleaner Production*, vol. 293, p. 126023, Apr. 2021.
16. Q. Zhang, "Opinion paper: Precision agriculture, smart agriculture, or digital agriculture," *Computers and Electronics in Agriculture*, vol. 211, p. 107982, Aug. 2023.
17. Z. Liu, J. Liu, and Y. Xu, "Research on the path of agriculture sustainable development based on the concept of circular economy and big data," *Acta Agriculturae Scandinavica Section B-Soil And Plant Science*, vol. 71, no. 9, pp. 1024–1035, Aug. 2021.
18. X. Xia and J. Ruan, "Analyzing Barriers for Developing a Sustainable Circular Economy in Agriculture in China Using Grey-DEMATEL Approach," *Sustainability*, vol. 12, no. 16, p. 6358, Aug. 2020.
19. K. Feng and C.-Y. Lam, "An Overview of Circular Economy in China: How the Current Challenges Shape the Plans for the Future," *Chin. Econ.*, vol. 54, no. 5, pp. 355–371, 2021.
20. M. A. Khan and R. Mahajan, "Exploring factors influencing circular economy adoption and firm-level practices in the agribusiness sector: an exploratory study of Indian firms," *Environ. Dev. Sustain.*, Dec. 2023.
21. N. N. Batova, I. E. Tochitskaya, and P. Sachek, "Circular Economy in Agriculture: Conceptual Basis and Implementation Possibilities in Belarus," *Proc. Natl. Acad. Sci. Belarus-Agrar. Ser.*, vol. 59, no. 3, pp. 277–291, 2021.
22. L. A. Zadeh, "Fuzzy sets," *Information and Control*, vol. 8, no. 3, pp. 338–353, Jun. 1965.
23. J. Michnik, "Weighted Influence Non-linear Gauge System (WINGS) – An analysis method for the systems of interrelated components," *European Journal of Operational Research*, vol. 228, no. 3, pp. 536–544, Aug. 2013.
24. J. Michnik, "The WINGS method with multiple networks and its application to innovation projects selection," *International Journal of Applied Management Science*, Jun. 2018, Accessed: Jan. 31, 2024. [Online]. Available: https://www.inderscienceonline.com/doi/10.1504/IJAMS.2018.092077

Tailoring Preference-based Food Recommendation for Diabetic Patients

M. J. Barranco*, R. Yera and L. Martínez

Computer Science Department, University of Jaén, Jaén, Spain
**barranco@ujaen.es*

Food recommendation is currently becoming an area of special interest regarding its high impact on the end users. In this direction, for individuals with non-communicable diseases such as diabetes, a tailored food recommendation according to the user's preferences and needs and the available nutritional knowledge could be derived to relevant benefits regarding the importance of an appropriate diet in such diseases. The current contribution presents then a novel preference-based food recommendation approach tailored to diabetes patients. An experimental analysis taking as base the USDA database suggests that in this context, it is reachable a balance between the sugar levels and the overall preferences of the generated menus.

Keywords: Food recommendation; diabetes; user preferences.

1. Introduction

In the context of e-health, the recommendation of dietary choices is gaining prominence as a significant advisory scenario, considering its close connection to non-communicable diseases and the concept of personalized nutrition.[1] Personalized nutrition is characterized as tailored healthy eating guidance based on individual factors, such as genetic data, personal health status, lifestyle, nutrient intake, and phenotypic data.[2] Due to the expenses associated with managing genetic data, recent research efforts have intensely focused on handling alternative data sources, leading to the proposal of various computational solutions.

Specifically addressing diabetes, numerous studies have demonstrated that adopting more favorable dietary patterns is linked to lower glycemic load in older adults.[3] Moreover, low-carbohydrate, Mediterranean, and high-protein diets have been recommended as effective in enhancing various cardiovascular risk markers in individuals with diabetes, warranting consideration in the overall diabetes management strategy.[4] Consequently, a tailored personalized nutrition approach in this context not only contributes directly to patient safety but also holds global significance. Diabetes is a prevalent global health crisis, estimated to incur an annual cost of US$827 billion. In 2021, it was reported that 537 million adults worldwide were living with diabetes.[a]

Despite numerous efforts in recent years concentrating on the creation of recommendations for healthy food, the motivation for this research work within the framework of existing knowledge stems from the observed absence of comprehensive approaches for food recommendation tailored to the domain of diabetes. As recently pointed out by Yera *et al.*[5] in a recent survey on this topic, it is necessary for a better exploitation of the knowledge domain for proposing computational solutions in this scenario, and in addition, it is necessary for a larger exploitation of the user preferences in the whole recommendation process.

This work is focused on these two directions, by tailoring a preference-based food recommendation model, to be used in the diabetes domain. Specifically, our goal is to exploit medical criteria used for managing nutrition in diabetic patients to accomplish our objective.

The contribution is structured as follows. Section 2 presents a brief background on food recommendations for diabetes patients. Section 3 discusses our approach for generating tailored menu recommendations for diabetic patients. Section 4 illustrates the performed experiments. Section 5 concludes the contribution.

2. Background

The generation of food recommendations for diabetic patients has been a research goal that have attracted attention in the last few years. Yera et al.[5] identifies four groups of works focused on this objective:

- Semantic-driven methodologies, encompassing those that involve the integration of various forms of semantic knowledge management[6].
- Approaches centered around optimization, which aim to incorporate methods utilizing restriction-based or optimization-based models for the generation of food recommendations[5].
- Approaches based on rules and classifications, presenting a diverse range of works incorporating techniques such as classification, mul-

[a]https://www.makingdiabeteseasier.com/es/diabetes-explicada/diabetes/panorama-mundial-diabetes-2021.

ticriteria decision-making, fuzzy inference, or production rules to support food recommendations[7].

- Interaction-centric approaches, characterized by a shared attribute of engaging in an interactive process with users to derive final recommendations[5].

Beyond the previous efforts done by the research community in this direction, Yera et al.[5] pointed out the following shortcomings:

- The lack of a a cohesive framework intended to serve as a foundation for subsequent research in generating nutritional recommendations for individuals with diabetes.
- The scarce utilization of the related knowledge domain to suggest computational solutions in this context.
- The necessity of a heightened integration of user preferences in the generation of recommendations.

Taking account this scenario, the current contribution attempts to performing the first steps towards the development of preference-based food recommendation approach, tailored to the diabetes domain.

3. Generating Tailored Menu Recommendations for Diabetic Patients

The overall objective of the current work is the development of computational approaches focused on reaching improved menu recommendations regarding their composition, tailored for diabetic patients, and keeping a high overall preference with them.

As starting point it is taken the previously proposed model focused on menu recommendation based on user preferences and nutritional information developed by Yera et al.[1], which is not focused on any specific disease:

$$Maximize \sum_{k \in A} w_k f_k \tag{1}$$

s.t.

$|\sum_j (nt_{kj} * f_k) - b_j| \leq \alpha,$ *for each nutrient* $1, 2, 3, ..., J$

$\sum_{k \in G_a} f_k = n_{G_a},$ *for each* $n_{G_a} \in \{n_{G_1}, n_{G_2}, n^l_{G_3}, n^l_{G_4}, n^l_{G_5}, n^d_{G_3}, n^d_{G_4}, n^d_{G_5}, n_{G_6}, \},$

being G_a *the groups in Table* 1.

The menu recommendation generation in this context is driven by an objective function (based on the user preferences), and two group of restrictions based on a) nutrients requirements (proteins, carbohydrates and lipids), and b) diversity of foods from different groups, that should contain the menu (milk, cereals, meat, fruits, etc).

According to Yera et al.[1], this model is focused on:

(1) **Maximizing the sum of preferences w_i of all the foods i included in the plan.** Herein, w_i depends on the frequent of consumption of the food i, as well as how recent was its last consumption (Equation 2).

$$w_k = \frac{N_k}{N}(e^{\theta(\frac{t_c - t_k}{t_c})} - 1) \tag{2}$$

(2) **Verifying that the nutrients of the generated plan are very close to the required nutrients for the current user profile.** This goal is verified by assuring that for each nutrient, the absolute difference between the required amount (b_j) and the final amount $\sum_j (nt_{kj} * f_k)$, is always under a threshold α.

(3) **Guaranteeing that the generated plan fills the menu templates presented in Table 1.** This table defines several groups of foods, considered in the composition of the generated menus.

Table 1. The template for the daily meal plan

Breakfast
n_{G_1} foods of group G_1 (Milk, yogurts)
n_{G_2} foods of group G_2 (Breakfast cereals)
n_{G_6} foods of group G_6 (Fruits)
Lunch
$n^l_{G_3}$ foods of group G_3 (Proteins)
$n^l_{G_4}$ foods of group G_4 (Carbohydrates)
$n^l_{G_5}$ foods of group G_5 (Vegetables)
n_{G_6} foods of group G_6(Fruits)
Dinner
$n^d_{G_3}$ foods of group G_3 (Proteins)
$n^d_{G_4}$ foods of group G_4 (Carbohydrates)
$n^d_{G_5}$ foods of group G_5 (Vegetables)
n_{G_6} foods of group G_6(Fruits)

Based on this framework, the goal of the current work is to tailoring the screened menu recommendation task, for the context of the diabetic patients.

To accomplish this task, primarily we have adopted the following premise that is supported by the related literature, as explained below.

Premise: *Reducing sugar consumption benefits diabetic patients.* It has been identified that diabetics should control in a special way the consumption of sugar (rapid-absorption carbohydrates), replacing them by low-absorption carbohydrates[8–10].

Related to this goal, the literature has identified that:

- Fructose consumed as "free fructose" (i.e., naturally occurring in foods such as fruit, (that also contain fiber) may result in better glycemic control compared with isocaloric intake of sucrose or starch[8].
- Gray et al.[8] also suggest the replacing of high GI (glycemic index) foods with low GI foods; as well as decreasing the consumption of sugary foods like cookies, cakes, candy, and soft-drinks.
- People with diabetes and those at risk are advised to avoid sugar-sweetened beverages (including fruit juices) in order to control glycemia and weight[8].
- Although all carbohydrates can be incorporated into carbohydrate counting, for good health, carbohydrates from vegetables, fruits, whole grains, legumes, and dairy products take priority over other carbohydrate sources, especially those that contain added fats, sugars, or sodium[9].

Based on this statement that supports the tailoring of the food recommendation considering preferences and nutritional knowledge for the diabetes domain, we screen the following issue:

Can we obtain menu recommendations in which sugar restrictions on the daily menu are applied with no (or minimal) reduction in user preference?

To solve this issue, we will consider a different objective function in relation to the formerly proposed model, focused on reducing the overall amount of sugar in the generated menu, without affecting the preference w_k of the user over the generated menu.

This objective function (Equation 3) is represented through a convex combination of the preference w_k of the user over the item k, and the amount of sugar that this item has. In the referred equation, such amount is represented through a normalize way by taking as reference the maximum amount of sugar that could be associated to an item.

$$Max \sum_{k \in A} ((1-\beta) w_k + \beta \frac{compo^{sugar}_{max} - compo^{sugar}_{k}}{compo^{sugar}_{max}}) f_k \quad (3)$$

4. Experiments

The evaluation of the presented framework uses the USDA National Nutrient Database [b], which characterizes each food according to their macronutrients and micronutrients. We link each food in this dataset with a food group of the menu template illustrated in Table 1, being considered six groups: 1) Milk, yogurt; 2) Breakfast cereals; 3) Proteins; 4) Carbohydrates, 5) Vegetables, and 6) Fruits.

For testing the proposed model, different types of user profiles are built taking into account a higher or lower sugar consumption. Three types are particularly built: 1) $Profile_{80}$, with users with a daily consumption of up to 80mg of sugar, 2) $Profile_{60}$, with users consuming up to 60mg, and 3) $Profile_{40}$, with users consuming up to 40mg. For each category 500 users profiles are created, each one with 10 associated menus generated in a random way satisfying the sugar-related constraints.

Based in this setting, we generate a menu considering the formerly presented model (Equation 1) and assuming the objective function presented in Equation 3. We consider several values of β in the range [0,1]. For the generated menus it will be considered as evaluation criteria: 1) the overall preference over the generated menu, and 2) the amount of sugar in the menu.

In the first case, Figure 1 illustrates how the overall user preference on the generated menu evolved with different values of β, considering the three mentioned types of users respectively for the subfigures a), b), and c). Here the X-axis illustrates the sugar restriction value (lower values mean greater restriction) and the Y-axis the preferences of the users.

Figure 2 follows a similar purpose and structure, with the particularity that in this case the Y-axis illustrates the actual amount of sugar in the generated menus.

The experiment shows that as the restriction on sugar in the optimization formula increases (i.e., as the value of x decreases), both overall preference and the actual amount of sugar in the menu decrease. However, a

[b]https://data.nal.usda.gov/dataset/composition-foods-raw-processed-prepared-usda-national-nutrient-database-standard-referen-10

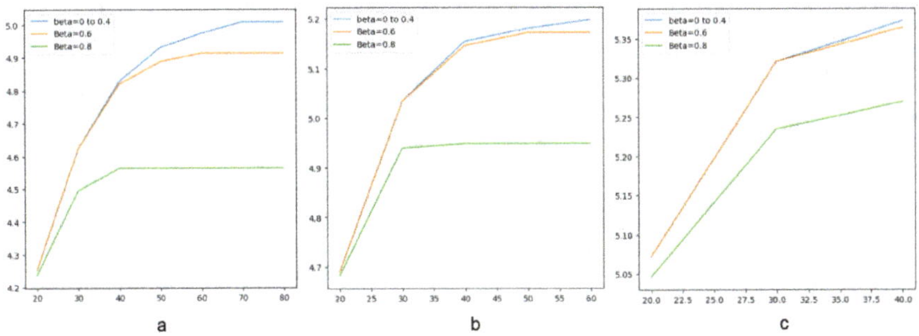

Fig. 1. Evolving of sugar restrictions in relation to the overall preferences of the generated menus. a) $Profile_{80}$ case. b) $Profile_{60}$ case. c) $Profile_{40}$ case.

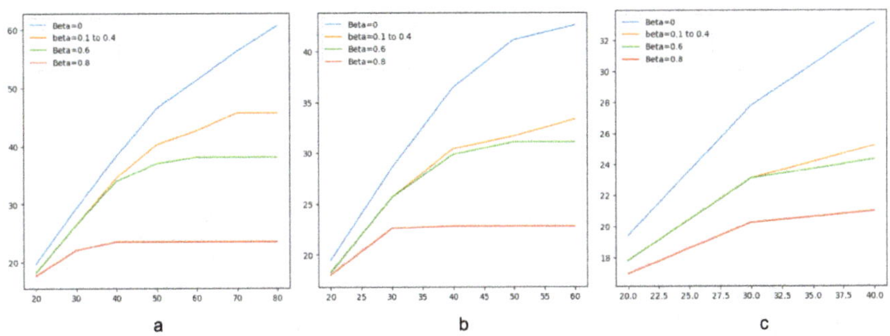

Fig. 2. Evolving of sugar restrictions in relation to actual amount of sugar in the generated menus. a) $Profile_{80}$ case. b) $Profile_{60}$ case. c) $Profile_{40}$ case.

greater decrease in the actual amount of sugar is observed than in the value of overall preference, particularly when the increase in restriction is slight. For example, in $profile_{80}$ (users who typically consume 80g of sugar daily), applying a value of $\beta = 0.4$ in the formula, we have a reduction of 30g of sugar per day, approximately, while only losing one-tenth in user preference value.

5. Conclusions

This research work has been focused on the exploration of a novel preference-based food recommendation approach tailored to diabetes patients. As a main contribution, the proposal suggests the reaching of an optimization criterion over the generated menus, which balances the sugar level and the overall user preferences. Several experimental settings were explored, identifying the peculiarites of each scenario. Future works will be focused on a more extensive exploitation of the medical knowledge in this scenario.

Acknowledgments

This research is supported by the Andalusian PAIDI 2020 Program under the project PROYEXCEL_00257.

References

1. R. Yera, A. A. Alzahrani and L. Martinez, A food recommender system considering nutritional information and user preferences, *IEEE Access* **7**, 96695 (2019).
2. M. J. Gibney and M. C. Walsh, The future direction of personalised nutrition: my diet, my phenotype, my genes, *Proceedings of the Nutrition Society* **72**, 219 (2013).
3. M. Davis, C. Miller and D. Mitchell, More favorable dietary patterns are associated with lower glycemic load in older adults, *Journal of the American Dietetic Association* **104**, 1828 (2004).
4. O. Ajala, P. English and J. Pinkney, Systematic review and meta-analysis of different dietary approaches to the management of type 2 diabetes, *The American Journal of Clinical Nutrition* **97**, 505 (2013).
5. R. Yera, A. A. Alzahrani, L. Martínez and R. M. Rodríguez, A systematic review on food recommender systems for diabetic patients, *International Journal of Environmental Research and Public Health* **20**, p. 4248 (2023).
6. N. S. Selvan, S. Vairavasundaram and L. Ravi, Fuzzy ontology-based personalized recommendation for internet of medical things with linked open data, *Journal of Intelligent & Fuzzy Systems* **36**, 4065 (2019).
7. O. M. Omisore, B. A. Ojokoh, A. E. Babalola, T. Igbe, Y. Folajimi, Z. Nie and L. Wang, An affective learning-based system for diagnosis and personalized management of diabetes mellitus, *Future Generation Computer Systems* **117**, 273 (2021).
8. A. Gray and R. J. Threlkeld, Nutritional recommendations for individuals with diabetes, in *Endotext [Internet]*, (MDText. com, Inc., 2019)
9. M. J. Franz, J. L. Boucher and A. B. Evert, Evidence-based diabetes nutrition therapy recommendations are effective: the key is individualization, *Diabetes, metabolic syndrome and obesity: targets and therapy*, 65 (2014).
10. M. E. Lean and L. Te Morenga, Sugar and type 2 diabetes, *British Medical Bulletin* **120**, 43 (2016).

"Unveiling the Invisible": Deep Learning-based Semantic Segmentation for Analyzing Activity Patterns

Gurkiran Kaur* and Li Zhang†

Department of Computer Science, Royal Holloway, University of London
Surrey, TW20 0EX, UK
**zkac303@live.rhul.ac.uk*
†li.zhang@rhul.ac.uk

The ubiquity of internet-enabled devices has led to a rapid increase in the use of connected cameras for real-time monitoring, creating a high demand for (automated) visual data analytics across various industries. The prospect of automating visual data analysis to drive positive change involves extracting actionable insights from data that will inform decision-making processes, improving efficiency, and contributing to evidence-based strategies across diverse applications and industries. This research explores and compares well-known semantic segmentation models such as DeepLabV3+ and UNet, determining the best-suited for use in a visual analytics and scene understanding, culminating in a proof of concept program capable of automating video analysis, plotting detections, average trajectories, and identifying outliers.

Keywords: DeepLabV3+; automated data analysis; semantic segmentation.

1. Introduction

Analysis of activity patterns within a scene using object tracking and visual data analytics have long been the focus of extensive research and development efforts, as evidenced by the multitude of seminal works across a range of industries [1]. These techniques have important applications across a wide range of industries and scenarios, from surveillance and autonomous vehicles to healthcare in the automated identification of cancerous tumors. The ability to monitor, analyze, and interpret visual data streams has revolutionized decision-making processes and driven advancements across diverse domains.

The research delves into the complex technical landscape of these technologies, exploring cutting-edge methods, models, and tools that fuel the nexus between computer vision and data analytics offering insight into fundamental components that will guide the development of our own proof of concept tool capable of creating modal average activity masks for a given scene.

This project aims to automate and streamline visual data analysis, addressing the tedious and error-prone nature of the task, while offering

*Corresponding author.

potential solutions for various industries, such as policing, surveillance, research, and traffic management, where large volumes of video data require efficient and accurate processing.

2. Related Work

The research holds significance due to its unique aims, particularly in the context of a scarcity of directly comparable endeavors, though this also presents challenges in terms of available support. Franklin et al. [2] demonstrated a highly effective anomaly detection method in video footage, closely aligned with our proposed approach, utilizing a Convolutional Neural Network (CNN) and the COCO dataset for real-time video data. Similarly, Bajestani et al. [3] presented a system for anomalous vehicle and pedestrian detection using object recognition techniques, contrasting with our proposed higher granularity semantic segmentation techniques. Santhosh et al. [5] conducted a comprehensive survey of existing literature, offering valuable insights into contextual anomalies and statistical methods for quantifying deviation, which will guide our research and serve as a reference point for industry standards and common practices.

3. Preliminaries

Remark 1. An anomaly is a data point which does not fit the patterns of normal behavior, which will be defined by our average heat map created from multiple segmented images of a scene over time.

Remark 2. Normal activity encompasses data points that conform to the established patterns of behavior as defined by the average heat map derived from multiple segmented images of a scene over time. This definition ensures that normal activity is reflective of the prevailing spatio-temporal patterns detected by the system.

Remark 3. Conformity is exhibited by data points or objects that align with the established patterns of behavior, as defined by the average heat map derived from multiple segmented images of a scene over time.

4. Methodology

4.1. *Proposed architecture*

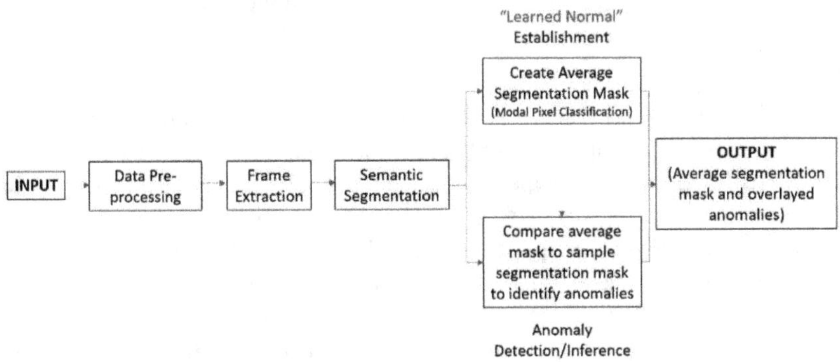

Fig. 1. The proposed pipeline structure of the visual data analytics tool

Table 1. Table describing the broad function of each described section in the pipeline architecture proposed in Fig 1.

Stage	Function
Input	Acquire initial raw data. The data should be from a single fixed camera and should be at least 30 seconds in length.
Data Pre-processing	Raw data should be standardized, addressing inconsistencies or errors in data.
Frame Extraction	Reduce the number of frames to reduce computational burden. Extract frames at timed intervals.
Semantic Segmentation	Create multiple segmentation masks for each given frame.
Average Heatmap ("Learned Normal")	Create average heatmap from the modal classification of each pixel within the segmented frames. This will be used as a reference point foı normal activity.
Anomaly Detection	Compare newly segmented frames to our 'learned normal' and identify potential outliers. Create a 'delta map' showing the differences in expected normal to the newly segmented scene.
Output	Output desired results via our average heatmap along with potential anomalies, that could be marked onto frames.

4.2. *Semantic segmentation model comparison and results: DeepLabV3+ vs UNet*

Both DeepLabV3+ and UNet were chosen for their prominence in semantic segmentation tasks and as models that continue to show high efficiency and promise in innovative machine learning applications. DeepLabV3+ employs a deep CNN with atrous convolutions to capture multi-scale contextual information, while UNet features a U-shaped encoder-decoder architecture for high-resolution feature extraction, commonly used in biomedical research.

4.2.1. *UNet*

Our U-Net Jupyter Notebook file documents our experimentation in creating a U-Net model, following resources and code from the pyimagesearch online tutorial [6], and trained on the 'Oxford iiit pet/3.2.0' dataset [7]. Despite the seeming irrelevance of this dataset to our use case, this experimentation serves solely for model comparison purposes, allowing us to assess the average performance of our chosen models, DeepLabV3+ and UNet, using a dataset of comparable size to those we anticipate using in our program.

After training for a suggested 20 epochs, the final statistics on the UNet model stood as follows:

- loss: 0.2892
- accuracy: 0.8884
- val loss: 0.339029
- val accuracy: 0.8757

A loss value of 0.2892 suggests good performance, while the 0.89 accuracy value indicates strong performance. Working on unseen data, or the validation set, a loss of ~0.34 suggests good performance on unseen data and as for accuracy, at ~0.88, this implies that the model predicts outcomes correctly about 88% of the time, which is an impressive result.

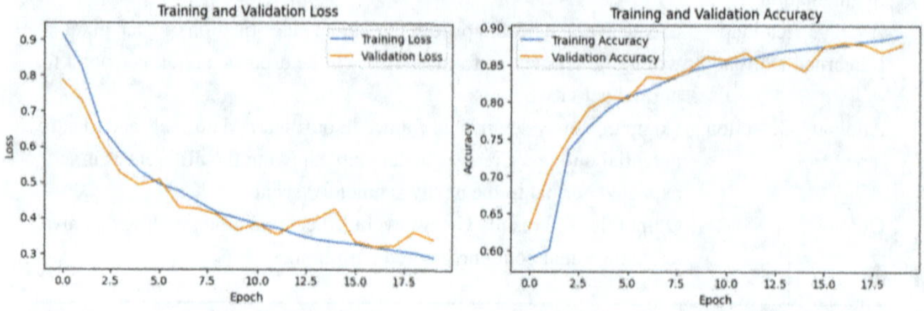

Fig. 2. Loss and Accuracy Graphs for both Validation and Training Data for UNet model

The graphs in Fig. 2 show expected curves whereby our accuracy increases as our loss decreases. Furthermore, the validation curve closely follows the training curve, indicating that the model performs well on unseen data (generalizes well) without overfitting.

4.2.2. *DeepLabV3+*

Our DeepLabV3+ Jupyter Notebook documents the creation and training of our DeepLabV3+ model, following steps and resources from the Keras website tutorial [8]. Differing from the UNet approach, we trained our DeepLabV3+ model on the Crowd Instance-level Human Parsing Dataset (CIHP) [9] due to its specific data input requirements.

After training for 20 epochs, as we did with our UNet model, these are the statistics we obtained:

- loss: 0.2214
- accuracy: 0.9339
- val loss: 1.0525
- val accuracy: 0.7517

A loss value of 0.2214 suggests good performance, while the 0.93 accuracy value indicates very good performance, suggesting that the model is correct around 93% of the time. The model's performance on the validation set, with a loss of 1.0525 and an accuracy of 0.75, suggests there is potential for improvement, possibly due to overfitting. Further optimization and regularization techniques could enhance the model's generalization ability and address these issues.

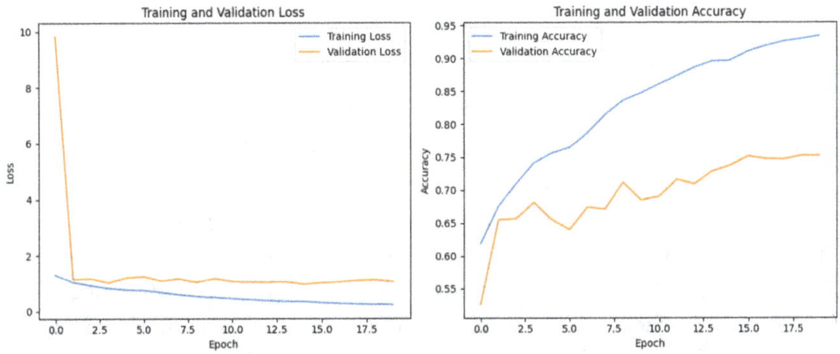

Fig. 3. Loss and Accuracy Graphs for both Validation and Training Data for DeepLabV3+ model

The graphs in Fig. 3 show expected curves whereby our accuracy increases as our loss decreases, however, there is a slight discrepancy between the validation and training set curves. This is likely attributable to model overfitting or a mismatch in dataset and model complexity. This should not be a cause for alarm as we will be using a different implementation and dataset in our tool. A good example of the output of our DeepLabV3+ model performing on unseen data can be seen in Fig. 4, which depicts the original image, alongside the overlaid mask and the mask alone.

Fig. 4: An example output of running the DeepLabV3+ model on an unseen image

4.2.3. *Summary*

DeepLabV3+ demonstrated superior performance with a final accuracy of 0.9339 and validation accuracy of 0.7517, outperforming UNet's final accuracy of 0.8884 and validation accuracy of 0.8757 after 20 epochs. DeepLabV3+ also exhibited lower validation loss, indicating better generalization. While acknowledging inherent complexities limiting perfect performance, such as label ambiguity and computational constraints, DeepLabV3+ remains the preferred model for advancing our project despite slight divergences in training and validation curves.

4.3. *Implementation*

We utilised the Pixellib [10] library for efficient and reliable implementation of our DeepLabV3+ model for semantic segmentation. Pixellib simplifies the integration of complex machine learning technologies into software solutions, addressing our project's challenge of limited time and technical expertise. Unlike other libraries, Pixellib leverages deep learning techniques like DeepLabV3+, making it suitable for our pipeline. It also supports future work development by facilitating the training of custom segmentation models and enabling real-time video segmentation. Our implementation uses the DeepLabV3+ model trained on the ADE20K dataset, which includes classes relevant to our project such as vehicles and pedestrians.

Frame taken from TfL JamCam

Frame with mask overlayed after segmentation using Pixellib library DeepLabV3+

Fig. 5: A TfL JamCam frame with and without the Pixellib segmentation mask

Utilizing the Pixellib library for semantic segmentation, we concentrate on generating average activity maps to pinpoint anomalous and conformal activity within scenes. By creating segmentation masks from frames at regular intervals, we derive an average activity heatmap (using the modal pixel classification), aiding in the identification of normal and anomalous behavior. Figs 5 and 6 shows more example segmentation outputs using DeepLabV3+.

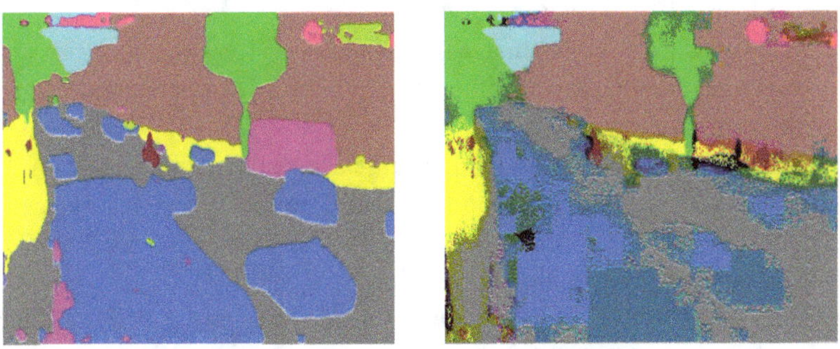

Fig. 6: (Left) Segmentation mask created using Pixellib DeepLabV3+ of a frame in the video referenced in above (Fig. 5), and (Right) Average activity mask generated from a collection of frame segmentation masks (like the one on the left) from our pipeline using DeepLabv3+ semantic segmentation. Judd Street (West) Transport for London JamCam

5. Conclusion

This research marks a successful exploration into novel computer vision techniques for automating decision-making and visual analysis, representing a valuable learning opportunity in my pursuit of a career in machine learning and mathematics. The creation of average activity masks demonstrates technical success, meeting our initial aims for a proof of concept tool using semantic segmentation. Moving forward, there is potential for further exploration, including delving into deep learning methodologies [11–13], researching data sanitization's impact, and exploring novel trajectory plotting methods.

References

1. E.R. Davies, Computer and machine vision: theory, algorithms, practicalities. Amsterdam; Boston: Elsevier, 4th ed., 2012.
2. R. J. Franklin, Mohana, and V. Dabbagol. Anomaly detection in videos for video surveillance applications using neural networks, in International Conference on Inventive Systems and Control, pp. 632–637, 2020.
3. M. F. Bajestani, S. S. H. R. Abadi, S. M. D. Fard, and R. Khodadadeh, AAD: Adaptive Anomaly Detection through traffic surveillance videos, Aug. 2018. arXiv:1808.10044 [cs].
4. S. Ren, K. He, R. Girshick, and J. Sun, Faster R-CNN: Towards Real-Time Object Detection with Region Proposal Networks, Jan. 2016. arXiv:1506.01497 [cs].
5. K. K. Santhosh, D. P. Dogra, and P. P. Roy, Anomaly Detection in Road Traffic Using Visual Surveillance: A Survey, ACM Computing Surveys, vol. 53, pp. 119:1–119:26, Dec. 2020.
6. M. Maynard-Reid, U-Net Image Segmentation in Keras, Feb. 2022.
7. Visual Geometry Group - University of Oxford. Dataset containing around 8000 images of dogs and cats.
8. K. Team, Keras documentation: Multiclass semantic segmentation using DeepLabV3+. Tutorial used for using DeepLabV3+ to Segment Images.
9. Papers with Code - CIHP Dataset, *paperswithcode.com*. https://paperswithcode.com/dataset/cihp
10. Papers with Code - Simplifying Object Segmentation with PixelLib Library. https://paperswithcode.com/paper/simplifying-object-segmentation-with-pixellib
11. S. Slade, L. Zhang, L., H. Huang, et al., (In Press). Neural Inference Search for Multiloss Segmentation Models. *IEEE Transactions on Neural Networks and Learning Systems.*
12. L. Zhang, S. Slade, C.P. Lim, et al., (2023). Semantic segmentation using Firefly Algorithm-based evolving ensemble deep neural networks. *Knowledge-Based Systems*, 277, p.110828.
13. L. Zhang and C.P. Lim, 2020. Intelligent optic disc segmentation using improved particle swarm optimization and evolving ensemble models. *Applied Soft Computing*, 92, p.106328.

Study of the Factors that Influence the Tourism Sector's Acceptance of the Use of ChatGPT

José Javier Galán Hernández*,‡ and Gabriel Marín Díaz†
*ENAE Business School, Murcia 30100, Spain
†Faculty of Statistics, Complutense University
Madrid 28040, Spain
‡josejgh@outlook.com

The interest in the use of the ChatGpt tool by professionals in the tourism sector is increasing, but the acceptance of this tool by these professionals is not the same for all, there being different factors that affect its acceptance. In this chapter, first, the interest in this topic is demonstrated by means of a bibliometric study formed by related scientific publications, second, a co-occurrence study is carried out on the publications found to determine the most significant variables, and, from them, lastly, to obtain the factors mentioned at the beginning, thus creating a theoretical adaptation of the technological adaptation model, that model how the tourism sector comes to accept and use ChatGpt. With this information, a company in the tourism sector has a theoretical evaluation model to measure the degree of acceptance of its workers toward ChatGpt tools, being able to distinguish the strongest points and having the opportunity to establish an adaptation strategy according to the weakest points.

Keywords: Technological factors; technological acceptance; ChatGPT; TAM; tourism; route; trip.

1. Introduction

In today's digital era, innovative technologies have transformed industries and customer engagement. Chatbots [1], like ChatGPT [2], are gaining traction in tourism to enhance customer experience and streamline processes. However, their successful adoption depends on the willingness and acceptance of tourism professionals, which varies due to numerous factors. As the tourism industry increasingly embraces digital solutions, understanding the factors influencing the adoption of ChatGPT becomes imperative. This chapter aims to fill this gap by investigating the adoption dynamics of ChatGPT within the tourism sector. Specifically, we seek to identify key factors influencing its adoption and adapt the

Technology Acceptance Model (TAM) [3] to elucidate its adoption process. By conducting a bibliometric analysis and applying co-occurrence techniques, we aim to provide insights into recent interest in ChatGPT in tourism and uncover factors driving its adoption.

The objective of this chapter is to investigate the adoption dynamics of ChatGPT within the tourism sector. Specifically, this chapter aims to understand the factors influencing the adoption of ChatGPT and to adapt the Technology Acceptance Model (TAM) to elucidate its adoption process. By conducting a bibliometric analysis and applying co-occurrence techniques, this chapter seeks to provide insights into recent interest in ChatGPT in tourism and uncover factors driving its adoption.

2. Methodology and State of the Art

To investigate the adoption dynamics of ChatGPT in the tourism sector, we conducted a comprehensive literature review using the Scopus database. We searched for literature related to the tourism sector, ChatGPT, and its acceptance up to and including the year 2023. The query used was: ("tourism industry" OR "tourism sector" OR "hospitality industry") AND ("chatgpt" OR "chat-gpt"). The search yielded a significant number of documents, particularly in the year 2023, indicating a growing interest in this topic within the scientific community.

The methodological approach employed in this study focuses on two main aspects: comprehensive literature search and the application of co-occurrence techniques to analyze scientific publications related to the adoption of ChatGPT in the tourism sector. These aspects are primarily detailed in sections 3 and 4 of the document.

2.1. *Discovery of External Factors:*

To begin with, a search was conducted in the Scopus database to find literature related to the tourism sector, ChatGPT, and its acceptance up to the year 2023. The query used included keywords such as "tourism industry," "ChatGPT," and "acceptance." The aim was to identify relevant publications that provide insights into the adoption dynamics of ChatGPT within the tourism sector.

2.2. *Proposed Theoretical Model:*

Once relevant publications were identified, co-occurrence techniques were applied using tools like VosViewer. Co-occurrence analysis allows visualizing the relationships between terms and identifying the key factors influencing the adoption of ChatGPT. This analysis facilitates the classification of variables into factors, which serve as input data for adapting the TAM model to the context of ChatGPT adoption in tourism.

By combining these methodological approaches, the study aims to provide a comprehensive understanding of the adoption dynamics of ChatGPT within the tourism sector and contribute to the existing body of knowledge in this area.

3. Discovery of External Factors

Having verified the interest of the topic of study, the search for information was expanded, now adding the necessary terms so that the search would include all the publications related to the concept of technological acceptance, obtaining a total of 26 publications through the query: (("acceptance of technology" OR "technology acceptance model" OR "user acceptance") AND ("tourism industry" OR "tourism sector" OR "hospitality industry") AND ("chatgpt" OR "chat-gpt") AND ("factors" OR "variables" OR "influences")).

Making use of the VosViewer tool [4], the co-occurrence technique [5, 6] is applied on the obtained publications, by which the bibliometric network [7,8] (Figure 1) of the most relevant terms is visualized. These terms are analyzed, thus obtaining the most used variables in relation to the topic of study, subsequently classifying them into the factors that we will use as input data in the TAM model (Table 1).

Table 1. Factors and influential variables in the proposed TAM theoretical model.

Technological Factors	Training	Behavior
advanced technologies	hallucination misinformation	behavioral intentions
challenges in technology usage in tourism	human resources	critical reflection
generative artificial intelligence (ai)	recruitment	tourism behavior
technology adoption		trust

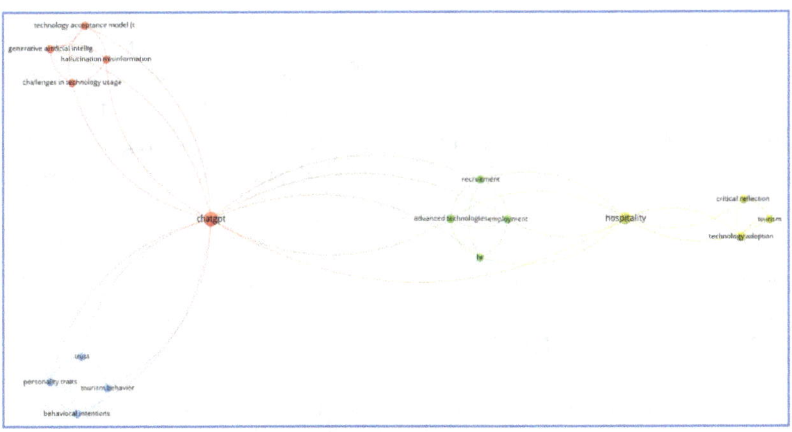

Figure 1. Bibliometric network

4. Proposed Theoretical Model

The factors determined in the previous section together with their variables as a source of inputs are used to form a theoretical model of technological adaptation. The original TAM model is used, which determines that any assessment of the acceptance of a system is based on the degree of acceptance of Perceived Usefulness and Perceived Ease of Use. As indicated in the previous section, when the objective is the use of the Chatgpt tool in the tourism sector, these two variables must be fed by Technological Factors, Training and Behavior. Thus we have the proposed model, Figure 2.

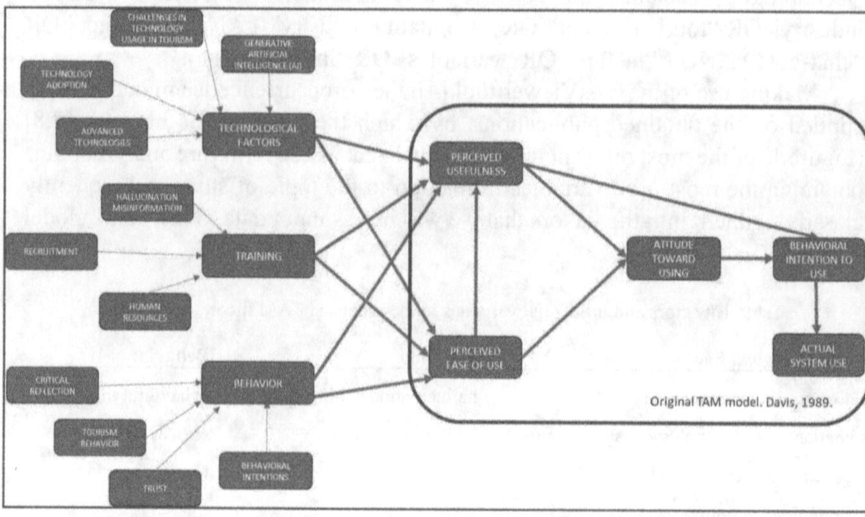

Figure 2. Proposed theoretical model.

- **Advanced technologies**

The perceived usefulness of ChatGPT is influenced by its ability to provide fast and accurate responses, as well as its ability to adapt to the changing needs of tourism. The ease of use of ChatGPT [9], facilitated by technological advances, also increases its acceptance by minimizing adoption barriers and improving operational efficiency in the tourism sector.

- **Challenges in technology usage in tourism**

Concerns about data privacy and security, as well as integration with existing systems [10], can generate resistance and skepticism in the tourism sector.

- **Generative artificial intelligence (ai)**

ChatGPT's ability to understand and respond to the specific needs of tourists, as well as to provide personalized recommendations, enhances the perceived usefulness of the technology. Confidence in the accuracy and relevance of

ChatGPT-generated responses may increase due to advanced AI capabilities, driving its acceptance in the tourism industry [11]. The ability to dynamically adapt to changing tourism demands also reflects the evolution of AI, which may increase its acceptance and adoption in the industry.

- **Technology adoption**

The willingness to adopt innovative technologies such as ChatGPT reflects the attitude of the tourism industry towards innovation and continuous improvement of the customer experience [12]. The perception of ChatGPT as an effective solution to optimize processes and improve customer service increases its acceptance and adoption in the tourism industry. The technology adoption strategy implemented by tourism companies can influence the speed and success of ChatGPT implementation in the industry.

- **Recluitment**

Hiring personnel with skills in artificial intelligence and natural language processing can facilitate the effective implementation and adoption of ChatGPT in the tourism industry [13].

- **Hallucination misinformation**

Lack of understanding about the capabilities and benefits of ChatGPT can lead to resistance and skepticism among tourism professionals. Dissemination of incorrect or misinterpreted information about ChatGPT can increase distrust and reduce the willingness of the tourism sector to adopt this technology [14,15].

- **Human Resources**

The existence of human resources trained in artificial intelligence and natural language processing is important. The continuous support of personnel in the use of ChatGPT are fundamental to increase its acceptance and use in the tourism industry [16].

- **Critical reflection**

Critical reflection can drive the evaluation of how ChatGPT can improve customer experience and internal processes in tourism [17]. In addition, it will encourage consideration of possible ethical and social implications of its implementation in the tourism industry.

- **Tourism behavior**

The positive response of tourists to ChatGPT interactions may increase the acceptance of this technology among tourism professionals. Alignment between ChatGPT functionalities and tourism behavior preferences can drive its adoption and utilization in the industry [18].

- **Trust**

The perception of reliability and consistency in interactions with ChatGPT strengthens confidence in its usefulness and effectiveness in the industry. Transparency in the operation and processes of ChatGPT contributes to building a trusting relationship between users and the technology. Confidence in the accuracy and relevance of the responses generated by ChatGPT is critical to its adoption and effective use in the tourism industry [19].

- **Behavioral intentions**

A higher intention to incorporate ChatGPT into tourism processes indicates a more positive acceptance of the technology. Favorable intentions may be driven by the perceived usefulness and ease of use of ChatGPT, as well as the positive experience of other users [20].

- **Technological Factors.**

They influence the acceptance of the tourism sector towards the ChatGPT tool by reflecting the willingness of professionals to adopt and use the technology. In addition, previous experience with similar technologies and observation of other users' behavior may also influence behavioral intentions and thus the acceptance of ChatGPT in the tourism sector [21].

- **Training**

Familiarity with the technology and confidence in its appropriate application in the tourism context are key to acceptance. In addition, a solid background in the capabilities and limitations of ChatGPT promotes its effective integration into tourism operations, thus influencing its acceptance in the industry [22].

- **Behavior**

Perceived usefulness and convenience in integrating ChatGPT into daily activities in the tourism industry also affect adoption behavior. In addition, positive user feedback and success stories influence collective behavior, promoting the acceptance and utilization of ChatGPT in the tourism industry [23].

5. Conclusions

Interest in integrating ChatGPT into the tourism sector has surged within the scientific community, particularly noticeable in the increased volume of related publications, notably in 2023. By employing co-occurrence techniques in scientific literature, pivotal factors impacting ChatGPT adoption in tourism have been discerned, including perceived utility, technological hurdles in tourism application, and generative artificial intelligence. The presence of AI and NLP-trained personnel, alongside conscientious deliberation on ethical and societal ramifications, emerges as pivotal considerations for successful ChatGPT integration. Confidence in the technology, its perceived reliability, and consistency in user interactions emerge as critical drivers shaping ChatGPT's embrace in tourism. Beyond perceived utility and usability, behavioral intentions and technological aspects also wield significant influence, reflecting the industry's stance and eagerness for innovation to enhance customer experiences. These findings underscore the necessity of comprehending the factors influencing ChatGPT adoption in tourism, laying a robust groundwork for future research and advancements in this domain.

Acknowledgment

Work partially supported by grant 2-4570.5 of the Swiss National Science Foundation.

References

1. Esiyok, E., Gokcearslan, S., & Kucukergin, K. G. (2024). Acceptance of Educational Use of AI Chatbots in the Context of Self-Directed Learning with Technology and ICT Self-Efficacy of Undergraduate Students. International Journal of Human–Computer Interaction, 1-10.
2. Solomovich, L., & Abraham, V. (2024). Exploring the influence of ChatGPT on tourism behavior using the technology acceptance model. Tourism Review.
3. Davis F. D.: Perceived usefulness, perceived ease of use, and user acceptance of information technology. MIS Quart. 13, 319–339 (1989)
4. Van Eck, N. J., & Waltman, L. (2013). VOSviewer manual. Leiden: Univeristeit Leiden, 1(1), 1-53.
5. Coulter, N., Monarch, I., & Konda, S. (1998). Software engineering as seen through its research literature: A study in co-word analysis. Journal of the American Society for Information Science, 49(13), 1206-1223.
6. Callon, M., Courtial, J. P., Turner, W. A., & Bauin, S. (1983). From translations to problematic networks: An introduction to co-word analysis. Social science information, 22(2), 191-235.
7. Galán, J. J., Carrasco, R. A., & LaTorre, A. (2022). Military applications of machine learning: a bibliometric perspective. Mathematics, 10(9), 1397.
8. Carrasco-Aguilar, A., Galán, J. J., & Carrasco, R. A. (2022). Obamacare: A bibliometric perspective. Frontiers in Public Health, 10, 979064.
9. Elmohandes, N., & Marghany, M. (2024). Effective or ineffective? Using ChatGPT for staffing in the hospitality industry. European Journal of Tourism Research, 36, 3617-3617.
10. Matosas-López, L. (2024). The influence of brand credibility and brand loyalty on customer satisfaction and continued use intention in new voice assistance services based on AI. Journal of Marketing Analytics, 1-22.
11. Christensen, J., Hansen, J. M., & Wilson, P. (2024). Understanding the role and impact of Generative Artificial Intelligence (AI) hallucination within consumers' tourism decision-making processes. Current Issues in Tourism, 1-16.

12. Hollebeek, L. D., Menidjel, C., Sarstedt, M., Jansson, J., & Urbonavicius, S. (2024). Engaging consumers through artificially intelligent technologies: Systematic review, conceptual model, and further research. Psychology & Marketing.
13. The impact of artificial intelligence (AI) on employees' skills and well-being in global labor markets: A systematic review
14. Kim, J. H., Kim, J., Kim, C., & Kim, S. (2023). Do you trust ChatGPTs? Effects of the ethical and quality issues of generative AI on travel decisions. Journal of Travel & Tourism Marketing, 40(9), 779-801.
15. Elmohandes, N., & Marghany, M. (2024). Effective or ineffective? Using ChatGPT for staffing in the hospitality industry. European Journal of Tourism Research, 36, 3617-3617.
16. Makkonen, M., Salo, M., & Pirkkalainen, H. (2023). The Effects of Job and User Characteristics on the Perceived Usefulness and Use Continuance Intention of Generative Artificial Intelligence Chatbots at Work. In CEUR Workshop Proceedings. RWTH Aachen.
17. Kim, J. H., Kim, J., Kim, S., & Hailu, T. B. (2023). Effects of AI ChatGPT on travelers' travel decision-making. Tourism Review.
18. Ariza-Montes, A., Quan, W., Radic, A., Yu, J., & Han, H. (2023). Human values and traveler behaviors: metaverse for conferences and meetings. Journal of Travel & Tourism Marketing, 40(6), 490-511.
19. Esiyok, E., Gokcearslan, S., & Kucukergin, K. G. (2024). Acceptance of Educational Use of AI Chatbots in the Context of Self-Directed Learning with Technology and ICT Self-Efficacy of Undergraduate Students. International Journal of Human–Computer Interaction, 1-10.
20. Rajaguru, R. (2023). Effects of contemporary technologies, such as blockchain and artificial intelligence (AI) in enhancing consumers' trustworthiness of online reviews. Journal of Hospitality Marketing & Management, 1-9.
21. Jia, S., Chi, O. H., Martinez, S. D., & Lu, L. (2023). When "Old" Meets "New": Unlocking the Future of Innovative Technology Implementation in Heritage Tourism. Journal of Hospitality & Tourism Research, 10963480231205767.
22. Cramarenco, R. E., Burcă-Voicu, M. I., & Dabija, D. C. (2023). The impact of artificial intelligence (AI) on employees' skills and well-being in global labor markets:A systematic review. Oeconomia Copernicana, 14(3), 731-767.
23. Mogaji, E., Viglia, G., Srivastava, P., & Dwivedi, Y. K. (2024). Is it the end of the technology acceptance model in the era of generative artificial intelligence?. International Journal of Contemporary Hospitality Management.

New Advances in Node Classification Problem: Feature Engineering in Decision Trees

E. Martínez-Tena[*,‡], J. Álvarez-Liébana[†], D. Gómez[†] and J. Castro[†]

[*] *Faculty of Statistics, Complutense University, Madrid, Spain*
[†] *Faculty of Statistics, Department Statistics and Data Science, Complutense University, Madrid, Spain*
[‡] *edumar09@ucm.es*

Motivated by the rise of social networks, as well as their applications in fields such as genetics or marketing, the node classification problem is one of the most actively researched topics within network-based machine learning. In contrast to methodologies based on neural networks or relational probabilistic algorithms, the main objective of this chapter is to provide an alternative by using classical algorithms (such as decision trees). This chapter aims to illustrate how, by using methodologies inherent to community detection, datasets can be enriched in such a way that the performance of these algorithms improves. A new measure of homophily may be proposed based on that improvement ratio.

Keywords: Community detection; feature engineering; graphs on machine learning; homophily; node classification problem.

1. Introduction

As has occurred in the fields of sociology or genetics, marketing and social media benchmarks have undergone a significant transformation in recent decades, becoming a critical part of connecting brands and consumers. Marketing, at its core, seeks to understand and satisfy market needs, but in the digital age, social media has emerged as a powerful channel to engage audiences more immediately. For that purpose, brands are looking to build more authentic relationships with their customers.[1] User-specific information, enhanced with network structures, is now crucial.

The incorporation of social media benchmarks in marketing strategies has evolved alongside advancements in classification and prediction techniques. Similar to breakthroughs in fields where understanding complex relationships appear, marketing now leverages social media's power to strengthen brand-consumer connections. While traditional methods focused on customer characteristics, the recognition of interconnections among individuals and their influences has become pivotal.[2]

Classical classification and prediction approaches in machine learning often assume that multidimensional items have no interrelationships. For instance, in customer churn problems, methods typically focus on individual characteristics like spending, satisfaction, and age to classify the clients according to the target class. However, the influence of relationships with others and personal opinions on the decision to leave may be overlooked. Specific classification problems, especially in image classification, semantic segmentation, and natural language processing, have been addressed by recurrent neural networks (RNNs[3]) or long short-term networks (LSTM[4]). These networks are employed for tasks, such as sentiment analysis.

Motivated by the success of the fusion of graph theory and statistical learning, particularly in node classification problem, the proposed methodology implies an extension of the concept of neighborhood by generalizing it from a community detection problem. Hence, new variables will be incorporated by aggregating each community's own information, providing a feature engineering to enhance the performance of classical machine learning algorithms, in contrast with the most used graph convolutional network approaches.[5,6] This chapter is organized as follows. Section 2 provides background information and notation. Section 3 outlines the fundamental concepts underlying the proposed methodology. Section 4 presents the classification results, while Section 5 exposes concluding remarks.

2. Preliminaries

2.1. *Graph theory*

Within the present paper, we shall denote by $G = (V, E)$ an undirected graph composed of a finite set of vertices or nodes V and a finite set of edges E (paired vertices). Henceforth, the set of nodes V will be made up of instances $V = \{v_1, v_2, \ldots, v_n\}$, where $n = |V|$ is the order of graph. Likewise, we shall denote the vertices as a set of pairs of distinct vertices $E = \{e_{ij} = (v_i, v_j)\,|v_i,\ v_j \in V\} \subset V \times V$, where $s = |E|$ is the size of graph. As usual, the graph structure G can be fully determined by its adjacency matrix $A = (a_{ij})_{i,j=1,\ldots,n}$ where $a_{ij} = 1$ if there exists an edge between v_i and v_j, and $a_{ij} = 0$ otherwise. This adjacency matrix characterizes the neighbourhood of v_i as the set of vertices $N(v_i) = \{v_j | a_{i,j} \neq 0,\ j = 1, \ldots, n\}$ such that its cardinality is denoted as the degree (or valency) of vertex given by $d(v_i) = |\{e_{ij} = (v_i, v_j)\,|v_j \in V\}| = \sum_{j=1}^{n} a_{ij}$, for each $i = 1, \ldots, n$.

From now on, we assume that the set of edges are constituted by unordered nodes such that $e_{ij} \in E$ if and only if $e_{ji} \in E$, with $a_{ii} = 0$ for each $i = 1, \ldots, n$, such that there exists at least one pair of connected nodes.

2.2. *Node classification problem*

Let us now consider that, in addition to G, we have at our disposal a labelled dataset $(\mathbf{X}, \mathbf{Y})$. On the one hand, $\mathbf{X} = (\overline{x}_i)_{i=1,\ldots,n}$ will denote a row-by-row dataset, where each observation is constituted by a p-dimensional vector $\overline{x}_i = (x_{i1}, \ldots, x_{ip}) \in \mathbb{R}^p$, for each $i = 1, \ldots, n$. On the other hand, $\mathbf{Y} = (y_1, \ldots, y_n)^T \in \mathbb{R}^n$ will represent a vector of labels. As is customary in the Node Classification Problem, each labelled pair $(\overline{x}_i, y_i)$, with $i = 1, \ldots, n$, is directly associated with a node $v_i \in V$, and then, related to each $\overline{x}_j$ for which $a_{ij} = 1$, violating the classical independent and identically distributed (i.i.d.) assumption. In the following, the set of nodes V will be splitted into $V = V_l \cup V_u$, with $V_l \cap V_u = \emptyset$, where V_l is a set of nodes with known labels. Hence, given a set of labelled observations $(\mathbf{X}, \mathbf{Y}_l)$, related by $G = (V, E)$, the goal is to predict the unknown targets of V_u. This general benchmark is commonly adopted in a wide variety of graph-related Machine Learning tasks, among which the following stand out as node classification, graph labeling ($p = 0$, just considering the relationships from the graph structure) and link prediction, among others (see Fig. 1).

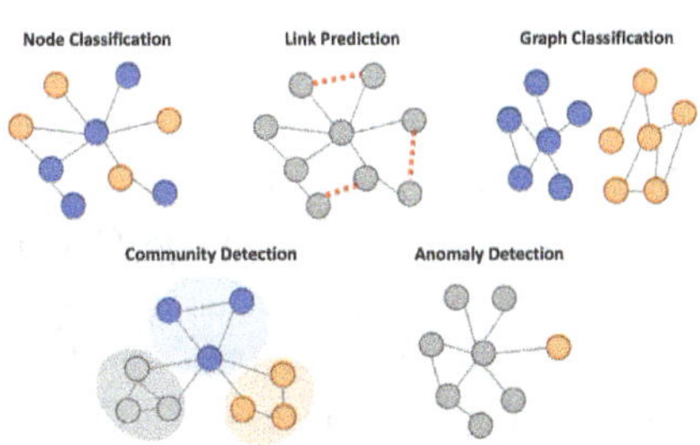

Fig. 1. Illustration about graph-related problems extracted from[7]

We will focus on node classification and graph labelling problem, in which homophily plays a crucial role. Homophily, also informally known as "birds of a feather flock together," is when a bond between individuals (such as friendship or other social connection) is correlated with those individuals being similar in nature. For example, friends often share similar characteristics such as age, social background, and educational level. If it can be argued that this phenomenon applies within an interest network, then information about neighboring nodes with similar attributes may be useful in helping to classify a node of interest.

As succinctly exposed, besides the emerging deep learning models developed in the literature (e.g., GNN or GCN techniques), our hypothesis is that the topological structure derived from the graph allows us to capture complex relationships, to be included in classical Machine Learning algorithms. In the following, the classifier to assign $\hat{Y}_i$, for each node V_i, will be defined as $\hat{Y}_i = k_j$ if and only $\hat{P}(Y_i = k_j | (\mathbf{X}, \mathbf{Y}_1; G_l)) = \arg\max_{k \in \mathcal{K}} \hat{P}(Y_i = k | (\mathbf{X}, \mathbf{Y}_1; G_l))$ where $G_l = (V_l, E)$, $\mathcal{K}$ is the unique set of labels to be classified and $\hat{P}$ the estimated probability of belonging. Note that, in the case of graph labelling problem, the estimated probability $\hat{P}$ can be simplified as $\arg\max_{k \in \mathcal{K}} \hat{P}(Y_i = k | (\mathbf{Y}_1; G_l))$. Our proposal relies on adopting a decision tree as classifier $\hat{P}$ considering the information about relationships in G_l.

Additionally, we will provide a performance comparison among a traditional classification tree using **i)** solely graph information (graph labelling problem, denoted as GL), **ii)** exclusively intrinsic variables from $\mathbf{X}$ (ignoring the graph-structure, ML), and **iii)** the integration of both sources of information (feature engineering, GL-ML). In the following, $\mathbf{Z} = (\overline{z}_i)_{i=1,\ldots,n}$, with $\overline{z}_i = (z_{i1}, \ldots, z_{iq}) \in \mathbb{R}^q$, denotes the set of exogenous predictors extracted from the graph, and $\tilde{\mathbf{X}} = (\mathbf{X}, \mathbf{Z}) \in \mathbb{R}^{p+q}$ the extended matrix of covariates. Note that in case **i)**, $p = 0$, whereas $q = 0$ in case **ii)**.

3. Methodology

We now summarise the main aspects related with our methodology, considering the GL, ML and GL-ML scenarios. All of them will be tested in some of the most used datasets in the Node Classification benchmark (see Table 2), regarding homophily measures as assortativity coefficient[8] and Gupta-Anderson-May (GAM)[9]. On the one hand, we consider the Cora[10] and PubMed[11] homophilic datasets. On the other hand, respect to sparse homophily, we include the Wikipedia graphs Chameleon[12], the co-occurrence graph Actor, and WebKB's Wisconsin and Cornell website graphs[13]. Covariates are 0/1-valued vectors indicating the absence/presence of a bag of words, except Pubmed dataset (covariates as inverse weighted frequencies).

3.1. *Feature engineering: Louvain variables*

Even though the inclusion of more variables could be studied (e.g., related to the centrality), in the current work we have just considered exogenous

Table 1. Summary of citation network datasets.

Dataset	Assortativity	GAM measure	Nodes	Edges	Classes	Features
Cora	0.775	0,612	2708	5429	7	1433
Pubmed	0.686	0.462	19717	44338	3	500
Chameleon	0.031	-0.089	2277	36101	5	3133
Actor	0.002	-0.108	7600	29926	5	931
Cornell	-0.143	-0.125	183	298	5	1703
Wisconsin	-0.193	-0.145	251	515	5	1703

variables derived from a community detection framework. For this purpose, we have adopted one of the most known greedy method, the Louvain's algorithm[14], with complexity of order $\mathcal{O}\left(n\log(n)\right)$[15]. The main idea revolves around the modularity, a scalar parameter aimed at quantifying the density of intra-community connections respect to connections with other communities, defined in[14] as $Q = \frac{1}{2m}\sum_{i,j=1}^{n}\left(w_{ij} - \frac{w_{i\cdot}w_{j\cdot}}{2m}\right)\delta(c_i, c_j)$, with $-1 < Q < 1$, where w_{ij} denotes the weight for edge $e_{i,j}$, c_i the assigned community for node v_i, $w_{i\cdot}$ the sum of weights for each $e_{i\cdot}$ and $m = \frac{1}{2}\sum_{i,j=1}^{n} w_{ij}$, for each $i = 1, \ldots, n$. Hence, the Louvain's method is based on a recursive agglomeration, quantifying the gain/loss in modularity for each community configuration. We have implemented it with the function `group_louvain()` from the `R` package `{tidygraph}`[16].

Despite the fact that the Louvain's algorithm generates communities that may be arbitrarily misconnected[17], we firstly consider it at the expense of alternatives as Leiden's algorithm[17]. The referred modularity can be generalized by replacing $\frac{w_{i\cdot}w_{j\cdot}}{2m}$ terms by $\gamma\frac{w_{i\cdot}w_{j\cdot}}{2m}$, depending on a resolution parameter $\gamma > 0$: if we maximize Q_γ, higher resolutions favors smaller communities (minimizing $w_{i\cdot}w_{j\cdot}$ terms), and derive to more communities.

In this preliminary work, we have considered a grid of 10 possible resolution parameters, adopting two different approaches for each scenario:

- **Resolution List (RL)**. All possible community configurations are checked, incorporated for each one a categorical variable with the assigned community and $k-1$ variables representing the relative frequency of each class in each one (i.e., $q = s(k-1)$ such that $\tilde{\mathbf{X}} = (\mathbf{X}, \mathbf{Z}) \in \mathbb{R}^{p+s(k-1)}$ with $s = 10$ and k the number of classes).
- **Resolution Tune (RT)**. Resolution parameters are incorporated as extra parameter to be tuned (i.e., $s = 1$).

3.2. *Validation strategy*

The developed methodology has been applied similarly to both RL and RT approaches. Within each of these scenarios, we consider 20 random instances train/test splits (80%-20%), stratified by target variable. The whole train/test subsets will be used for scoring our best models. Subsequently, a repeated 5-fold cross-validation process (with 3 replicates) is conducted on the train subset, involving a second split stratified by the target variable. Aiming to be fair in the learning process, target values of the validation subsets should not be used, and then, additional $10(k-1)$ Louvain variables are generated a new in each iteration, recomputing the relative frequency, just considering as known the target values for train subsets (see Fig 2). This fact represents a significant shift in the data engineering paradigm during the model training process, as the intrinsic variables of each observation remain unchanged. In other words, the new variables derived from the graph structure depend on their environment. Finally, the optimal configurations of the trained tree model are evaluated in terms of accuracy on the respective test sets separated at the beginning of the workflow.

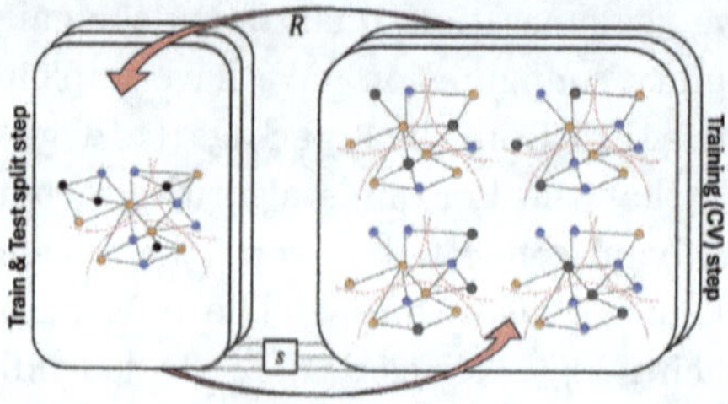

Fig. 2. On the left, black nodes represents test observations (unknown labels). On the right, for each iteration, grey nodes represent the observations which are excluded from the training (validation nodes).

4. Results

First of all, concerning the different approaches about how to consider the resolution parameter (RL vs RT approach), the inclusion of the last one was motivated by the importance of variables illustrated in Fig. 3. As shown, regardless of the level of homophily, simulation studies suggest that, when different resolution are included, a particular community configuration is always selected. This fact supports the idea of regarding resolution parameter as a parameter to be tuned (RT approach).

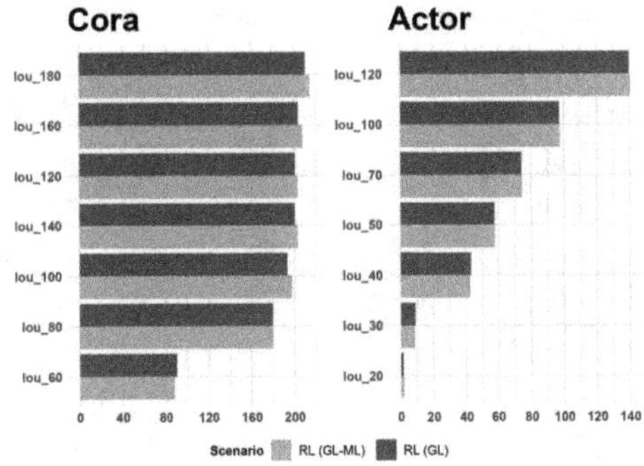

Fig. 3. Average of importance of variables (goodness of split provided by vip package), comparing RL-GL-ML (orange) and RL-GL (blue) scenarios.

As shown in Table **??**, RT's results improve also slightly the RL's performance, in terms of the average accuracy for each of the scenarios discussed. Note also that the performance is directly related to the graph homophily of each dataset. Hence, initial results display that the incorporation of the aforementioned Louvain's variables leads to noticeable improvements under high levels of homophily. In contrast, in the case of heterophilic data, the results show a slight, though not remarkable, worsening.

Table 2. Test results in terms of assortativity (Ass), mean accuracy and improvement ratio (IR) between RT-GL-ML and RT-ML scenarios.

Dataset	Ass	ML	RL GL	RL GL-ML	RT GL	RT GL-ML	IR
Cora	0.775	**0.65**	0.79	0.79	0.82	**0.82**	**25.37%**
Pubmed	0.686	**0.83**	0.81	0.84	0.81	**0.86**	**2.83%**
Chameleon	0.031	**0.52**	0.45	0.45	0.52	**0.57**	**7.83%**
Actor	0.002	0.32	0.23	0.23	0.26	0.30	-3.81%
Cornell	-0.143	0.65	0.27	0.43	0.45	0.62	-4.29%
Wisconsin	-0.193	0.76	0.30	0.69	0.42	0.73	-2.97%

5. Conclusions and Open Research Lines

In conclusion, we appreciate how the incorporation of graph structure, under the hypothesis of homophily, may help improve the results with respect to classical machine learning approaches. Contrarily, in the case of high levels of heterophily, the same result may not be guaranteed. Our hypothesis is based on the idea that heterophilic data provide us with poorer information about its nearby nodes, and hence, considering the information coming from the graph structure does not substantially improve a classical decision tree. Motivated by the exposed conclusions, currently, we are working on considering the mentioned improvement ratio as an alternative homophily measure. Our proposal is now being tested in the most-known machine learning classification benchmarks (support vector machine, nearest neighbors, etc.). Some improvements are being studied, implementing a weighted-based community detection, a Bayesian perspective considering as prior probabilities those provided by probabilistic relational models, or adopting a fuzzy graph perspective, to mention some of them.

Acknowledgments

This work was supported in part by project PID2021-122905NB-C21, of the DGI, MINECO, Spain.

References

1. J. H. McAlexander, J. W. Schouten and H. F. Koenig, Building brand community, *Journal of marketing* **66**, 38 (2002).
2. S. Bhagat, G. Cormode and S. Muthukrishnan, Node classification in social networks, *Social network data analytics* , 115 (2011).
3. F. J. Pineda, Generalization of back-propagation to recurrent neural networks, *Physical review letters* **59**, p. 2229 (1987).
4. S. Hochreiter and J. Schmidhuber, Long short-term memory, *Neural computation* **9**, 1735 (1997).
5. T. N. Kipf and M. Welling, Semi-supervised classification with graph convolutional networks, *arXiv preprint arXiv:1609.02907* (2016).
6. J. Bruna, W. Zaremba, A. Szlam and Y. LeCun, Spectral networks and locally connected networks on graphs, *arXiv:1312.6203* (2013).
7. T. Masui, *Graph neural networks with PyG on Node Classification, link prediction and anomaly detection*, (2022). https://t.ly/BPoHA.
8. M. E. J. Newman, Mixing patterns in networks, *Physical review E* **67**, p. 026126 (2003).
9. S. Gupta, R. M. Anderson and R. M. May, Networks of sexual contacts: implications for the pattern of spread of hiv, *Aids* **3**, 807 (1989).

10. A. K. McCallum, K. Nigam, J. Rennie and K. Seymore, Automating the construction of internet portals with machine learning, *Information Retrieval* **3**, 127 (2000).
11. Z. Yang, W. Cohen and R. Salakhudinov, Revisiting semi-supervised learning with graph embeddings, in *International conference on machine learning*, 2016.
12. B. Rozemberczki, C. Allen and R. Sarkar, Multi-scale attributed node embedding, *Journal of Complex Networks* **9**, p. cnab014 (2021).
13. H. Pei, B. Wei, K. C.-C. Chang, Y. Lei and B. Yang, Geom-gcn: Geometric graph convolutional networks, *arXiv:2002.05287* (2020).
14. V. D. Blondel, J. L. Guillaume, R. R. Lambiotte and E. Lefebvre, Fast unfolding of communities in large networks, *J. Statistical Mechanics: Theory and Experiment* **10** (2008).
15. A. Lancichinetti and S. Fortunato, Community detection algorithms: a comparative analysis, *Phys. Rev. E* **80 (5)**, p. 056117 (2009).
16. T. L. Pedersen, *tidygraph: a tidy API for Graph Manipulation*, (2024). https://github.com/thomasp85/tidygraph.
17. V. A. Traag, L. Waltman and N. J. van Eck, From louvain to leiden: guaranteeing well-connected communities, *Sci Rep* **9 (5233)** (2019).

Navigating the Landscape of Artificial Intelligence: Computing with Words in Marketing

Ziwei Shu[*,‡] and Ramón Alberto Carrasco[†]

[*]*Department of Statistics and Data Science, Faculty of Statistics, Complutense University of Madrid, Madrid 28040, Spain*

[†]*Department of Marketing, Faculty of Statistics, Complutense University of Madrid, Madrid 28040, Spain*

[‡]*ziweishu@ucm.es*

Artificial Intelligence (AI) is increasingly playing a vital role and transforming our lives. Explainable AI is promoted by the concept of Computing with Words (CWW), which uses a human-like analytical framework to interpret and process information. In marketing, it is crucial to transform ambiguous information from customer feedback or expert knowledge into interpretable information for effective decision-making and the creation of marketing strategies. This chapter aims to explore the application of the 2-tuple linguistic model, a widely used methodology for CWW, in the field of marketing. Through a systematic exploration of articles published from 2000 to 2003, this chapter identifies areas of investigation worthy of further exploration and analysis.

Keywords: Computing with words; artificial intelligence; 2-tuple linguistic model; aggregation operators; marketing.

1. Introduction

In the dynamic landscape of the digital era, Artificial Intelligence (AI) drives the transformation of companies and individuals. Computing with Words (CWW) is a concept enabling thinking, reasoning, decision-making, and problem-solving through a human-like analytical framework. This promotes the development of explainable AI. In marketing, converting vague, semantically uncertain information from customer feedback or expert knowledge into easily understandable data facilitates more effective and informed marketing strategies for businesses.

This chapter aims to explore the application of the 2-tuple linguistic model, one of the most popular methodologies for CWW, in the field of marketing. Articles applying the 2-tuple linguistic model in marketing from 2000 to 2023 are reviewed. Through systematic exploration, areas worthy of further investigation are identified.

The remainder of this chapter is organized as follows. Section 2 offers a brief overview of the 2-tuple linguistic model, introducing its fundamental elements and concepts. Section 3 analyzes the results of a systematic search conducted on the Web of Science Core Collection (WoSCC) for articles related to the application of the 2-tuple linguistic model in marketing. Finally, Section 4 presents conclusions and outlines potential directions for future research.

2. The 2-Tuple Linguistic Model

2.1. *A brief overview of the 2-tuple linguistic model*

The 2-tuple linguistic model, founded on fuzzy logic and linguistic variables, was introduced to tackle information loss in the linguistic information fusion of other linguistic computing models [1]. Drawing inspiration from symbolic models used in decision-making, the 2-tuple linguistic model consists of linguistic terms and a symbolic translation, each represented by a numeric value [2]. Characterized by representing linguistic information in a continuous domain, this model streamlines the CWW process while preserving the linguistic basis.

The 2-tuple linguistic model excels among existing computational models for CWW [3], thanks to its easy-to-understand interpretation and capability to express any linguistic data in a continuous domain. This model, along with its extensions, has been applied in various fields, including finance industry [4, 5], healthcare industry [6-8], hospitality industry [9, 10], among others. Figure 1 shows the evolution of the number of published articles and the total citations related to the 2-tuple linguistic model from 2000 to 2023 on the WoSCC.

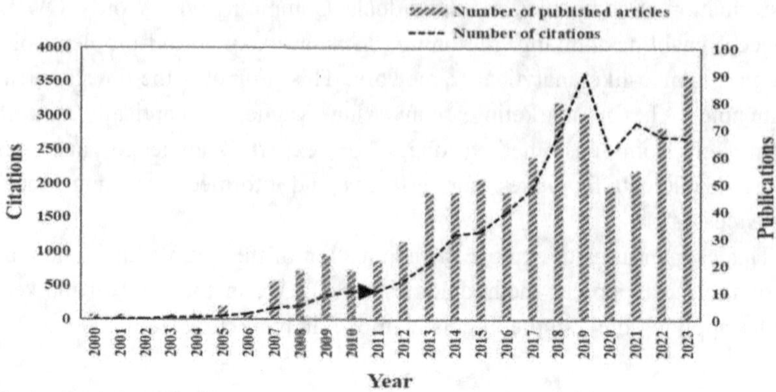

Fig. 1. Trends in publications and citations related to the 2-tuple linguistic model (2000-2023).

Figure 1 indicates that between 2000 and 2006, articles on the 2-tuple linguistic model were consistently below 10, with a low number of citations. From 2007 onward, the number of publications and citations started to increase, although some years experienced decreases (e.g., production dramatically decreased in 2020 due to the pandemic). Overall, there is a trend of increasing articles and total citations related to the 2-tuple linguistic model.

2.2. *Fundamental elements and concepts of the 2-tuple linguistic model*

The 2-tuple linguistic model represents the linguistic information using a pair of values known as a 2-tuple value (s_i, α), where $s_i \in S$ is a linguistic term, and $\alpha \in [-0.5, 0.5)$ is a numerical value representing the distance to the central value of s_i. The following defines the 2-tuple linguistic model.

Definition 1. Let $S = \{s_0, \ldots, s_g\}$ be a linguistic term set, whose cardinality is $g + 1$. $\beta \in [0, g]$ is a value that represents the result of a symbolic aggregation operation. The function Δ: $[0, g] \rightarrow \langle S \rangle = S\ x\ [-0.5, 0.5)$ is used to convert β to 2-tuple value (s_i, α) as shown in Eq. (1):

$$\Delta(\beta) = (s_i, \alpha), with \begin{cases} i = round(\beta) \\ \alpha = \beta - i, \quad \alpha \in [-0.5, 0.5) \end{cases} \tag{1}$$

where *round*(·) is the rounding operation; s_i has the nearest index label to β; and α represents a numerical value of the symbolic translation. As the function Δ is bijective, its inverse function Δ^{-1}: $\langle S \rangle = Sx[-0.5, 0.5) \rightarrow [0, g]$ can convert the 2-tuple value into its equivalent numerical value as $\Delta^{-1}(s_i, \alpha) = i + \alpha = \beta$.

The comparison between 2-tuple values, the negation operator of a 2-tuple value applied to handle inverse dimensions, and aggregation operators for 2-tuple linguistic computing are further detailed in [11]. The example illustrating the calculation of the 2-tuple linguistic model can be consulted in [10].

3. Application of the 2-Tuple Linguistic Model in Marketing

The 2-tuple linguistic model has found applications in various fields. This paper focuses on its application in marketing, conducting a systematic search on the WoSCC for relevant articles from 2000 to 2023. The literature search used the following terms in the query: "2-tuple linguistic model," "marketing," and their corresponding abbreviations (e.g., "2TL" for "2-tuple linguistic model"), along with their synonyms (e.g., "promotion," "advertising," "sales," "branding" for "marketing," etc.). The initial search produced 112 results. However, by applying inclusion and exclusion criteria to pinpoint pertinent and valid papers for the research, the literature scope was narrowed down to 93 articles. Figure 2 shows the literature review process used to identify relevant and valid papers.

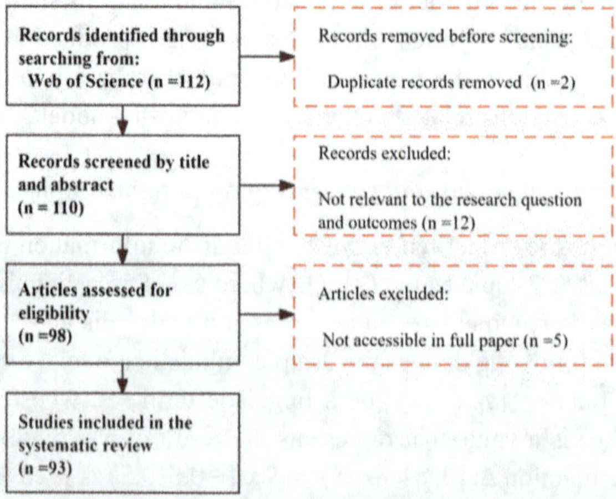

Fig. 2. Flow diagram of literature review process.

Bibliometric analysis was conducted on 93 selected articles using *Biblioshiny* software. The following are the interesting findings:

- Three-fields plot

Figure 3 shows the relationship among the title (words in the title), sources, and countries in the application of the 2-tuple linguistic model in marketing. The three words that appear most frequently in the title are “2-tuple” (indicated as ‘_tuple’ in Figure 3 due to a word limit in the software), “linguistic”, and “fuzzy”. These words are associated with four sources, namely the Journal of Intelligent & Fuzzy Systems, Kybernetes, Engineering Applications of Artificial Intelligence, and Applied Soft Computing. Moreover, these publications exhibit a significant connection with two countries: China and Spain.

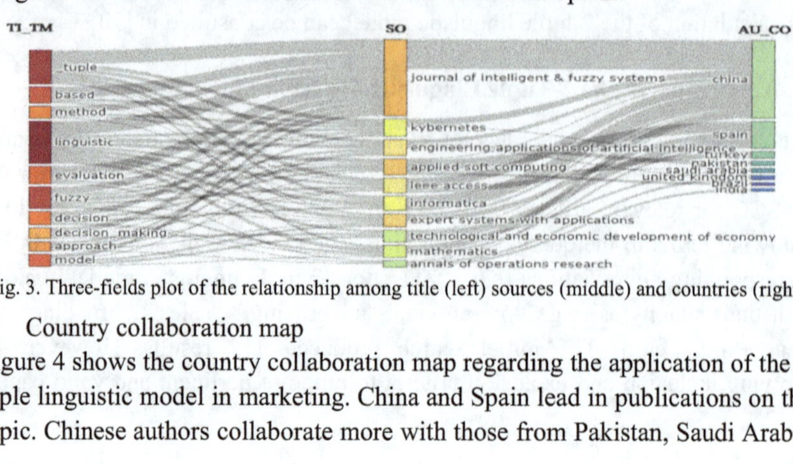

Fig. 3. Three-fields plot of the relationship among title (left) sources (middle) and countries (right).

- Country collaboration map

Figure 4 shows the country collaboration map regarding the application of the 2-tuple linguistic model in marketing. China and Spain lead in publications on this topic. Chinese authors collaborate more with those from Pakistan, Saudi Arabia, and Turkey, while Spanish authors have stronger collaborations with authors from Australia and Saudi Arabia.

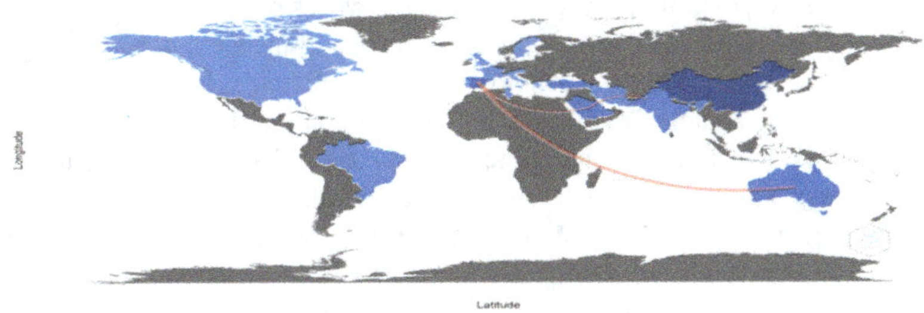

Fig. 4. Country collaboration map.

- Analyses of author keywords and keywords plus

Table 1 displays occurrences of author keywords and keywords plus, each with a frequency of at least 5. As shown in Table 1, author keywords have a significant relationship with the terms used in the literature research query, such as "2-tuple linguistic model," "marketing," etc. Therefore, it becomes more meaningful to work with keywords plus to explore topics closely associated with the application of the 2-tuple linguistic model in marketing.

Table 1. Author keywords and keywords plus occurrences.

Author Keywords	Occurrences	Keywords Plus	Occurrences
"2-tuple linguistic model"	30	"decision-making"	26
"multiple attribute decision making (madm)"	9	"aggregation operators"	25
"multiple attribute group decision making (magdm)"	8	"model"	25
"quality function deployment"	7	"information"	18
"2-tuple linguistic neutrosophic sets (2tlnss)", "neutrosophic numbers", "fuzzy linguistic approach"	5	"selection"	13
		"sets"	11
		"customer satisfaction"	10
		"fuzzy"	7
		"design", "management", "similarity measures", "stock-market", "words"	6
		"operators", "todim method"	5

Observing the occurrences of keywords plus, it is evident that words or phrases that frequently appear in the titles of cited references in articles related to the application of the 2-tuple linguistic model in marketing are "decision-making," "model," and "aggregation operators." This implies that articles in this domain not only focus on the 2-tuple linguistic model but also integrate models associated with decision-making. For instance, Table 1 shows that the TODIM method (i.e., an acronym in Portuguese for Interactive Multi-criteria Decision Making) appears with a frequency of 5 in the keywords plus.

Decision-making holds paramount significance in the field of marketing, as it profoundly shapes strategic choices and significantly influences overall business success. Articles exploring the application of the 2-tuple linguistic model in marketing frequently integrate it with various decision-making models. In this context, aggregation operators emerge as indispensable components, as they are effective tools widely used in addressing decision-making problems. These operators essentially enable the amalgamation of disparate inputs, helping to distill valuable insights and streamline the decision-making process.

- Thematic map

The thematic map is generated based on density and centrality to capture deeper variations, classifying the results into four regions: motor themes (upper-right), niche themes (upper-left), emerging or declining themes (lower-left), and basic themes (lower-right). Figure 5 shows a thematic map constructed based on keywords plus.

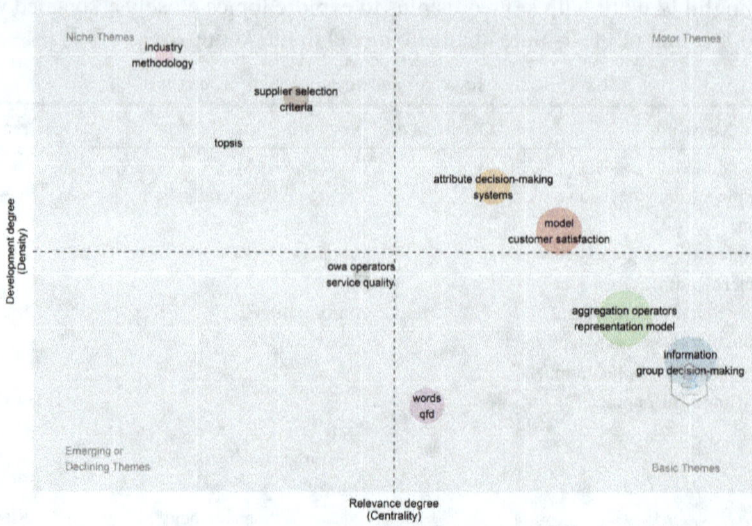

Fig. 5. Thematic map for keywords plus.

As seen in Figure 5, motor themes include “attribute decision-making,” “systems,” “model,” and “customer satisfaction.” They are well-developed and warrant further development in future research due to their significance. Niche themes, including “industry,” “methodology,” “supplier selection,” “criteria,” and “topsis,” are identified in the upper-left quadrant. These specific, under-represented topics, despite their rapid development, exhibit low centrality. Regarding emerging or declining themes, analyzing the frequency of specific keywords facilitates the determination of their status (i.e., whether they are emerging or declining). For example, in conjunction with Figure 5 and Table 1, it can be observed that "owa operator" (i.e., an acronym for Ordered Weighted Averaging operator) is weakly developed regarding the application of the 2-tuple linguistic model in marketing. However, as one of the types of aggregation operators, it represents an emerging theme that could be intriguing to explore further in its combination with the 2-tuple linguistic model.

Finally, basic themes include "aggregation operators," "representation model," "information," "group decision-making," "words," and "qfd" (i.e., an acronym for Quality Function Deployment). They are important for research as general topics but haven't been extensively developed as motor themes. In motor and basic themes, two types of decision-making have been mentioned: "attribute decision-making" and "group decision-making. Both approaches merit further exploration in applying the 2-tuple linguistic model within marketing.

4. Conclusion

The 2-tuple linguistic model, known for its accuracy and simplicity, has found extensive application across various fields. This chapter briefly introduces its concepts and analyzes articles related to the application of the 2-tuple linguistic model in marketing from 2000 to 2023 on the WoSCC. China and Spain stand out as countries with a higher number of publications related to the application of the 2-tuple linguistic model in marketing. Moreover, the 2-tuple linguistic model is applied in marketing primarily to enhance decision-making. It is often integrated with tools like aggregation operators to address decision-making problems, benefiting from its ability to fuse linguistic information without any loss. In the application of the 2-tuple linguistic model in marketing, areas warranting further exploration include "attribute decision-making," "systems," "model," and "customer satisfaction." Additionally, themes like "aggregation operators" are vital for research, with the potential for extensive development.

Acknowledgments

The authors would like to acknowledge the financial support from Universidad Complutense de Madrid and Banco Santander (Project CT58/21-CT59/21), the European Project KA220-HED - Cooperation partnerships in higher education in the Erasmus+ Program in 2022 (Ref: 2022-1-IT02-KA220-HED-000090206) and the grant PID2022-139297OB-I00 funded by MICIU/AEI/10.13039/501100011033 and by ERDF/EU+. They also express gratitude to the Faculty of Statistics at Universidad Complutense de Madrid.

References

1. F. Herrera and L. Martinez, A 2-tuple fuzzy linguistic representation model for computing with words, *IEEE Transactions on Fuzzy Systems* **8(6)**, 746–752 (2000).
2. F. Blanco-Mesa, E. León-Castro, and J. M. Merigó, A bibliometric analysis of aggregation operators, *Applied Soft Computing* **81**, 105488 (2019).
3. T. Malhotra and A. Gupta, A systematic review of developments in the 2-tuple linguistic model and its applications in decision analysis, *Soft Computing* **27(4)**, 1871–1905 (2023).
4. Z. Li, X. Zhang, W. Wang, and Z. Li, Multi-criteria probabilistic dual hesitant fuzzy group decision making for supply chain finance credit risk assessments, *Expert Systems* **39(8)**, e13015 (2022).
5. W. Wang, L. Ding, X. Liu, and S. Liu, An interval 2-Tuple linguistic Fine-Kinney model for risk analysis based on extended ORESTE method with cumulative prospect theory, *Information Fusion* **78**, 40–56 (2022).
6. L. Huang, L.-X. Mao, Y. Chen, and H.-C. Liu, New method for emergency decision making with an integrated regret theory-EDAS method in 2-tuple spherical linguistic environment, *Applied Intelligence* **52(11)**, 13296–13309 (2022).
7. M. Sarwar, Improved assessment model for health-care waste management based on dual 2-tuple linguistic rough number clouds, *Engineering Applications of Artificial Intelligence* **123**, 106255 (2023).
8. J. Serrano-Guerrero, M. Bani-Doumi, F. P. Romero, and J. A. Olivas, A 2-tuple fuzzy linguistic model for recommending health care services grounded on aspect-based sentiment analysis, *Expert Systems with Applications* **238**, 122340 (2024).
9. I. Bueno, R. A. Carrasco, C. Porcel, G. Kou, and E. Herrera-Viedma, A linguistic multi-criteria decision making methodology for the evaluation of tourist services considering customer opinion value, *Applied Soft Computing* **101**, 107045 (2021).
10. Z. Shu, R. A. Carrasco, M. Sánchez-Montañés, and J. P. García-Miguel, A Multi-Criteria Decision Support Model for Restaurant Selection Based on Users' Demand Level: The Case of Dianping.com, *Information Processing & Management* **61(3)**, 103650 (2024).
11. E. Herrera-Viedma, F. Herrera, L. Martínez, J.C. Herrera, and A. López, Incorporating filtering techniques in a fuzzy linguistic multi-agent model for information gathering on the web, *Fuzzy Sets and Systems* **148(1)**, 61–83 (2004).

Carbon Emission Reduction and Pricing under Timely Payment and Bank Loan Financing Strategy in Dual-Channel Bilateral Capital Constrained Supply Chain of Manufacturers

Limin Du*, Yuhao Zhang, Qing Yu, Mengjia Zhang and Jie Zhao

International Business School, Henan University,
Zheng Zhou, Henan, China
**dulimin@vip.henu.edu.cn*

Aiming at the dual-channel low-carbon supply chain system in which upstream and downstream enterprises composed of a manufacturer and a retailer are constrained by funds, a financing decision model of timely payment and bank loan strategy is constructed through the Stackelberg game model, and the carbon emission reduction rate and profit involved in the carbon emission reduction decision of the model are analyzed. Finally, the numerical analysis is combined to verify and expand the conclusions. The analysis finds that the increase in the cost coefficient of carbon emission reduction and the increase in loan interest rate will hurt the carbon emission reduction behavior of enterprises when they choose the strategy of timely payment and bank loans. It is worth noting that for retailers, the profit level of the just-in-time payment strategy at low-interest rates shows a temporary upward trend with the increase in interest rates.

Keywords: Bilateral financial constraints; dual-channel low-carbon supply chain; timely payment and bank loan portfolio financing strategies.

1. Introduction

As industrialization advances, global warming accelerates, drawing attention to the importance of low-carbon supply chains. Many domestic enterprises are actively carrying out online and offline dual-channel low-carbon supply chain construction. However, it is difficult for manufacturers and retailers to solve the financing problem. This chapter analyzes the joint decision-making of pricing and carbon abatement in dual-channel low-carbon supply chains under bilateral financial constraints, which will provide support for supply chain construction.

There have been several studies on low-carbon supply chain financing strategies. Shen studied the optimal payment method for the manufacturer's unilateral capital constraint in a low-carbon supply chain.[1] Cai *et al.* studied the financing and operation decisions of a supply chain consisting of a manufacturer with a unilateral capital constraint.[2] Dai *et al.* discussed different payment modes for the financing problem of capital-constrained suppliers in a low-carbon supply chain.[3] Sun *et al.* explored the supply chain financing strategy of bilateral capital constraints and analyzed the impact.[4] Therefore, scholars have not sufficiently studied the dual-channel supply chains.

There is much literature on the strategies of supply chain financing. Zhang *et al.* explored financing methods and found that supplier financing is more dominant.[5] Wang *et al.* found that the core enterprise suppliers in the supply chain can provide commercial credit financing for capital-constrained retailers.[6] Zou *et al.* studied financing strategy considering unilateral financial constraints and compared the optimal decisions.[7] Cao *et al.* considered the impact of advance payment and bank loan financing strategy on optimal pricing and carbon abatement of dual-channel supply chains under financial constraints.[8]

In summary, the innovation of this chapter mainly focuses on the bilateral financial constraints in the dual channels of a manufacturer and a retailer in a low-carbon supply chain and proposes a combination of timely payment and bank loan financing strategies. This will extend the research on low-carbon supply chains and is also important for realizing the goals of carbon neutrality.

2. Assumptions and Description of the Problem

This study examines a supply chain with a manufacturer and retailer both under capital constraints. The manufacturer produces eco-friendly products and sells them through two channels: directly to consumers online and indirectly through the retailer offline. The focus is on how they manage financing, combining instant payments with bank loans to facilitate sales in this dual-channel system.

Assuming that the initial funds of both manufacturer and retailer are zero, information symmetry exists between manufacturer and retailer, and the market can clear completely, carbon emissions are mainly generated during the production process, and consumers have low-carbon preferences. The manufacturer is the leader, carbon abatement rate, wholesale price and online selling price of the products are its decision variables, and offline selling price is a decision variable for the retailer. The relevant symbols in this paper are shown in Table 1.

Table 1. Related symbols and their meanings

Symbols	Meanings
β	Cross-price elasticity between offline and online channel
λ	Low carbon preference coefficient for consumers
c	Production cost per unit product for manufacturer
w	Wholesale price per unit product
A	Potential market demand
θ	Proportion of consumers who prefer offline retail channels
$1-\theta$	Proportion of consumers who prefer online sales channels
τ	Carbon abatement rate. Under this rate, the carbon abatement technology investment[9] is $C(\tau) = 0.5\,\eta\tau^2$
η	Carbon abatement cost coefficient
P_d	Manufacturer's online selling unit product price
P_r	Retailer's offline selling unit product price
r_1	Interest rate at which a manufacturer borrowings from a bank for financing
r_2	Interest rate at which a retailer borrowings from a bank for financing
π_m	Manufacturer's profit
π_r	Retailer's profit
Q_d	Online sales channel market demand
Q_r	Offline sales channel market demand

Based on Table 1, taking into account carbon emission reduction, the market demand assumption regarding the online direct sales channel and offline retail channel of the bilateral capital-constrained dual-channel supply chain[8], the online demand is $Q_d = (1-\theta)A - P_d + \beta P_r + \lambda\tau$, the offline demand is $Q_r = \theta A - P_r + \beta P_d + \lambda\tau$.

3. Model Building and Analysis

Decision-making process are illustrated in Fig. 1.

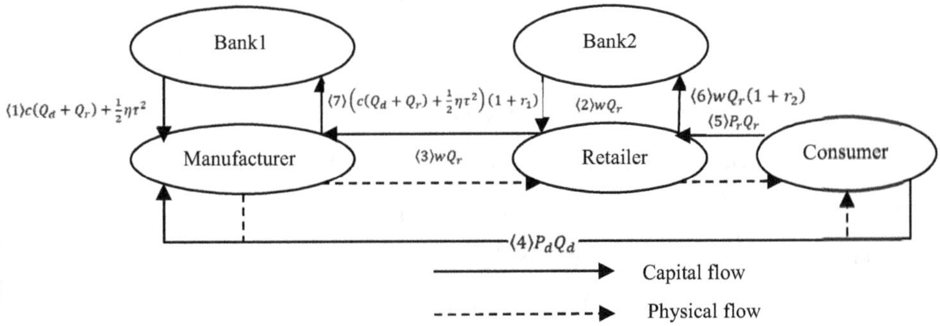

Fig. 1. Decision-making process illustration

The profit equation of the manufacturer and retailer are

$$\pi_m = (p_d - c)Q_d + (w - c)Q_r - \frac{1}{2}\eta\tau^2 - r_1\left((Q_d + Q_r)c + \frac{1}{2}\eta\tau^2\right). \quad (1)$$

$$\pi_r = (P_r - w)Q_r - r_2 w Q_r. \quad (2)$$

The manufacturer is the leader, adopting backward induction, let

$$\frac{d\pi_r}{dP_r}=0$$

$$P_r=\frac{\theta A+\beta P_d+\lambda\tau+w+r_2 w}{2}. \tag{3}$$

Putting Eq. (3) into Eq. (1) we get

$$\pi_m=-\frac{r_1\eta\tau^2}{2}-(P_d-c)\left(P_d-\lambda\tau+A(\theta-1)-\beta\frac{w+P_d\beta+A\theta+\lambda\tau+r_2 w}{2}\right)+r_1 c\left(P_d+\frac{w-P_d\beta-A\theta-3\lambda\tau+r_2 w}{2}+A(\theta-1)-\beta\frac{w+P_d\beta+A\theta+\lambda\tau+r_2 w}{2}\right). \tag{4}$$

Calculating the second-order partial derivatives of P_d, w, τ in Eq. (4) respectively, the Hessian matrix of the manufacturer's profit π_m can be obtained

$$\mathrm{H}=\begin{pmatrix}\frac{\partial^2\pi_m}{\partial P_d^2} & \frac{\partial^2\pi_m}{\partial P_d\partial w} & \frac{\partial^2\pi_m}{\partial P_d\partial\tau}\\ \frac{\partial^2\pi_m}{\partial w\partial P_d} & \frac{\partial^2\pi_m}{\partial w^2} & \frac{\partial^2\pi_m}{\partial w\partial\tau}\\ \frac{\partial^2\pi_m}{\partial\tau\partial P_d} & \frac{\partial^2\pi_m}{\partial\tau\partial w} & \frac{\partial^2\pi_m}{\partial\tau^2}\end{pmatrix}=\begin{pmatrix}\beta^2-2 & \beta\left(1+\frac{r_2}{2}\right) & \lambda+\frac{\beta\lambda}{2}\\ \beta\left(1+\frac{r_2}{2}\right) & -r_2-1 & \frac{\lambda}{2}\\ \lambda+\frac{\beta\lambda}{2} & \frac{\lambda}{2} & -\eta-\eta r_1\end{pmatrix}$$

When $\beta^2<2$, $(1-\beta^2)(1+r_2)-\frac{\beta^2 r_2^2}{8}>0$, $2\beta^2\eta-2\eta r_1-2\eta r_2-2\eta+2\beta\lambda^2+\lambda^2 r_2+\frac{1}{2}(3\lambda^2+\beta^2\lambda^2+3\beta\lambda^2 r_2+2\beta^2\eta r_2^2)+\beta^2\lambda<0$, The matrix H is negative definite, the manufacturer's profitπ_m under the strategy of combining immediate payment and bank loan is a concave function about the online direct sales price P_d, the retailer's offline wholesale price$w$, and the carbon abatement rate τ, so there are optimal solutions: $P_d^{\#}$, $w^{\#}$, $\tau^{\#}$.

$$\xi=2(1+\beta)\lambda^2(3+\beta+(2+\beta)r_2)+\eta(1+r_1)\left(-8(1+r_2)+\beta^2(8+r_2(8+r_2))\right)$$

$$P_d{}^{\#}=c(1+r_1)\left((1+\beta)\lambda^2(6+5r_2+2\beta(1+r_2))\right)/\xi+c(1+r_1)\left((-1+\beta)\eta(1+r_1)(1+r_2)(4+\beta(4+r_2))\right)/\xi$$

$$+A\left(\lambda^2(1-2\theta)-\eta(1+r_1)(4+4(-1+\beta)\theta+r_2(4-4\theta+3\beta\theta))\right)/\xi. \quad (5)$$

$$w^{\#}=$$

$$-A(\lambda^2(2-4\theta)+\beta\lambda^2(1-2\theta)+4\eta\theta(1+r_1)+\beta^2\eta\theta(1+r_1)r_2)/\xi$$
$$+A(2\beta\eta(-1+\theta)(1+r_1)(2+r_2))/\xi$$
$$-c(1+r_1)(-1+\beta)\eta(1+r_1)(-4(1+\beta)+(-4+(-2+\beta)\beta)r_2)/\xi$$
$$+c(1+r_1)(1+\beta)\lambda^2(2(3+\beta)+(2+\beta)r_2)/\xi. \quad (6)$$

$$\tau^{\#}=$$

$$-\lambda(1+\beta)c(1+r_1)(-6(1+r_2)+\beta(2\beta(1+r_2)+(2+r_2)^2))/\xi$$
$$-\lambda A\left(4+4\beta+4r_2+3\beta r_2+2\left(-1-2r_2+\beta^2(1+r_2)\right)\theta\right)/\xi. \quad (7)$$

Putting Eqs. (5) (6) (7) into Eq. (3), we can get the retailer's optimal offline sales price $P_r^{\#}$

$$P_r^{\#}=(1+r_2)\left(-A(3\lambda^2+2\beta\lambda^2+\beta\eta(1+r_1)(4+r_2))\right)/\xi$$
$$+(1+r_2)c(1+r_1)((1+\beta)\lambda^2(6+2\beta+r_2))/\xi$$
$$+(1+r_2)c(1+r_1)(-1+\beta)\eta(1+r_1)(2(1+\beta)^2+(2+\beta)r_2)/\xi$$
$$+(1+r_2)A\theta\left(2(3+2\beta)\lambda^2+\eta(1+r_1)(-6+\beta(4+2\beta+r_2))\right)/\xi. \quad (8)$$

Theorem 4.1: *When the initial funds of the manufacturer and retailer are zero,* $\beta^2<2$, $(1-\beta^2)(1+r_2)-\frac{\beta^2 r_2^2}{8}>0$, $2\beta^2\eta-2\eta r_1-2\eta r_2-2\eta+2\beta\lambda^2+\lambda^2 r_2+\frac{1}{2}(3\lambda^2+\beta^2\lambda^2+3\beta\lambda^2 r_2+2\beta^2\eta r_2^2)+\beta^2\lambda<0$ *the optimal price and carbon abatement rate exist when manufacture and retailer adopt the combination of immediate payment with bank loans.*

4. Numerical Analysis

This paper will analyze the financing strategies of manufacturers and retailers in a low-carbon supply chain with bilateral financial constraints and dual-channel.

In order to simplify the calculation according to Cao et al. [8] and combine with the method of Bai et al. [10] to set the basic operating parameters Table 2, numerical simulation analysis.

Table 2. Basic operating parameters

β	θ	A	r_1	r_2	η	λ	c
0.5	0.4	500	0.2	0.2	6.5×10^4	30	20

4.1. *Analysis of financing strategies*

Fig. 2 shows the impact of carbon abatement cost coefficient η on the optimal carbon abatement rate τ, manufacturer's profit π_m and retailer's profit π_r .

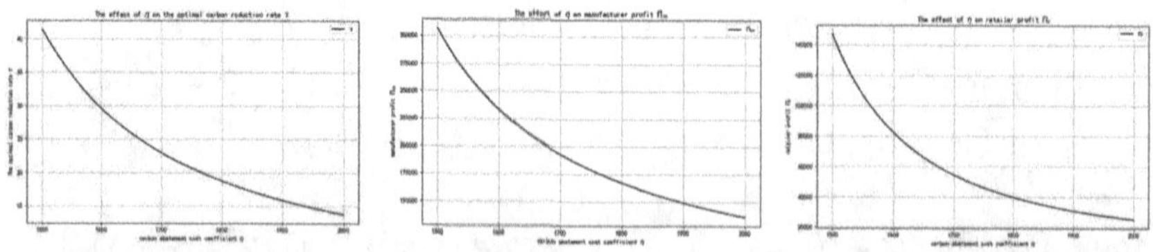

Fig. 2. Impact of carbon abatement cost coefficient η on optimal carbon abatement rate τ, manufacturer's profit π_m, and retailer's profit π_r

Figure 2 illustrates that under the combination of prompt payment and bank loan financing, the carbon abatement cost coefficient has a negative relationship with the optimal carbon abatement rate, manufacturer's profit and retailer's profit. This means that a higher carbon reduction cost will have a negative impact on profitability. Therefore, companies need to consider the balance between cost and profit when formulating their carbon reduction strategies to achieve their carbon reduction targets.

When the manufacturer's lending rate r_1 is the same as the retailer's lending rate r_2, the impact of the manufacturer/retailer lending rate r on the optimal carbon emission reduction rate τ, the manufacturer's profit π_m, and the retailer's profit π_r is shown in Fig.3.

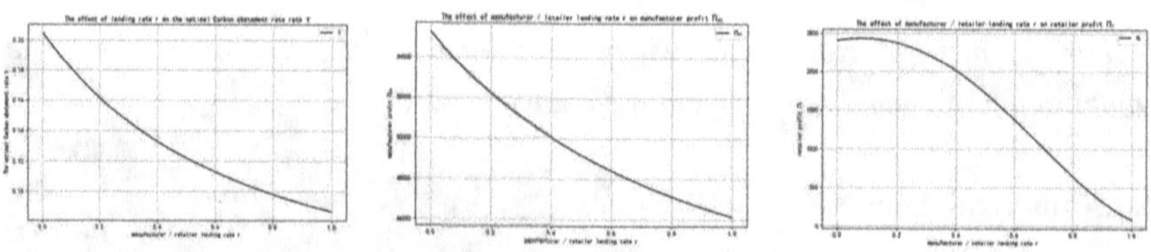

Fig. 3. The effect of manufacturer/retailer lending rate r on the optimal carbon abatement rate τ, manufacturer's profitπ_m, and retailer's profit π_r

Figure 3 shows that under the combination strategy of prompt payment and bank loan financing, the loan interest rate and retailer's profit first show a weak positive correlation and then switch to a negative correlation. This may be due to the fact that when the loan interest rate is low, retailers use loan financing strategy to support business expansion. However, as lending rates rise, retailers face higher borrowing costs and may need to cover the costs by increasing retail prices, which will lead to a negative correlation between lending rates and retailers' profits.

The following paper will study when the manufacturer and retailer loan interest rates change at the same time, how the manufacturer's lending rate r_1 and the retailer's lending rate r_2 affect the optimal carbon abatement rate τ, the manufacturer's profit π_m and the retailer's profit π_r, and the result was shown in Fig. 4.

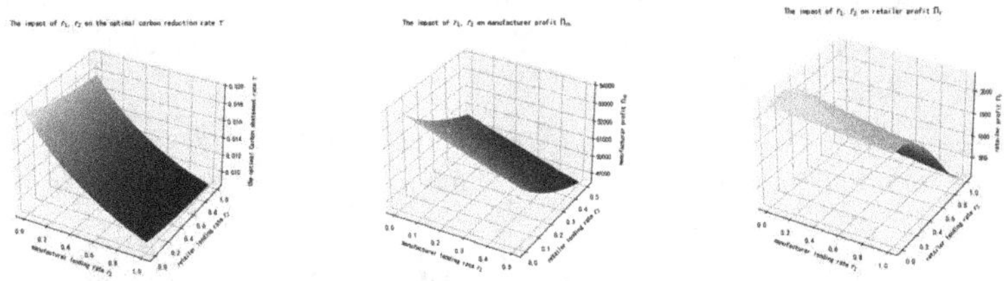

Fig. 4. Impact of Manufacturer and Retailer Lending Rates r_1, r_2 on the Optimal Carbon Emission Reduction Rate τ, Manufacturer's Profit π_m and Retailer's Profit π_r.

Figure 4 reflects that under the combination strategy of prompt payment and bank loan financing, when the manufacturer/retailer loan interest rate increases, its corresponding optimal carbon emission reduction rate, and the manufacturer's profit decrease accordingly. In contrast, the retailer's profit first increases with the increase of the loan rate and then decreases. This may be because, the retailer may increase the price of its products as the lending rate increases initially, leading to a rise in profits. However, as the lending rate increases further, the retailer may need to cover the costs by increasing the price of the product, which may lead to a decline in sales and ultimately hurt profits.

5. Conclusions and Recommendations

In conclusion, this study investigates the bilateral capital-constrained dual-channel low-carbon supply chain, employing a Stackelberg game model to analyze the impact of carbon emission reduction cost coefficients and loan interest rates on optimal strategies. The findings underscore the significance of considering financing strategies in low-carbon supply chains, revealing that higher carbon emission reduction costs and loan interest rates lead to decreased emission reduction behaviors and financial performance for enterprises. Additionally, this chapter highlights the complex interplay between lending rates and profits, indicating a negative correlation between optimal emission reduction rates and manufacturer profits as lending rates increase.

Future research avenues could explore diverse financing strategy choices, consider non-zero initial capital for manufacturers, and examine the effects of financial constraints on sales strategies amidst online and offline retail expansions.

References

1. J. N. Shen and X. F. Shao. A study on operation strategy and payment selection of capital-constrained manufacturers under product competition. Journal of Management Engineering, vol. 35, no. 05, pp. 184-201, 2021.
2. M. Cai and J. W. Luo. *Optimal financing and quality decisions for financially constrained manufacturers.* Journal of Systems Management, vol. 31, no. 02, pp. 230-240, 2022
3. J. S. Dai and G. Li. *A co-operation and coordination model for capital-constrained supply chain.* China Management Science, vol. 31, no. 09, pp. 83-93, 2023.
4. F. S. Sun, X. Song and M. Jian. Analysis of bilateral capital constraint supply chain financing strategy based on technological innovation. Operations Research and Management, pp, 1-9, December, 2023.
5. Y. L. Zhang and X. J. Hu. *A study on supply chain decision-making considering manufacturers' financial constraints under carbon cap policy.* Journal of Hefei University of Technology (Natural Science Edition), vol. 45, no. 10, pp. 1413-1421, 2022
6. D. P. Wang, M. Y. Zhu and H. X. Dong. Research on supply chain operation decision-making based on blockchain technology under capital constraint. China Management Science, pp. 1-13, November, 2023.
7. Q. M. Zou, C. Liu and T. J. Zou. Research on financing and coordination of low-carbon supply chain with uncertain output under financial constraint. Operations Research and Management, vol. 31, no. 12, pp. 76-85, 2022
8. X. Y. Cao, Y. H. Qin and X. Y. Yu. A study on dual-channel supply chain pricing and carbon emission reduction decision-making considering manufacturer's financial constraint. Journal of Central China Normal University (Natural Science Edition), vol. 57, no. 06, pp. 859-869, 2023.
9. C. Q. Xu, D. Z. Zhao, B. Y. Yuan, et al. Differential game modeling of joint upstream and downstream emission reduction and low-carbon publicity. Journal of Management Science, vol. 19, no. 02, pp.53-65, 2016.
10. S. Z. Bai and X. L. Jia, Research on the operation strategy of dual-channel supply chain of fresh agricultural products with capital constraint. China Management Science, pp. 1-13, November, 2023.

Pricing and Carbon Emission Reduction of Delayed Payment and Bank Loan Combination Strategies in Manufacturers' Dual-Channel Supply Chains under Bilateral Funding Constraints

Limin Du*, Mengjia Zhang, Qing Yu, Jie Zhao and Yuhao Zhang

International Business School, Henan University,
Zheng Zhou, Henan, China
**dulimin@vip.henu.edu.cn*

This chapter focuses on the dual-channel low-carbon supply chain system, comprising upstream and downstream enterprises, consisting of a manufacturer and a retailer, facing capital constraints. A financing decision model employing a delayed payment and bank loan strategy is formulated using the Stackelberg game model. The model analyzes the carbon emission reduction rate and its impact on the profits of manufacturers and retailers. Additionally, numerical analysis is conducted to validate and extend the conclusions. The findings suggest that increasing the cost coefficient of carbon emission reduction and the loan interest rate negatively influence enterprise carbon emission reduction behavior when employing the deferred payment and bank loan strategy.

Keywords: Bilateral financial constraints; dual-channel low-carbon supply chain; deferred payment and bank loan financing strategies.

1. Introduction

Currently, the world is facing serious climate change, and countries are committed to reducing carbon emissions. Therefore, the low-carbon supply chain has received widespread attention. China's low-carbon supply chain construction is not mature enough, and manufacturers and retailers are prone to face financial constraints. This chapter studies the joint decision-making on pricing and carbon abatement in dual-channel low-carbon supply chains under bilateral financial constraints to provide useful references for the theory and practice.

At present, much domestic literature has explored the financial constraints and financing strategies of low-carbon supply chains. Wang *et al.* studied the decision-making problem in a low-carbon supply chain and found that the more adequate the supplier's funds, the more favorable.[1] Liu solved the optimal wholesale price of risk-neutral suppliers for a low-carbon supply chain.[2] Zhang *et al.* studied the financing decision of retailers with insufficient purchasing funds in a government-subsidized low-carbon supply chain.[3] Xu studied the impact of different financing methods in a low-carbon supply chain.[4] We find that the discussion on the financial constraints of dual-channel supply chains is relatively insufficient.

To solve the problem, academics have conducted extensive research on the methods and strategies of financing. Zhou *et al.* found that the adoption of retailer sharing or bank loan financing strategy by manufacturers can effectively improve performance.[5] Zhang *et al.* investigated the choices of different financing options for financially constrained manufacturers in low-carbon supply chains.[6] Ren focused on the choice of optimal financing strategies for manufacturers and retailers in different scenarios.[7] Bai *et al.* argued that financially challenged suppliers may be able to achieve a win-win situation with the retailer by adopting an internal financing model.[8] It can be seen that the bilateral financial constraint problem of the low-carbon supply chain still needs more in-depth analysis.

The innovation of this chapter is to combine the bilateral financial constraints and dual channels to study the pricing and carbon abatement decisions of low-carbon supply chains, and to propose a financing combination strategy of bank loans and deferred payment. This provides new perspectives and methods for the financial management of low-carbon supply chains.

2. Assumptions and Description of the Problem

Considering a two-stage supply chain consisting of a single manufacturer and retailer. The manufacturer is green products producer, facing capital constraints, and retailers, who sell green products, also face capital constraints. Manufacturers can sell green products to retailers offline, or they can sell green products to consumers online. We define this phenomenon as that both manufacturer and retailer have capital constraints and manufacturers can carry out dual-channel sales as a bilateral capital-constrained dual-channel supply chain.

For the bilateral capital-constrained dual-channel supply chain problem, this paper mainly considers the combination of external financing and internal financing. This combines deferred payment with bank loan financing, which means the manufacturer lends money from the bank and allows the retailer to delay payment.

Assuming that the initial funds of both manufacturer and retailer are zero, information symmetry exists between manufacturer and retailer, and the market can clear completely, carbon emissions are mainly generated during the production process, and consumers have low-carbon preferences. The manufacturer is the leader, carbon abatement rate, wholesale price and online selling price of the products are its decision variables, and offline selling price is a decision variable for the retailer. The relevant symbols in this paper are shown in Table 1.

Table 1. Related symbols and their meanings

Symbols	Meanings
β	Cross-price elasticity between offline and online channel
λ	Low carbon preference coefficient for consumers
c	Production cost per unit product for manufacturer
w	Wholesale price per unit product
A	Potential market demand
θ	Proportion of consumers who prefer offline retail channels
$1-\theta$	The proportion of consumers who prefer online sales channels
τ	Carbon abatement rate. Under this rate, the carbon abatement technology investment[9] is $C(\tau) = 0.5\,\eta\tau^2$
η	Carbon abatement cost coefficient
P_d	Manufacturer's online selling unit product price
P_r	Retailer's offline selling unit product price
r_1	Interest rate at which a manufacturer borrowings from a bank for financing
π_m	Manufacturer's profit
π_r	Retailer's profit
Q_d	Online sales channel market demand
Q_r	Offline sales channel market demand

Based on Table 1, taking into account carbon emission reduction, the market demand assumption regarding the online direct sales channel and offline retail channel of the bilateral capital-constrained dual-channel supply chain [10], the online demand is $Q_d = (1-\theta)A - P_d + \beta P_r + \lambda\tau$, the offline demand is $Q_r = \theta A - P_r + \beta P_d + \lambda\tau$.

3. Model Building and Analysis

Decision-making process are illustrated in Fig. 1.

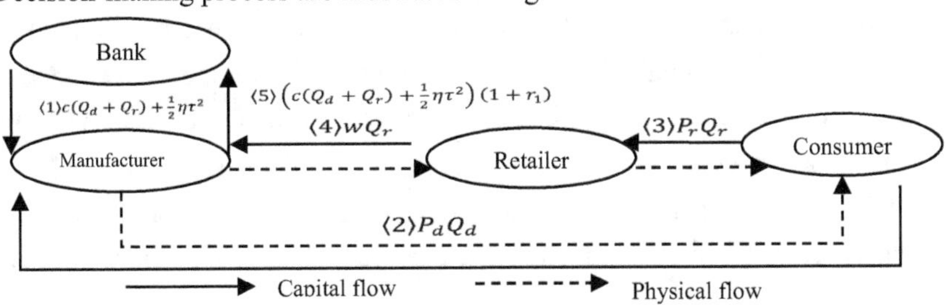

Fig.1. Decision-making process illustration

The profit equation of the manufacturer and retailer are

$$\pi_m = (P_d - c)Q_d + (w - c)Q_r - \frac{1}{2}\eta\tau^2 - r_1\left((Q_d + Q_r)c + \frac{1}{2}\eta\tau^2\right). \quad (1)$$

$$\pi_r = (P_r - w)Q_r. \quad (2)$$

The manufacturer is the leader, adopting backward induction, let $\frac{d\pi_r}{dP_r} = 0$

$$P_r = \frac{\theta A + \beta P_d + \lambda\tau + w}{2}. \quad (3)$$

Putting Eq. (3) into Eq. (1) we get

$$\pi_m = -(P_d - c)\left(P_d - \lambda\tau - \beta\frac{w + P_d\beta + A\theta + \lambda\tau}{2} + A(\theta - 1)\right) - (c - w)\frac{P_d\beta - w + A\theta + \lambda\tau}{2} - \frac{\eta\tau^2}{2} - r_1\left(\frac{\eta\tau^2}{2} + c\left(\frac{P_d\beta - w + A\theta + 3\lambda\tau}{2} - P_d\right)\right) - r_1\left(\frac{\eta\tau^2}{2} + c\left(\beta\frac{w + P_d\beta + A\theta + \lambda\tau}{2} - A(\theta - 1)\right)\right). \quad (4)$$

Calculating the second-order partial derivatives of P_d, w, τ in Eq. (4) respectively, the Hessian matrix of the manufacturer's profit π_m can be obtained

$$H = \begin{pmatrix} \frac{\partial^2\pi_m}{\partial P_d^2} & \frac{\partial^2\pi_m}{\partial P_d\partial w} & \frac{\partial^2\pi_m}{\partial P_d\partial\tau} \\ \frac{\partial^2\pi_m}{\partial w\partial P_d} & \frac{\partial^2\pi_m}{\partial w^2} & \frac{\partial^2\pi_m}{\partial w\partial\tau} \\ \frac{\partial^2\pi_m}{\partial\tau\partial P_d} & \frac{\partial^2\pi_m}{\partial\tau\partial w} & \frac{\partial^2\pi_m}{\partial\tau^2} \end{pmatrix} = \begin{pmatrix} \beta^2 - 2 & \beta & \lambda + \frac{\beta\lambda}{2} \\ \beta & -1 & \frac{\lambda}{2} \\ \lambda + \frac{\beta\lambda}{2} & \frac{\lambda}{2} & -\eta - \eta r_1 \end{pmatrix}$$

When $\beta^2 < 1$ and $2\beta^2\eta r_1 + 2\beta^2\eta + \frac{1}{2}(\beta^2\eta^2 + 3\lambda^2) + 2\beta\lambda^2 - 2\eta r_1 - 2\eta < 0$, The matrix H is negative definite, the manufacturer's profitπ_m under the strategy of combining deferred payment and bank loan is a concave function about the online direct sales price P_d, the retailer's offline wholesale price w, and the carbon abatement rate τ, so there are optimal solutions: $P_d^{\#}$, $w^{\#}$, $\tau^{\#}$.

Derive the Eq. (4) for the online direct sales price P_d, the wholesale price w, and the carbon emission reduction rate τrespectively, and set them 0, we get

$$P_d^{\#} = \frac{2(1+\beta)c(1+r_1)((3+\beta)\lambda^2 + 2(-1+\beta)\eta(1+r_1))}{2(1+\beta)((3+\beta)\lambda^2 + 4(-1+\beta)\eta(1+r_1))} + \frac{A(\lambda^2(1-2\theta) - 4\eta(1+r_1)(1+(-1+\beta)\theta))}{2(1+\beta)((3+\beta)\lambda^2 + 4(-1+\beta)\eta(1+r_1))}. \tag{5}$$

$$w^{\#} = \frac{2(1+\beta)c(1+r_1)((3+\beta)\lambda^2 + 2(-1+\beta)\eta(1+r_1))}{2(1+\beta)((3+\beta)\lambda^2 + 4(-1+\beta)\eta(1+r_1))} + \frac{A\left(-2\lambda^2 - \beta(\lambda^2 + 4\eta(1+r_1))\right)}{2(1+\beta)((3+\beta)\lambda^2 + 4(-1+\beta)\eta(1+r_1))} + \frac{A(2((2+\beta)\lambda^2 + 2(-1+\beta)\eta(1+r_1))\theta)}{2(1+\beta)((3+\beta)\lambda^2 + 4(-1+\beta)\eta(1+r_1))}. \tag{6}$$

$$\tau^{\#} = -\left(\frac{\lambda((-1+\beta)(3+\beta)c(1+r_1) + A(2+(-1+\beta)\theta))}{(3+\beta)\lambda^2 + 4(-1+\beta)\eta(1+r_1)}\right). \tag{7}$$

Putting Eqs. (5) (6) (7) into Eq. (3), we can get the retailer's optimal offline sales price $P_r^{\#}$

$$P_r^{\#} = \frac{2(1+\beta)c(1+r_1)((3+\beta)\lambda^2 + (-1+\beta^2)\eta(1+r_1))}{2(1+\beta)((3+\beta)\lambda^2 + 4(-1+\beta)\eta(1+r_1))} + \frac{A\left(-3\lambda^2 - 2\beta(\lambda^2 + 2\eta(1+r_1))\right)}{2(1+\beta)((3+\beta)\lambda^2 + 4(-1+\beta)\eta(1+r_1))} + \frac{A(2((3+2\beta)\lambda^2 + (-1+\beta)(3+\beta)\eta(1+r_1))\theta)}{2(1+\beta)((3+\beta)\lambda^2 + 4(-1+\beta)\eta(1+r_1))}. \tag{8}$$

Theorem 4.1: *When the initial funds of the manufacturer and retailer are zero,* $\beta^2 < 1, 2\beta^2\eta r_1 + 2\beta^2\eta + \frac{1}{2}(\beta^2\eta^2 + 3\lambda^2) + 2\beta\lambda^2 - 2\eta r_1 - 2\eta < 0,$ *the optimal online sales price, offline sales price, the wholesale price and carbon abatement rate exist when manufacture and retailer adopt the combination of deferred payment with bank loans.*

4. Numerical Analysis

This paper will use MATLAB and Python software to build two-dimensional and three-dimensional models to sequentially analyze the financing strategies of manufacturers and retailers in a low-carbon supply chain with bilateral financial constraints and dual-channel sales since the optimal solution formula is too complex.

Retailers are more focused on their earnings than manufacturers are on the pace of reduction of carbon emissions in the green supply chain of dual-channel sales with bilateral financial limitations. Table 2 numerical simulation analysis was used to define the fundamental operating parameters to simplify the calculation using the approach of Bai Shizhen et al. [8] and in accordance with Cao Xiyu et al. [10].

Table 2. Basic operating parameters

β	θ	A	r	η	λ	c
0.5	0.4	500	0.2	6.5×10^4	30	20

4.1. *Analysis of financing strategies*

The effect of carbon abatement cost coefficient η on the optimal carbon abatement rate τ, manufacturer's profit π_m and retailer's profit π_r is shown in Fig.2.

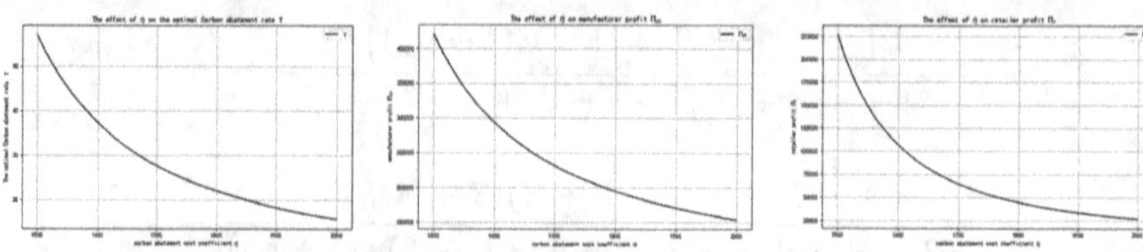

Fig.2 Impact of carbon abatement cost coefficient η on optimal carbon abatement rate τ, manufacturer's profit π_m, and retailer's profit π_r

Figure 2 shows that under the combination strategy of delayed payment and bank loan financing, the carbon abatement cost coefficient is negatively correlated with the optimal carbon abatement rate, manufacturer's profit, and retailer's profit. This negative correlation also serves as a reminder that manufacturers and retailers need to carefully weigh economic benefits and environmental impacts when pursuing carbon reduction targets to ensure economic viability in the context of sustainable development.

The effect of the manufacturer's lending rate r on the optimal carbon reduction rate τ, the manufacturer's profit π_m, and the retailer's profit π_r is shown in Fig.3.

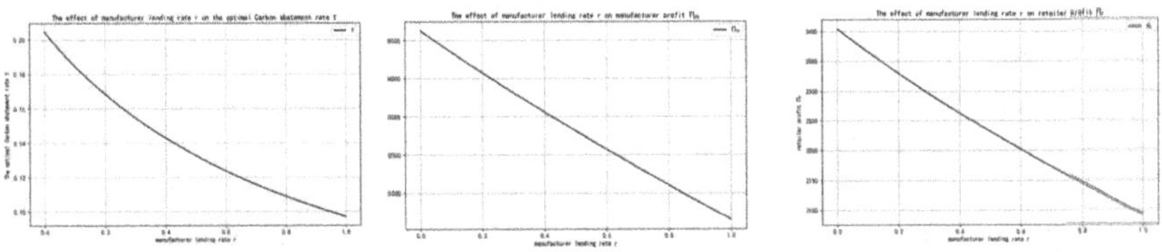

Fig.3 The effect of manufacturer's lending rate r on the optimal carbon reduction rate τ, manufacturer's profit π_m and retailer's profit π_r.

Figure 3 demonstrated that under the combination strategy of delayed payment and bank loan financing, the manufacturer loan interest rate is negatively correlated with the optimal carbon emission reduction rate, the manufacturer's profit and the retailer's profit. This suggests that higher lending rates may negatively affect firms' environmental investments, profits and profitability. Firms may need to weigh economic and environmental objectives and seek more appropriate financing strategies to mitigate the impact of lending rates on their sustainability efforts.

5. Conclusions and Recommendations

This chapter investigates the bilateral capital-constrained dual-channel supply chain, analyzing the impact of carbon emission reduction costs and manufacturers' loan interest rates on optimal strategies. Findings reveal that in adopting deferred payment and bank loan financing, increasing carbon emission reduction costs negatively affects both emission reduction behavior and financial performance of enterprises. Moreover, simultaneous increases in loan interest rates lead to decreased optimal carbon emission reduction rates and profits for manufacturers and retailers.

Future research avenues should explore innovative financing methods, including timely disbursement and bank lending strategies. Additionally, investigations should consider the impact of capital constraints on financing decisions and sales strategies, especially in scenarios where initial capital for manufacturers and retailers is non-zero, and in contexts where retailers expand both online and offline.

References

1. M. Wang and L. Y. Li. *Research on the decision-making of low-carbon supply chain with both supply and demand being constrained by capital.* Computer Science, vol. 1, no. 02, pp. 989-997, 2023.

2. G. A. Liu, *Research on low carbon supply chain decision making considering risk appetite and financial constraints.* Chongqing Jiaotong University, 2022.
3. K.Y. Zhang, C. X. Li, J. M. Yao et al. *A study on low-carbon supply chain financing decision-making under retailers' purchasing capital constraint.* Operations Research and Management, vol. 30, no. 08, pp. 108-116, 2021.
4. S. Xu. *Research on low carbon supply chain operation and financing strategy under financial constraints.* Nankai University, 2020.
5. T. Zhou, M. Tao and X. Q. Meng. *A study on manufacturers' low-carbon financing strategies considering capital constraints under carbon trading regulations.* Financial Theory and Practice, vol. 1, no. 09, pp. 44-56, 2022.
6. C. Zhang, H. M. Ma and Z. Guo. *Supply chain emission reduction and financing strategies under carbon cap-and-trade mechanism and consumer low-carbon preference.* Control and Decision Making, vol. 1, no. 11, pp. 3271-3278, 2023.
7. K. Ren. *Research on Green Supply Chain Financing Strategies Considering Capital Constraints.* Hangzhou University of Electronic Science and Technology, 2023.
8. S. Z. Bai and X. L. Jia, *Research on the operation strategy of dual-channel supply chain of fresh agricultural products with capital constraint.* China Management Science, pp. 1-13, November, 2023.
9. W. L. Wang, F. L. Wang and S.X. Zhang. *A study of dual-channel supply chain coordination contract considering low-carbon efforts.* Management Review, vol. 33, no. 04, pp. 315-326, 2021.
10. X. Y. Cao, Y. H. Qin and X. Y. Yu. *A study on dual-channel supply chain pricing and carbon emission reduction decision-making considering manufacturer's financial constraint.* Journal of Central China Normal University (Natural Science Edition), vol. 57, no. 06, pp. 859-869, 2023.

A Control Method for Power Robot Operation Process Based on Job Files

Zhong Liangmin[*,†,§], Hua Dingzhong[*,†], Yang Mingxin[*,†], WEN Xiangqing[*,‡],

Xie Yonglin[*,†], and Peng Yi[*,†]

[*]*NARI Group Corporation, State Grid Electric Power Research Institute, Nanjing, Jiangsu, China*

[†]*Shenzhen NARI Technologies Co., Ltd., Shenzhen, Guangdong, China*

[‡]*NARI Technology Co., Ltd, Nanjing, Jiangsu, China*

[§]*mail_zlm_0597@163.com*

To address the problems of customized design, unclear functional interfaces between modules, and complex process control logic in the autonomous operation process control of power robots, this chapter proposes a standardized job file based on XML format for power robot operation task process control, From the perspective of general generality and top-level unified management, a universally applicable control method for the operation process of electric robots was studied. Through hierarchical design, job tasks are subdivided into task items, job items, action items, and other subcategories in standardized format, which improves the flexibility of job sequence configuration and forms job files with complete task information. Corresponding to the hierarchy of job files, a multi-layer finite state machine software framework is designed, combined with job file information, to achieve clear interactive scheduling management of upstream and downstream multi-modules in the job process. This method solves the problems of weak adaptability, low flexibility, and complex control logic in the task flow control components and methods of power robots, and improves the reusability and portability of software functional modules. The application analysis of the autonomous operation of transmission line inspection robots shows that compared with conventional methods, this method can effectively improve the efficiency of operation process control and has good generalization ability for different types of job tasks.

Keywords: Power robot; job files; finite state machine; process control; hierarchical design.

1. Introduction

Domestic and foreign researchers have conducted relevant research on the autonomous operation task control of power robots. However, such research generally targets specific and concrete job tasks. For example, transmission line/substation inspection, mobile robot obstacle avoidance planning, robotic arm grabbing jobs, and patrol jobs in distribution rooms are customized designs for predetermined scenarios,[1,2,3,4] lacking research on generic and universally applicable job process control methods with top-level management and dispatching functionality abstracted from scenario types, robot types, and job types.

This chapter proposes a power robot job task process control method based on XML format standardized job files. Through standardized and hierarchical design, combined with finite state machine technology, it solves the problems of poor adaptability, low flexibility, and complex control logic in job task process control components and methods of power robots. This enhances the reusability and portability of software functional modules and achieves efficient and reliable control of the job processes of various types of power robots.

2. Problems in Power Robot Job Task Process Control

Power robots, facing diverse job task logics and different electricity production scenarios, raise higher requirements for the adaptability and reliability of job task process control methods, as well as the standardization and flexibility of control interfaces[5]. To accelerate robot industrialization, major manufacturers have developed power robots capable of performing specific job tasks for various application scenarios. Currently, the development of various types of robots mainly focuses on solving safety and reliability issues in jobs[6,7], and improving job accuracy and success rates[8,9]. However, there are still problems in job task process control, such as:

(1) Weak Adaptability. Difficult to quickly deploy control modules to other power robots, achieving module reusability and functional generalization.
(2) Low Flexibility. Customized job process control methods make jobs like adding or deleting steps, adjusting sequences, etc., more complex.
(3) Complex Control Logic. In the job task process, there is a lack of hierarchical design for tasks and robot actions. The interface between the control module and modules for job task analysis, logic judgment, actuator scheduling, data transmission, and state monitoring is not clear enough.

3. Robot Job Process Control Based on Job Files

3.1. *Structure design of job task files*

This paper designs a form of job content organization based on general job task files to sequentially and structurally describe the job process.

The specific hierarchical division is shown in Figure 1. Through self-developed job files configuration tool software, jobs such as adding, deleting, querying, and modifying job files are possible, allowing job task files to fully describe the job process and conditions.

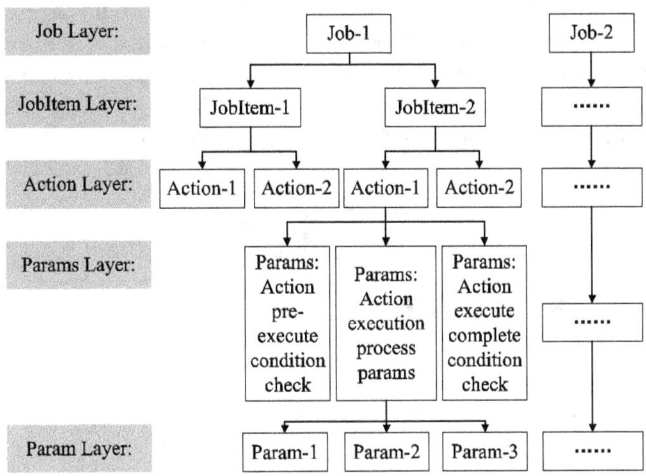

Fig. 1. Job File Structure

The structure of the job file is divided into five layers:

Layer 1 - Job Task: This includes the attributes of the job task. The attributes of the job task are used to describe the basic information of the job task, like the type of the task. Layer 2- Job Item: This contains the attributes of the job item and the actuator that completes it. A job item is a logical or job task meaningful division of the job task, used to describe information about a specific set of job actions. Layer 3- Actuator Action: This includes the attributes of the actuator and the actions of the actuator. The attributes of the actuator are used to describe the basic information of the specific actuator to be operated. Layer 4- Action Execution Process Parameters: This contains the attributes of execution and execution parameters. Execution attributes set the basic information for current action execution, such as preparation conditions, execution conditions, completion judgment, etc. Layer 5- Action Execution Parameter Items: This contains specific detailed indicators of a particular action execution parameter item, which is an instantiation of the fourth layer's parameter types.

3.2. *Job task functional interaction design*

The execution of job tasks is managed by the job task control component, which is responsible for process progression and closed-loop status management. This component encompasses the main functional modules of the robot software platform, and the process of job task execution, i.e., the interaction between various modules within the job task management component to complete job task functions, is a unified process of interaction and integration of job task flows, equipment flows, and data flows in robot software.

As shown in Figure 2, the job task function framework for the execution of job task processes is presented. Within this framework, the job task control component can be broadly divided into three categories of main functional modules/devices: remote terminal device modules, job task management modules, and execution and computing service modules.

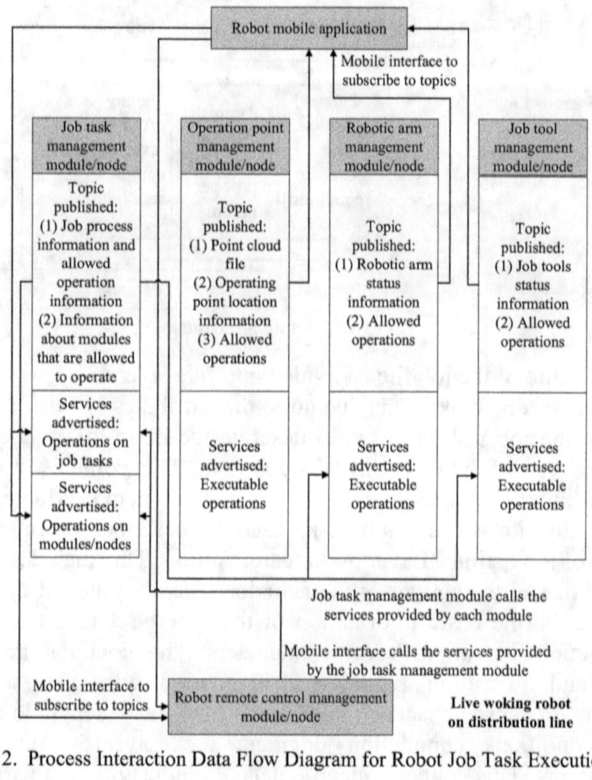

Fig. 2. Process Interaction Data Flow Diagram for Robot Job Task Execution

The job task functional process supported by the aforementioned three types of main device modules can be described as follows:

The upstream remote control terminal device initiates the job task execution process, sending the task command to the job task management module.

After receiving the command, the job task management module starts the execution process of the job task. Specifically, the job task management module forms an job task file with specific logical significance based on the content of the job task, containing specific execution instructions. The job task management node sequentially retrieves execution instructions from the file and sends them to different downstream execution and computing service modules according to different application objects.

The downstream execution and computing service modules, upon receiving specific action execution instructions, immediately execute the corresponding instructions.

Once all subdivided actions/computing instructions under the job task logic are completed, the execution of the job task is finished. The above process is initiated by the remote control terminal module, with the job task management module responsible for process control and status management, and the execution and computing module responsible for specific execution implementation and status feedback. This constitutes a fully self-consistent entire process of job task management for live working robots, realizing the autonomous execution mechanism of robots for complex job tasks.

3.3. *Job task execution process control design*

To meet the needs of autonomous job for power robots, this paper designs a general job task execution process based on Finite State Machine (FSM). By standardizing the data interaction interfaces of each component, it achieves autonomous execution control throughout the entire job task process.

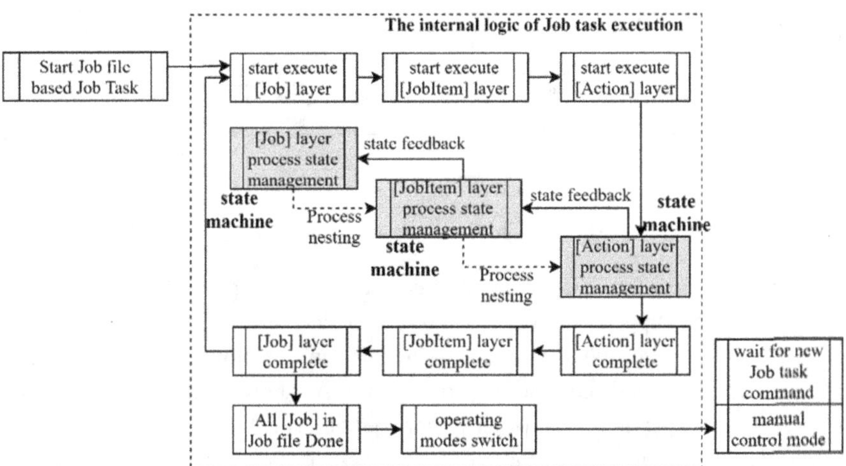

Fig. 3. Internal Logic of Job Execution based on Job File

As shown in Figure 3, when initiating the autonomous job task, the robot enters automatic job mode. The job task control module will sequentially retrieve each job content from the job task files according to its arrangement and progressively advance until the entire job task is completed.

The process state management is realized by three-layered state machines corresponding to the Job (task layer), JobItem (job item layer), and Action (action execution layer) in the job file. These three-layered state machines are nested within each other, with the innermost action execution layer state machine providing outward state feedback, forming the state representation of the job item layer and task layer. For example, in the event of an job anomaly, there is both a macro-level job task anomaly alarm and specific information on the anomaly in the job item or action.

4. Application Analysis and Results

To validate the application effects of the job file-based job process control method, this method was applied in the autonomous inspection tasks of transmission line inspection robots, aiming to achieve robot job process management in unsupervised situations.

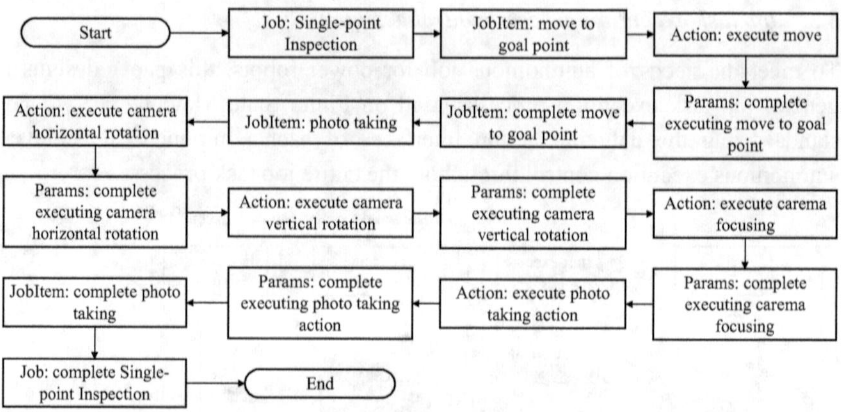

Fig. 4. Job Task Flow of Line Inspection Robot Inspection

Transmission line inspection robots use overhead ground wires as their travel tracks. Equipped with chassis motion mechanisms, lifting arms, and pan-tilt cameras, these robots autonomously perform functions such as walking, obstacle crossing, and fixed-point photo-taking. As shown in Figure 4, a simple single-point inspection task process can be described as follows in normal execution: the robot receives the job task, drives the chassis mechanism to move to a designated docking point, aims the pan-tilt camera at a preset inspection target for photo-taking at the docking point, and then concludes the job task.

Figure 5 shows the XML format file of a single-point inspection task. The file consists of five hierarchical labels: Job (task item), JobItem (subdivided job item), Action (action item), Params (action execution process parameters), and Param (action execution parameter). The task item describes the task type as a single-point detailed inspection task; the job item includes two items, which are two job task-significant projects in the task, namely walking to the target point and conducting inspection (photo-taking) at the target point. The action items further break down and specify the robot's behaviors in the job items, with walking involving the movement of the chassis and inspection being a combination of camera rotation, focusing, and photo-taking. Action process parameters are used to categorize different types of action execution parameters, generally divided into pre-condition, perform, and complete parameters. Action execution parameters are preset for specific actions, such as the number of photos, movement speed limit, timeout durations for each action, etc. Action parameters are essential elements of the Action layer (Action).

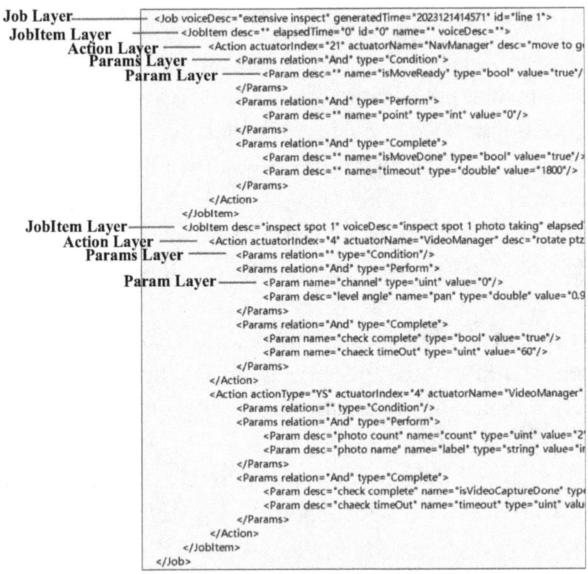

Fig. 5. Example of Line Inspection Robot Job File

Clearly, the design of the Job, JobItem, and Action in the job file corresponds with the three-layer state machine described in the previous section. As shown in Figure 4, the execution process of an action item layer (Params/Param) manages the execution process, thereby advancing the job item and task item processes until all tasks are completed. If the movement mechanism gets stuck, it may cause the movement execution conditions (pre-execution/during execution/post-execution) to be unsatisfied, preventing the completion of the movement action, leading to the failure of the move-to-target-point job item, and ultimately causing the entire inspection task to end anomalously.

5. Summary

This chapter presents a method of job task process control for power robots based on standardized XML format job files. The transmission line inspection robot, distribution line live working robot, and substation firefighting robot auto-operation software system based on this method have been successfully piloted in practical applications. Test results show that it achieves high software module reusability, flexible configuration methods, standardized job file interfaces, and layered internal state control. It can accelerate software update iteration and the application of software modules on multiple types of electric robots.

References

1. BAI Guangrui, ZHENG Yuxiang, WU Kai, WU Shaolei, et al. Power Line Identification and Localization for Live Working Robot of Distribution Line[J] (in Chinese) Power System Technology, 2023, 47(6):2604-2611.
2. Zhu, Jinsong, et al. "Intelligent fire monitor for fire robot based on infrared image feedback control." Fire Technology 56 (2020): 2089-2109.
3. Zengin, Aydin Tarik, et al. "Measurement of power line sagging using sensor data of a power line inspection robot." Ieee Access 8 (2020): 99198-99204.
4. Li, Hong Jun, et al. "Robust motion control for multi-split transmission line four-wheel driven mobile operation robot in extreme power environment." Industrial Robot: the international journal of robotics research and application 47.2 (2020): 219-229.
5. CHENG Jin, et al." Design of online multitask scheduling system for robot based on PLC" (in Chinese) Manufacturing Automation 45.12(2023):127-130+149.
6. Yan, Yu, et al. "Research on mechanism configuration and coordinated control for power distribution network live working robot." Industrial Robot: the international journal of robotics research and application 47.3 (2020): 453-462.
7. Mao Xiaoliang, et al. Research on Robot System for Centrally Installed Switchgear Based on ROS [J] (in Chinese). Drive System Technique, 2023, 37(2): 20-25.
8. LI Chengxi, et al." An Object Detection Method for Electricity Service Robot" (in Chinese) Modern Computer. 18(2021):109-112.
9. Wei Haibin. Semi-autonomous teleoperated robot system for power grid operations[D]. Master thesis (in Chinese), Guangdong University of Technology, 2023. DOI:10.27029/d.cnki.ggdgu.2022.000296.

Rule Acquisition of Semantic Decision Based on Description Logics

Yanling Wang[*,‡], Zheng Yang[*,§], Xi Zhang[*,¶], Lixian Xu[†,‖], Hengfei Li[*,**] and Li Zou[*,††]

[*] *School of Computer and Technology, Shandong Jianzhu University, Jinan 250101, China*
[†] *School of Computing, Ulster University, Northern Ireland, BT37 0QB, UK*
[‡] *yanlingwang_99@163.com*
[§] *yzharder@163.com*
[¶] *Zhangxi_33@163.com*
[‖] *Xu-L2@ulster.ac.uk*
[**] *lihengfei@sdjzu.edu.cn*
[††] *zoulicn@163.com*

The proposition of semantic formal context, especially the reasoning mechanism in description logics, provides a new formal tool for semantic search and data analysis. However, as a favorable tool for knowledge discovery, formal decision context mainly focuses on the syntax point, and the research on semantic information is still insufficient. Considering semantic formal context, we propose semantic formal decision context, integrating semantic information into formal decision context based on ontology. And we introduce an algorithm about rule acquisition of semantic formal decision context. By comparing with the original formal decision context, the advantages of semantic formal decision context in semantic search are verified.

Keywords: Formal concept analysis; description logics; formal decision context; rules acquisition.

1. Introduction

Since formal concept analysis[1,2] (FCA) was proposed, it has gradually become a powerful formal tool for data analysis and knowledge discovery and applied to many fields. Among them, formal decision context[3] (FDC) was proposed on the basis of formal context, which has a set of conditional attributes and a set of decision attributes. The knowledge hidden in the data can be found in the FDC and expressed in the form of "if-then". Its proposal not only expands the conceptual level of formal context but also provides a formal tool for rule acquisition in knowledge discovery.

[‡]Corresponding author.

An Ontology[4,5] is defined as "*formal, explicit specifications of a shared conceptualization*", where it was referred to the study of the essence of existence originally. Description Logics (DLs) is mainly used to implement ontologies, a family of logic for representing structure knowledge. Jiang[6] proposed semantic formal context (SFC) by investigating background knowledge and the semantic of attribute for FCA. On this basis, Zhang *et al.*[7] proposed a semantic formal context based on knowledge graph and applied it to semantic search. Wei *et al.*[8] combined FCA with knowledge graph to provide a semantic basis for related semantic technologies and proposed a new semantic recommendation algorithm.

The FDC is still limited to data itself and lacks background knowledge. Similar to the SFC, we use DLs to semanticize FDC. The advantage of this semanticization is mainly reflected in the fact that as long as the ontology of the extracted decision formal background is coordinated (without logical errors), FDC is correct (without contradictory binary relationships between attributes and objects). And by utilizing the logical equivalence in the ontology, richer decision rules are obtained. In Section 2, we introduce some basic knowledge. In Section 3, we combine the SFC with FDC to propose the semantic formal decision context (SFDC) and rule acquisition based on SFDC. In Section 4, by comparing with the original decision formal context, the effectiveness of SFDC and its advantages in semantic search are verified. Then we conclude this chapter and introduce possible trend of the research in Section 5.

2. Preliminaries

In this section, we review some basic concepts briefly, some basic definition has been proposed in Reference[1,2].

Definition 2.1.[3,9] Let (U, M, R), (U, N, J) be two formal contexts, if $\forall (Y, B) \in \mathcal{L}(U, N, J)$, there exist a concept $(X, A) \in \mathcal{L}(U, M, R)$, satisfied $X = Y$. Then we say $\mathcal{L}(U, M, R)$ is finer than $\mathcal{L}(U, N, J)$, written as $\mathcal{L}(U, M, R) \leq \mathcal{L}(U, N, J)$. A five-tuple (U, M, R, N, J) is called formal decision context, which consist of two formal contexts (U, M, R) and (U, N, J). Further, if $\mathcal{L}(U, M, R) \leq \mathcal{L}(U, N, J)$, then we say (U, M, R, N, J) is strongly consistent formal decision context.

Definition 2.2.[10] Let (U, M, R, N, J) be strongly consistent formal decision context, if there exist two concepts $(X, A) \in \mathcal{L}(U, M, R)$ and $(Y, B) \in \mathcal{L}(U, N, J)$, satisfied $X = Y$,then we have $A \to B$, which is called a decision rule, read if A, then B.

Definition 2.3.[6] Let $\mathcal{DL}$ be a DLs($\mathcal{ALC}, \mathcal{SHOJQ}$, etc.),$\mathcal{DO} =< \mathcal{AB}, \mathcal{TB}, \mathcal{RB} >$ be a domain ontology (i.e., DL-ontology) expressed in $\mathcal{DL}$. Π is a set of $\mathcal{DL}-concept$, T is a set of individual of $\mathcal{DO}$, $I = (\triangle^I, \bullet^I)$ is a model of $\mathcal{DO}$(i.e., $\mathcal{DL}$-knowledge base). A semantic formal context (SFC) is a triplet $G = (U, M, R)$ w.r.t. $\mathcal{DO}$, which consist of two nonempty set U, M and a binary relation R between two set U and M, where $U \subseteq T$, $M \subseteq T$ and $R = I$(i.e., I is the restriction on $\{M, U\}$)

Obviously, there exist semantic formal concept and lattice, which have already defined in Reference 6. So $\mathcal{SL}(U, M, R)$ is used to denote semantic formal concept lattice of semantic formal context (U, M, R).

3. Semantic Formal Decision Context

In this section, we semantify the formal decision context by combining formal decision context with DLs. We introduce some basic concepts and an algorithm of rule acquisition.

Definition 3.1. Let (U, M, R), (U, N, J) be two semantic formal contexts w.r.t. $\mathcal{DO}$. Π is a set of $\mathcal{DL}-concept$, T is a set of individual of $\mathcal{DO}$, $I = (\triangle^I, \bullet^I)$ is a model of $\mathcal{DO}$(i.e., $\mathcal{DL}$-knowledge base). If $\forall (Y, B) \in \mathcal{SL}(U, N, J)$, there exist a concept $(X, A) \in \mathcal{SL}(U, M, R)$, satisfied $X = Y$. Then we say $\mathcal{L}(U, M, R)$ is finer than $\mathcal{SL}(U, N, J)$, written as $\mathcal{SL}(U, M, R) \leq \mathcal{SL}(U, N, J)$.

Definition 3.2. A semantic formal decision context is five-tuple (U, M, R, N, J), which consist of two semantic formal contexts $(U, M, R)$and (U, N, J), where $U = \{x_1, x_2, ..., x_n\}$ is a finite and nonempty object set, $M = \{m_1, m_2, ..., m_n\}$ and $N = \{n_1, n_2, ..., n_n\}$ are finite and nonempty attribute sets and $R \subseteq U \times M$ and $J \subseteq U \times N$ are binary relation. If $\mathcal{SL}(U, M, R) \leq \mathcal{SL}(U, N, J)$, then we say (U, M, R, N, J) is strongly consistent semantic formal decision context, where M is called condition attribute set, and N decision attribute set.

In semantic formal decision context(SFDC) $G = (U, M, R, N, J)$, M, N are the set of $\mathcal{DL}-concept(attribute)$. Because SFDC is constructed based on domain ontology, the semantic decision context is correct if a domain ontology $\mathcal{DO} =< \mathcal{AB}, \mathcal{TB}, \mathcal{RB} >$ is consistent. The concept is interpreted as a subset of domain knowledge, and the relationship is interpreted as the binary relationship between individuals and concepts in this domain. Obviously, SFDC can obtain more semantic concepts than original formal decision context. For example, we will obtain more semantic information, if we use Web search based on FCA. This kind of advantages can help us obtain more decision rules. Analogously to Definition 2.2, we study semantic formal decision rule.

Definition 3.3. Let $G = (U, M, R, N, J)$ be strongly consistent formal decision context w.r.t. $\mathcal{DO}$, where $\mathcal{DL}$ is a kind of description logics expression and $\mathcal{DO} =< \mathcal{AB}, \mathcal{TB}, \mathcal{RB} >$ is a domain ontology in $\mathcal{DL}$. If there exist two semantic formal concepts $(X, A) \in \mathcal{SL}(U, M, R)$ and $(X, A) \in \mathcal{SL}(U, M, R)$, satisfied $X = Y$,then we have $A \to_{\mathcal{R}} B$, which is called a semantic decision rule, read if A, then B.

Definition 3.4. Suppose $G = (U, M, R, N, J)$ be strongly consistent formal decision context w.r.t. $\mathcal{DO}$. If there exist two semantic decision rules $A \to_{\mathcal{R}} B$ and $A^{'} \to_{\mathcal{R}} B^{'}$, satisfied $A \subseteq A^{'}, B \supseteq B^{'}$,then we say $A \to_{\mathcal{R}} B$ implying $A \subseteq A^{'}, B \supseteq B^{'}$, and $A \subseteq A^{'}, B \supseteq B^{'}$ is redundant.

Algorithm 3.1 Rule acquisition in semantic formal decision

Input: the semantic formal decision context $G = (U, M, R, N, J)$ based on $\mathcal{DO}$

Output: the number of the semantic formal decision rules, and all of semantic decision rules

$num = 0, Rule = \phi$;
for all semantic concept in $\mathcal{SL}(U, M, R)$ **do**
  if there exist two semantic formal concepts $(X, A) \in \mathcal{SL}(U, M, R)$ and $(Y, B) \in \mathcal{SL}(U, N, J)$, satisfied $X = Y$ **then**
$Rule = Rule + \{(A \to_R B)\}, num + +$;
    if there exist $A \equiv_{\mathcal{DO}} C$ in $\mathcal{DO}$ **then**
$Rule = Rule + \{(C \to_R B)\}, num + +$;
    end if
  end if
end foroutput $num, Rule$;
end.

In addition, Algorithm 3.1 is an approach of creating semantic decision rules. We consider some equality information in the Terminological Box ($\mathcal{TB}$) of description logics. And it is not influenced by the Assertion Box ($\mathcal{AB}$). So we can obtain more semantic decision rules.

4. Motivating Examples

In order to illustrate the advantage of semantic formal decision context, we provide an example.

Example 4.1. Assume $U = \{x_1, x_2, x_3, x_4\}$ is a set of objects(i.e. patients), $M = \{a, b, c, d, e\}$ is a set of condition attribute (i.e. symptoms) and $N = \{g, h, k\}$ is a set of decision attribute (i.e. diagnosis), where a is *LackingStrength*, b is *Fever*, c is *Cough*, d is *NasalObstruction*, e is *SoreThroat*, g is *Influenza*, h is *Cold*, k is *Pneumonia*. Taking medical decision-making as an example, supposing a medical domain ontology $\mathcal{DO} =< \mathcal{AB}, \mathcal{TB}, \mathcal{RB} >$ expressed in a kind of description logic $\mathcal{DL}$ as follows:

$\mathcal{RB} = \phi$; $\mathcal{TB} = \{InfluenzaA \equiv Fever \sqcap Headache \sqcap LackingStrength \sqcap Cough, InfluenzaB \equiv Fever \sqcap Headache \sqcap LackingStrength, Cold \equiv Headache \sqcap Cough, Pneumonia \equiv Fever \sqcap Cough \sqcap Dyspnea, Influenza \equiv InfluenzaA \sqcup InfluenzaB, LackingStrength \equiv BodyAche, Cough \equiv SoreThroat\}$; $\mathcal{AB} = \{LackingStrength(x_3), LackingStrength(x_4), Fever(x_1), Fever(x_3), Cough(x_1), Cough(x_3), NasalObstruction(x_2), NasalObstruction(x_3), NasalObstruction(x_4), SoreThroat(x_2), SoreThroat(x_4), Influenza(x_3), Cold(x_2), Cold(x_4), Pneumonia(x_1), Pneumonia(x_3), Headache(x_2), Headache(x_3), Headache(x_4), Dyspnea(x_1), Dyspnea(x_3)\}$.

Suppose that we have interpretation I, where $I \models \mathcal{TB}$ and $I \models \mathcal{AB}$, as follows:

$LackingStrength^I = \{x_3, x_4\}, Fever^I = \{x_1, x_3\}, Cough^I = \{x_1, x_3\}, NasalObstruction^I = \{x_2, x_3, x_4\}, Headache^I = \{x_3, x_4\}, SoreThroat^I = \{x_1, x_3\}, Dyspnea^I = \{x_1, x_3\}, Influenza^I = \{x_3\}, Cold^I = \{x_2, x_4\}, Pneumonia^I = \{x_1, x_3\}$

Based on interpretation I, we can obtain the following binary relation R between U and M and binary relation J between U and N by Definition 2.6, as follows:

$LackingStrength^R = \{x_3, x_4\}, Fever^R = \{x_1, x_3\}, Cough^R = \{x_1, x_3\}, NasalObstruction^R = \{x_2, x_3, x_4\}, SoreThroat^I = \{x_1, x_3\}, Influenza^J = \{x_3\}, Cold^J = \{x_2, x_4\}, Pneumonia^J = \{x_1, x_3\}$

According to these interpretation, we can construct the medical semantic formal decision context $G = (U, M, R, N, J)$, which is shown in Table 1. Then we create the semantic formal decision context by using Algorithm 3.1. Figs. 1–4 are the semantic formal concepts.

Table 1. The semantic formal decision context w.r.t. medical domain ontology.

	a	b	c	d	e	g	h	k
x_1	0	1	1	0	0	0	0	1
x_2	0	0	0	1	1	0	1	0
x_3	1	1	1	1	0	1	0	1
x_4	1	0	0	1	1	0	1	0

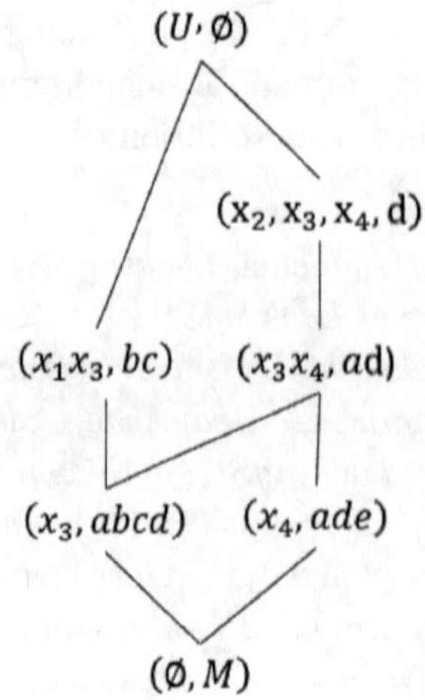

Fig. 1. $L(U, M, R)$

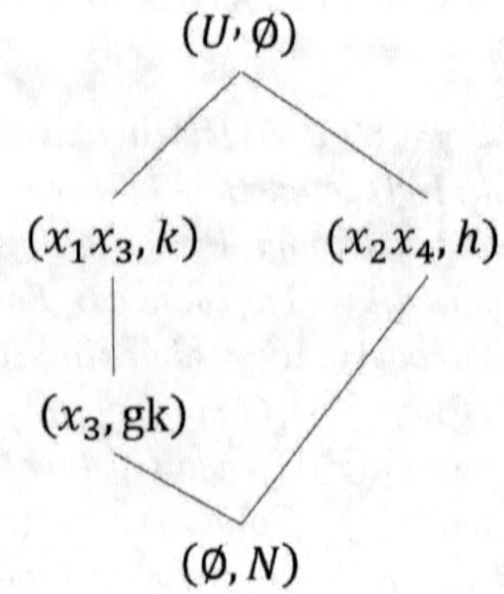

Fig. 2. $L(U, N, J)$

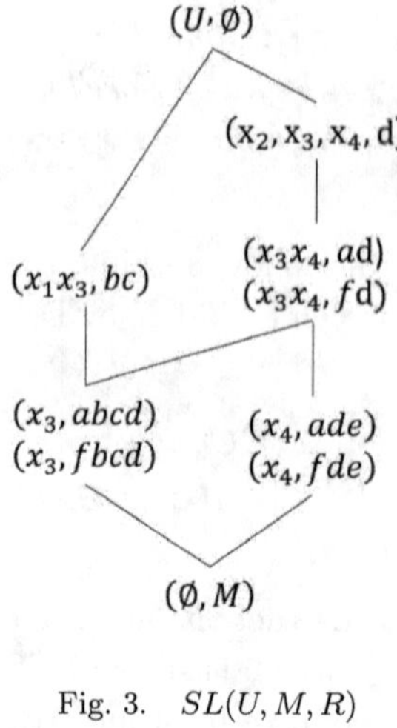

Fig. 3. $SL(U, M, R)$

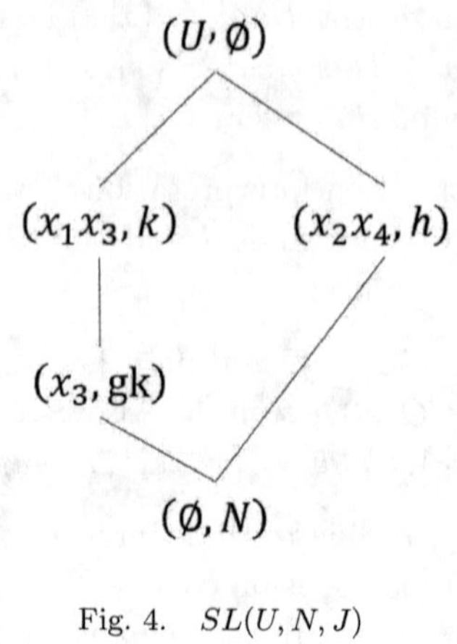

Fig. 4. $SL(U, N, J)$

According to Definition 3.4, there are seven the semantic decision rules of G: $ad \rightarrow_R k, fd \rightarrow_R k, abcd \rightarrow_R gk, fbcd \rightarrow_R gk, ade \rightarrow_R h, fde \rightarrow_R h, bc \rightarrow_R k$. At same time, there are two redundant semantic decision rules: $abcd \rightarrow_R k, fbcd \rightarrow_R k$. If we consider its syntax level, we can only obtain four decision rules: $abcd \rightarrow k, abcd \rightarrow gk, ade \rightarrow h, bc \rightarrow k$. Obviously, semantic formal decision context add semantic information, which make the obtained rules more comprehensive than original decision rules.

5. Conclusion

As we know, the formal decision concept analysis is an effective parameterization tool that possesses decision-making data, and the semantic information based on the ontology and the reasoning mechanism based on description logics provide strong support for rule extraction, and rule acquisition can discover the rule relationships contained between individual in the domain ontology, which is necessary for updating the terminology of the ontology. This chapter utilizes description logics to semanticify the FDC based on ontology and defines the rule acquisition on the formal decision context. Finally, through examples, it is demonstrated that SFDC can obtain richer semantic information. In terms of future research directions, the semantic decision rules can be integrated with the rules contained under domain ontology for knowledge inference. At the same time, we apply SFDC to different fields, including semantic web search and decision systems.

Acknowledgments

This paper is partially supported by the National Natural Science Foundation of China (62176142, 62177031), the Natural Science Foundation of Shandong Province (ZR2021MF099, ZR2022MF334), and Undergraduate Education Reform Project of Shandong Province (M2021130, M2022245), the high quality Professional degree teaching case base construction project of Shandong Province(SDYAL2022155) and Special Foundation for Distinguished Professors of Shandong jianzhu University.

References

1. W. Rudolf, Restructuring lattice theory: An approach based on hierarchies of concepts, in *Ordered Sets*, 1982.
2. B. Ganter, R. Wille and C. Franzke, *Formal Concept Analysis: Mathematical Foundations*, (1998).

3. L. Wei, J. Qi and W. Zhang, Attribute reduction theory of concept lattice based on decision formal contexts, *Science in China Series F: Information Sciences* **51**, 910 (2008).
4. R. Studer, V. Benjamins and D. Fensel, Knowledge engineering: Principles and methods, *Data and Knowledge Engineering* **25**, 161 (1998).
5. F. Bobillo, M. Delgado and J. Gómez-Romero, Representation of context-dependant knowledge in ontologies: A model and an application, *Expert Systems with Applications* **35**, 1899 (2008).
6. Y. Jiang, Semantifying formal concept analysis using description logics, *Knowlege-Based System* **186**, p. 104967 (2019).
7. Fusing semantic aspects for formal concept analysis using knowledge graphs, *Multimedia Tools and Applications* **83**, 16763 (2023).
8. L. Wei and X. Zhu, Semantic recommendation model via fusing knowledge graph and formal concept analysis, *IEEE Access* **11**, 62337 (2023).
9. W. Zhang, L. Wei and J. Qi, Attribute reduction theory and approach to concept lattice, *Science in China Series F: Information Sciences* **48**, 713 (2005).
10. L. Wei, L. Liu, J. Qi and T. Qian, Rules acquisition of formal decision contexts based on three-way concept lattices, *INFORMATION SCIENCES* **516**, 529 (2020).

A One-stage Method for Style Recognition from Fashion Images By Using YOLOv5 Network

Ruolin Wang*, Kaixuan Liu†, Kim Phuc Tran* and Xianyi Zeng*,‡

*University of Lille, ENSAIT, GEMTEX, F-59056 Roubaix, France

†Apparel and Art Design College, Xi'an Polytechnic University, Xi'an 710048, China

‡xianyi.zeng@ensait.fr

With the quick development of e-shopping for fashion products, finding a method to help consumers identify their preferred style becomes a significant challenge in the field of clothing industry. Moreover, consumers' fashion requirements have become more and more diversified to enhance their respective personalities. Therefore, fashion designers should master a large number of design styles and associated fashion images describing these styles. In this context, this chapter proposes a new study for style recognition from fashion images using the YOLOv5 network. Through iterative learning from experimental data, effective fashion image recognition is achieved. The efficacy of the model is assessed through a comprehensive validation process utilizing confusion matrix and mean Average Precision (mAP). The proposed recognition method has been validated through several cases of Cheongsam styles (traditional Chinese women's costume), revealing that the network has good performance in fashion image recognition. An object detection system is constructed based on YOLOv5 model predictions, capable of recognizing cheongsam styles in dynamic and static settings. It visualizes the results of recognition, offering technical support and a reference for intelligent classification of garments. Additionally, it serves as a guide for the intelligent advancement of the clothing industry.

Keywords: Fashion style recognition; fashion image; one-stage detection; YOLOv5 network.

1. Introduction

The fashion industry, constituting an important part of the consumer industry, is playing a key role in the worldwide economy. In the competition of the international fashion markets, consumer preferences have become more diversified.[1] Purchasing clothing is no longer solely for warmth and coverage. Instead, clothing has become a form of self-expression and charm. Simultaneously, with the development of digitization and artificial intelligence, consumers' lives, work, and entertainment behaviors have increasingly been digitized.[2, 3] Traditional shopping methods have undergone significant changes, giving rise to new business models, and e-commerce has become a new driving force leading economic and social development.[4] The focus of clothing consumption has shifted from offline to online.[5, 6] In the process of online clothing shopping, a search engine is crucial. Currently, there are two categories of search engines used in online fashion shopping: text search-based and image search-based. The results enable us to provide a multitude of fashion items to consumers in order to choose them according to their personalized fashion requirements. However, consumers invest significant time and effort during selection. Therefore, finding a method to help consumers quickly identify their preferred fashion style becomes a significant challenge in the field of clothing industry.[7]

In recent years, with the rapid development of information technology, artificial intelligence (AI) has been continuously evolved and improved. It has been widely applied in various fields, promoting industrial reform and upgrading, and becoming the driving force for enterprise development.[8] AI has infused data intelligence into the fashion industry, being applied in production planning,[9] fashion recommendations,[10] personalized customization,[11] sustainable development,[12] and more. It has effectively tackled issues in the traditional clothing industry and made valuable contributions to the field.

‡Corresponding author.

In this context, the proposed study applies deep learning technology to classify fashion images according to their representative styles. This research enhances understanding of designers and consumers on garment styles and provides a reference library for future design. The YOLOv5 model has a small size, high network flexibility, fast algorithm processing speed, and can be detected in real time. By utilizing the YOLOv5 network, the automatic classification of fashion images demonstrating various garment fitting effects is achieved, enabling consumers in e-shops to select the most favorable garment styles by means of garments shown on the most relevant fashion images. This study will contribute to fashion companies by reducing manual costs and enhancing their overall competitiveness. The proposed study has been effectively validated through a business case of cheongsam design.

2. Methodology

2.1. Basic concept

2.1.1. Garment style

Style is a concept formed by the integration of historical, intellectual, and cultural elements. Unlike fashion, it is a highly abstract and ambiguous concept that considers social, emotional, political, cultural, regional, physical characteristics, and other factors to classify garment. Garment style is an important attribute of apparel, representing the overall visual perception of clothing. Garment elements are the essential components of garment style. The elements are typically categorized into aspects such as design, color, fabric, accessories, structure, craftsmanship, patterns, etc. The combination of these elements forms the garment style.

2.1.2. Fashion image recognition

Fashion image recognition is a common and important task in fashion industry, which has attracted widespread attention due to its significant value in many e-commerce applications such as apparel retrieval and recommendation. Fashion images always contain rich information, and sometimes there may even be deformation and occlusion, which poses a huge challenge for efficient and accurate fashion recognition[13]. To address this challenge, early methods primarily emphasized hand-crafted features. However, given the limited amount of annotated data, it became challenging to effectively tackle the diverse issues encountered in online fashion stores. The growing abundance of fashion images on the internet has promoted the recent advances in fashion image recognition through deep learning technology. At present, relying on powerful deep learning algorithms such as the YOLO series, online fashion stores can manage products faster and more effectively.

2.2. YOLO algorithm

At present, the main object recognition algorithms include the R-CNN series and the YOLO series. The R-CNN series is a two-stage algorithm with higher accuracy, it first proposes regions through selective search algorithms, and then uses convolutional neural networks to extract the features of these proposed regions and classify and refine the bounding boxes, its network structure is complex and the detection speed is slow. The You Only Look Once (YOLO) is one of the famous one-stage object detection algorithms. The YOLO series algorithms use the idea of regression to complete the target detection and feature extraction directly, reducing the computational workload of the detection process and greatly improving the speed of object detection. At present, the YOLO series algorithms have been widely applied due to their simple network structure and extremely fast detection speed. YOLOv1 is built upon GooLeNet, offering rapid processing but lower recognition accuracy compared to two-stage algorithms. YOLOv2, based on YOLOv1, enhances both accuracy and processing speed with the Darknet-19 framework, yet struggles with small object recognition. YOLOv3 addresses this issue by introducing the Neck module for feature fusion, elevating the output dimension to 3 and improving small object recognition. YOLOv4, combined with technical studies and experimental results, provides better object detection accuracy[14, 15]. YOLOv5 incorporates the concept of anchor boxes to improve the speed of the R-CNN algorithm, eliminating the manual anchor box selection. The YOLOv5 model has four different scales, namely s, m, l, and x, representing Small, Medium, Large, and Xlarge, respectively. Each scale applies different multipliers to the depth and width of the model. Larger models achieve higher accuracy but may result in slower processing speeds. Given the requirements for accuracy and speed in the cheongsam clothing detection algorithm, this paper selects the small (s) version from YOLOv5 as the network model. The network structure is shown in Figure 1, which includes four parts: the image input module (Input), the backbone network module (Backbone), the feature fusion module (Neck), and the prediction module (Head).

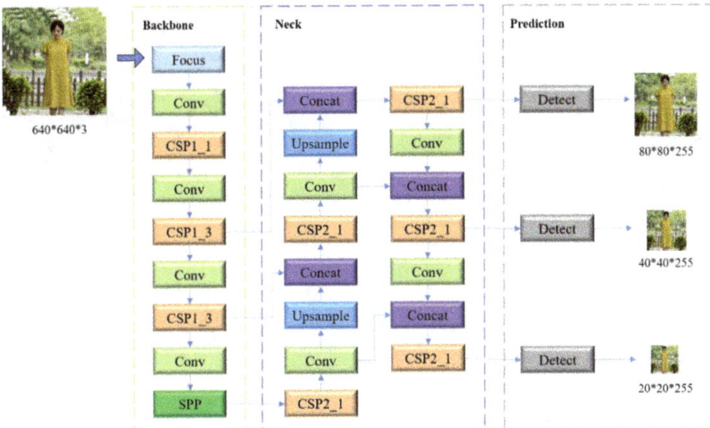

Fig. 1. YOLOv5 Network.

2.2.1. *Input section*

In this section, operations involve mosaic data augmentation, adaptive anchor box calculation, and image scaling. YOLOv5s training goes beyond the original dataset to improve the generalization of the network. The model expands the training dataset scale using the mosaic data augmentation algorithm. It connects four images in a random scaling manner, which can increase small target diversity, enhance the overall network robustness. Image scaling involves resizing to a standard size (3 × 640 × 640) before feeding it into the network for detection.

2.2.2. *Backbone section*

The YOLOv5s backbone comprises the Focus and Cross Stage Partial (CSP) structures. The original 608×608×3 image is processed through the Focus structure, creating a 304×304×12 feature map. Further convolution with 32 kernels results in a final 304×304×32 feature map. CSP structure, inspired by CSPNet, addresses computational challenges during inference. It maps the base layer, divides it into two parts, and fuses them through a cross-stage hierarchical structure. Moreover, YOLOv5s introduces the Spatial Pyramid Pooling (SPP) layer between the backbone and neck, enabling efficient feature fusion across scales without compromising speed.

2.2.3. *Neck section*

This section combines image features and transmits them to the prediction layer, integrating abstract semantic information with shallow features. It employs Feature Pyramid Networks (FPN) and Path Aggregation Network (PAN) structures. FPN upsamples high-level features using a top-down approach, while PAN, acting as the bottom-level pyramid, fuses features from FPN with positional features in a bottom-up manner. This effectively addresses information preservation issues for small targets in object detection.

2.2.4. Prediction section

The loss function used by the YOLOv5s in the prediction section consists of three parts, including class prediction loss (L_{class}), geometric intersection-over-union loss (L_{GIoU}), and confidence prediction loss (L_{conf}), defined as Eq. (1).

$$L_{total} = L_{conf} + L_{class} + L_{GIoU} \tag{1}$$

The loss calculation formula for geometric intersection-over-union is as Eq. (2).

$$L_{GIoU} = 1 - IoU + \frac{|A - u)|}{|A|} \tag{2}$$

where, IoU (Intersection-over-Union) refers to the intersection rate of the candidate box and the real box, A is the intersection area of the predicted box and the real box, and u is the merging area of the predicted box and the real box.

Class prediction loss and confidence prediction loss are calculated using the binary cross entropy loss function, which are defined as Eq. (3). and Eq. (4).

$$\begin{aligned} L_{conf} = \lambda_{obj} \sum^{S^2} \sum^{B} I_{ij}^{obj} [-\widehat{C}_i \ln C_i - (1 - \widehat{C}_i) \ln(1 - C_i)] \\ + \lambda_{nobj} \sum_{i=0}^{S^2} \sum_{j=0}^{B} I_{ij}^{nobj} [-\widehat{C}_i \ln C_i - (1 - \widehat{C}_i) \ln(1 - C_i)] \end{aligned} \tag{3}$$

$$L_{class} = \sum_{i=0}^{S^2} \sum_{j=0}^{B} \sum_{c \subset classes} I_{ij}^{obj} [-\widehat{p}_i(c) \ln(p_i(c)) - (1 - \widehat{p}_i(c)) \ln(1 - p_i(c))] \tag{4}$$

where, S represents the size of the grid, B represents the number of predicted boxes per grid, I_{ii}^{obj} indicates whether the j-th predicted box in the i-th grid contains a predicted target. If the overlap between the predicted box and the truth box exceeds a certain threshold, indicating the need to predict a target, I_{ii}^{obj} is set to 1, and it contributes to the calculation of the loss function. Otherwise, I_{ii}^{obj} is set to 0. Conversely, I_{ii}^{nobj} represents whether the j-th predicted box in the i-th grid contains a non-target. If the overlap between the predicted box and the truth box is below a certain threshold, I_{ii}^{nobj} is set to 1, otherwise, it is set to 0. λ_{obj}、λ_{nobj} are coefficients used to adjust confidence loss. C_i represents the confidence of the predicted box, $\widehat{C}_i$ represents the confidence of the truth box. $\widehat{p}_i(c)$ is the predicted probability of belonging to class c when the target is detected in the i-th grid, and $p_i(c)$ is the actual probability of belonging to class c when the target is detected in the i-th grid.

3. Result and Discussion

This paper focuses on cheongsam as the research subject, utilizing a Python web scraping tool to collect fashion images. After filtering and removing irrelevant images, a total of 2600 cheongsam images were retained. Based on the visual attributes, cheongsam categories were annotated according to patterns, resulting in five categories. They are geometry, scenery, flowers and plants, animal, and solid color. The distribution includes 497 images in the geometry, 500 in scenery, 560 in flowers and plants, 441 in animal, and 602 in solid color, as shown in Table 1. The images were divided into training and validation sets in a 7:3 ratio, comprising 1795 images in the training set and 805 in the validation set. Figure 2 illustrates examples of cheongsam images collected from the dataset. The images were annotated using the LabelImg software, a graphical image annotation tool written in Python with a Qt-based graphical interface. The labeled files can be saved in three formats: PASCAL VOC, YOLO, and CreateML.

The confusion matrix, depicted in Figure 3, visually illustrates the prediction accuracy of the model. In this matrix, "True" indicates the true distribution of a category, and "Predicted" indicates the predicted distribution of a category. It can be seen that the probability of correct predictions for each category. The highest prediction accuracy is achieved for geometry and solid color, both at 91%. Next is the scenery with a prediction accuracy of 80%, followed by the animal with 72%, and finally the flowers and plants with a prediction accuracy of 70%. The confusion matrix results indicate that the trained model exhibits good classification performance, and can mostly achieve accurate classification of input fashion images.

Table 1. Cheongsam dataset.

Category	Quantity (count)
Geometry	497
Scenery	500
Flowers and plants	560
Animal	441
Solid color	602

Fig. 2. Examples of cheongsam images from the dataset.

This paper introduces precision (P), recall (R), average precision (AP) and mean average precision (mAP) to evaluate the performance of the detection model. The formulas are shown in Eq. (5) to Eq. (8).

$$AP = \int_1^0 PdR \tag{5}$$

$$mAP = \frac{1}{Q}\sum_{q\in Q} AP(q) \tag{6}$$

$$P = \frac{TP}{TP + FP} \tag{7}$$

$$R = \frac{TP}{TP + FN} \tag{8}$$

where TP(true positives) represents positive samples with correct classification, FP(false positives) represents negative samples with correct classification, FN(false negatives) represents positive samples with incorrect classification, AP represents the average of all precision values obtained under the assumption of all possible recall values, Q is the number of categories.

Figure 4 is the PR curve, precision is plotted on the vertical axis, and recall on the horizontal axis. The area enclosed by the PR curve and the coordinate axes represents the value of AP, and mAP represents the average AP for all categories in the dataset. The PR curve provides a comprehensive view of the model performance at different classification thresholds. The curve being closer to the upper-right corner indicates that the model achieves a better balance between high precision and high recall. From the training results, the $mAP0.5$ is 0.859. Among them, $mAP0.5$ indicates the average detection accuracy of all detection categories when the IoU threshold is 0.5. The model exhibits the highest ability to balance precision and recall in the geometry, reaching 0.929, the balance ability on animals is the worst, with 0.744. Among them are solid color, scenery, flowers and plants, reaching 0.912, 0.898, and 0.813, respectively. Therefore, it can be concluded that the model achieves a relatively balanced performance between precision and recall, and has good classification ability.

The confidence threshold is a crucial parameter in object detection, influencing the recognition criteria of the detector for the target. It impacts the precision and recall of detection results. It can be considered as the harmonic mean of precision and recall. Analyzing the curve provides insights into how the F1 score changes across various categories at different confidence thresholds. The F1 curve is valuable for selecting optimal thresholds and assessing overall performance. The F1 score ranges from 0 to 1, where 1 is the best and 0 is the worst. Figure 5 shows the F1 curve, from which it can be seen that when the confidence level is 0.290, the F1 score of YOLOv5s reaches the maximum value of 0.80.

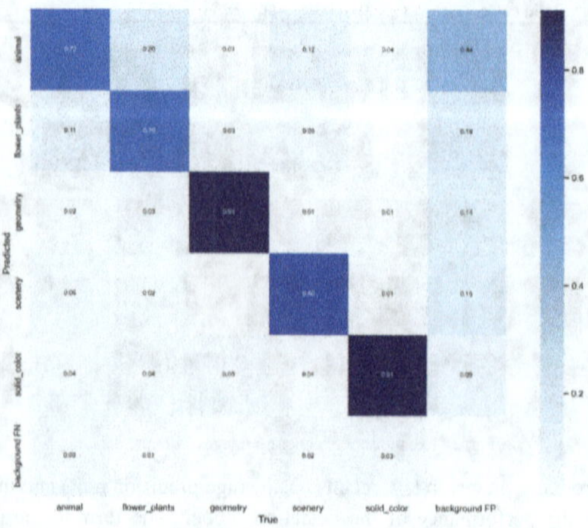

Fig. 3. Confusion matrix.

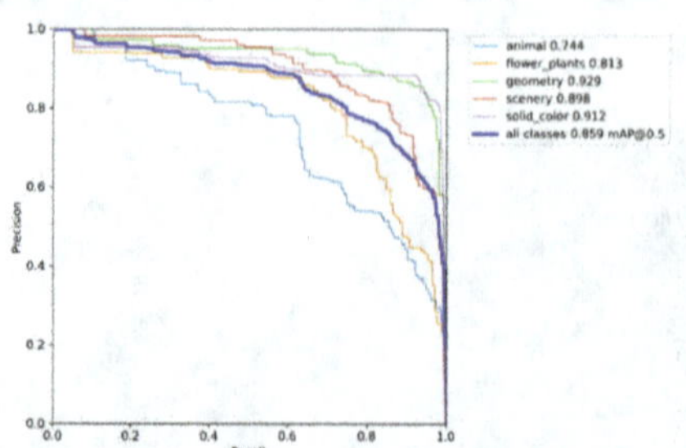

Fig. 4. PR curve.

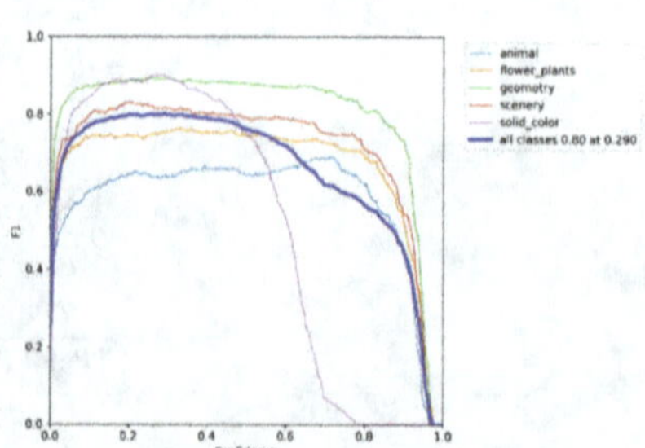

Fig. 5. F1 curve.

This article studies the style recognition methods for fashion images. Firstly, we have constructed a novel dataset focused on garment styles. Given the limitations of the Imagenet dataset in fashion style recognition, where differences in feature extraction exist, our dataset is designed with a more professional and practical approach to accurately capture various garment styles. In order to achieve better processing speed and performance, we have chosen the lightweight YOLOv5 network. Through improvements to the network structure tailored to our dataset, we have achieved favorable performance during training. This demonstrates the significant practical potential of our approach in fashion image style recognition.

4. Application

With the rapid development of e-commerce, an increasing number of consumers are choosing online shopping platforms to purchase clothing products. Consumers often encounter challenges such as difficulty determining the style of a desired garment. In this paper, an object detection model for cheongsam recognition is developed. The system analyzes input images, compares them with cheongsam information from the dataset, and ultimately provides the style to which the target object belongs. The system is efficient and accurate, completing recognition tasks in a short time while ensuring the accuracy and reliability of the results. The developed object detection system features both image and video recognition function. Users can detect cheongsam by uploading pictures or selecting the option to open the camera or uploading video files. When users upload a picture, the system will automatically detect the style of cheongsam in the image and visualize them. Similarly, when users choose to open the camera or upload video files, the system can also provide results. Taking image detection as an example, as shown in Figure 6, a randomly cheongsam image from the internet is uploaded to the system, and the system correctly identified it as belonging to the flowers and plants with a detection accuracy of 92%. The object detection system can be applied in many aspects, such as enabling e-shops to identify cheongsam product categories, automate labeling, and manage cheongsam images. Consumers can use it to quickly find desired garment styles, making their shopping experience more convenient and efficient.

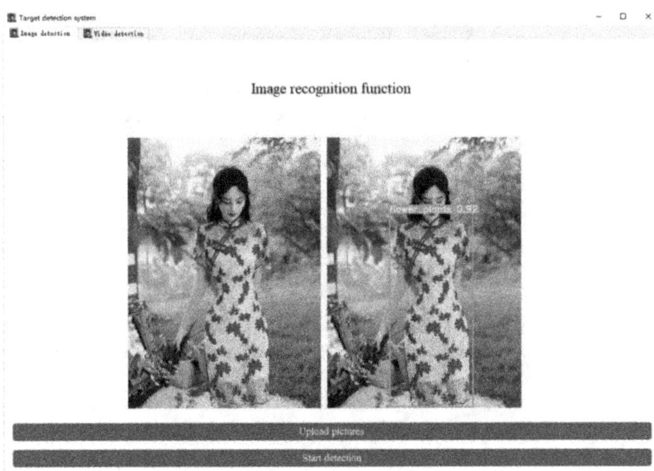

Fig. 6. Interface of the object detection system.

5. Conclusion and Perspective

This chapter applies artificial intelligence technology to research the recognition of cheongsam. The main conclusions are summarized as follows:

(1) This chapter applies a lightweight learning algorithm, YOLOv5s, for cheongsam recognition. YOLOv5s is a one-stage object method that primarily focuses on the superiority of detection speed and model size. It outperforms other network models in terms of recognition and detection performance, making it suitable for deployment on embedded platforms and mobile terminals for efficient detection.

(2) Images such as confusion matrix and precision-related indicator curves were provided, which can provide a more intuitive understanding and analysis of the results of the experiment. The experimental results show that YOLOv5s performs well in indicators, such as confusion matrix, mAP, and F1 score, meeting the accuracy requirements of cheongsam recognition.

(3) An object detection system has been constructed. Built on the dataset created in this chapter, the system has learned extensive features from cheongsam images. It can recognize target objects from input images, compare them with cheongsam information in the dataset, and finally provide the apparel category to which the target belongs. This system serves as a technological reference for clothing recognition research.

Due to the limited supervision information of deep learning models, such as clothing bounding boxes and apparel types, such lightweight image-based style recognition system is difficult to meet the needs of the rapidly developing and changing fashion industry. This forces us to shift our focus from the field of artificial intelligence to a more fine-grained side that is closer to the fashion industry. In the future, we first address the connection between garment and basic human attributes, such as skeletons and muscles, so that people of all body types can have a definite and appropriate style representation in the fashion field. Second, for different body postures, we need to consider how fashion styles are presented from multiple angles of view. Finally, based on the fine-grained style representation in the fashion field mentioned above, we need to model the invariance of individual fashion styles from changing images in deep learning models to achieve a more accurate and efficient style recognition system for the fashion industry.

Acknowledgments

The work was financially supported by China Scholarship Council (No. 202308610054).

References

1. Sumarliah E, Usmanova K, Mousa K, Indriya I. E-commerce in the fashion business: the roles of the COVID-19 situational factors, hedonic and utilitarian motives on consumers' intention to purchase online. *International Journal of Fashion Design, Technology and Education* 2022; 15(2):167-177.
2. Yan W-J, Chiou S-C. Dimensions of customer value for the development of digital customization in the clothing industry. *Sustainability* 2020; 12(11):4639.
3. Noor A, Saeed MA, Ullah T, Uddin Z, Ullah Khan RMW. A review of artificial intelligence applications in apparel industry. *The Journal of The Textile Institute* 2022; 113(3):505-514.
4. Koszewska M. Circular economy—Challenges for the textile and clothing industry. *Autex Research Journal* 2018; 18(4):337-347.
5. Petrova S, Ivanova Z, Marinov I. Impact of information and communication technology on the retail industry: changes in online clothing shopping in the European Union. *Economic and Social Development: Book of Proceedings* 2023:62-72.
6. García-Salirrosas EE, Acevedo-Duque Á. PERVAINCONSA scale to measure the consumer behavior of online stores of MSMEs engaged in the sale of clothing. *Sustainability* 2022; 14(5):2638.
7. Akhtar WH, Watanabe C, Tou Y, Neittaanmäki P. A New Perspective on the Textile and Apparel Industry in the Digital Transformation Era. *Textiles* 2022; 2(4):633-656.
8. Nouinou H, Asadollahi-Yazdi E, Baret I, Nguyen NQ, Terzi M, Ouazene Y, et al. Decision-making in the context of Industry 4.0: Evidence from the textile and clothing industry. *Journal of Cleaner Production* 2023; 391:136184.
9. Mohiuddin Babu M, Akter S, Rahman M, Billah MM, Hack-Polay D. The role of artificial intelligence in shaping the future of Agile fashion industry. *Production Planning & Control* 2022:1-15.
10. Kotouza MT, Tsarouchis SF, Kyprianidis A-C, Chrysopoulos AC, Mitkas PA. Towards fashion recommendation: an AI system for clothing data retrieval and analysis. In: *Artificial Intelligence Applications and Innovations: 16th IFIP WG 125 International Conference, AIAI 2020, Neos Marmaras, Greece, June 5–7, 2020, Proceedings, Part II 16*: Springer; 2020. pp. 433-444.
11. Sharma S, Koehl L, Bruniaux P, Zeng X, Wang Z. Development of an intelligent data-driven system to recommend personalized fashion design solutions. *Sensors* 2021; 21(12):4239.
12. Ahmad S, Miskon S, Alabdan R, Tlili I. Towards sustainable textile and apparel industry: Exploring the role of business intelligence systems in the era of industry 4.0. *Sustainability* 2020; 12(7):2632.
13. Seo Y, Shin K-s. Hierarchical convolutional neural networks for fashion image classification. Expert *systems with applications* 2019; 116:328-339.
14. Liu K, Lin K, Zhu C. Research on Chinese traditional opera costume recognition based on improved YOLOv5. *Heritage Science* 2023; 11(1):1-14.
15. Chang Y-H, Zhang Y-Y. Deep learning for clothing style recognition using YOLOv5. *Micromachines* 2022; 13(10):1678.

Advancing Forest Management: Integrating LiDAR and RGB Imagery through Ordered Weighted Averaging Operators

P. A. Flores-Vidal* and D. Gómez†

*Instituto de Estadística y Ciencia de los Datos, Universidad complutense de Madrid, Avda. Puerta de Hierro, s/n, 28040, Madrid, España

†Facultad de Estudios Estadísticos, Universidad complutense de Madrid, Avda. Puerta de Hierro, s/n, 28040, Madrid, España

*pflores@ucm.es

†dagomez@estad.ucm.es

This chapter introduces a novel approach to tree detection by fusing LiDAR (Light Detection and Ranging) and RGB imagery, leveraging Ordered Weighted Averaging (OWA) aggregation operators to improve image fusing. It focuses on enhancing tree detection and classification by combining LiDAR's structural data with the spectral details from RGB images. The fusion methodology aims to optimize information retrieval, employing image segmentation and advanced classification techniques. The effectiveness of this method is demonstrated on the PNOA dataset, highlighting its potential for supporting forest management.

Keywords: Tree detection; remote sensing; LiDAR; aggregation operator; OWA operators; information fusion; RGB.

1. Introduction: Advances in Remote Sensing for Tree Detection

Tree detection using remote sensing has evolved significantly, initially relying on either LiDAR (Light Detection and Ranging) or optical imagery. LiDAR excels in structural data, potentially ideal for capturing height and canopy density, while lacking the spectral data crucial for tree detection and species identification, a strength that RGB imagery can reach.[1] Fusing LiDAR with RGB imagery enhances both tree detection and classification.[2] This fusion not only enhances the accuracy of tree detection but also supports effective forest management and ecological monitoring, improving individual tree delineation in dense forests and facilitating species classification.

*Corresponding author.

In this chapter, we introduce a novel approach for tree detection by integrating LiDAR and RGB imagery, leveraging Ordered Weighted Averaging (OWA) aggregation operators for effective image fusion. This method distinctively separates trees (foreground) from non-trees (background) by combining the high-resolution 3D structural data from LiDAR with the contextual details like texture, color, and edges present in RGB images. Our fusion method aggregates these data and optimizes the information retrieval process by employing specific image segmentation methods and mathematical morphology. Further, we apply an advanced object classification strategy to enhance the accuracy and reliability of tree detection. The efficacy of our approach is tested on a dataset belonging to the PNOA project, a public data of National Georgraphic Institute of Spain.[3]

The rest of this chapter follows with a theoretical framework section focused on the key concepts and topics. After this, we find the third section related to our proposal and preceded by the employed dataset explanation. Finally, the results are followed by the conclusions.

2. Theoretical Framework

2.1. *Image processing basics*

Let us denote a digital image by I, and its pixel coordinates of the *spatial domain* by (i, j). For clarity's sake, the coordinates are integers, where each point (i, j) represents a pixel with $i = 1, \ldots, n$ and $j = 1, \ldots, m$. Thus, the image's size is equal to $n \times m$. The next image types are used.

- *binary map*: (I_{bin}): $I_{i,j} \in \{0, 255\}$ (as well it is usually expressed as $\{0, 1\}$).
- *grayscale image*: (I_{gray}): $I_{i,j} \in \{0, 1, \ldots, 255\}$.
- *normalized image*: (I): $I_{i,j} \in [0, 1]$.
- *RGB image*: (I_{RGB}): (I^R, I^G, I^B) $I_{i,j} \in \{0, 1, \ldots, 255\}^3$. (R=*Red*; G=*Green* and B=*Blue*).

RGB images are usually transformed to grayscale for simplification purposes, what is also done in this work.

2.2. *OWA operators*

An Ordered Weighted Averaging (OWA) aggregator is a mathematical operator used in decision-making processes and data fusion. It was introduced by Ronald R. Yager[4] as a generalized approach to aggregating information that allows for a flexible compromise between the minimum and maximum operators, among others. The OWA operator is particularly significant in scenarios involving imprecise information, and its key feature is its ability to aggregate multiple inputs by considering both its magnitude and the importance (weights) assigned to them.

$$OWA(a_1, a_2, \ldots, a_n) = w_1 \cdot a_{(1)} + w_2 \cdot a_{(2)} + \ldots + w_n \cdot a_{(n)}$$

Following this expresion some well-known statistics could be achieved, for example if $w_1 = 1$ and the rest of $w_i = 0$ the maximum is obtained (assuming that $a_{(1)}$ corresponds to the highest value). A common restriction for OWA operators is thay they sum up to one: $\sum_{i=1}^{n} w_i = 1$. OWA have been employed in different disciplines, IP being one of them[5]. Yager also developed a special version of OWA that allowed prioritizing[6,7]. Inspired by Yager's, a special version of it was developed in[8].

Definition 2.1. *The Yager-inspired hierarchical prioritized aggregation operator is defined as:*

$$V(y_1, \cdots, y_n) = \sum_{i=1}^{n} w_i \prod_{k=1}^{i} y_{n-k+1} \tag{1}$$

Where 'V' stands for vertical, as every hierarchized set of data is placed in a different box inside a vertical structure that shows the hierarchy between the clusters 'y_i', and 'n' is the number of different clusters

Inspired by Yager's, in this research we propose a novel approach utilizing prioritization-focused OWA operators.

2.3. *LiDAR data*

In Remote Sensing, LiDAR sensors are commonly employed. The data captured by them consists of a collection of points created by sending laser pulses towards a target area and measuring the time it takes for each pulse to return to the sensor. LiDAR data is represented as a set of points in a 3D space: (X, Y, Z), with X and Y representing respectively the horizontal (longitude) and vertical (latitude) locations, and Z is the elevation above a reference plane.

Definition 2.2. *Formally, the raw LiDAR dataset can be defined as a set $\mathcal{P} = \{P_1, P_2, \ldots, P_n\}$, where each point $P_i = (X_i, Y_i, Z_i)$ for $i = 1, 2, \ldots, n$, and n is the total number of points captured by the LiDAR system.*

This detailed 3D model of the environment is crucial for many applications, from digital elevation modeling in Archaeology[9], or Agriculture[10], to the analysis of urban landscapes and forest canopies[11], etc. LiDAR data can capture more information, as the intensity value of the reflected laser, its direction, number of returns, etc. Moreover, when a LiDAR system is equipped with a RGB camera, it can capture color images of the surveyed area, allowing each point in the dataset $\mathcal{P}$ to be augmented with RGB values, thus providing a color representation of the surface. Alternatively, RGB information can be integrated into the LiDAR dataset through post-processing techniques by aligning the two data sources based on their spatial coordinates[3]. In this research, we work with RGB layers added to the main LiDAR's coordinates resulting in $P_i = (X_i, Y_i, Z_i, R_i, G_i, B_i)$ corresponding the extended channels to red, green and blue channels respectively.

2.3.1. *From 3D to 2D: The raster image*

The transformation of a LiDAR point cloud (3D data) into a raster image I (2D data) involves several steps, including point cloud filtering, ground point identification, creation of the elevation model (aka DEMs) and interpolation. For mathematical analysis, LiDAR data is often processed into a raster image, which is a digital image that translates the 3D information of LiDAR's into a 2D 'gridded' format. This process can be considered an interpolation done by assigning pixel values from the LiDAR point cloud. These pixel values can be assigned in different ways, being the most common asssigning the Z value. This is useful for creating digital elevation models (DEMs). In this context, let I_{LiDARZ} denote the raster image derived from the Z layer of LiDAR's. Each pixel in I_{LiDARZ} corresponds to a specific area on the ground and contains a value representing the elevation (Z) of that area. The raster image I_{LiDARZ} is in fact an $I^{gray}_{i,j}$ image (see Section 2.1), and its value at (i, j) corresponds to the Z elevation. In paralell to the creation of raster I_{LiDARZ} another version is created for RGB channels. This raster image is an I_{RGB} colored image (see 2.1), and its spectral information is luminosity in threee different colors: red, green and blue.

3. Methodology Overview

Our study introduces a comprehensive methodology for enhanced tree detection by combining LiDAR and RGB images. The core idea is based on the employment of OWA operators inspired by Yager prioritized OWA operator[6,7] that has already found applications in different fields[8,12].

3.1. *Data adquisition and LiDAR data preprocessing*

The data belongs to PNOA-LiDAR dataset[3], an Open data project who's responsible is the National Geographic Institute of Spain. The chosen area is placed near the city of Pamplona in Navarra region and it covers about one squared kilometer. The reason to choose this specific region is that the resolution of the images is higher (10 points/m^2) than in the rest of regions (0.5 to 2 points/m^2). The point cloud preprocessing has been made with CloudCompare 2.12, a specilized software for this task. This preprocessing can be divided in these steps:

(1) The point cloud data undergoes a preliminary cleaning process, which includes removing outliers and resizing to ensure data quality. The cloud is double-checked to assure that has not anomalous data.
(2) Ground and non-ground elements within the point cloud are differentiated using the Cloth Simulation Filter (CSF)[13].
(3) A mesh model is generated from the ground points, providing a structured representation of the terrain surface (aka DTM or Digital Terrain Model).
(4) The vertical distance between the original point cloud and the generated mesh is calculated, forming the basis for CHM (Canopy Height Model) creation.
(5) The CHM is then produced, along with its corresponding raster images: I_{LiDARZ} and I_{RGB}, both in TIFF format, which assures the highest quality.

3.2. *Images transformation*

After creation of both I_{LiDARZ} and I_{RGB} images, they are imported into the Python environment. Empty pixels (NaN values) are a common problem when creating the raster images from the 3D LiDAR point cloud. Due to this, the I_{LiDARZ} is processed to fill NaN values. A custom median filter, specifically crafted to this task, is employed. It consists in applying a median filter over the 3x3 neighbourhood of NaN values. The filter is iteratively applied until all NaN values are replaced. This approach preserves essential edge details crucial for tree crown identification.

Parallely, I_{RGB} is transformed into grayscale using the luminance formula ($I_{gray} = 0.299 * I_{Red} + 0.587 * I_{Green} + 0.114 * I_{Blue}$), capturing crucial textural and color information. A normalization is applied to I_{gray} to enhance contrast and detail, preparing it to be fused with I_{LiDARZ}.

3.3. *Fusion of image data with prioritized OWA operators*

Fusing such different images like LiDAR's and RGB's, we encounter a scenario akin to multicriteria decision making (MCDM) where certain types of information may have prioritization. In this scenario, structural data from LiDAR images may take precedence due to their detailed $3D$ representation of objects, crucial for canopy measurement. Conversely, RGB images provide essential spectral information, valuable for identifying features like vegetation health or land cover classification. However, in some scenarios, spectral data should not compromise the structural accuracy provided by LiDAR. To address this, we apply a hierarchical approach to data fusion, using importance weights of the OWA operators. Here, the weights assigned to the RGB data are conditioned on the level of detail or confidence provided by the LiDAR data. Below, the equation that condenses this idea:

$$I_{\text{fused}} = \text{Yager}(I_{\text{LiDAR}Z}, I_{\text{RGB}}, w_1, w_2) = w_1 \cdot I_{\text{LiDAR}Z} + w_2 \cdot I_{\text{LiDAR}Z} \cdot I_{\text{RGB}} \tag{2}$$

This fused image is applied following two different methodologies:

- **Method A ("Pre-aggregation"):** Both images are fused prior to the binarization process that leads to the segmentation (the creation of objects representing potential trees):
  (1) I_{fused} is created.
  (2) I_{fused} is binarized and the objects are created from there.
- **Method B ("Post-aggregation"):**
  (1) $I_{\text{LiDAR}Z}$ is binarized to $I_{\text{bin}Z}$.
  (2) I_{RGB} is binarized to I_{binRGB}.
  (3) $I_{\text{fused}} = \text{Yager}(I_{\text{bin}Z}, I_{\text{binRGB}}, w_1, w_2) = w_1 \cdot I_{\text{bin}Z} + w_2 \cdot I_{\text{bin}Z} \cdot I_{\text{binRGB}}$.

For simplification purposes the I_{RGB} image was transformed into a grayscale's using the 'luminance' formula explained before. Different w_i were used: $w_1 = \{0.5, \ldots, 0.9, 1\}, w_2 = \{0, 0.1, \ldots, 0.5\}$ and $w_1 + w_2 = 1$. The prioritization of LiDAR information against RGB's was assured with this restriction: $w_1 \geq w_2$.

3.4. *Tree detection from the fused image*

Noise produced after Otsu binarization[14] is addressed through morphological opening, employing a circular structural element to maintain the tree crowns shape. Subsequently, dilation is applied to enhance the distinction between trees and their surroundings. The sure foreground areas are created using a distance transform and thresholding technique. This step is pivotal in differentiating between tree crowns and non-tree regions, aided by the subtraction of the sure foreground from the background. The Watershed algorithm[15] is then applied, using the markers derived from the Otsu's binarization[14]. Finally, the potential 'tree-objects' are identified.

4. Results

After studying the outputs for all different weights, $w_1 = 0.7$ was chosen as an appropriate value for the fusion image. We can see in Figure 4 that 'Pre-aggregation' Method A captured more forestry areas (i.e. 'tree-objects') than 'Post-aggregation' Method B (277 against 233). Moreover, 'tree-objects' detected by method B tend to be smaller than in A's.

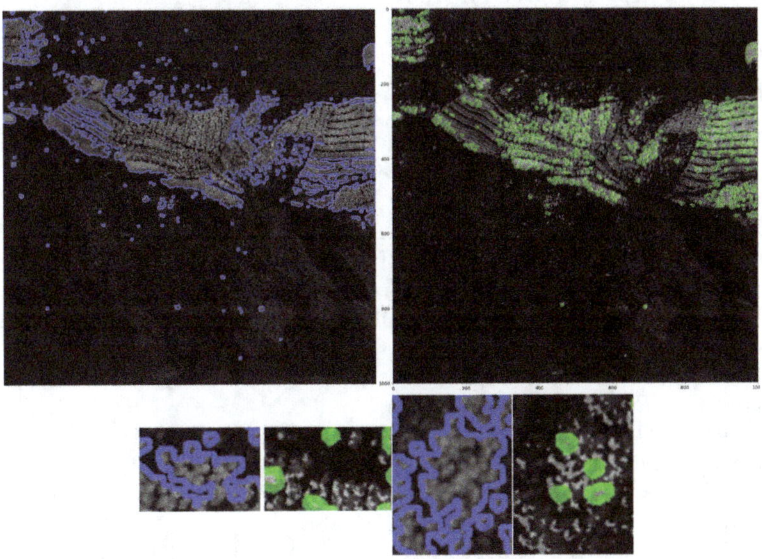

Fig. 1. Identidied tree-objects with Method A (blue) and B (green) with w1=0.7

5. Conclusions

The results of this chapter show that the fusion of information of RGB and LiDAR has promising future for tree detection. The use of Yager-inspired prioritized has been proven effective, but more research and evaluation are needed. As a preliminar work, this research can go further. For instance, a more sophisticated technique could complement the final classification step for improving the identification of individual trees, and this could be done employing clustering techniques. It could be also interesting to improve the fused image by using edge detection. Moreover, more channels could be included in the fused image, such as the Near infrared or LiDAR intensity.

Acknowledgments

This research has been partially supported by the Spanish Ministry of Science, grant PGC2018-096509-BI00.

References

1. M. A. Lefsky, W. B. Cohen, G. G. Parker and D. J. Harding, Lidar remote sensing for ecosystem studies: Lidar, an emerging technology that directly measures the 3-d distribution of plant canopies, *BioScience* **52**, 19 (1 2002).
2. D. Gatziolis and H. E. Andersen, A guide to lidar data acquisition and processing for the forests of the pacific northwest, *USDA Forest Service - General Technical Report PNW-GTR* (2008).
3. S. L. Martínez, J. C. O. Manrique, B. Rodríguez-Cuenca, E. G. Cristóbal and P. Muñoz, Procesado y distribución de nubes de puntos en el proyecto pnoa-lidar, *Proceedings of the XVII Congreso de la Asociación Española de Teledetección, Murcia, Spain* **2015** (2017).
4. R. R. Yager, On ordered weighted averaging aggregation operators in multicriteria decisionmaking, *IEEE Transactions on Systems, Man and Cybernetics* **18** (1988).
5. H. Bustince, J. Fernández, A. Kolesárová and R. Mesiar, Generation of linear orders for intervals by means of aggregation functions, *Fuzzy Sets and Systems* **220**, 69 (2013).
6. R. R. Yager, Modeling prioritized multicriteria decision making, *IEEE Transactions on Systems, Man, and Cybernetics, Part B: Cybernetics* **34** (2004).
7. R. R. Yager, Prioritized aggregation operators, *International Journal of Approximate Reasoning* **48**, 263 (2008).
8. K. Rojas, D. Gómez, J. Montero, J. T. Rodríguez, A. Valdivia Barrios and F. Paiva, Development of child's home environment indexes based on consistent families of aggregation operators with prioritized hierarchical information, *Fuzzy Sets and Systems* **241**, 41 (2014).
9. B. J. Devereux, G. S. Amable, P. Crow and A. D. Cliff, The potential of airborne lidar for detection of archaeological features, *Antiquity* **79**, 648 (2005).
10. U. Weiss and P. Biber, Plant detection and mapping for agricultural robots using a 3d lidar sensor, *Robotics and autonomous systems* **59**, 265 (2011).
11. D.-A. Kwak, W.-K. Lee, J.-H. Lee, G. S. Biging and P. Gong, Detection of individual trees and estimation of tree height using lidar data, *Journal of Forest Research* **12**, 425 (2007).
12. P. Flores-Vidal, D. Gómez, J. Castro and J. Montero, New aggregation approaches with hsv to color edge detection, *International Journal of Computational Intelligence Systems* **15** (2022).
13. W. Zhang, J. Qi, P. Wan, H. Wang, D. Xie and G. Yan, An airborne lidar data filtering method based on cloth simulation, *Remote Sensing* **8** (2016).
14. N. Otsu, Threshold selection method from gray-level histograms., *IEEE Trans Syst Man Cybern* **SMC-9** (1979).
15. L. Vincent and P. Soille, Watersheds in digital spaces: an algorithm based on immersion simulations, *IEEE Transactions on Pattern Analysis & Machine Intelligence* **13**, 583 (1991).

Horn Clause First in Ordered Linear Resolution for the CDCL Solver

Z. H. Li*,‡, G. F. Wu*,§, Y. Xu*,¶ and J. Liu†,‖

*School of Mathematics, Southwest Jiaotong University,
Chengdu 610031, China

†School of Computing, Ulster University,
Belfast, BT15 1AP, Northern Ireland, UK

‡lizhihui@swjtu.edu.cn

§wgf1024@swjtu.edu.cn

¶xuyang@swjtu.edu.cn

‖j.liu@ulster.ac.uk

This chapter presents the Horn-First Combinatorial Strategy (HFCS) within the Conflict-Driven Clause Learning (CDCL) solver framework, which leverages unit propagation to resolve conflicts via a directed acyclic entailment diagram. HFCS boosts the role of Horn clauses in clause management and their strategic use in solving, emphasizing learning clauses generated from their involvement. Experimental results confirm HFCS significantly enhances solver performance.

Keywords: Horn clause; resolution; CDCL; SAT solver.

1. Introduction

The SAT solver stands as a pivotal tool in addressing substantial real-world challenges. The efficacy of contemporary Conflict-Driven Clause Learning (CDCL) solvers is attributed to a synergy of critical components, such as Boolean Constraint Propagation (BCP) and the 2-watched literal scheme, orchestrated under the aegis of diverse heuristic strategies. The essence of the CDCL method lies in its systematic identification and resolution of conflicts throughout an efficient exploration of the entire search space, subsequently refining this space with learned clauses derived from conflict analysis.[1] This process of conflict analysis essentially involves the strategic dissection of the BCP entailment diagram to extract informative learned clauses, grounded in a generalized resolution principle focused on the propagation of literal sequencess.[2]

It has been proved that a broad class of CDCL SAT solvers corresponds to proof systems as powerful as general resolution and that general

‡Corresponding author.

resolution has more powerful reasoning.[3] The literals and clauses involved in the dissemination constitute the basic object used to resolve it. The results show that the solver's performance can be significantly improved by prioritizing specific literals as decision-making literals and preferentially propagating within the range of specific clauses.

The Horn clause is a special type of clause that contains at most one positive literal, which has the characteristics of concise form and clear logical meaning and has an important position in automatic reasoning proofs. In this context, we introduce the Horn-first combinatorial strategy (*HFCS*), alongside two innovative algorithms that elevate the involvement of Horn clauses in the CDCL framework through a prioritized order of linear resolution. Our findings demonstrate that this novel approach significantly augments the efficiency of cutting-edge solvers, showcasing the potential of HFCS in advancing the state of the art in SAT solving technology.

2. Preliminaries

We assume that you are familiar with the basic concepts of propositional logic and the principles and functions of the basic modules of a complete SAT solver. Robert Anderson and W. W. Bledsoe present one technique for establishing completeness proofs that can be used to establish a wide variety of deductive systems based on resolution [4] . The literature [5] improve on existing results on the relationship between proof systems obtained from CDCL SAT solvers and general resolution.

Theorem 2.1. ***The principle of reduction in propositional*** *Let C_1 and C_2 be two clauses, if one of the disjunctive terms L_1 in C_1 is complementary to one of the disconjunctions L_2 in C_2, then L_1 and L_2 are eliminated from C_1 and C_2 respectively, and two simple clauses $C_1^{'}$ and $C_2^{'}$ are obtained. $C_1^{'} \vee C_1^{'}$ is called the reduction of C_1 and C_2.*

Theorem 2.2. ***OL resolution*** *Let S be a finite set of ordered clauses, $C_0 \in S$. C_n is called the linear inference of S with C_0 vertex, if the following conditions are true:*

(i) The existence of the central clause $C_0, C_1, \ldots, C_{n-1}$ and the edge clause $B_0, B_1, \ldots, B_{n-1}$ is such that:

(a) When C_i can be reduced, and C_{i+1} is the reduction of C_i.
(b) When C_i irreducible, $B_i \in S$, C_{i+1} is C_i's ordered reduction of B_i, and the eliminated literal is the last literal of C_i.

(c) if $B_i \notin S$, $C_j (j < i)$, then B_i can be used as a candidate for C_i.

(ii) Logically valid formulas do not appear in reasoning.

This is followed by an introduction to two basic lemma that prioritize the provision of ordered queues after the .definition of the relevant ordinal linear reduction.

Lemma 2.1. *High-quality atductive literal sequences drive linear reductive deduction and corresponding deductive results.*

Lemma 2.2. *When the literal l is assigned to the propagation queue trail, the reason clause where the literal l is located is satisfied, and the literal order of the reason clause has been adjusted.*

In the CDCL solver, the search is a step-by-step selection of branch decision literals, and the linear deduction process is synchronized with the CDCL clause learning process, which can effectively solve the problem of uncertain reduction literal and deduction depth in the linear deduction. So we come up with the following inference Lemma 2.1.

When the literal l is assigned to the propagation queue *trail*, the *reason clause* where the literal l is located is satisfied, and the literal order of the *reason clause* has been adjusted. The order of the literals is dynamically ordered in the propagated clause during the CDCL-solving process. Specific clauses, including Glue clausess[6] and Core clauses[7], as well as Duplicate Learnt Clauses[8] have been experimentally proven to play an essential role in improving the ability to solve. So, the horn clause is also studied as an entry point from the perspective of ordered linear reduction in the next chapter.

3. Horn First Strategy

In this section, we give the theorem of HFCS and two improved algorithms based on it, that promote the priority participation of Horn clauses in the CDCL framework.

3.1. *Theorem*

In the CDCL framework, the Horn clauses have the properties of Lemma 2.1 and Lemma 2.2. Based on this, we propose a new idea: the *Horn First* strategy, which preferentially uses Horn clauses to scan sequentially

Algorithm 1: unitPropagation_HFCS( )

```
Input: (Φ ∪ Γ, α_p)
Output: c /* a conflict clause or null. */
α_p: trail stack of decision literals and implications;
c := null;
for q := index of list unvisited literal in α_p to α_p.size do
    l := α_p[q];
    W[l]: set of clauses watched by literal l;
    /* Simplify the expression of the length of W[l] with W[l].size,
    the last index of Horn clause in W[l] with W[l]_hornIndex = 0 */
    for k := 0 to W[l].size do
        if W[l][k] is unit then
            u := the unassigned literal of W[l][k];
            Push u to the end of α_p;
            if W[l][k] is horn clause and k > W[l]_hornIndex then
                Swap(W[l][W[l]_hornIndex], W[l][k]);
                W[l]_hornIndex ++;
        elseif W[l][k] is falsified then
            c := W[l][k];
            break;
    if c ≠ null then
        break;
return c;
```

every clause over the watcher list in BCP, and retains Horn learning clauses throughout the learning clause management.

Theorem 3.1. ***HFCS***: *in BCP, the 2-Watched Literal order is adjusted to give priority to the use of the Horn clause, and the Horn learning clause is retained throughout the learning clause management process.*

3.2. *Improved algorithms*

In line with *HFCS* we have made two improvements to CDCL-related components : *unitPropagation()* and *reduceDB()*, to give full play to the ability of the Horn clause to participate in the OL resolution in the BCP and conflict analysis phases.

We implement to select core clauses first by moving Horn clauses ahead of general clauses during unit propagation. When watchlists are built initially, Horn clauses are not in front of the clause corresponds to the watcher element. Like the standard BCP, the goal of *unitPropagation_HFCS( )* is to search for all unit clauses. The Horn First strategies embodies in the update of watchlists. Algorithm 1 shows unitPropagation_HFCS( ).

Algorithm 2: reduceDB_HFCS()

```
if (γ is Horn Clause) then
    add γ to the Core set;
else
    add γ to one of three sets (Core, Iter2, Local).
    /* according to the original quality evaluation criteria (divided by
    LBD value or Size of the learn clause.)*/
```

In the conflict analysis phase we get the learning clauses, and then we assign them to different sets of learning clauses according to the quality evaluation criteria of the learning clauses. The learning clauses in different sets have different flow patterns and deletion strategies.

In Algorithm 2, we add the horn clause to the Core set, which is permanently retained, and is not subject to the phased deletion operation, so that it can participate in the BCP preferentially.

4. Experimental Results

In order to evaluate the practical efficiency of HFCS, we implemented HFCS as a modification of state-of-the-art solver Relaxed_LCMDCBDL[9], cadical-rel1.5.3, Kissat_MAB-HyWalk [10], which are the leaders of the Glucose, Cadical and Kissat series of solvers. We refer to those who use the Horn First strategy as Horn First. Due to space limitations, this paper focuses solely on the impact of deploying HornFirst within the context of Relaxed LCMDCBDL.

The experiments were executed on Sugon cloud computing platform cluster where processoris 1*7185 32C 2.5GHz with 128GB RAM and 200Gb Computing network in standard-mode.

We evaluated our implementation on the benchmarks from the main track of the last three years of competition. The results of experiments for the base solvers and their corresponding updated versions are reported in Table 1. Detailed statistics on solved instances from SC2021, SC2022, and

SC2023. S and U stand for the number of solved satisfiable and unsatisfiable instances, respectively. SCR stands for the total number of instances resolved. The best SCR scores for SC2021, SC2022, and SC2023 are marked with bold.

Table 1. Solved instances of the SAT Competitions 2021-2023

Configure	2021SC			2022SC			2023SC		
	SCR	S	U	SCR	S	U	SCR	S	U
Original	237	123	114	197	113	84	211	115	96
HornFirst	**245**	129	116	**214**	118	96	**212**	113	99

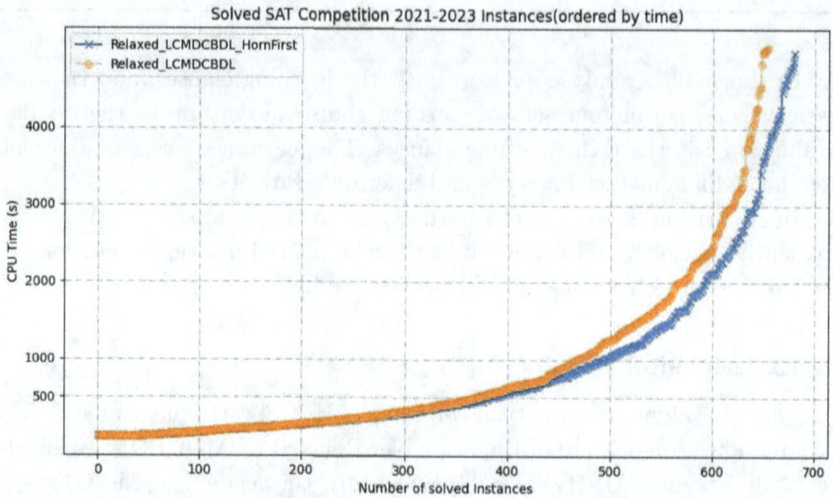

Fig. 1. Cactus plot for benchmarks of the main track of the SAT Competitions 2021–2023. One point (x, y) indicates that x instances were solved within y seconds.

The solver after adopting the improved *HFCS* algorithm has improved the solution performance of 1200 instances in the past three years, especially the solution speed has been improved on a considerable number of instances. As you can see from Fig. 1, the blue line is always below the yellow line. This shows that the improved version has a significant improvement in solution speed and the 26 more solved instances prove the effectiveness of the Horn clause first in ordered linear resolution for the CDCL solver. In Fig. 2, the

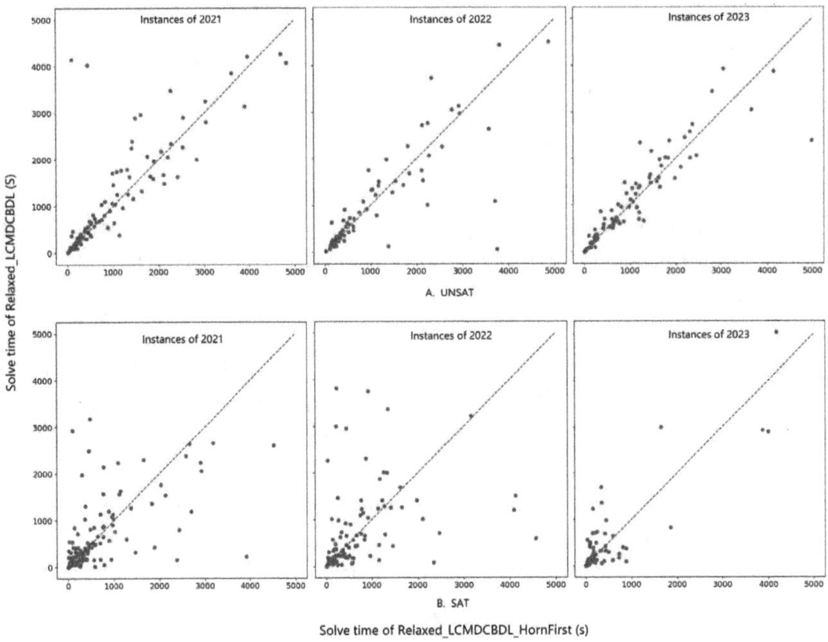

Fig. 2. Comparison chart of category results on benchmarks from SAT Competitions 2021–2023.

comparison between A.UNSAT and B.SAT outcomes highlights the Horn clause's consistent role in enhancing solver performance, without favoring results towards SAT or UNSAT specifically. UNSAT outcomes tend to group close to the diagonal, indicating predictability, while SAT outcomes are more spread out, showing a wider variability. This demonstrates that the HFCS strategy significantly affects SAT solutions, with minimal impact on the predictability of UNSAT outcomes.

5. Conclusion

In this chapter, we introduced the Horn-First Combinatorial Strategy (HFCS) and two novel algorithms to enhance Horn clauses' role in the CDCL framework through priority-based resolution. This approach optimizes clause management and the utilization of learned clauses from Horn clause involvement, significantly improving solver performance, as confirmed by our experiments.

Acknowledgments

This work is supported by National Natural Science Foundation of China (Grant No: 62106206).

References

1. S. Cai and X. Zhang, Deep cooperation of CDCL and local search for SAT (extended abstract), in *Proceedings of the Thirty-First International Joint Conference on Artificial Intelligence, Vienna, Austria, 23-29 July 2022*,
2. G. Audemard, L. Bordeaux, Y. Hamadi, S. Jabbour and L. Sais, A generalized framework for conflict analysis, in *Theory and Applications of Satisfiability Testing - SAT 2008, 11th International Conference, Guangzhou, China, May 12-15, 2008. Proceedings*,
3. C. Li, F. Xiao and F. Manyà, A resolution calculus for minsat, *Log. J. IGPL* **29**, 28 (2021).
4. R. Anderson and W. W. Bledsoe, A linear format for resolution with merging and a new technique for establishing completeness, *J. ACM* **17**, 525 (1970).
5. Y. Ajani and C. Bright, A hybrid SAT and lattice reduction approach for integer factorization, in *Proceedings of the 8th SC-Square Workshop co-located with the 48th International Symposium on Symbolic and Algebraic Computation, SC-Square@ISSAC 2023, Tromsø, Norway, July 28, 2023*,
6. M. S. Chowdhury, M. Müller and J. You, Exploiting glue clauses to design effective CDCL branching heuristics, in *Principles and Practice of Constraint Programming - 25th International Conference, Stamford, CT, USA, September 30 - October 4, 2019, Proceedings*, , Lecture Notes in Computer Science Vol. 11802
7. J. Chen, Core first unit propagation, *CoRR* **abs/1907.01192** (2019).
8. S. Kochemazov, O. Zaikin, A. A. Semenov and V. Kondratiev, Speeding up CDCL inference with duplicate learnt clauses, in *ECAI 2020 - 24th European Conference on Artificial Intelligence, 29 August-8 September 2020, Santiago de Compostela, Spain*,
9. X. Zhang and S. Cai, Relaxed backtracking with rephasing, *Proceedings of SAT competition* , 15 (2020).
10. J. Zheng, K. He, Z. Chen, J. Zhou and C.-M. Li, Combining hybrid walking strategy with kissat mab, cadical, and lstech-maple, *SAT COMPETITION 2022* , p. 20 (2022).

Temporal Knowledge Graph Embedding for Metro Flow Analysis

Dexian Wang[*,§], Pengfei Zhang[*,¶], Ping Deng[†,‖] and Tianrui Li[‡,**]

[*] *School of Intelligent Medicine, Chengdu University of Traditional Chinese Medicine, Chengdu 611137, China*

[†] *School of Computer and Software Engineering, Xihua University Chengdu 610039, China*

[‡] *School of Computing and Artificial Intelligence, Southwest Jiaotong University Chengdu 611756, China*

[§] *wangdexian@cdutcm.edu.cn*

[¶] *feifeihappy55@163.com*

[‖] *dengping609@gmail.com*

[**] *trli@swjtu.edu.cn*

The study of metro flow has become a hot topic and an important research element in urban computing. Metro flow is affected by the topology of metro network and the POI around metro stations. It is a challenge to extract the effective patterns of metro flow from the complex data. In this chapter, we construct metro flow knowledge graph to depict the topological relations between metro stations, the relations between metro stations and POIs, and the relations between metro stations and flows. Since metro flow changes over time, we add the corresponding time constrain to each relation. In addition, due to the complexity of relations, the time granularity of the constraints on the relations varies. Therefore, for effective knowledge graph representation learning, we propose a multi-temporal granularity metro flow temporal knowledge graph (MGMF-TKG) embedding method. Specifically, first, we construct a metro flow temporal knowledge graph from metro flow data and use a combination of sinusoidal waves to represent temporal information. Then, a multi-time granularity knowledge graph representation learning framework is constructed to realize the representation of relations between entities under complex time information constraints. Finally, experiments are conducted using Chongqing metro flow data, and the experimental results show that the method outperforms the benchmark methods.

Keywords: Temporal knowledge graph; multiple-time granularity; representation learning; spatio-temporal data.

[§]Corresponding author.

1. Introduction

The study of traffic flow patterns has been a hot topic in the field of urban computing. Since urban metro flow has obvious spatio-temporal characteristics, metro flow-based models are constructed from both temporal and spatial perspectives. Chen *et al.*[1] modeled the metro network as a structured graph with a graph convolutional network (GCN) and a stacked long-short time memory (LSTM) network to capture the dynamic spatial correlation and complex temporal features among metro stations, respectively. Ye *et al.*[2] constructed the metro system as a directed weighted graph and proposed a new spatio-temporal deep learning framework for metro passenger flow prediction. Guo *et al.*[3] designed a self-attention mechanism in the temporal dimension to capture the temporal dynamics of flow data using local contexts. A dynamic graph convolution module was developed in the spatial dimension to capture spatial correlations in a dynamic manner using self-attention. In these studies, the researchers basically used recurrent neural networks and attention-based mechanisms to obtain dynamic evolutionary features of the data over time. In addition, the metro network was constructed as graphs and used graph convolutional networks to learn spatial features.

In this chapter, entities and relations in the metro system are transformed into time-constrained knowledge quaternions to construct the metro flow TKG. The main contributions of this chapter are follows:

(1) Using urban metro flow data, we construct a temporal knowledge graph for urban metro flow. The knowledge graph is used to describe the topology relation of the metro network, the flow of metro stations, and the number of POIs around the metro stations.
(2) We propose a multi-time granularity knowledge graph embedding framework. For the representation of a time, we represent the different time units of a time separately and then accumulated the representations of each unit to achieve the representation of a time. For the representation of time periods, we achieve this by accumulating the representations of time periods corresponding to units.
(3) Experiments are conducted on Chongqing metro flow data, and the experimental results show that the proposed method outperformed some benchmark methods.

2. Our Method

We obtain knowledge quaternions based on POI data around metro stations, flow information of metro stations and topology information of metro network to build metro flow temporal knowledge graph.

2.1. *Multi-time granularity embedding framework*

We adopt the diachronic embedding[4] method to implement the metro flow temporal knowledge graph embedding. First, a knowledge quaternion $q = (h, r, t, \tau)$ is divided into the knowledge triplet (h, r, t) and time τ. Then, the time representation is spliced with the representations of entities and relations in the triplet, respectively, to obtain the representation of the knowledge quaternion. Finally, the DistMult method is used to learn the knowledge quaternion representation. However, there is multi-time granularity of time in the metro flow time knowledge graph in this paper. Therefore, we propose a multi-time-granularity knowledge graph embedding framework as shown in Fig. 1. For the representation of time, the framework first represents each time unit in a time separately, and then accumulates the representations of each time unit to achieve a time representation.

According to the diachronic embedding method, the knowledge quaternion $q = (h, r, t, \tau)$ in the embedding space is shown as follows.

$$\kappa_q = (\kappa_h \oplus \kappa_\tau, \kappa_r \oplus \kappa_\tau, \kappa_t \oplus \kappa_\tau), \tag{1}$$

where κ_q is the representation of the fact quaternion q, $\oplus$ denotes the splicing operation, which splices the embedding vectors of $h/r/t$ and time τ in the embedding space. $\kappa_h \oplus \kappa_\tau$ denotes the representation of entity h at time τ.

In this paper, we combine the diachronic embedding method with the spectral analysis method to design a novel time representation method as follows.

$$\begin{cases} F_1(\theta_k) = W_a^k \cdot \sin\left(W_b^k \cdot \theta_k + W_c^k\right), \\ F_2(\theta_k) = W_s^k \cdot \cos\left(W_e^k \cdot \theta_k + W_l^k\right), \\ \kappa_\tau^k = \left(F_1(\theta_k) + F_2(\theta_k)\right)/2, \theta_k \in X, \end{cases} \tag{2}$$

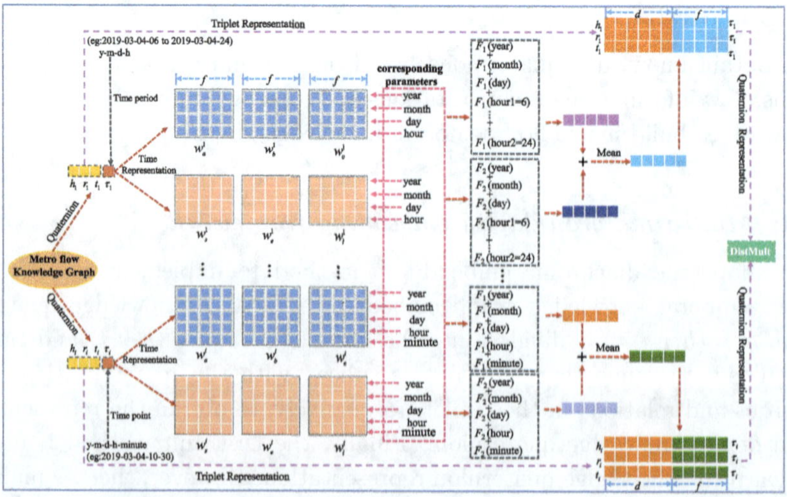

Fig. 1: Multi-time granularity knowledge embedding framework.

where $W_a^k, W_b^k, W_c^k, W_s^k, W_e^k, W_l^k \in \mathbb{R}^{f^*|X|}$ are vectors with learnable parameters for k-th time unit presentation. κ_τ^k denotes the representation of the time in the k-th time unit by combination of sinusoidal waves. Moreover, in Eq.(2) all vectors are calculated in an element-wise manner. To represent a time, we represent each time unit of that time separately according to Eq.(2). By fusing the individual time unit embedding vectors, we can obtain the time representation as follows.

$$\kappa_\tau = \sum_{k=1}^{|X|} \left(F_1(\theta_k) + F_2(\theta_k)\right), \theta_k \in X. \tag{3}$$

It is worth noting that Eq.(2) and Eq.(3) can only be represented for time points. To achieve the representation of time periods, we obtain the following time representations.

$$\begin{cases} F_1(\theta_k) = \sum_{n=1}^{N_k} \left(W_a^k * \sin\left(W_b^k * \theta_{k,n} + W_c^k\right)\right), \\ F_2(\theta_k) = \sum_{n=1}^{N_k} \left(W_s^k * \cos\left(W_e^k * \theta_{k,n} + W_l^k\right)\right), \\ \kappa_\tau^k = \left(F_1(\theta_k) + F_2(\theta_k)\right)/2, \theta_k \in X, \end{cases} \tag{4}$$

where $\theta_{k,n}$ denotes the n-th value of the k-th time unit. When $N_k > 1$, Eq.(4) is a representation for a period time.

2.2. *Loss function*

The positive samples are the knowledge quaternions directly observable in the temporal knowledge graph i.e. $q = (h, r, t, \tau) \in G$, $G \subset S$. The negative samples $(h', r, t, \tau) \in G_{h'}$ and $(h, r, t', \tau) \in G_{t'}$ are generated by randomly replacing the heads and tails of the positive samples. In this paper, we use cross entropy[5–7] to construct the loss function. The representation vector of the knowledge quaternion is learned by minimizing the cross entropy loss.

$$\xi = -\left(\sum_{q=(h,r,t,\tau)\in G} \frac{\exp(\phi(h,r,t,\tau))}{\sum_{(h',r,t,\tau)\in G_{h'}} \exp(\phi(h',r,t,\tau))} + \frac{\exp(\phi(h,r,t,\tau))}{\sum_{(h,r,t',\tau)\in G_{t'}} \exp\left(\phi\left(h,r,t',\tau\right)\right)} \right) \tag{5}$$

3. Experiments

3.1. *Dataset*

The operating hours of the metro are 6:00 - 24:00, and there are 169 metro stations and 8 metro lines. The traffic road network information is used to obtain information about which line the station belongs to, the station's neighbors and the station's transfer information. Our dataset contains 184,448 fact quaternions, 1,937 relations and 217 entities.

3.2. *Parameter setting*

In this paper, we set $d = 64$, $f = 36$, which is the dimension of 100 in our embedding space. the paper uses gradient descent method with min-batches for training. Where the batch size is 128 and the gradient descent optimization method is adam[8]. We set the learning rate to 0.001. Furthermore, the prediction is performed based on the metro outflow flow for the experiments.

3.3. *Evaluation methodology*

In this paper, to evaluate the performance of the constructed model, we use link prediction and metro flow prediction to evaluate the model.

3.4. *Baseline method*

In terms of link prediction, our proposed method MGMF-TKG is compared with TransE-based[4,9] and DistMult-based[4,10] methods.

For the flow prediction. First, we will obtain the relation representation in $q = (h, r, t, \tau) \in P$ based on method MGMF-TKG and all the comparison methods. Then, they are input into the LSTM for flow prediction, respectively. In addition, to prove the validity of method MGMF-TKG, we input the raw time series flow data into the LSTM for comparison.

3.5. *Link prediction results analysis*

Table 1 shows the link prediction results for all knowledge graph embedding methods. From the Table 1, we can find the proposed embedding method has superior link prediction results. Therefore, it can perform more effective embedding for multi-time granularity knowledge graphs. As shown in Table 1, our method is higher than other embedding methods in metrics hit@1, hit@3, hit@10, MRR, respectively. Specifically, our method outperforms TransE, DistMult, DE-TransE, DE-DistMult by 36.81%, 33.81%,15%, and 7.92% in metric hit@1, respectively.

Table 1: Link prediction results

Models	Hit@1	Hit@3	Hit@10	MRR
TranE	0.269	0.536	0.610	0.425
DistMult	0.275	0.535	0.608	0.426
DE-TransE	0.320	0.559	0.650	0.464
DE-DistMult	0.341	0.558	0.646	0.474
MGMF-TKG	**0.368**	**0.574**	**0.663**	**0.495**

3.6. *Flow prediction results analysis*

Fig.2 shows the flow prediction accuracy obtained by feeding the representation of flow r and the raw time series data into the LSTM separately. We use two evaluation metrics, RMSE and MAE, to evaluate the prediction performance, and we can find the effective knowledge graph embedding method can better obtain the features of the data. As shown in Fig.2(a), we input the representation of the relation r obtained by all embedding methods into LSTM, our proposed algorithm has the lowest MAE, followed by DE-DistMult, DE-TransE, DistMult, TransE. The prediction results obtained by inputting the raw time-series data into LSTM are the highest. Therefore, our proposed method has an advantage in metro station prediction. The same situation can be seen in Fig.2(b), where our method also achieves the lowest accuracy on RMSE.

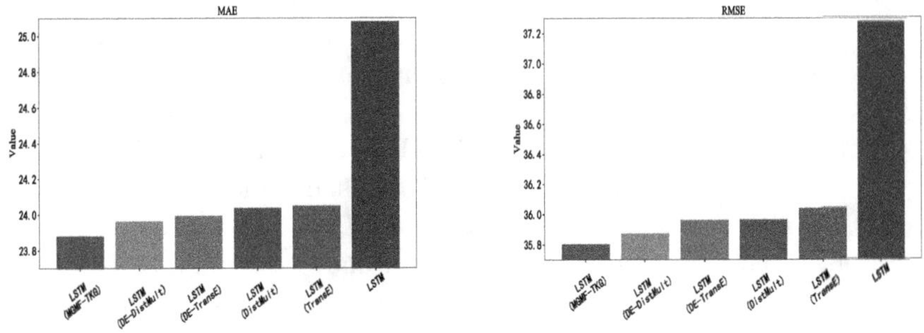

(a) (b)

Fig. 2: Metro station flow prediction.

4. Conclusion

In this chapter, we proposed a multi-time granularity temporal knowledge graph embedding algorithm MGMF-TKG. We utilized the temporal knowledge graph to simultaneously describe the spatial topology of the metro network and the time-evolution-based flow data to achieve the modeling of the spatio-temporal characteristics of metro flow. In addition, we constructed a multi-time granularity representation framework to realize the representation learning of multi-time granularity metro flow temporal knowledge graph. Finally, we conducted experiments on the Chongqing metro flow dataset, and the experimental results show that our method outperforms other algorithms in both link prediction and flow prediction. In addition, based on this method, we can use methods such as classification and clustering[11–13] to realize applications such as traffic pattern analysis and business location selection in the future.

Acknowledgments

This work was supported by the Postdoctoral Fellowship Program of CPSF (Nos. GZB20230091 and GZB20230092) and the China Postdoctoral Science Foundation, China (Nos. 2023MD744127 and 2023M740383).

References

1. P. Chen, X. Fu and X. Wang, A graph convolutional stacked bidirectional unidirectional-lstm neural network for metro ridership prediction, *IEEE Transactions on Intelligent Transportation Systems* , 1 (2021).

2. J. Ye, J. Zhao, K. Ye and C. Xu, Multi-stgcnet: A graph convolution based spatial-temporal framework for subway passenger flow forecasting, in *2020 International Joint Conference on Neural Networks (IJCNN)*, 2020.
3. S. Guo, Y. Lin, H. Wan, X. Li and G. Cong, Learning dynamics and heterogeneity of spatial-temporal graph data for traffic forecasting, *IEEE Transactions on Knowledge and Data Engineering* (2021).
4. R. Goel, S. M. Kazemi, M. Brubaker and P. Poupart, Diachronic embedding for temporal knowledge graph completion, in *Proceedings of the AAAI Conference on Artificial Intelligence*, (04)2020.
5. D. Ruffinelli, S. Broscheit and R. Gemulla, You can teach an old dog new tricks! on training knowledge graph embeddings, in *International Conference on Learning Representations*, 2019.
6. K. Zhou, W. X. Zhao, S. Bian, Y. Zhou, J.-R. Wen and J. Yu, Improving conversational recommender systems via knowledge graph based semantic fusion, in *Proceedings of the 26th ACM SIGKDD International Conference on Knowledge Discovery & Data Mining*, 2020.
7. H. He, A. Balakrishnan, M. Eric and P. Liang, Learning symmetric collaborative dialogue agents with dynamic knowledge graph embeddings, *arXiv preprint arXiv:1704.07130* (2017).
8. D. P. Kingma and J. Ba, Adam: A method for stochastic optimization, *arXiv preprint arXiv:1412.6980* (2014).
9. A. Bordes, N. Usunier, A. Garcia-Duran, J. Weston and O. Yakhnenko, Translating embeddings for modeling multi-relational data, in *Advances in Neural Information Processing Systems*, eds. C. Burges, L. Bottou, M. Welling, Z. Ghahramani and K. Weinberger (Curran Associates, Inc., 2013).
10. B. Yang, W.-t. Yih, X. He, J. Gao and L. Deng, Embedding entities and relations for learning and inference in knowledge bases, *arXiv preprint arXiv:1412.6575* (2014).
11. D. Wang, T. Li, P. Deng, F. Zhang, W. Huang, P. Zhang and J. Liu, A generalized deep learning clustering algorithm based on non-negative matrix factorization, *ACM Trans. Knowl. Discov. Data* **17** (may 2023).
12. D. Wang, T. Li, W. Huang, Z. Luo, P. Deng, P. Zhang and M. Ma, A multi-view clustering algorithm based on deep semi-nmf, *Information Fusion* , p. 101884 (2023).
13. D. Wang, T. Li, P. Deng, J. Liu, W. Huang and F. Zhang, A generalized deep learning algorithm based on nmf for multi-view clustering, *IEEE Transactions on Big Data* **9**, 328 (2023).

Heterogeneity-Aware Personalized Federated Learning for Mobile Edge Computing

Ximing Chen*, Yue Cong*, Chengliang Gao*, Rongrong Chen*, Fei Tang*, Yongxin Cai* and Jing Qiu[†,‡]

*The Cyberspace Institute of Advanced Technology, Guangzhou University, Guangzhou 510006, China

[†]Pengcheng Laboratory, Shenzhen, China

[‡]qiujing@gzhu.edu.cn

With advances in the Internet of Things and communication technologies, the diversity of edge devices has attracted excessive computational loads to these technology systems while causing device heterogeneity. In the context of mobile edge computing, differences in communication and computing resources of edge devices may affect the performance of the model. In personalized federated learning, the differences among different devices can further widen the convergence effect differences. Thus, we propose a multi-exit selection mechanism based on limited resource allocation, which allows devices with weaker computing power to end training early. Subsequently, we propose a hierarchical personalized federated learning algorithm called FedMEM, which is based on the multi-exit selection mechanism. The algorithm divides the local model into a global layer and a personalized layer, with the global layer participating in federated model aggregation while the personalized layer is retained locally. Simulation results show that compared with FedAvg and FedPre algorithms, the proposed algorithm achieves better results and considerably reduces transmission delay.

Keywords: Mobile computing; personalized federated learning; wireless communication; system heterogeneity; multi-exit mechanism.

1. Introduction

With the advancement of mobile communication technology and the advent of the Internet of Things (IoT) era, more and more mobile devices are connected to the Internet through wireless networks.[1] Sensors and mobile devices send the collected data to the cloud servers via wireless networks and train AI models, which greatly facilitates the development of applications, such as intelligent driving, intelligent transportation, and whole-house intelligence. However, the traditional cloud computing paradigm faces challenges in terms of communication quality, security, and latency. To address these challenges, researchers have proposed the Mobile Edge Computing (MEC),[2] which aims to effectively reduce the latency and communication cost by sinking efficient computation services to the edge of the network, and realize the edge collaborative computation in a distributed approach.

As large amounts of sensitive data are transmitted and processed on cloud servers, the data privacy and security of mobile devices are facing serious threats.[3] As a privacy-preserving distributed learning process, federated learning addresses these challenges to an extent. In federated learning, mobile devices perform training and model updates locally and transfer the updated model parameters to servers for aggregation. The aggregated model parameters are then distributed to mobile devices to update the global model. Since training data are not required to leave the device, federated learning effectively ensures user privacy and data security by avoiding the potential risks associated with uploading data to the cloud. However, federated learning still faces two major types of challenges: system heterogeneity and non-independently and identically distributed (non-IID) data.

Concerning system heterogeneity,[4] Rui *et al.*[5] proposed a model compression algorithm based on model pruning and clustering. However, the data on different nodes may vary significantly, which may lead to unfair training on devices with limited computing resources and thus affect the performance to some extent. Concerning non-IID data,[6] in FedBN,[7] the authors added normalization layers to the neural network and kept the layers local, that is, not participating in the federated learning aggregation operation. Although improving the personalization performance of devices has achieved better results in enhancing the generalization effect of the global model in federated learning, however, local model personalization methods ignore the computational resources of devices. In this chapter, a multi-exit selection mechanism for neural networks based on computational resources is first proposed. In addition, based on the multi-exit selection mechanism, a personalized federated learning training framework FedMEM is proposed, which focuses on the personalized performance of each edge device while preserving the global model performance. The main contributions are as follows:

(1) A neural network multi-exit selection mechanism is developed to allow devices with less computational resources to finish training and output inference results earlier, reducing computational delay for some devices.

(2) A personalized federated learning framework is proposed, which divides the model into a global layer and a personalized layer. Only global layer performs model aggregation, allowing the model to benefit from learning from other devices while maintaining the personalization capability of the local model.

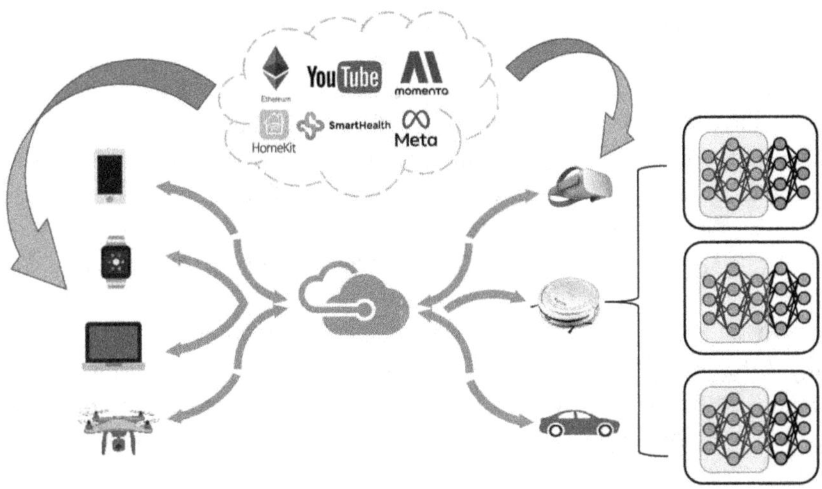

Fig. 1. The architecture of FedMEM

2. Proposed FedMEM Algorithm

In this section, we proposed a multi-exit mechanism for federated learning in MEC systems. The mechanism considers the differences in device hardware computing resources, allowing devices with insufficient resources to terminate training early to overcome system heterogeneity and fault-tolerance challenges. Then, a hierarchical personalized federated learning framework (FedMEM) based on the multi-exit mechanism is proposed, the overall flow is shown in Fig. 1

2.1. *Neural network multi-exit selection mechanism*

Considering a multi-user MEC system includes an edge server and N edge devices $\mathcal{U} = \{1, ..., N\}$ with non-id data distribution and heterogeneous computational resources. To ensure that all local models are trained in a limited time period, we consider a multi-exit selection mechanism for neural networks based on communication and computational resources. This mechanism allows devices with poor computational resources to terminate training early by shallow exit of deep models, an effective exits selection and bandwidth allocation strategy allows the devices to transmit the models to the edge servers within a specified time frame and ensures that the devices participate in model aggregation. Fig. 2 illustrates an schematic of the mechanism.

First, define K model exits as $\theta(x|1), ..., \theta(x|K)$. Our intuition is that deeper models tend to realize better performance and also imply more inference latency. Define the optimization objective of the multi-exits mechanism as minimizing the sum of local training loss and system latency of all devices. Where the training loss of device n can be denoted as $\mathcal{L}(y, \theta(x|k))$. Latency T_{total} is mainly divided into two parts: transmission latency T^t and task computation latency T^c. The transmission latency depends on the bandwidth b_n, the signal-to-noise ratio of the transmission link $SNR_n = \frac{P_n|h_n|^2}{\sigma_n^2}$ and the model size s_k, while the computation latency depends on the CPU cycles per second c_n and the amount of computation g_k. Therefore, the cost function of the multi-exit mechanism can be expressed as:

$$J(\mathbb{B}, \mathbb{K}) = \sum_{i=1}^{N} \mathcal{L}(y, \theta(x|k)) + \sum_{i=1}^{N} \left(\frac{s_k}{b_n log_2(1 + \frac{P_n|h_n|^2}{\sigma_n^2})} + \frac{g_k}{c_n} \right) \tag{1}$$

To construct the optimization equations for the multi-outlet selection mechanism, delay and bandwidth constraints are introduced. In order to motivate more devices to participate in the model aggregation, the maximum delay γ of a round of model aggregation is limited, which is the reason why we only consider local reasoning delay and unilateral transmission delay in our cost function. Meanwhile, by using power control technique, the appropriate carrier bandwidth is dynamically assigned to different devices while the total bandwidth B remains fixed. Therefore, the optimization equation of the multi-exit selection mechanism can be described as:

$$\begin{aligned} \mathbf{P1}: \min_{\mathbb{B}, \mathbb{K}} \quad & J(\mathbb{B}, \mathbb{K}) \\ \text{s.t.} \quad & T^t + T^c \leq \gamma \\ & \textstyle\sum_{1 \leq n \leq N} b_n \leq B \end{aligned} \tag{2}$$

By decomposing the optimization, each client can follow the exit selection strategy to guide the model to output results from different early exits and complete the model's inference in advance.

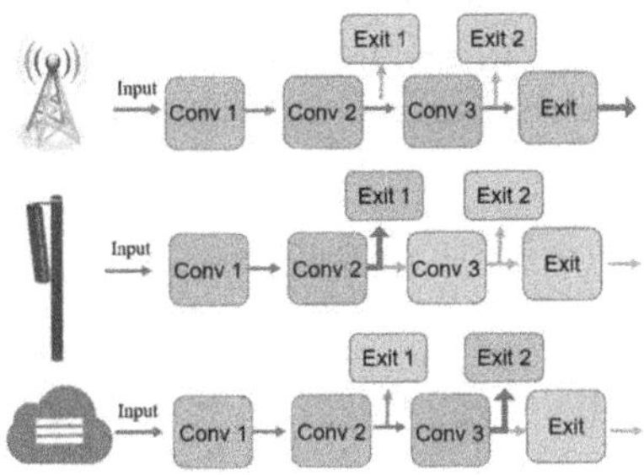

Fig. 2. Multi-exit selection mechanism

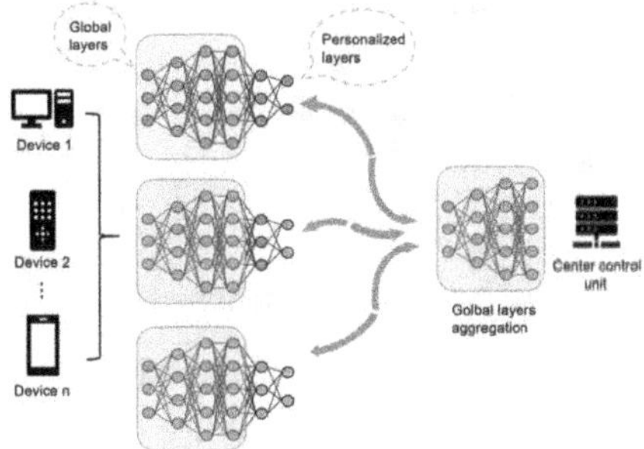

Fig. 3. Personalized federated learning

2.2. *Personalized federated learning framework*

To achieve better personalization performance of local models in federated learning, we divide the local neural network into the global layer W_g and personalized layer W_p, which are inspired by FedPre[8] and FedVF[9], respectively. The global layer is closer to the input and is mainly used for training the global model in federated learning. The personalized layer is closer to the exits and mainly relies on the training effect of local data. For a better division of the global layer and the personalized layer and to mitigate the adverse effects of differences in device hardware resources on federated learning, we utilized the minimum exit layer $l_{min}^{k_n}$ as the division criterion for the global layer and personalized layer. This personalized federated learning method based on the division of the local node computing resources is called FedMEM.

The workflow of FedMEM is shown in Fig. 3, which is mainly divided into four phases: multi-exit selection, local training, edge server aggregation, and local personalized model fine-tuning. First, the edge server computes the exit selection scheme and bandwidth allocation scheme for each device using the multi-exit mechanism. In the local training phase, each device starts training according to the initialized model w_0 and trains the local model according to the multi-exit mechanism. After training is completed, the global layer parameters are uploaded to the edge server and the personalized layer parameters are stored at the device side. In the edge server aggregation phase, the server receives the aggregated global layer model parameters uploaded by the devices, updates the unified global layer model using the weighted average algorithm, and sends the aggregated global model parameters back to each device. Finally, in the local personalized model phase, the local model is updated by combining the aggregated global layer parameters with the locally stored personalized layer parameters.

3. Analysis of Simulation

In this section, we first introduce the simulation settings and implementation details; the performance and communication efficiency of the proposed algorithm are then analyzed and discussed.

3.1. *Dataset and parameter setting*

The dataset used in this paper is CIFAR-10, which contains a total of 60,000 color images, including 50,000 training images and 10,000 testing images. To simulatenon-IID features of the device data, we used the CIFAR10 dataset as an example and assigned each device image from at most 4 categories. In order to validate the effectiveness of the proposed method in different types of neural networks, in this study, we choose a residual neural network, *i.e.* ResNet-34; and lightweight convolutional neural network, *i.e.* MobileNet V1. Models were locally trained with an epoch of 4, learning rate of 0.01, and batch size of 128. For the wireless MEC system, we assumed that there were $N = 10$ edge devices, with each device serving in a dedicated resource block.

Table 1. Experimental result

Model	Method	Accuracy (Epoch)					Inference duration(ms)
		1	10	20	30	40	
ResNet-34	fedavg	61.92	66.20	76.30	81.86	82.31	8.03
	fedpre	64.29	71.50	79.58	82.13	82.45	8.03
	fedmem	**66.49**	**72.70**	**80.03**	**82.75**	**83.16**	**5.18**
MobileNet-V1	defavg	68.60	70.15	73.53	76.03	76.75	4.83
	fedpre	71.24	73.49	75.82	76.67	76.89	4.83
	fedmem	**71.74**	**73.77**	**76.03**	**76.94**	**77.13**	**3.50**

3.2. *Evaluation of model*

We examined the influence of device heterogeneity on model training and aggregation. Devices with lower computing power may perform local training slowly, which affects the efficiency of model aggregation. To simulate this real scenario, we assumed that computing power differs with node device, with hardware parameters being as described earlier. Therefore, we selected fixed models, ResNet and MobileNet, with model depths of 34 and 28 layers, respectively. Early exits were set at layers 20 and 28 for the ResNet model and layers 15 and 24 for the MobileNet model. We documented the training accuracies of the two models on the proposed method versus the BASELINE method, as well as the average inference latency of a single round, and the corresponding results are shown in Table 1. The comparative analysis shows that the proposed personalized federated learning framework consistently outperforms fedavg and fedpre over the 40-epoch training process. Furthermore, a comparison of the average inference time shows an average reduction of 36% for ResNet and 28% for MobileNet. This highlights that the proposed method not only achieves significant performance gains in the training phase, but also offers significant efficiency advantages in the inference phase.

4. Conclusion

In this chapter, we investigated the problem of using personalized MEC in heterogeneous IoT networks. We considered the impact of communication and computation resource differences among different devices on personalized federated learning. To address these challenges, we proposed a multi-exit selection mechanism based on a resource allocation neural network. Subsequently, we proposed a hierarchical personalized federated learning method called FedMEM. The simulation results indicated that FedMEM could improve the personalized performance of federated learning in MEC scenarios while ensuring global performance and significantly reducing communication overhead.

Acknowledgments

This work was supported in part by the National Science and Technology Major Project (2022ZD0119602), National Natural Science Foundation of China (62272114), Major Key Project of PCL (PCL2022A03), the Science and Technology Program of Guangzhou (No. 2024A03J0399), the CCF-

NSFOCUS 'Kunpeng' Research Fund (CCF-NSFOCUS 2023003), Joint Research Foundation of Guangzhou University (202201020380), Guangdong Higher Education Innovation Group (2020KCXTD007), Pearl River Scholars Funding Program of Guangdong Universities (2019).

References

1. S. Iftikhar, S. S. Gill, C. Song, M. Xu, M. S. Aslanpour, A. N. Toosi, J. Du, H. Wu, S. Ghosh, D. Chowdhury *et al.*, Ai-based fog and edge computing: A systematic review, taxonomy and future directions, *Internet of Things* **21**, p. 100674 (2023).
2. B. Ji, Y. Wang, K. Song, C. Li, H. Wen, V. G. Menon and S. Mumtaz, A survey of computational intelligence for 6g: Key technologies, applications and trends, *IEEE Transactions on Industrial Informatics* **17**, 7145 (2021).
3. W. Lu, Y. Ding, Y. Gao, Y. Chen, N. Zhao, Z. Ding and A. Nallanathan, Secure noma-based uav-mec network towards a flying eavesdropper, *IEEE Transactions on Communications* **70**, 3364 (2022).
4. X. Du and Y. Xiao, Energy efficient Chessboard Clustering and routing in heterogeneous sensor networks, *International Journal of Wireless and Mobile Computing* (2007).
5. L. Rui, S. Yang, S. Chen, Y. Yang and Z. Gao, Smart network maintenance in an edge cloud computing environment: An adaptive model compression algorithm based on model pruning and model clustering, *IEEE Transactions on Network and Service Management* **19**, 4165 (2022).
6. Y. Tan, Y. Liu, G. Long, J. Jiang, Q. Lu and C. Zhang, Federated learning on non-iid graphs via structural knowledge sharing, *Proceedings of the AAAI conference on artificial intelligence* **37**, 9953 (2023).
7. X. Li, M. Jiang, X. Zhang, M. Kamp and Q. Dou, Fedbn: Federated learning on non-iid features via local batch normalization, *arXiv preprint arXiv:2102.07623* (2021).
8. M. G. Arivazhagan, V. Aggarwal, A. K. Singh and S. Choudhary, Federated learning with personalization layers, *arXiv preprint arXiv:1912.00818* (2019).
9. Y. Mei, B. Guo, D. Xiao and W. Wu, Fedvf: Personalized federated learning based on layer-wise parameter updates with variable frequency, *2021 IEEE International Performance, Computing, and Communications Conference (IPCCC)*, 1 (2021).

The Influence of the Electric Vehicle Start-stop and Turning Process on Human Perception

Hong Li[*,‡], Yanlin Song[*], Zhong Li[*] and Zhong Li[†,§]

[*]*Beijing Jiaotong University, Beijing 100044, China*

[†]*FernUniversität in Hagen, Hagen, 58097, Germany*

[‡]*hli@bjtu.edu.cn*

[§]*zhong.li@fernuni-hagen.de*

With the increasing development of electric vehicle technology, people have a growing pursuit for the riding comfort of electric vehicles. Compared to fuel vehicles, the driving motor of the electric vehicle has the mechanical characteristics of low-speed constant torque and high-speed constant power, which leads to their stronger accelerations, which also makes the development of electric vehicles face the risk of reduced riding comfort. In order to better explain this issue, this chapter starts from the human body comfort and analyzes the relationship between acceleration during the starting, breaking and turning processes of electric vehicles and the subjective feelings of the human body. Besides, an acceleration model for electric vehicles is established, which provides a detailed analysis for the riding comfort of electric vehicles.

Keywords: Electric vehicles; riding comfort; acceleration model.

1. Introduction

As a mainstream product of new energy vehicles, the high instantaneous acceleration of electric vehicles is still a crucial factor causing discomfort to the human body. Therefore, how to balance the optimization of acceleration performance while ensuring riding comfort in electric vehicles is an important consideration in the field of new energy vehicles nowadays.

Methods for evaluating and analyzing the riding comfort of electric vehicles are constantly being proposed. Reference [1] introduces the heart rate variability as an evaluation indicator for riding comfort, comparing subjective and objective evaluations of riding comfort for passengers in different physiological

[‡]Corresponding author.

states; Ref. [2] proposes a novel evaluation index for bus ride comfort: comprehensive comfort, which can effectively reflect the ride comfort of buses; Ref. [3] obtains a ranking of the advantages and disadvantages of ride vibration comfort for different public transportation vehicles and determines a ride vibration acceleration threshold of 1.071m/s^2; Ref. [4] introduces subjective and objective evaluations of driving performance of automobiles and develops a software for calculating driving performance indicators.

2. Comparison and Analysis of Electric Vehicles and Fuel Vehicles

This section first conducts a comparison of electric vehicles and fuel vehicles, which makes it more accurate to identify the key factors different from fuel vehicles that affect the human body.

2.1. *Power sources*

The starting torque during the starting process of an internal combustion engine is relatively small, which takes some time to increase to the maximum torque; however, electric vehicles are driven by electric motors and have mechanical characteristics of low-speed constant torque and high-speed constant power.[5] Therefore, during the starting process of an electric vehicle, the motor first works in the constant torque area. At this point, the electric cvehicle has the maximum starting torque, which can achieve rapid acceleration.

The mechanical characteristics of motors can be represented by Eq. (1), where T is the motor output torque and n is the motor speed.

$$P = \frac{\pi}{30000} T * n. \tag{1}$$

2.2. *Acceleration performance*

Shock and Vibration Handbook mentions that people will experience discomfort when the acceleration reaches 0.45g and lasts for 3-5 seconds,[6] and when the average acceleration of vehicles exceeds 0.45g, the corresponding 0-100km/h acceleration time should not exceed 6.3 seconds.[7] Meanwhile, It can be concluded that the average 0-100km/h acceleration time of electric vehicles is about 6 seconds, with an average acceleration of 0.46g; and the average acceleration time of fuel vehicles is about 10 seconds, corresponding to an average acceleration of 0.28g.[7] According to the above data, it can be seen that electric vehicles have better acceleration performance then fuel vehicles, while are also more likely to cause discomfort to the human body.

3. Comfort Evaluation Methods for Electric Vehicles

3.1. *Evaluation indicators for acceleration of starting process*

ISO2631-1-1997[8] proposes a riding comfort standard to determine the subjective perception of the human body based on the root mean square value of the total weighted acceleration obtained from the frequency weighting function of each axis.[1] The specific correspondence is shown in Table 2.

Table 2. The relationship between total weighted acceleration root mean square value av and human subjective sensation.

Total weighted acceleration root mean square value a_v (m/s^2)	Subjective sensation of the human body	Total weighted acceleration root mean square value a_v (m/s^2)	Subjective sensation of the human body
Less than 0.315	Comfortable	0.8~1.6	Uncomfortable
0.315~0.63	Slightly uncomfortable	1.25~2.5	Very uncomfortable
0.5~1	Quite uncomfortable	More than 2	Entirely uncomfortable

3.2. *Evaluation indicators for acceleration of turning process*

During the turning process, vehicles will experience axial and lateral acceleration, which will significantly affect the subjective perception of the human body and thus affect the ride comfort of vehicles. A common evaluation method is to use lateral acceleration as indicator, as shown in Table 3.[9]

Table 3. The relationship between lateral acceleration ah and subjective sensation of the human body.

Lateral acceleration a_h (m/s^2)	Subjective sensation of the human body
Less than 1.8	Comfortable
1.8~3.6	Slightly uncomfortable
3.6~5.0	Very uncomfortable
More than 5.0	Entirely uncomfortable

3.3. *Evaluation indicators for acceleration of braking process*

Table 4.[10] shows the deceleration of a 1340kg electric vehicle from an initial speed of 50 km/h at different maximum regenerative braking forces, which have extremely low riding comfort, usually used in emergency situations.

The larger the maximum regenerative braking power of an electric vehicle, the faster the average braking deceleration during deceleration and the worse the riding comfort. For passenger cars, the deceleration acceleration is suggested not to exceed 1.0m/s^2, that is, the deceleration action under the condition that the maximum recuperated power does not exceed 10 kW belongs to smooth braking.

Table 4. The relationship between the maximum regenerative braking power and the braking deceleration during the braking process of electric vehicles.

Maximum regenerative braking power (kW)	Deceleration time (s)	Maximum deceleration (m/s^2)
5	6.9	59
10	10.0	35
15	16.5	33

4. The Acceleration Model for Electric Vehicles

Electric vehicles meet the EV driving force balance equation during operation:[12]

$$F_t = F_f + F_w + F_i + F_j. \tag{2}$$

F_t *is* the driving force for electric vehicles, and its calculation formula is:

$$F_t = \frac{T_e i_t \eta_t}{r}, \tag{3}$$

F_w is the air resistance ,and its calculation formula is:

$$F_w = \frac{C_D A u^2}{21.15}, \tag{4}$$

F_j is the acceleration resistance, and its calculation formula is:

$$F_j = \delta m \frac{du}{dt}, \tag{5}$$

F_f and F_i are respectively the rolling resistance and the slope resistance, and their calculation formula is:

$$F_f = mgf \cos \alpha_G, \tag{6}$$

$$F_i = mg \sin \alpha_G, \tag{7}$$

Since this model considers only the acceleration process of the electric vehicle on the horizontal road, so it can be considered that α_G=0, so the ramp resistance during the driving of the electric vehicle F_i=0.

Then the acceleration model can be established by Eq. (2)-(7):

$$a_j = \frac{du}{dt} = \frac{F_t - F_f - F_w}{\delta m}. \tag{8}$$

The symbols in Eq. (2)-(8) are explained in Table 5:

Table 5. Summary of the parameters in 4th section.

Symbol	Meaning	Symbol	Meaning
T_e	Output torque	u	Speed
i_t	Total transmission ratio	δ	Conversion coefficient for rotational mass
η_T	Transmission efficiency	m	Overall mass
r	Rolling radius	f	Rolling resistance coefficient (generally 0.01)[11]
C_D	Air resistance coefficient	g	Gravitational acceleration
A	Windward area	α_G	*Slope angle*

5. Simulation

5.1. *Simulation results of the electric vehicle's dynamic characteristics*

The program design and model building are completed on MATLAB, and the balance diagram of driving force and driving resistance, speed-time diagram and acceleration-time diagram of the electric vehicle are shown in Figure 1, Figure 2, and Figure 3.

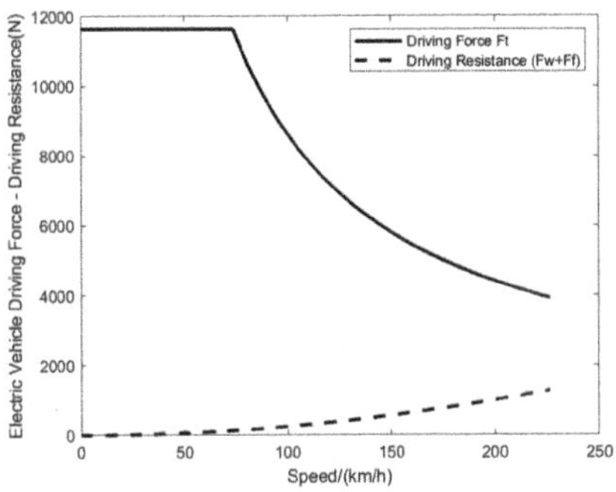

Fig. 1. Driving force-driving resistance balance diagram.

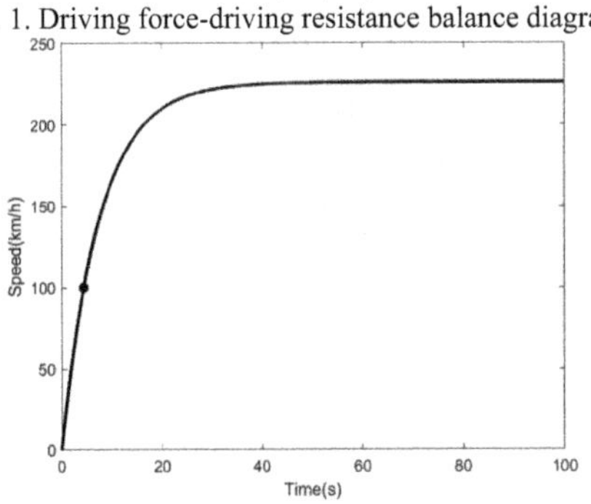

Fig. 2. Speed-time relationship diagram.

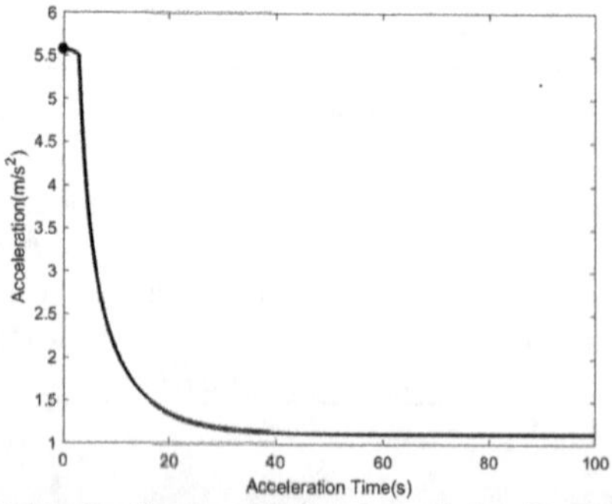

Fig. 3. Acceleration-acceleration time relationship diagram.

The 0-100km/h acceleration time and maximum instantaneous acceleration of the electric vehicle are related to the setting of parameters such as the rotational inertia, damping coefficient, and motor gain. As shown in the figures, the maximum speed of the electric vehicle calculated under the simulation parameters in Table 6 is 226.200 km/h, with a 0-100km/h acceleration time of 4.435s, and a maximum instantaneous acceleration of 5.576 m/s^2.

5.2. *Factors influencing the maximum instantaneous acceleration of the electric vehicle*

The simulation results show that the acceleration of the electric vehicle has a maximum value at the initial moment. At this time, the output torque should be equal to the peak torque T_m. The air resistance experienced F_w is zero. According to (2) - (8), the calculation formula for the maximum instantaneous acceleration a_{max} of the electric vehicle can be expressed as follows:

$$a_{\max} = \frac{F_t - F_f}{\delta m} = \frac{1}{\delta m}\left(\frac{T_m i_t \eta_t}{r} - mgf\right) \cdot \tag{9}$$

From this, the relationship curves between the maximum instantaneous acceleration of the electric vehicle and the peak torque or vehicle mass can be obtained, as shown in Fig. 4 and 5.

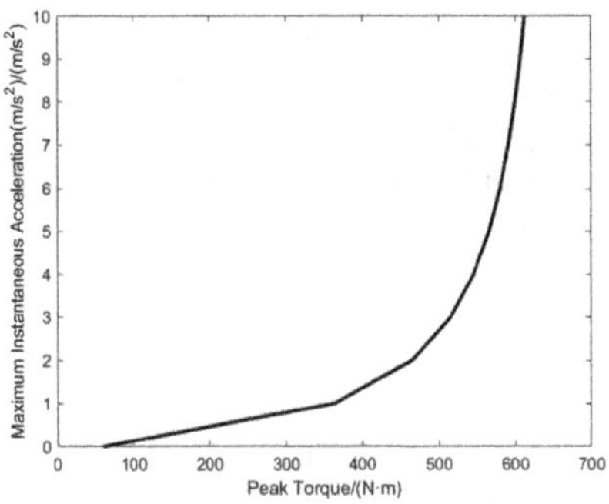

Fig. 4. Peak torque-the maximum instantaneous acceleration relationship diagram.

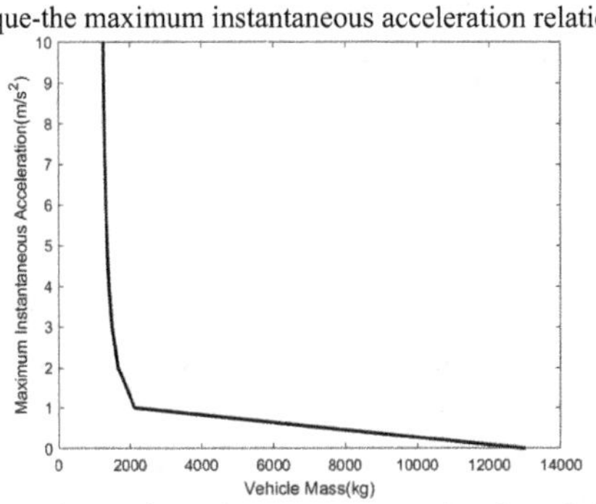

Fig. 5. Vehicle mass-the maximum instantaneous acceleration relationship diagram.

Therefore, the riding comfort of electric vehicles is influenced by factors such as peak torque, vehicle mass, and total transmission ratio. The specific correlations are shown in Table 6.

Table 6. Factors affecting the maximum instantaneous deceleration of electric vehicles.

Factor	Correlation
Vehicle mass	Negative
Peak torque	Positive
Total transmission ratio	Positive

6. Conclusion

This chapter has studied the impact of the start-stop and operation process of electric vehicles on human perception and analyzed the differences in power sources and acceleration performance between electric vehicles and fuel vehicles. Meanwhile, the comfort evaluation indicators of electric vehicles in different speed-changing situations have been proposed, providing a basis for evaluating the comfort of electric vehicle riding. Besides, the results of the electric vehicle acceleration model remarked that the maximum instantaneous acceleration of electric vehicles is related to parameters such as peak torque, vehicle mass and total transmission ratio, which also affect the riding comfort of electric vehicles. This model has a wide range of applications and provides certain reference significance for the design and parameter setting of electric vehicles.

References

1. Y. Li, Research on Ride Comfort of Pure Electric Bus Based on Occupant' s Heart Rate Variability. Chang' an University, 2022. DOI:10.26976/d.cnki.gchau.2021.001535.
2. W. Ren, Research and Application of Evaluation Method for Ride Comfort of Bus. Beijing Jiaotong University, 2012.
3. C. Yan, Study on Vibration Riding Comfort of Public Transportation Based on Physiology and Psychology Indicators. Beijing Municipal Engineering Research Institute, 2015.
4. Y. Li, C. Ye, W. Mao, Drivability evaluation of vehicle acceleration conditions, 2021(04):84-87. DOI:10.13612/j.cnki.cntp.2021.04.027.
5. M. Zhou, Study on Control Strategy for Electric Vehicle at Drive Condition. Chongqing University, 2013.
6. C.M.Harris.C.E.Crede.Shock and Vibration Handbook. New York. McGraw-Hill, 1976.
7. J. Zhang, S. Lu, X. Zhang, Application and Strategy Research of Acceleration in Electric Vehicles. 2022(18):118-120.
8. ISO 2631-1:1997. Mechanical vibration and shock-Evaluation of human exposere to whole-body vibration-Part 1: General requirements.
9. H. Gu, The Study of Highway Alignment Comfortableness Evaluation Method Based on the Lateral Acceleration Rate. South China University of Technology, 2011.
10. Vyacheslav Rakov. Determination of optimal characteristics of braking energy recovery system in vehicles operating in urban conditions. Vologda State University, 2020.
11. GB/T18385-200. Test Methods for Power Performance of Electric Vehicles
12. LI Kai. Simulation and Analysis of Power Performance of Pure Electric Vehicles Based on MATLAB. Automobile Applied Technology, 2023, 48(9) : 1-5.

Harmonic State-Space Modeling and Frequency Coupling Analysis for PLL-tied VSCs

Shengyu Chen*, Samson Shenglong Yu†, Guidong Zhang*, Zhong Li‡ and Yun Zhang*

*School of Automation, Guangdong University of Technology, Guangzhou 510006, China

†School of Engineering, Deakin University, Melbourne, Australia

‡Faculty of Mathematics and Computer Science, FernUniversitaet in Hagen Hagen, Germany

To mitigate unpredictable stability challenges arising from frequency coupling phenomena in voltage-source converter (VSC) systems within renewable energy sources, this chapter employs an innovative Harmonic State-Space (HSS) modeling approach based on linear time-periodic (LTP) theory. The proposed model can capture the complexities of a VSC incorporating phase-locked loop (PLL) and time-delay components. By considering the multi-frequency response properties of input and output signals, the HSS model facilitates a comprehensive analysis of the interplay among various harmonic frequencies. Through the derivation of the harmonic transfer function matrix for the VSC system, the findings unveil noteworthy harmonic coupling components evident in the non-diagonal elements. This underscores the substantial influence of harmonic coupling admittance and emphasizes potential inaccuracies in harmonic response analyses employing traditional linear time-invariant (LTI) models. Finally, the model's precision is validated through time-domain simulations.

Keywords: Harmonic state-space model; voltage-source converter; linear time-periodic; harmonic coupling.

1. Introduction

With the increasing growth of renewable energy sources like wind and solar power, VSCs are commonly used in current power systems as typical power electronics equipment. Electronic power technology guarantees superior conversion efficiency and precise control over loads and sources, but it frequently displays unpredictable behavior. Oscillation problems in VSC grid-connected systems have increased in recent years, presenting a significant risk to the system's secure and stable functioning.[1,2]

Rygg *et al.*[3] identified the mirror frequency coupling phenomenon in grid-connected inverter systems. This phenomenon shows that a sinusoidal disturbance signal at a particular frequency produces current responses at the same and different frequencies. Nian *et al.*[4,5] stated that the design of PLL and the structure of unbalanced current controllers provide substantial off-diagonal components in the converter impedance matrix, resulting in a frequency coupling impedance model. However, these models do not account for the interaction of harmonics of varying orders.

HSS modeling expands the variables of the state equation in the form of a fourier series, transforming the time-domain model into a LTP system in the frequency domain using Euler's formula and harmonic balance processing. Kwon *et al.*[6] employed HSS theory to model and analyze conventional AC/DC converters, investigating the harmonic relationship between DC side voltage, AC side current, and grid voltage. However, previous studies of the HSS model have been relatively limited in addressing the effect of the PLL, a key nonlinear control factor that significantly contributes to the frequency coupling phenomena. Moreover, existing studies have mainly focused on the frequency coupling patterns of individual converters, neglecting the frequency coupling issues on the renewable side.

In this chapter, we address the integration of VSC for renewable energy into the grid by investigating the frequency coupling model using the HSS theory. We present an HSS model that incorporates PLL and time-delay components. The model illustrates the coupling between input and output variables and quantitatively analyzes the interaction effects of frequency coupling. Through comparative simulations and mathematical model validation, we affirm the accuracy of the proposed model.

2. Principles of Harmonic State-Space Modeling

The goal of HSS theory is to establish a LTP dynamic system analogous to a LTI system. This is achieved by transforming the time domain into the frequency domain, thereby achieving the steady-state representation of time-varying models. The state-space equations for a LTP system are expressed as follows.

$$\begin{cases} \dot{x}(t) = A(t)\,x(t) + B(t)\,u(t) \\ y(t) = C(t)\,x(t) + D(t)\,u(t) \end{cases} \tag{1}$$

where, $x(t)$, $u(t)$, and $y(t)$ denote the state, input, and output variables of the system, respectively. $A(t)$, $B(t)$, $C(t)$ and $D(t)$ are the coefficient matrices of the equation of state, which are time-dependent and periodic. For any periodically varying signal, it can be expressed by a fourier transform as follows.

$$\begin{aligned}&\sum_{k=-\infty}^{\infty} (jk\omega + s) X_k e^{(jk\omega+s)t} \\ &= \sum_{k,m=-\infty}^{\infty} A_{k-m} X_m e^{(jk\omega+s)t} + \sum_{k,m=-\infty}^{\infty} B_{k-m} U_m e^{(jk\omega+s)t}\end{aligned} \tag{2}$$

According to the harmonic balance principle, integrating the same frequency components on both sides of the equation yields the HSS model as follows.

$$\begin{cases} sX = (\Gamma(A) - N) X + \Gamma(B) U \\ Y = CX + DU \end{cases} \tag{3}$$

$$\begin{cases} X = [X_{-k}, \ldots, X_{-1}, X_0, X_1, \ldots, X_k] \\ U = [U_{-k}, \ldots, U_{-1}, U_0, U_1, \ldots, U_k] \\ N = diag[-jk\omega, \ldots, -j\omega, 0, j\omega, \ldots, -jk\omega] \end{cases} \tag{4}$$

Where, X represents the frequency domain information of the state variable, U represents the frequency domain information of the input variable, and N is a diagonal matrix. $\Gamma(A)$ and $\Gamma(B)$ represent Toeplitz matrices constructed from the fourier decomposition of matrices $A(t)$ and $B(t)$ These matrices are the harmonic coefficient matrices obtained through fourier decomposition.

3. HSS Modeling of VSC Grid-connected Systems

The typical topology and control structure of a VSC grid-connected system are illustrated in Fig.1. The renewable energy station is connected to the infinite power grid through inverters and transmission lines. In the figure, P_{in} denotes the active power input of the renewable energy station, P_{g} denotes the active power output of the renewable energy station, u_{dc} denotes the DC bus voltage, C_{dc} denotes the DC-side stabilizing capacitor, C_f and L_f denote the output filtering capacitor and damping resistor, respectively. ω_0 is the fundamental frequency of the grid.

3.1. *Modeling of main circuit*

For the DC bus section of the inverter, the active power flow into and out of the VSC's DC bus determines the dynamics of the DC bus voltage.

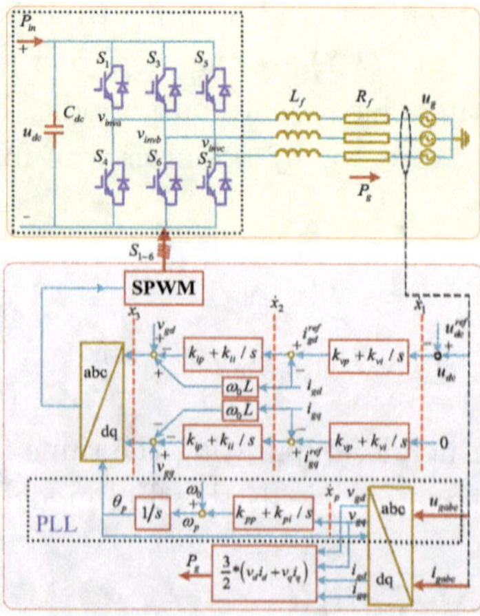

Fig. 1. Topology and control structure of a VSC grid-connected system

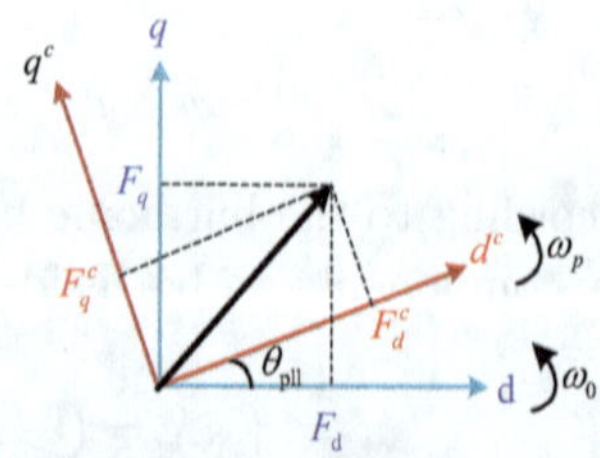

Fig. 2. The Relationship between the synchronous dq-coordinate system and the control dq-coordinate system

Therefore, the relationship for the VSC's DC bus section can be expressed as follows.

$$\begin{cases} C_{dc}U_{dc}\frac{dU_{dc}}{dt} = P_{in} - P_g \\ P_g = \frac{3}{2}\left(u_{gd}i_{gd} + u_{gq}i_{gq}\right) \end{cases} \tag{5}$$

The VSC is connected to the grid through series inductors and resistors. The dynamic model can be expressed as.

$$\begin{cases} L_{fa}\frac{di_{ga}}{dt} = v_{inva} - v_{ga} - R_{fa}\cdot i_{ga} \\ L_{fb}\frac{di_{gb}}{dt} = v_{invb} - v_{gb} - R_{fa}\cdot i_{gb} \\ L_{fc}\frac{di_{gc}}{dt} = v_{invc} - v_{gc} - R_{fa}\cdot i_{gc} \end{cases} \tag{6}$$

Define E_n, Z_n, and N_n to denote the (2k + 1)-dimensional unit, zero, and diagonal matrices, respectively. Following the modeling approach introduced in Section 2, $X_m = \left[U_{dc}\ i_{ga}\ i_{gb}\ i_{gc}\right]^{\mathrm{T}}$, and $U_m = \left[P_{in}\ P_g\ u_{inva}\ u_{inva}\ u_{inva}\ u_{ga}\ u_{gb}\ u_{gc}\right]^{\mathrm{T}}$.Eqs. (5)-(6) can be organized into the form of HSS equations and expressed in the form of matrices as follows.

$$\dot{X}_{\mathrm{m}} = A_{\mathrm{m}}X_{\mathrm{m}} + B_{\mathrm{m}}U_{\mathrm{m}} \tag{7}$$

$$A_m = \begin{bmatrix} E_n - N_n & Z_n & Z_n & Z_n \\ Z_n & -\frac{R_{fa}}{L_{fa}}E_n - N_n & Z_n & Z_n \\ Z_n & Z_n & -\frac{R_{fb}}{L_{fb}}E_n - N_n & Z_n \\ Z_n & Z_n & Z_n & -\frac{R_{fc}}{L_{fc}}E_n - N_n \end{bmatrix} \tag{8}$$

$$B_m = \begin{bmatrix} \frac{1}{C_{dc}U_{dc0}}E_n & -E_n & Z_n & Z_n & Z_n & Z_n & Z_n & Z_n \\ Z_n & Z_n & \frac{1}{L_{fa}}E_n & Z_n & Z_n & \frac{1}{L_{fa}}E_n & Z_n & Z_n \\ Z_n & Z_n & Z_n & \frac{1}{L_{fb}}E_n & Z_n & Z_n & -\frac{1}{L_{fb}}E_n & Z_n \\ Z_n & Z_n & Z_n & Z_n & \frac{1}{L_{fc}}E_n & Z_n & Z_n & -\frac{1}{L_{fc}}E_n \end{bmatrix} \tag{9}$$

3.2. *PLL and coordinate transformation links*

The synchronization between the VSC and the power grid is achieved through a conventional PLL structure, as illustrated in Fig.1. The system employs a d-axis voltage-oriented control strategy, and its mathematical relationship can be expressed as follows.

$$\begin{cases} \dot{x}_p = v_{gq} \\ \dot{\theta}_p = \Delta\omega_p = k_{pp}v_{gq} + k_{pi}x_p \end{cases} \tag{10}$$

Define $X_{pll} = [X_p, \theta_p]^{\mathrm{T}}$,$U_{pll} = [V_{gq}]$, The HSS formula can be stated as

$$\dot{X}_{pll} = A_{pll}X_{pll} + B_{pll}U_{pll} \tag{11}$$

where, $A_{pll} = [E_n - N_n, Z_n; k_{pi}E_n, -N_n]$,$B_{pll} = [E_n; k_{pp}E_n]$.

v_{gq} is obtained from v_{gabc} by a series of coordinate transformations. First, the collected three-phase voltages v_{gabc} are transformed by the classical park transformation. To represent matrix T_{abc-dq} as a Toeplitz matrix, it can be expressed as follows:

$$\Gamma\left(T_{abc-dq}\right) = \frac{2}{3}\begin{bmatrix} \Gamma\left[\cos\omega_0 t\right] & \Gamma\left[\cos\left(\omega_0 t - \frac{2}{3}\pi\right)\right] & \Gamma\left[\cos\left(\omega_0 t + \frac{2}{3}\pi\right)\right] \\ \Gamma\left[\sin\omega_0 t\right] & \Gamma\left[\sin\left(\omega_0 t - \frac{2}{3}\pi\right)\right] & \Gamma\left[\sin\left(\omega_0 t + \frac{2}{3}\pi\right)\right] \end{bmatrix} \tag{12}$$

The main network and control system of the VSC are modeled in different coordinate systems. As shown in the Fig.2, the main network is represented in the system's *dq* coordinate system, which is synchronized with the ideal AC power grid. Meanwhile, the control system is modeled in the control *dq* system, which is synchronized with the PLL. F is the voltage or current vector and θ_{pll} denotes the phase difference between coordinate systems. The superscript c indicates the control coordinate system.

$$\begin{bmatrix} F_d^c \\ F_q^c \end{bmatrix} = \begin{bmatrix} \cos\theta_{pll} & \sin\theta_{pll} \\ -\sin\theta_{pll} & \cos\theta_{pll} \end{bmatrix} \begin{bmatrix} F_d \\ F_q \end{bmatrix} \tag{13}$$

The grid current i_{gabc} sampled by the controller also needs to be transformed into the control dq coordinate system using the same method, employing the T_{abc-dq} and T_{m-c} transformations.

$$\begin{bmatrix} V_{gd} \\ V_{gq} \end{bmatrix} = \Gamma\left(T_{m-c}\right) \begin{bmatrix} V'_{gd} \\ V'_{gq} \end{bmatrix}, \begin{bmatrix} V'_{gd} \\ V'_{gq} \end{bmatrix} = \Gamma\left({T_{m-c}}^{-1}\right) \begin{bmatrix} V_{gd} \\ V_{gq} \end{bmatrix} \tag{14}$$

3.3. *Time delay of digital control system*

In this paper, we use the first-order Pade approximation to represent the time-delay link $e^{-\tau \cdot s}$, and obtain the following state-space expression for the time-delay link of the modulated signal. $\tau = 1.5T_s$ is the delay time generated by the digitally computed delay (T_s) and the pulse width modulation delay ($0.5T_s$) and T_s is the inverter sampling period.

$$\dot{v}^d_{invabc} = -\frac{2}{1.5 \cdot T_s} v^d_{invabc} + v'_{invabc}, v_{invabc} = \frac{4}{1.5 \cdot T_s} v^d_{invabc} - v'_{invabc} \tag{15}$$

v^d_{invabc} and v_{invabc} are the state variables and modulated signal after delay, respectively. To reveal the interactions between the delay and other internal states, the delay element can be converted to HSS form, as shown below.

$$\dot{X}_d = A_d X_d + B_d U_d \tag{16}$$

where $X_d = \left[V^d_{inva}, V^d_{invb}, V^d_{invc}\right]^{\mathrm{T}}$, $U_d = \left[V'_{inva}, V'_{invb}, V'_{invc}\right]^{\mathrm{T}}$

$$A_d = \begin{bmatrix} \frac{-2}{1.5T_s}E_n - N_n & Z_n & Z_n \\ Z_n & \frac{-2}{1.5T_s}E_n - N_n & Z_n \\ Z_n & Z_n & \frac{-2}{1.5T_s}E_n - N_n \end{bmatrix}, B_d = \begin{bmatrix} E_n \, Z_n \, Z_n \\ Z_n \, E_n \, Z_n \\ Z_n \, Z_n \, E_n \end{bmatrix} \tag{17}$$

4. Simulation Results

This section simulates the various components of the HSS model described in Section 3 and concatenates these components into a full HSS model based on the relationship between input and output variables. The above procedure was implemented using MATLAB with the m-code. For the matrix forms of the HSS given by Eq.(3), we can obtain the following expression.

$$Y = HU = \left[C(sI - A)^{-1}B + D\right]U \tag{18}$$

The matrix H has a Toeplitz matrix form and is an n-order square matrix. The main diagonal elements of the matrix represent the LTI components, while the off-diagonal elements represent the coupling frequency components. Fig.3 and Fig.4 illustrate the impact of grid-side perturbation volt-

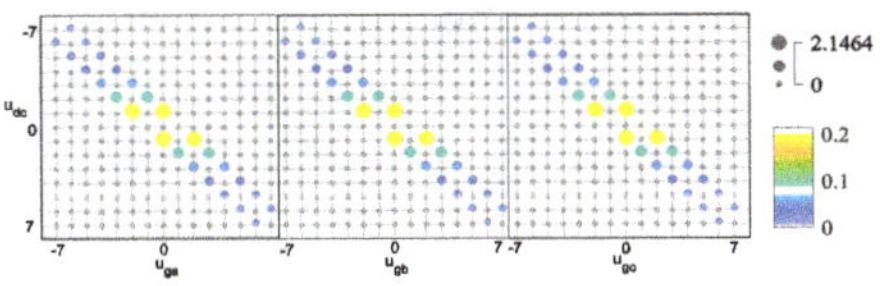

Fig. 3. Diagram of harmonic coupl -ing strength between u_g and u_{dc}

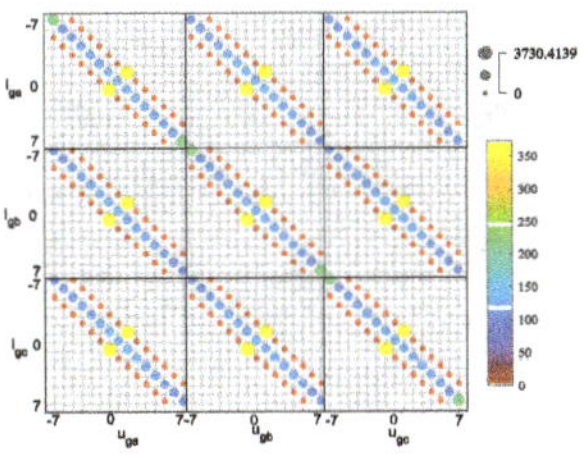

Fig. 4. Diagram of harmonic coupling strength between u_g and i_g

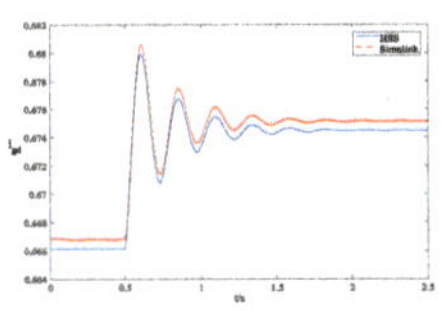

Fig. 5. Diagram of harmonic coupling strength between u_g and i_g

age on DC-side current and the effect of the grid-side interference voltage on the grid-side current, respectively.The axes indicate the kth harmonics. It is apparent that when there is a voltage perturbation with an angular frequency ω_n at the grid side, harmonic coupling components with an angular frequency $\omega_n \pm 2\omega_0$ appear on the DC side. As the frequency increases, the strength of the harmonic coupling gradually decreases, but the harmonic coupling around the main diagonal remains non-negligible. Especially near the fundamental frequency, significant harmonic coupling components are generated. This suggests that the conventional LTI approach may lead to considerable errors during the analysis.

A time-domain simulation model of a three-phase VSC system is built in Simulink to compare the accuracy of the HSS model. The time-domain response of the AC-side grid current i_{gd} in the three-phase VSC system is depicted in Fig. 5 when the active power increases by 10 MW. It can be seen that the values found using the HSS model agree well with the time-domain waveforms found using Simulink simulations.

5. Conclusion

In this chapter, we utilize the HSS approach grounded in LTP theory to conduct a comprehensive mathematical modeling of VSC within the realm of renewable energy grid-connected systems. Notably, the study fully incorporates the effects of PLL and time delay components. By deriving the harmonic transfer function matrix, our research characterizes the harmonic coupling relationships intrinsic to the VSC, offering a detailed analysis of its harmonic coupling mechanism. Remarkably, the HSS model, by accounting for coupling frequency components across various harmonic orders, exhibits superior modeling accuracy when compared to conventional LTI models.

References

1. J. Shair, X. Xie, L. Wang, W. Liu, J. He and H. Liu, Overview of emerging subsynchronous oscillations in practical wind power systems, *Renewable and Sustainable Energy Reviews* **99**, 159 (2019).
2. S. Zhao, N. Wang, R. Li, B. Gao, B. Shao and S. Song, Sub-synchronous control interaction between direct-drive pmsg-based wind farms and compensated grids, *International Journal of Electrical Power & Energy Systems* **109**, 609 (2019).
3. A. Rygg, M. Molinas, C. Zhang and X. Cai, A modified sequence domain impedance definition and its equivalence to the dq-domain impedance definition for the stability analysis of ac power electronic systems, *IEEE Journal of Emerging & Selected Topics in Power Electronics* **4**, 1383 (2016).
4. H. Nian, L. Chen, Y. Xu, H. Y. Huang and J. C. Ma, Sequences domain impedance modeling of three phase grid-connected converter using harmonic transfer matrices, *IEEE Transactions on Energy Conversion* **PP**, 1 (2017).
5. H. Nian, B. Hu, C. Wu, L. Chen and F. Blaabjerg, Analysis and reshaping on impedance characteristic of dfig system based on symmetrical pll, *IEEE Transactions on Power Electronics* **PP**, 1 (2020).
6. J. B. Kwon, X. Wang, F. Blaabjerg, C. L. Bak, V. S. Sularea and C. Busca, Harmonic interaction analysis in a grid-connected converter using harmonic state-space (hss) modeling, *IEEE* (2017).

A New Fuzzy Hypergeometric Naive Bayes Network and Its Application for User's Assessment in Virtual Reality Simulators

Isaac S. L. Araújo[*,‡], Jodavid A. Ferreira[*,§], Liliane S. Machado[†,¶], and Ronei M. Moraes[†,||]

[*]*Laboratory of Technologies for Virtual Teaching and Statistics, Federal University of Paraíba João Pessoa, Paraíba, Brazil*
[†]*Graduate Program in Decision Models and Health, Federal University of Paraíba João Pessoa, Paraíba, Brazil*
[‡]*isaac.sebastian@academico.ufpb.br*
[§]*jodavid@protonmail.com*
[¶]*liliane@di.ufpb.br*
[||]*ronei@de.ufpb.br*

Assessment methods have been used to identify the quality of procedures performed by human beings. In virtual reality simulators, user's interaction data can be collected and used by systems named Single User Assessment Systems (SUAS) in order to assess the user's performance. In particular, some procedures can be considered well performed when the same objective is achieved v times within a limited universe of attempts. In this case, the data can be modeled by a hypergeometric distribution. This chapter presents the proposal of a new fuzzy hypergeometric naive Bayes network as basis for an SUAS, which can be used in virtual reality simulators. The results showed that this new SUAS was able to achieve better results when compared to other SUASs based on different naive Bayes networks. That new proposed network can also be used for data classification and machine learning applications.

Keywords: Fuzzy hypergeometric naive Bayes; hypergeometric distribution; user's assessment; virtual reality.

1. Introduction

Assessment plays an important role in identifying the quality of procedures performed by human beings. In a scenario in which technologies allow the reproduction of real situations in simulated environments, training procedures in virtual reality has become an important resource in the educational process. Virtual reality (VR) simulators can collect user's interaction data by devices that read eye movements, get user's spatial position or capture the number of virtual object selections, among other

[‡]Corresponding author.

possibilities. Therefore, the different particularities of a procedure require the choice and use of specific devices and data collection can occur at different granularities of time, depending on the device used and its software interfaces. Some devices offer a sampling rate of 1000Hz, which can result in millions of information collected in just a few minutes of simulation.[4]

The use of assessment systems coupled to these simulators makes it possible to provide a response about the quality of the procedure carried out in the simulated environment.[8] These systems use data from the virtual environment and user's interactions and they are named Single User Assessment Systems (SUAS). Considering the need to provide real-time responses, several assessment systems have been proposed in recent years.[3,5–7] However, the choice of a method depends on the nature of the procedure since each procedure generates datasets that follow specific statistical distributions. By considering the statistical distribution of data, it is possible to provide more accurate assessment systems.

This chapter contributes by proposing a methodology based on fuzzy Bayesian networks that use hypergeometric distribution for data modeling. This methodology considers that the procedure is well performed when the same objective is achieved v times within a limited universe of attempts but with the probability of success changing at each try. In this case, the data follow a hypergeometric distribution.[2] The following sections describe the theoretical basis for that new SUAS, the results obtained by its application and a brief comparison with other SUASs.

2. Statistical Modeling

2.1. *Hypergeometric distribution*

Consider a set with N objects of which r are of type I and $N - r$ are of type II. A sample is randomly chosen without replacement, with size n ($n < N$), and we define X as the number of objects with characteristic I in the sample. In this case, we say that the random variable X follows the hypergeometric distribution with parameters r, N, n, denoted $X \sim Hgeo(r, N, n)$[2].

Thus, the hypergeometric distribution is a discrete probability distribution that describes the probability of v successes in n draws from a finite population without replacement, where each draw is either a success or a failure. This distribution has applications in scenarios involving sampling without replacement from a finite population that contains two types of elements. The probability function of the hypergeometric distribution is given by:

$$P(X = v) = \frac{\binom{r}{v}\binom{N-r}{n-v}}{\binom{N}{n}} \tag{1}$$

where, N is the population size, r is the number of success states in the population, n is the number of draws, v is the number of observed successes.

2.2. *Mathematical properties*

In this subsection, important results for the hypergeometric distribution are presented, such as the expectation and variance of the distribution. Thus, let $X \sim Hgeo(r, N, n)$, its expectation is given by:

$$E(X) = n\frac{r}{N} = np, \tag{2}$$

where $p = \frac{r}{N}$, and the variance of hypergeometric distribution is given by

$$V(X) = E(X^2) - (E(X))^2 = n\frac{r}{N}\left(1 - \frac{r}{N}\right)\frac{N-n}{N-1} = npq\frac{N-n}{N-1}, \tag{3}$$

where, $q = 1 - p = (1 - \frac{r}{N})$.

2.3. *Parameters estimation*

Let $X_1, \ldots, X_n$ a random sample i.i.d. of a random variable $X \sim Hgeo(\theta)$, such that $\theta = [r, N, n]^\top$ is a hypergeometric distribution parameter vector. Assuming the model is being used in a simulator with information obtained during training time, then in this way, we have that the value of the parameter N is always known, being represented by the maximum number of attempts made by an experienced professional within certain evaluation levels. Thus, we have that $\hat{\theta} = [\hat{r}, \hat{n}]^\top$ is obtained by[11]

$$\hat{r} = \frac{\overline{X}(N+1)}{\hat{n}} \tag{4}$$

where $\overline{X} = \sum_{i=1}^{t} X_i/t$ is the expectation estimator $(E(X))$, where t is the sample size, and $\hat{n}$ is estimated obtaining the equation roots, by

$$-\overline{X}\hat{n}^2 + \left(\overline{X}(\overline{X} + N) - S^2(N+1)\right)\hat{n} - \overline{X}^2 N = 0 \tag{5}$$

The roots of equation (5) represent the estimate of the parameter n. It is important to remember that equation 5 normally has two roots that are complex conjugates, thus, the estimative to n will be represented by the modulus of a complex number of one of the roots.

3. A New SUAS based on Fuzzy Hypergeometric Naive Bayes Network

A Naive Bayes network assumes the hypothesis that information about an event is not dependent of other events, i. e., the variables are conditionally independent. A Naive Bayes network assumes multinomial distribution for its variables.

Formally, let $\mathbf{x}_i = \{X_{i1}, X_{i2}, \ldots, X_{ik}\}$ be a random vector of data in the i-th sample with k-information (dimension/variables) obtained when training is performed. Let $w_j, j \in \Omega$ be the performance class most likely to be chosen, where the set $\Omega = 1, ..., M$ and M is the total number of performance classes for the assessment of a user in the virtual reality simulator.

The probability of the class of performance w_j assuming the Naive Bayes hypothesis, i. e., each variable X_{it} is conditionally independent of any other variable X_{il} for all $t \neq l \leq k$, is given by:

$$P\left(w_j | X_{i1}, X_{i2}, ..., X_{ik}\right) = \frac{1}{S} P(w_j) \prod_{t=1}^{k} P(X_{it} | w_j) \qquad (6)$$

where S is a correction factor in order to maintain P as a measure of probability. Thus, the equation (6) represents a Naive Bayes Network.

In this paper, the variables present in the training simulation follows the Hypergeometric distribution. Then, a network that considers this statistical distribution could be more suitable for the problem.

3.1. *The hypergeometric naive Bayes network*

For better understanding, we present first the classical Hypergeometric Naive Bayes Network. In this case, it is assumed for $P(X_{it}|w_j)$ in equation (6) the conditional probability using Hypergeometric distribution. Thus, after mathematical manipulations, the equation (6) can be rewritten as a discriminating function g, as follows:

$$
\begin{aligned}
g(w_j|\mathbf{x}_i) =& \log P(w_j|\mathbf{x}_i) \\
=& \log P(w_j) + \left\{\sum_{v=1}^{r_t}\log v - \sum_{v=1}^{X_{it}}\log v - \sum_{v=1}^{r_t-X_{it}}\log v\right\} \\
&+ \left\{\sum_{v=1}^{N_t-r_t}\log v - \sum_{v=1}^{n_t-X_{it}}\log v - \sum_{v=1}^{(N_t-r_t)-(n_t-X_{it})}\log v\right\} \\
&- \left\{\sum_{v=1}^{N_t}\log v - \sum_{v=1}^{n_t}\log v - \sum_{v=1}^{N_t-n_t}\log v\right\}
\end{aligned} \tag{7}
$$

The decision rule for the vector $\mathbf{x}_i$ is given by

$$\hat{w}_j = \arg \underset{j\in\Omega}{max}\, g(w_j|\mathbf{x}_i).$$

3.2. *The fuzzy hypergeometric naive Bayes network*

A new network named fuzzy hypergeometric naive Bayes network is proposed from union of the Zadeh's definition of probability of fuzzy events[10] and hypergeometric distribution. Formally, let ($\mathbb{R}^k$, B, P) a probability space, where B is σ-field of Borel subsets in $\mathbb{R}^k$ and P is probability measure of $\mathbb{R}^k$. Let F in B a fuzzy event with membership function μ_F: $\mathbb{R}^k \to [0,1]$, then the probability of a fuzzy event F is defined by the Lebesgue-Stieltjes integral presented in the equation (8).

$$P(F) = \int_{F\subseteq\mathbb{R}^n} \mu_F(x)dP = E(\mu_F) = \int_{F\subseteq\mathbb{R}^n} \mu_F(x)f(x)dP \tag{8}$$

where $f(x)$ is a distribution function of a random variable X[10]. So, the probability of a fuzzy event F is the mathematical expectation of its membership function. Thus, let a random vector $\mathbf{x}_i = \{X_{i1}, X_{i2}, \ldots, X_{ik}\}$, such that each X_{it}, $t = 1, \ldots, k$ is a fuzzy random variable with membership function $\mu_j(X_{it}), j = 1, \ldots, M$, so we have

$$P\left(w_j|X_{i1}, X_{i2}, ..., X_{ik}\right) = \frac{1}{S}P(w_j)\prod_{t=1}^{k} P(X_{it}|w_j)\mu_j(X_{it}). \tag{9}$$

As in the case of the Hypergeometric Naive Bayes Network, assuming that $P(X_{it}|w_j)$ follows a Hypergeometric distribution, and applying some mathematical manipulations, the Fuzzy Naive Bayes Hypergeometric Network can be described as a fuzzy discriminating function g_f:

$$
\begin{aligned}
g_f(w_j|\mathbf{x}_i) &= \log P(w_j|\mathbf{x}_i) \\
&= \log P(w_j) + \sum_{t=1}^{k} \log \mu_{\omega_j}(X_{it}) \\
&\quad + \left\{ \sum_{v=1}^{r_t} \log v - \sum_{v=1}^{X_{it}} \log v - \sum_{v=1}^{r_t - X_{it}} \log v \right\} \\
&\quad + \left\{ \sum_{v=1}^{N_t - r_t} \log v - \sum_{v=1}^{n_t - X_{it}} \log v - \sum_{v=1}^{(N_t - r_t)-(n_t - X_{it})} \log v \right\} \\
&\quad - \left\{ \sum_{v=1}^{N_t} \log v - \sum_{v=1}^{n_t} \log v - \sum_{v=1}^{N_t - n_t} \log v \right\}.
\end{aligned} \tag{10}
$$

The vector $\mathbf{x_i}$ will be assigned to the class that

$$\hat{w}_j = \arg \max_{j \in \Omega} g_f(w_j|\mathbf{x}_i).$$

This new network is the core of the new SUAS based on Fuzzy Naive Bayes Hypergeometric.

4. Results

The tests carried out considered medical procedures, such as the collection of material for gynecological examination offered by SITEG[9] and the bone marrow harvest offered by BMHSim[4], both VR simulators. These procedures are well executed when the harvest is achieved within a limited universe of attempts. However, the probability of success changes at each try. In both cases, the interaction data follows a hypergeometric statistical distribution. Thus, the evaluation of the new Fuzzy Naive Bayes Hypergeometric Network proposed, as the base of a SUAS, was performed by 1000 replicas in a Monte Carlo simulation replicas in order to obtain its best result. In each replica, two samples were generated with 300 observations each and 100 observations per class. There are 3 classes of performance, representing three real possible assessments for a procedure: Class 1: "the procedure was performed well", Class 2: "the user needs more training" or Class3: "the user needs much more training".

The first sample was used for training and the second one for testing of the new SUAS based on Fuzzy Hypergeometric Naive Bayes Network. The samples were randomly generated with parameters $[r, N, n]^\top$ as follows: for the class 1 the parameters were $[10, 20, 8]$; for the Class 2 were $[20, 40, 14]$ and for the Class 3 $[30, 60, 20]$. For each Monte Carlo replica, the metrics of accuracy (A), kappa coefficient (κ) and kappa variance (σ^2_κ) were calculated. The results were obtained for all samples. This application used a Intel Core i7-1165G7, 2.80GHz with 8 cores, with 16GB of RAM, running Ubuntu Linux, and it was parallelized among the CPU cores, using 5 out of the 8. The code of the Fuzzy Hypergeometric Naive Bayes is available in the FuzzyClass package[1] for R software *(https://cran.r-project.org/web/packages/FuzzyClass/index.html).*

The same samples were used to compare the SUAS based on the proposed method with two other SUAS based on Naive Bayes and on Hypergeometric Naive Bayes methods. In the Table 1 are presented the results of assessments.

Table 1. Simulation results by Monte Carlo, where A is accuracy, κ is kappa coefficient and σ^2_κ is kappa coefficient variance.

Methods	A (%)	κ (%)	σ^2_κ
Naive Bayes	31.11	0	0.0271
Hyper. Naive Bayes	78.89	68.39	0.0039
Fuzzy Hyper. Naive Bayes	85.56	78.21	0.0029

The SUAS based on Fuzzy Hypergeometric Naive Bayes method was superior to the other SUAS in accuracy (85.76%), kappa (78.65%) and lower in kappa variance (0.0029). The Hypergeometric Naive Bayes was the second one most accurate SUAS (78.89%). Both SUAS achieved superior results with accuracy and kappa when compared to the Naive Bayes SUAS. A higher accuracy is important when considering the assessment of future professionals for health area in training simulators.

5. Conclusion

In this chapter was proposed a new naive Bayes network, named fuzzy hypergeometric naive Bayes network, which is able to treat data that follow hypergeometric distributions. This network was used as basis for an SUAS for virtual reality simulators as, for instance, SITEG or BMHSim.

Simulations were performed using data that followed a hypergeometric distribution. A brief comparison among three SUASs based on the new fuzzy hypergeometric naive Bayes network, hypergeometric naive Bayes network and naive Bayes network was provided. The simulation results showed that the SUAS based on the hypergeometric distribution (the new fuzzy hypergeometric naive Bayes network and the hypergeometric naive Bayes network) were able to discriminate the assessments, surpassing the classic naive Bayes. In adittion, the SUAS based on the new fuzzy hypergeometric naive Bayes network was able to provide the best results according to accuracy, kappa coefficient and by the lower kappa variance.

The fuzzy hypergeometric naive Bayes network proposed in this chapter can also be used for data classification and machine learning applications, when data follow a hypergeometric distribution.

References

1. J. Ferreira & R. Moraes, in *J of Open Source Software*, **8** 5613 (2023).
2. N. Johnson, A. Kemp, & S. Kotz, *Univariate discrete distributions*, 444. John Wiley & Sons (2005).
3. A. Lopes, J. Ferreira, L. Machado, & R. Moraes, A New Fuzzy Trapezoidal Naive Bayes Network as Basis for Assessment in Training Based on Virtual Reality, in *Proc. FLINS* (2022).
4. L. Machado, A. Mello, R. Lopes, V. Odone Fo, & M. Zuffo, in *Studies in Health Technology and Informatics* (2001).
5. R. Moraes, J. Ferreira, & L. Machado, in *Int J of Fuzzy Systems*, **23** 849 (2021).
6. R. Moraes, E. Soares, & L. Machado, in *J of Intelligent & Fuzzy Systems*, **38** 577 (2020).
7. R. Moraes, I. Silva, & L. Machado, Online Skills Assessment in Training Based on Virtual Reality Using a Novel Fuzzy Triangular Naive Bayes Network, in *Proc. FLINS* (2020).
8. R. Moraes & L. Machado, A New Architecture for Assessment of Multiple Users in Collaborative Medical Training Environments Based on Virtual Reality, in *Proc. FLINS* (2012).
9. R. Moraes, D. Souza, M. Valdek, & L. Machado, A VR Based Simulator for Gynecologic Exam Training, in *Proc. ITHET* (2006).
10. L.A. Zadeh, in *J of Mathematical Analysis and App.*, **23** 421 (1968).
11. H. Zhang, *A note about maximum likelihood estimator in hypergeometric distribution.* Comunicaciones en Estadistica (2009).